GENETICS AND THE LAW II

GENETICS AND THE LAW II

EDITED BY

AUBREY MILUNSKY, MB.B.Ch., M.R.C.P., D.C.H.

Harvard Medical School
Eunice Kennedy Shriver Center and
Massachusetts General Hospital
Boston, Massachusetts

AND

GEORGE J. ANNAS, J.D.

Boston University Schools of Law, Medicine, and Public Health
Boston, Massachusetts

PLENUM PRESS • NEW YORK AND LONDON

Library of Congress Cataloging in Publication Data

National Symposium on Genetics and the Law, 2d, Boston, 1979.
  Genetics and the law II.

  Proceedings of the symposium held in Boston, Mass., May 21–23, 1979, and co-sponsored by the American Society of Law and Medicine and the Eunice Kennedy Shriver Center.
  Includes index.
  1. Medical genetics–Law and legislation–United States–Congresses. 2. Medicine, Experimental–Law and legislation–United States. I. Milunsky, Aubrey. II. American Society of Law and Medicine. III. Eunice Kennedy Shriver Center. IV. Title. [DNLM: 1. Genetics–United States–Legislation. 2. Legislation, Medical–United States–Legislation. W3 NA677IH 1979g/QZ50 N28 1979g]
  KF3827.G4N38 1979   344.73′0419   80-16179

ISBN-13:978-1-4613-3080-6   e-ISBN-13:978-1-4613-3078-3

DOI: 10.1007/978-1-4613-3078-3

Contributors

George J. Annas, J.D., M.P.H.
Associate Professor of Law and Medicine
Boston University Schools of Medicine and Public
Health
Boston, Massachusetts 02118

Nicholas A. Ashford, Ph.D., J.D.
Associate Professor of Technology and Policy
Assistant Director, Center for Policy Alternatives
Massachusetts Institute of Technology
Cambridge, Massachusetts 02139

Mary Ellen Avery, M.D.
Thomas Morgan Rotch Professor of Pediatrics
Harvard Medical School
Boston, Massachusetts 02115
Physician-in-Chief
Children's Hospital Medical Center
Boston, Massachusetts 02115

John R. Ball, M.D., J.D.
Office of Science and Technology Policy
Executive Office of the President
Washington, D.C. 20500

Charles H. Baron, LL.B., Ph.D.
Professor of Law
Boston College Law School
Newton Center, Massachusetts 02159

Arthur D. Bloom, M.D.
Professor of Pediatrics and of Human Genetics and
Development
College of Physicians and Surgeons
Columbia University
New York, New York 10032
Director of Clinical Genetics
Columbia–Presbyterian Medical Center
New York, New York 10032

Robert A. Burt, J.D.
Professor of Law
Yale Law School
New Haven, Connecticut 06520

Daniel Callahan, Ph.D.
Director, The Hastings Center
Institute of Society, Ethics and the Life Sciences
Hastings-on-Hudson, New York 10706

Alexander Morgan Capron, J.D.
Professor of Law
University of Pennsylvania Law School
Philadelphia, Pennsylvania 19104

Barton Childs, M.D.
Professor of Pediatrics
Department of Pediatrics
Johns Hopkins University School of Medicine
Baltimore, Maryland 21205

Catherine J. Damme, J.D.
Assistant Professor
Department of Community Medicine
University of Texas Medical School
Houston, Texas 77025

Bernard D. Davis, M.D.
Adele Lehman Professor of Bacterial Physiology
Bacterial Physiology Unit
Harvard Medical School
Boston, Massachusetts 02115

John C. Fletcher, Ph.D.
Assistant for Bioethics to the Director
Clinical Center
National Institutes of Health
Bethesda, Maryland 20205

JOSEPH F. FLETCHER, Ph.D.
Visiting Professor in Biomedical Ethics
School of Medicine
University of Virginia
Charlottesville, Virginia 22903

CHARLES FRIED, J.D.
Professor of Law
Harvard Law School
Cambridge, Massachusetts 02138

PAUL R. FRIEDMAN, J.D.
Director
Mental Health Law Project
Washington, D.C. 20036

LEONARD H. GLANTZ, J.D.
Assistant Professor of Law and Medicine
Department of Socio-Medical Sciences and
    Community Medicine
Boston University School of Medicine
Boston, Massachusetts 02118

GREGORY T. HALBERT, J.D.
Enforcement Attorney
Enforcement Division
United States Environmental Protection Agency
New York, New York 10007

JOSEPH M. HEALEY, JR., J.D.
Assistant Professor
Department of Community Medicine and Health
    Care
University of Connecticut Medical School
Farmington, Connecticut 06032

NEIL A. HOLTZMAN, M.D.
Associate Professor of Pediatrics
The Johns Hopkins University School of Medicine
Baltimore, Maryland 21205

Chief, Division of Hereditary Disorders
Preventive Medicine Administration
State of Maryland Department of Health and
    Mental Hygiene
Baltimore, Maryland 21201

Y. EDWARD HSIA, M.R.C.P., D.C.H.
Professor of Genetics and Pediatrics
Director, Medical Genetic Services
John A. Burns School of Medicine
University of Hawaii
Honolulu, Hawaii 96826

BARBARA F. KATZ, J.D.
Office of General Counsel
Massachusetts Department of Public Health
Boston, Massachusetts 02111

JAY KATZ, M.D.
Adjunct Professor of Law and Psychiatry
Yale Law School
New Haven, Connecticut 06520

SANFORD N. KATZ, J.D.
Professor of Law
Boston College Law School
Newton Centre, Massachusetts 02159

SEYMOUR LEDERBERG, Ph.D.
Professor of Biology
Division of Biology and Medicine
Brown University
Providence, Rhode Island 02912

CHARLES U. LOWE, M.D.
Special Assistant to the Director
National Institutes of Health
Bethesda, Maryland 20205

AUBREY MILUNSKY, MB.B.Ch., M.R.C.P.,
    D.C.H.
Assistant Professor of Pediatrics
Harvard Medical School
Boston, Massachusetts 02115

Director, Genetics Division
Eunice Kennedy Shriver Center
Waltham, Massachusetts 02154

Medical Geneticist
Massachusetts General Hospital
Boston, Massachusetts 02114

ARNO G. MOTULSKY, M.D.
Professor of Medicine and Genetics
Director, Center for Inherited Diseases
University of Washington
Seattle, Washington 98195

GILBERT S. OMENN, M.D., Ph.D.
Associate Director of Human Resources and Social
    and Economic Services
Office of Science and Technology Policy
Executive Office of the President
Washington, D.C. 20500

STANLEY JOEL REISER, M.D., Ph.D., M.P.A.
Assistant Professor of History of Medicine
Co-Director of Harvard Interfaculty Program in
 Medical Ethics
Harvard Medical School
Boston, Massachusetts 02115

ARNOLD S. RELMAN, M.D.
Professor of Medicine
Harvard Medical School
Boston, Massachusetts 02115
Editor, New England Journal of Medicine
Boston, Massachusetts 02115

JOHN A. ROBERTSON, J.D.
Professor of Law
Law School and Medical School
University of Wisconsin
Madison, Wisconsin 53706

MARGERY W. SHAW, M.D., J.D.
Professor of Medical Genetics
Director, Medical Genetics Center
The University of Texas Health Science Center at
 Houston
Houston, Texas 77025

JAMES R. SORENSON, Ph.D.
Associate Professor
Department of Socio-Medical Sciences and
 Community Medicine
Boston University School of Medicine
Boston, Massachusetts 02118

DEWITT STETTEN, JR., M.D., Ph.D.
Senior Scientific Advisor
Deputy Director for Science
National Institutes of Health
Bethesda, Maryland 20205

RICHARD B. STEWART, J.D.
Professor of Law
Harvard Law School
Cambridge, Massachusetts 02138

JUDITH P. SWAZEY, Ph.D.
Professor
Department of Socio-Medical Sciences and
 Community Medicine
Boston University School of Medicine
Boston, Massachusetts 02118
Executive Director
Medicine in the Public Interest, Inc.
Boston, Massachusetts 02110

MELVIN L. TAYMOR, M.D.
Clinical Professor of Obstetrics and Gynecology
Harvard Medical School
Boston, Massachusetts 02115
Chief, Division of Infertility and Reproductive
 Endocrinology
Beth Israel Hospital
Boston, Massachusetts 02115

PETER V. TISHLER, M.D.
Assistant Professor of Medicine and Medical
 Geneticist, Channing Laboratory
Department of Medicine
Harvard Medical School
and the Peter Bent Brigham Hospital Division of
 the Affiliated Hospitals Center, Inc.
Boston, Massachusetts 02115

ROBERT M. VEATCH, Ph.D.
Senior Associate
The Hastings Center
Institute of Society, Ethics and the Life Sciences
Hastings-on-Hudson, New York 10706

NANCY S. WEXLER, Ph.D.
Psychologist
Neurological Disorders Program
National Institute of Neurological, Communicative
 Disorders and Stroke
National Institutes of Health
Bethesda, Maryland 20205

PREFACE

The law is a mandate and a mirror; it both commands and reflects. It should not come as a shock that scientists and physicians often prefer the mirror at times when society seems to be demanding a mandate. This may be especially true in the rapidly advancing field of medical genetics, where recent discoveries leading to potentially startling applications have raised old questions of law in a new light. Nevertheless, we believe that in general the conflict between the law and science, as illustrated in the field of genetics, is embroidered with exaggeration.

The Chief Justice of the United States Supreme Court, Warren Burger, has noted that "the prime function of the law is to protect basic human values—individual human values—sometimes even at the expense of scientific progress"; and that "it is not the function of the law to keep pace with science." While both of these statements are true as far as they go, we believe the law must make an affirmative effort to anticipate scientific developments so that those beneficial to society can be nurtured rather than stultified. It was to nurture cooperation and understanding that we brought together a distinguished faculty of internationally known experts on law and genetics to discuss their fields in 1975. The proceedings of that symposium were published as *Genetics and the Law* in January, 1976. The volume evidenced more controversy than concordance, and it was not uncommon for news accounts, like that from the International Medical News Service, to characterize the conference by noting: "courts of law and the science of medicine are on a collision course and probably will clash on a number of issues in the genetics field . . . medicine is racing forward into the future while the courts base their decisions on past events. . . ." That symposium was a general introduction to the field, but also concentrated on specific questions of the day such as the XYY controversy, fetal experimentation, the rights of the fetus, and cloning.

In the fall of 1978 we began to ask if anything had changed. Were lawyers and scientists growing closer or moving further apart? Would another national conference on the subject be fruitful? After some initial skepticism on our own part, we decided that there had been sufficient activity to justify a second national conference on Genetics and the Law. The proceedings of that conference, which was held in Boston in May, 1979, are presented in this volume. We have attempted to avoid repetition from the first volume, and to concentrate on legal and scientific developments of the past five years.

Since the first symposium, considerable heat, and some light, has been generated by genetic issues confronting the law. We have seen attempted incursions by government into the control and regulation of science. Much of the debate over recombinant DNA research regulation, for example, has taken place over the past five years. It is exemplified in this volume as essays on scientific freedom. Some scientists continue to invoke the name of Galileo as a talisman to ward off the evil spirits of restrictive legislation. Others, however, recognize that rights do not exist in a vacuum, and that society has the right to be free from unreasonable risks. Political, ethical, and constitutional dimensions of this debate are discussed. The passage of the National Genetic Disease Act has also spawned individual state genetic disease programs with many societal implications. Public funding for abortion and the "right to treatment" doctrine are also among issues discussed herein that were not debated in the first symposium.

The treatment of defective newborns, the retarded, and the "incompetent" has remained a major issue, but the focus of the debate has shifted radically. In 1975 it was on criminal law and legal liability for nontreatment. In 1979 the focus of both the courts and the conference was on protecting the rights of the defective newborn and "incompetent" patients in general, with physicians and lawyers debating which method could best insure such protection. Here an intense distrust between physicians and lawyers has often been in evidence. Physicians believe they and the families can make decisions in the best interests of the patients, while some lawyers believe that only in a court of law can the "best interests" of the newborn and others be protected.

Artificial insemination by donor is again dealt with in this volume, but from an entirely new perspective. For example, while the legal basis for artificial insemination was outlined in the previous symposium, the paper in this volume questions the policy behind current laws and practices, and calls for new procedures designed to protect the child rather than the sperm donor. Likewise, we have focused debate not on the fetus *per se* in this volume, but on the emerging debate concerning "wrongful life," i.e., in what circumstances can a child sue his mother's physician or his parents for being born defective? All sides of this debate are presented, with advocates for permitting children to sue their parents, and those who believe such suits are neither likely nor proper.

In vitro fertilization has been achieved since the theoretical discussions of the first symposium, making the need to resolve ethical, theological, genetic, and legal implications more cogent. Adopted persons continue to have few rights regarding access to their medical genetic history despite the sometimes crucial personal need for such information. A model approach to this dilemma is debated in this volume.

The 1970s have seen an escalation in litigation concerning genetic diseases. This increase has undoubtedly occurred *pari passu* with mounting public education and awareness of the real options available for the avoidance of mental retardation and certain genetic diseases. Not unexpectedly, suits have largely focused on physician errors of omission.

Eugenics and social policy remain a pervasive topic, and the Rev. Joseph Fletcher presents his argument in favor of limiting the right to reproduce to the genetically fit. In his words, which have already aroused editorial anger in the popular press, "children are often abused preconceptively and prenatally—not only by mothers drinking alcohol, by smoking, and by using drugs medicinally but also by their *knowingly* passing on or risking passing on genetic diseases. There are more Typhoid Marys carrying genetic disease than infectious diseases." We may question Fletcher's approach, but we must face the issues he raises head on.

If there is an emphasis in this volume, it is prevention. Prevention has become the leitmotiv of medicine, and medical genetics is no exception. People are urged to obtain genetic counseling; those providing genetic information to society are urged to be more active; and efforts to prevent the birth (and sometimes to end the life) of genetically defective children are becoming more vigorous.

A full half-day of the conference was devoted to what is emerging as the major issue of the future in the prevention of genetic damage: the regulation of environmental mutagens and teratogens. It is known that cancer has environmental origins in 80–90% of cases, and that some unknown number of these cases may be induced in embryonic or fetal life. The grave problems of industrial pollution require rigorous efforts to protect the human species from potentially lethal or damaging consequences. The message emphasized again is that the best approach to genetic disease is prevention.

We have made no effort to present a unified view of any particular case or statute in this volume or to delete comments made by participants outside their areas of expertise. Instead, we have left the interpretations of all writers, scientists, physicians, and lawyers intact. We have elected this route for four reasons: (1) opinions sometimes differ, even among legal scholars; (2) it is important for members of both professions to know how each views the other; (3) it is important for lawyers to know something about science and for scientists to know something about law—and the attempt of each to understand the other discipline should be encouraged; and (4) useful insights often emerge from such interdisciplinary musings.

We look forward to the Third National Symposium in the mid-1980s, when we anticipate a society further altered by advances in technology. Our hope is that the price will not be the diminishing of human values. Genetic progress and human values must coexist. We believe that the dialogue developed in these symposia will facilitate the translation of genetic advances into a legal framework for the public good.

AUBREY MILUNSKY
GEORGE J. ANNAS

Acknowledgments

We are most grateful to the National Endowment for the Humanities and the National Foundation—March of Dimes for grants that made this National Symposium possible. This meeting was held under the cosponsorship of the Eunice Kennedy Shriver Center and the American Society of Law and Medicine. Our sincere thanks are due to Dr. Raymond D. Adams, Director of the Shriver Center, and Dr. Elliot L. Sagall, President of the American Society of Law and Medicine, for their strong support of our undertaking.

The Program Committee was invaluable in drawing up an extremely effective and balanced meeting. We most sincerely thank Professor Margery W. Shaw and Professor Charles Fried for their help in planning the program.

We owe a debt of gratitude to the Planning Committee, chaired by Mrs. Babette Milunsky. Ms. Judy Heck, Mrs. Marie B. Jardine, Attorney Edward Doudera, and Dr. Elliot L. Sagall all contributed to the impeccable organization that characterized the symposium.

Finally we deeply appreciate the willing participation of outstanding scientists, lawyers, physicians, and ethicists who shared their wisdom and expertise, not only with the conference participants through their oral presentations, but also with many thousands more through their excellent manuscripts contained in this volume.

CONTENTS

ETHICS, EUGENICS, LAW, AND SOCIETY

LAW AND THE CONTROL OF GENETIC DISEASE

GENETICS AND FAMILY LAW

GENETICS, LAW, AND ENVIRONMENTAL MUTAGENS/TERATOGENS

GOVERNMENT CONTROL OF SCIENCE

MODERATOR: CHARLES FRIED

1

THREE SPECTERS
Dangerous Products, Powers, or Ideas

BERNARD D. DAVIS
Bacterial Physiology Unit
Harvard Medical School
Boston, Massachusetts 02115

Most of this volume will be concerned with obviously real moral and legal problems, generated by the impact of genetic research on medical practice and on social policy. But this first part is concerned with moral and legal problems that may be generated by the research itself rather than by its applications. In this area it is important to scrutinize the relevant technical facts carefully in order to dissect real or probable hazards from far-out fantasies. I shall try to do so in three distinct areas of public concern: potentially dangerous products, powers, or ideas created by modern genetic research. For the first two of these the recent excitement over recombinant DNA will serve as a paradigm, from which we can perhaps learn some lessons.

As a brief historical background, let me first note that for centuries the scientific community enjoyed virtually complete autonomy in directing and regulating its research, but in recent years demands for external regulation have mounted. These demands are rather different in the physical and in the biological sciences. In the former they arose primarily from belated recognition that large-scale technology has yielded not only benefits but also costs and dangers, ranging from the threat of nuclear catastrophe to despoliation of the environment. Accordingly, attention has been focused on controlling harmful technological applications. In biology, in contrast, the deepest concern is reserved for possible dangers in the basic research itself, and in its future consequences, rather than in its present applications.

Let us now consider three major areas of concern over genetic research.

DANGEROUS PRODUCTS

The first concern, over possibly dangerous products of genetic research, might have arisen, but failed to do so, during the past two decades of isolation of mutants and recombinants of various bacteria and viruses. It exploded, however, with the development of molecular recombination in the test tube, which made it possible to insert small blocs of DNA from any source into bacteria. A decade ago such a powerful tool for studying living systems would have been greeted solely as a remarkable breakthrough. In the altered current atmosphere, however, public discussion has focused much more on a conjectural danger: the creation of novel organisms that might harm the public or the environment.

The possibility of creating such hazards was initially raised in 1974 by a group of molecular geneticists, and while their concern was very much in the tradition of a responsible scientific community, they departed from tradition—perhaps in reaction to years of criticism of scientific elitism—by making their initial concern public before the matter had been explored in private. Although this action, and the request for governmental regulations, at first received acclaim, a handful of other scientists, for various reasons, expressed great alarm over such research. This man-bites-dog news inevitably had great appeal for the news media, and widespread public anxiety soon followed.

Meanwhile, apprehension has largely subsided. Let me briefly summarize some of the reasons. (1) After several years of work with such chimeric bacteria in many dozens of laboratories, the hazards remained entirely conjectural: no illness and no environmental damage could be traced to this source. (2) Mutant host strains have been developed that provide a huge increment in safety: they can be grown in the laboratory but they self-destruct rapidly under conditions encountered in nature. (3) The increasing evidence for promiscuous uptake and transfer of DNA in the bacterial world makes it extremely likely that in nature *E. coli* cells in the mammalian gut occasionally pick up DNA from surrounding host cells; hence the laboratory recombinants would not be a novel class of organisms after all. (4) Evolutionary principles, entering the discussion late, emphasize that spread of any harmful products would involve much more than their ability to multiply: for only an infinitesimal fraction of genetic novelty in nature survives. Thus natural selection of an organism depends on its adaptation to the environment; adaptation, in turn, depends not on the properties of a single gene but on a balanced set of genes; and that balance is hardly likely to be improved by insertion of DNA from a distant source. (5) It was also recognized belatedly that the problem is fundamentally one in epidemiology, and the input of persons knowledgeable in this area turned out to be highly reassuring. Thus recombinant *E. coli*, with 0.1% inserted foreign DNA, will retain the limited habitat of *E. coli* (i.e., the vertebrate gut) and the mode of spread of *E. coli*. It is therefore pertinent that spread of strains of this organism, and of related enteric pathogens such as the typhoid bacillus, is easily controlled by modern sanitation. Moreover, even with kinds of pathogens whose spread is much harder to control, such as respiratory viruses, no large epidemic has ever arisen from a laboratory. In a word, no expert in infectious disease held that any recombinant *E. coli* could be as probable a source of spread of disease as the many well-adapted, highly virulent organisms that may be encountered in any diagnostic laboratory at any time—and that have caused several thousand recorded laboratory infections in this century.

Eventually, then, reason prevailed, and it was recognized that as in every other interaction with the real world, a rational response must deal with probabilities and not with the misguided, absolute question: Can you prove that the following catastrophe could not occur? Public anxiety subsided, the NIH guidelines were relaxed a good bit, and Congress concluded that the issue was too fluid and complex for rigid legislative regulation. Nevertheless, an inexpensive and hobbling bureaucracy has been set up. And although more and more scientists close to the problem are now publicly questioning the need for any guidelines at all, the bureaucracy may be with us for a long time, and it may provide a precedent for similar overreaction to other potentially dangerous products of research.

What lessons can we learn from the episode? First, it is clearly legitimate for the public to be concerned about the production and use of dangerous materials in research. Where the dangers are well defined, as with inflammable or toxic chemicals, there has been no real tension between the public and the scientific community, or within the latter group. But where the potential dangers are matters of judgment rather than demonstration, the value of informed judgment must weigh heavily. Moreover, there is bound to be dissension in the scientific community during the early phase of digesting the relevant facts and principles. I would question whether the public benefits from participating, or from having the media present, during the highly technical phase, as opposed to participating in the later phase of risk–benefit analysis—a phase that must follow when risks are judged to be significant.

Second, while we are often faced today with the assertion that no group can be trusted to police itself, I would point out that in certain areas the scientific community has long been given that trust: the advancement of its goals was believed to coincide, and not to conflict, with the benefit of society, and policing could only hinder that advancement. In the recombinant DNA issue, where this trust was lost, it is now clear, in retrospect, that the scientific community had indeed acted thoroughly within its tradition of social responsibility, and the recent public mistrust resulted in a futile and enormously expensive exercise—expensive not only in time and money but also in terms of the morale of the scientific community and the public image of science. Accordingly, while there is great appeal in the principle of public participation in any discussions involving the public interest, perhaps this principle, like other principles of social action, has its limits.

I would conclude, then, that there clearly is room for procedural improvements in separating the technical assessment of the hazards of research from the political pressures that attend the subsequent process of decision making. There also is need for improved communication with the public—but this problem is complicated by the way the media select news. For I can attest that during the long period of intense publicity over the alleged dangers of recombinant DNA it proved impossible to get even some of the most responsible news organizations to pay attention to the reassuring arguments from epidemiology and evolution.

Dangerous Powers

Our second area of interest is knowledge that might give us dangerous powers. In the physical sciences such concerns have ranged from military applications of nuclear energy to electronic invasions of privacy. In the biomedical sciences similar concern has arisen over genetic engineering. While this phrase has frightening overtones, we must recognize that in medicine genetic engineering means gene therapy: the replacement of the single, biochemically defined defective gene in various hereditary diseases.

Such gene replacement, considered by itself, is surely as legitimate as the replacement of gene products, such as insulin. But 10 years ago a group of political activists working in science asserted that if advances in molecular genetics achieve this goal, the same techniques will be used to manipulate personalities. This widely accepted proposition clearly added to public anxiety in the recombinant DNA debate—though more as an undercurrent than as a central issue.

Such a prospect would indeed be frightening, if it were realistic. However, the relevant technical facts are highly reassuring. First, therapy even of single-gene defects still seems, unfortunately, very far off, except for defects in those cells that function in widely distributed, loosely organized locations (i.e., the precursor cells in the bone marrow that give rise to blood cells). Second, even if the more difficult goal of gene therapy for defects in localized organs (e.g., the liver) should eventually be achieved, an enormous technical gap would still separate it from any useful, predictable transfer of behavioral genes. For the organ of behavior, the brain, is at the other end of the spectrum from the blood cells: its characteristics depend primarily on the specific connections between its 10 trillion switches. And although we know nothing in molecular terms about the genes that contribute, via differences in brain structure, to individual differences in behavioral potentialities, we can be sure that an enormous number of genes are involved in guiding the development of the circuitry of the individual's brain. Moreover, these genes will have done their work before birth: gene transfer could not conceivably rewire an already developed brain.

For these reasons genetic manipulation of personalities will no doubt long remain remote; meanwhile, many other means of manipulation are already at hand. Nevertheless, if attacks on genetic engineering should continue to arouse public apprehension, the extremely desirable long-term medical goal of gene therapy may be threatened.

In answer to such a threat scientists have traditionally invoked the principle of freedom of inquiry, and in open societies they have interpreted this principle as a facet of the principle of freedom of expression. But perhaps it is no longer enough to wave the flag of Galileo: it is not inconceivable that as science digs deeper we might indeed acquire knowledge that is too hot to handle. However, if we accept this principle and are therefore prepared to proscribe knowledge that clearly is predominantly dangerous, we must also recognize that no such knowledge has yet been demonstrated. And in the absence of a convincing demonstration, the guiding principle of free inquiry must still be defended. For virtually any basic knowledge about nature is ambivalent—it can be applied in both good and bad ways—and we have very limited capacity to foresee the full range of uses. We have even less capacity to foresee the social consequences. The operational conclusion, then, would be that we can best serve society's interests not by blocking knowledge itself but by being quicker to recognize specific harmful applications, and to prevent or to halt them.

Socially Dangerous Knowledge

The third concern, over scientific knowledge that shakes the foundations of public morality or the social order, began with Galileo. It now seems restricted, however, to biology, which impinges much more directly on our views on human nature. In the 19th century it was the conservative establishment that saw a threat in Darwin's evidence against special creation; but today, ironically, it is the left that is most concerned with the possible social implications of a related kind of knowledge: that resulting from the modern fusion of evolutionary theory with genetics. There are three fears: that research in human behavioral genetics will distract attention from the effort to provide social solutions to social problems, that the results of the research might conflict with egalitarian preconceptions about the amount and the distribution of variation in genetic potentialities in our species, and that the implications of results may be distorted for political purposes.

The attack on this area of research has paralleled the frightening example of Lysenkoism, in subordinating genetic science to political dogma; yet it has evoked wide sympathy. The reasons are evident: our guilt over the inequities of slavery and race discrimination, and the all too real history of past abuses of evolutionary and genetic concepts, by Social Darwinists and by racists, to rationalize discriminatory practices. But in trying to prevent a repetition of this earlier subordination of science to reactionary political purposes, we must ask who are its present spiritual heirs: those who would now subordinate science to progressive political purposes or those who oppose its subordination to any politics, whether of the left or of the right?

Hitler's ghost thus continues to haunt us and to color our views on human genetics. And although we are unlikely to develop in this country the overt mechanisms of Lysenkoism, with rigid governmental proscription of a politically offensive field, we must recognize that effective suppression does not require this mechanism: in the face of the attacks of the last decade within universities, granting agencies are likely to be uncomfortable with such sensitive areas, and few graduate students are likely to enter the field. Indeed, several medical investigators have abandoned studies on the still undefined effects of an extra X or Y chromosome. Yet as J. B. S. Haldane emphasized in the 1930s, increased awareness and understanding of human genetic diversity offers great long-term promise for education, as well as more immediate promise for medicine.

The breadth of the issue and the price of a victory for ideology over science should not be underestimated. The problem will not remain confined to behavioral genetics: in time neurobiology, and evolutionary studies on social behavior, will surely also provide challenges to our preconceptions about human nature. Even today, efforts to inquire into biological and medical aspects of violence and crime, not necessarily genetic, encounter intense opposition from those who insist that the sickness lies entirely in society.

What is at stake is the intellectual freedom painfully acquired over past centuries—not freedom to make or to do something potentially harmful but freedom to know.

Of course, some would counter that it is callous to wish to unearth knowledge regardless of its political consequences. But we must recognize that the truths about human nature, both its universals and its diversity, will be there whether or not scientists discover them, and this reality will affect the success of those social policies that depend on assumptions about these matters. Moreover, if we recognize justice as a constantly evolving social construct, it is difficult to see how any valid new knowledge can itself threaten justice. On the contrary, as we deepen our understanding of the interaction of inborn and social factors that influence human behavior, we should be able to build more effective institutions of justice.

Summary

On purely technical grounds it seems clear that recent fears of biology have been greatly exaggerated. In the area of dangerous products the history of the controversy over recombinant DNA research suggests that the scientific community did behave responsibly, and that excessive involvement of the general public in the phase of technical assessment of risks impeded rather than advanced the process of arriving at reasonable judgments. In a second area, knowledge that may give us dangerous powers, restrictions on advances in the knowledge, as opposed to restrictions on its applications, are difficult to justify, simply because we are not able to foresee the full range of good and bad consequences of such advances.

A third concern, over increased insights into human nature and human diversity, is more complex. On the one hand, evidence that conflicts with treasured preconceptions may be painful. On the other hand, science gives us knowledge as well as power, and while technological powers have generated the growing crisis of mankind, insight into the biological roots of our behavior may be what is most needed to meet that crisis. If we should cut off the flow of such insights to avoid immediate problems, we may pay dearly in the long term—and we will also set a precedent that is inimical to an open, democratic society.

2

RESEARCH AND REGULATION

DEWITT STETTEN, JR.
National Institutes of Health
Bethesda, Maryland 20205

In 1553 Michael Servetus paid the supreme penalty. He was burned at the stake in Champel, France, having earlier been burned in effigy for daring to disagree with the establishment. The establishment, in those days, was more likely to be the church than the state, and the court that tried and convicted him was dominated by John Calvin. Prominent among his heresies was his description—in fact, his discovery—of the pulmonary circulation of blood. The punishment meted out involved not merely burning of the author but apparently also the burning of his book, of which only very few copies survived. Almost a half century later, Giordano Bruno paid the same price, and among his crimes was an acceptance of many of the precepts of Copernicus, including the heliocentric theory. Later, in the 17th century, we read about the trial of Galileo, whose heresies also included a belief in the Copernican hypothesis. Either due to his great prestige as a scholar or due to the changing times of the developing Renaissance, Galileo escaped with his life on the condition that he publicly retract his heretical beliefs. He was, however, forbidden further to explore or to teach these radical ideas. Recently, Jacob Bronowski identified this specific judgment of the Roman Court of Inquisition as the immediate cause for the collapse of Italian leadership in the scientific Renaissance. Up to that moment in history, Italy enjoyed unquestioned dominance in the process of the rebirth of the scientific enterprise, and immediately thereafter this leadership was lost, to be promptly assumed by the more northerly states of continental Europe and England.

When, in that particular layer of purgatory that is reserved for dissident scientists, the souls of Servetus, Bruno, and Galileo convene with the many others who have come into conflict with the laws and regulations imposed upon them by their church or by their state, one may suppose that they arrive at a consensus that may be expressed, "Regulation is antithetical to creativity." I do not deny that I have a bias in favor of this epigrammatic statement. When I first tested it on a scientific audience, it gained immediate acceptance and drew favorable comments from several auditors. It appeared to me that the scientist understands immediately what is meant by the antithesis of creativity and regulation, and also recognizes that creativity is the very stuff of which good scientific research is made. It is, of course, also the substance of literature, of art, of music, and of philosophy. There are many kinds of regulations. There are those that regulate ideas, those that regulate actions, and those that regulate things. Regulations that dictate the characteristics of Grade A beef or the level of pollutants in the discharge from automobiles are certainly less damaging to creativity than are regulations directed against ideas and thoughts. While it is true that courts of inquisition do not figure prominently in our culture today, it should not be supposed that attempts to regulate our ways of thought have entirely disappeared. By way of example, may I cite a statute that, for all I know, is still on the books of the state of Kansas. This statute sets pi equal to 3, a devaluation that, were it to be enforced, would wreak havoc with geometry and other branches of mathematics. In a more serious vein, however, we have ample evidence from what happened in the Soviet Union under the dictatorship of Stalin. Not only were art and literature subject to stern regulation but even scientific thought was regulated. Thus to support any views of genetics that were in conflict with those proposed by Lysenko was regarded as heresy and was punished as such. The punishments that were applied involved failure of academic promotion, blockage of normal channels of publication, eviction from learned societies, and on occasion imprisonment. It should be recalled that all of this occurred in a large, powerful, and developed country, and as recently as 30 years ago. We are reminded by the infamous Scopes trial and its derivative decisions that in this country also the legislative–judicial process tampers with and attempts to regulate the dissemination of ideas.

During the interval 1974–1978, I served as chairman of the Recombinant DNA Molecule Program Advisory Committee of the National Institutes of Health. It was during that period that I became acutely aware of the adverse effects that might result from excessive regulation of the scientific research process, and it was in summarizing the reasons for my resignation from that chairmanship that I referred to the antithetical nature of regulation and creativity. I was startled and somewhat grieved when a distinguished lawyer who had peripheral contact with the problem of recombinant DNA labeled my epigram as "absurd," pointing out that traffic regulations did not impede creativity and that regulations can be designed to be flexible. Clearly, traffic regulations will not impede creativity as long as driving a car is not creative. The case is, in my opinion, trivial. I was, however, intrigued with the concept of "flexibile regulations." This appears to me to be the kind of play on words much favored by members of the legal profession and often incomprehensible to those lacking legal training. It is my contention that regulations are designed to be *in*flexible, much as a ruler is designed to be inelastic. By this I mean that when we wish to make a measurement employing a ruler, we normally use a ruler made of a material with a modulus of elasticity sufficiently large so that stretching the ruler cannot introduce significant error into the measurement. We can, of course, print inches or centimeters on a rubber band and the product might resemble a ruler. However, it will fail in the functions of a ruler and is, I believe, essentially no ruler at all. Similarly, we can introduce flexibility into regulations. The suggestion has been made that regulations may be rendered flexible by the simple device of providing an easy means for amending or modifying the regulations. This suggestion, however, does not in my opinion withstand critical review. Easy amendment provides some degree of useful flexibility to the regulator and may be of itself useful. However, the regulator is characteristically not the party who is constrained by the regulations. The person who is regulated—that is, the regulatee—derives little or no benefit from the fact that the regulator now may easily change the regulations. On the contrary, it makes the position of the regulatee somewhat more difficult since he cannot be certain whether the regulations in effect yesterday will still be in effect tomorrow. In a community in which the chief of police drafts the traffic regulations, the provision of easy modification of these regulations may help the chief of police but it certainly does not help the driver of the car. Therefore, the flexibility that results from this maneuver will not fill the present requirement. As a ruler loses the characteristics of a ruler when it becomes elastic, so a regulation loses the characteristics of a regulation when it becomes flexible.

Since science would probably not prosper in complete anarchy, some degree of regulation is required. The real questions are: What should be regulated and what should be left unregulated? and: Of that which is to be regulated, to what degree is regulation useful? Let us first consider certain areas in which regulation should not be attempted or should be approached with great reluctance. In this country, to take an example, it is regarded as inappropriate for the government to regulate the nature of religious worship by its citizens. The First Amendment to the Constitution also guarantees some degree of freedom of speech and of the press. We cherish the Jeffersonian rights to life, liberty, and the pursuit of happiness. To these freedoms I should like to add, for purposes of the present discussion, the freedom of inquiry. This freedom, although not explicitly discussed in the Constitution, appears to me to be of the same quality as freedom of speech. It is a basic part of our political philosophy. It should be enjoyed by every citizen. And it should be abrogated by legislative or judicial process only when and if its continued license would lead to serious jeopardy to the people or to the body politic. The regulation of otherwise free inquiry should be measured by the same standards as we use to assess the abridgment of the freedom of speech or of the press, where the test, as I understand it, is that of "real and present danger."

The development of nuclear chemistry and physics during World War II resulted in a clear and present danger, both to the scientists who wished to continue this exploration and to their neighbors and their environment. Appropriate regulations were therefore drafted and administered initially by the Atomic Energy Commission and more recently by the Nuclear Regulatory Commission. Over the many ensuing years, my colleagues and I have conducted research using radionuclides. I have heard very little complaint directed toward the regulation of the use of these materials, and I believe that any negative impact that the regulations might have produced upon the research process has been modest. It must further be understood that the hazard to life and to health from exposure to radioactive materials was first observed within a few years after the discovery of radioactivity. We have repeatedly been reminded of this hazard as a result of a variety of accidents, and the whole matter was vividly and intensely tragically called to our attention in 1945 at Hiroshima and Nagasaki. The danger of radioactivity was well documented and never in question.

This situation stands in marked contrast to the problems raised by recombinant DNA molecules in 1974 and 1975. At that time, the anxiety developed in the minds of a number of very responsible scientists that these novel manipulations might contain significant dangers, not only to individuals but to the entire ecological system in which we reside. Once the scientists' imagination was fired, there appeared to be no limit to the horror stories that could be concocted, and by the time the scientists met with the press and representatives of the law at Asilomar in February 1975, a considerable wave of fear had been generated. Neither at that time nor at any time since, despite the fact that thousands upon thousands of recombinant DNA experiments have been conducted, has any evidence been presented that indicates that there truly is a hazard in this technology beyond that inherent in the microorganisms that are handled. In other words, the hazard of manipulating recombinant DNA molecules remains at this time entirely imaginary. This does not mean, of course, that no hazard exists. However, with the passage of time, in the absence of adverse findings, it would appear to me that the likelihood of serious hazard progressively diminishes. Since the risk remains to be demonstrated, no risk–benefit analysis can be conducted. The benefits derived from this research approach are already imposing and promise far greater things in the future. Yet this research is today being conducted under a bewildering mass of regulations, the drafting, interpretation, and application of which have consumed inordinate amounts of the time of some of our finest scientific intellects. I refer to them as regulations because, *de facto*, this is what they have become. They were drafted as guidelines by a committee, which I chaired, intending to give guidance both to investigators and to reviewing committees. However, in language and in substance, they have come to be interpreted both by governmental agencies and by scientists as regulations, not to be violated under threat of sanction.

It appears to me that in this instance we have erected a bastion against what may prove to be pure fantasy. Curiously, this has occurred at a time when we are surrounded by a number of very real dangers. These include the danger of nuclear holocaust, depletion of energy sources, inadequate food supplies, insufficient potable water, the several meteorological catastrophes that have been forecast, and self-inflicted hazards such as overpopulation, cigarettes, hand guns, and automobiles. Our problem, it seems to me, is that our legislatures have acted and our courts have ruled on the basis of anxiety rather than of hazard. Whenever a sufficient number of persons become sufficiently aroused by a fear, we hear the cry "There ought to be a law!" and if we are not very careful, shortly thereafter there is.

The resources of the federal government are very large but they are finite. The fraction of these resources that may be spent in protecting the people of the United States against future possible hazards is politically determined; but whatever its value, it clearly must be a fraction of the total. Common sense requires that we invest these funds to maximal effect. Three questions must be explored: (1) What is the likelihood of occurrence of the catastrophe? (2) How much damage will it produce if it does occur? (3) What is our capacity to reduce the likelihood of occurrence or the ensuing damage?

It should be on the basis of the answers to questions of this sort that funds to combat catastrophe would most wisely be invested. Analogously, in drafting of regulations designed to curb hazards, we would be well advised to concentrate upon the hazards that are real and of significance and to postpone drafting regulations against hazards that cannot be substantiated.

DISCUSSION

DR. DANIEL CALLAHAN (Hastings Center): Let me just say to Dr. Stetten that he might have remarked in a Marxist vein to his lawyer interlocutor that, after all, we lawyers make our living by regulation, even the best of us, and scientists, the best of you, make your living by creativity. And indeed the parallel with clerics at all times might be at hand as well. There is a certain professional bias involved, no doubt.

DR. JONATHAN BECKWITH (Bacterial Geneticist, Harvard Medical School): I am mainly responding to some of the comments of Dr. Davis. A few of his comments largely misrepresent the issues at least in terms of our position on many of them. First, on the recombinant DNA issue, I personally don't feel very strongly about the health hazards of recombinant DNA, although I think there are certain areas where one has to be very cautious, and I think that the process that has been pursued has been very good. I believe in the present guidelines.

It was suggested that there have been no accidents of people coming down with diseases from working with these organisms. The work has been done under the guidelines and that hopefully has provided a measure of safety. But, secondly, a comment like that seems to me amazingly like the assurances that are given by people around nuclear power plants that no one has suffered from the radiation at Three Mile Island.

We have a long history in this society of introducing agents into the atmosphere and then finding out only much later that they cause serious problems (such as asbestos and other things which cause cancer). We now have tests, for instance, which allow us to look at these agents and to determine whether there is a prospect for some of these agents to cause cancer. I think in the case of many recombinant DNA strains, we also have the possibility of doing tests to see whether they might be dangerous before we go ahead and use them.

A second specific technical comment. It was mentioned that there is now a bacterial strain which supposedly disintegrates once it enters the human body. But, in fact, a researcher at Tufts has shown that this particular strain actually survives for many days in the human body after it has been introduced. I think that many of the people who suggested that there are no problems and that we now have all the knowledge which indicates that recombinant DNA is perfectly safe have their objectivity somewhat clouded by other factors. In certain cases, many of the people involved in the research have the prospects of prizes before them. There has already been a Nobel prize awarded in this field.

In addition, many of the researchers in various universities are now involved as directors of various corporations or have stock in corporations and certain connections with a number of these corporations that have been set up to exploit recombinant DNA research. I think that we can't trust people with these interests to make decisions about what is safe and what isn't safe.

Finally, I think that in general the public should participate in decision making. The public pays for the research; the public will reap the benefits or suffer the consequences of it. We already have a whole variety of ways in which science is regulated. In many cases the decisions about science are certainly not determined by the public, not even by scientists themselves, but by corporate interests in society.

Now, one point of view that I believe is not represented here and I feel should be was presented by the Cambridge Experimental Review Board. People from the city of Cambridge have made conclusions about DNA research. I think it represents a model about how the public can participate in decision making about science.

PROF. BERNARD D. DAVIS: I am at least as concerned as Jonathan Beckwith about the failure of our society to do something much earlier than it did about such demonstrable hazards as exposure to asbestos. But there is a very large difference, I would suggest, between stimulating action in response to a very real hazard like that and arousing anxiety over hypothetical hazards, especially ones that would require for a catastrophe a long series of compounded events each of very low probability.

Now, it is very well for Dr. Bekcwith to indicate that the whole process really cannot be trusted to scientists because some of them are after prizes in this area. But there are many scientists who are not after prizes in this area, and there are some of us who aren't in the area at all but feel we know something about problems of infectious diseases and spread of organisms, an area whose relevance has gradually appealed to those responsible for legislative decisions.

Implicit in Dr. Beckwith's approach is another problem that I mentioned briefly in my talk: the search for absolute security. Can you prove that the following scenario couldn't happen? The scientific method cannot prove that something cannot happen. We have to deal with probabilities. And I think the main reason that anxiety has finally subsided is recognition that we are talking about hazards whose total probability, compounded from a series of very low-probability events, is too low to merit the amount of attention they have had.

DR. RICHARD RATZAN (University of Connecticut Medical School): What bothers me about the reasons why recombinant DNA has proven not to be dangerous is Dr. Beckwith's approach. I think the Asilomar conference was exactly the time to make such a warning. If there were to be a foreseeable catastrophe, it would be too late to go back and undo it.

I think at the time the feeling at the conference was that it was timely to raise the issue in an anticipatory way. It was not considered acceptable to wait to see if work in private laboratories turned out to be problem-free. I think that was the whole idea of the conference. Because you can now say five years later that it turned out to be an unwarranted fear does not detract from the fact that at the time it was a real fear. I think that is a retrospective judgment which is not justified.

My second point: Is it the fault of the people who made the warning that there was this process? I agree with Dr. Beckwith that it turned out to be a very healthy process, not only for heightening the awareness of the public but for establishing some kind of responsibility of scientists to the future consequences of their work.

But is it the fault of the scientists who raised the question that all these regulations came into effect and that there is a huge, expensive, time-consuming bureaucracy? Is it their fault that there is this effect, or is it the fault of the people who responded the way they did and the people who made the rules?

I don't think it is fair to blame the bureaucratic response, in effect, on the people who raised the issue.

PROF. DAVIS: Nowhere in my talk, or anywhere else, have I ever blamed the scientists who raised the question. In fact, I stated explicitly that I thought they acted extremely responsibly.

I do raise the question in retrospect, and I think any serious student of social policy must ask the question, looking back: Did it work out well? Now, what they did was unusual. It was equivalent to a physician facing a patient with some symptom that is not yet explained and instead of quietly going about testing various possibilities before he discusses with the patient whether this lump is due to some harmless lesion or is due to a cancer, speaks out the differential diagnosis that is going through his mind. The patient would inevitably become very anxious.

We have here a very serious problem. The scientists who see potential hazards have a real obligation to discuss and to analyze them. But I would suggest that if we do too much of the differential diagnosis out loud in public at the very beginning, before things have been narrowed down, we are bound, with the interest of the media, to arouse a great deal of excessive anxiety.

DR. DEWITT STETTEN: I would like to concur with Dr. Beckwith. I too am troubled by the scientists who have affiliated with or created entrepreneurial industries for profit to exploit the research which has been initiated at least largely with federally supported dollars. I'm worried about that.

I would, however, like to support the notion that Dr. Davis has stated. Absolute safety is of course not available anywhere and absolute safety is, therefore, a futile goal. We can reduce hazards, we can assess hazards. We cannot eliminate hazards in hardware or perhaps in any field of endeavor.

As regards the Asilomar conference, I did attend it. I attended it relatively cold. This was not my field of investigation. I had been briefed by colleagues at the National Institutes of Health before I went and I listened, and I was persuaded.

However, in retrospect, I realize that I had heard at this purportedly scientific meeting virtually no data. It was different from any other scientific meeting I had ever attended. And the more I thought about it, the more convinced I became that actually I had been the unwitting attendee of a revival meeting rather than a scientific conference. It took me about a year to wake up to that. And actually what alerted me to it was when in the process of writing the guidelines, I received an angry letter from one of the organizers of the Asilomar meeting who told me that I had violated the spirit of Asilomar. I have been attending scientific meetings for fifty years, most of them in Atlantic City, and I have never heard anyone refer to the spirit of Atlantic City.

I suggest this was a spiritual adventure, not a scientific one. It was very persuasive at the time, I can assure you.

DR. DANIEL STOWENS (Utica College and St. Luke's Memorial Hospital, Utica, New York): Scientists, lawyers, and other interested people at this meeting are talking about things in relationship to society. We have heard asbestos mentioned twice, as if this were a curse on humankind. On one side, we have the number of people who unfortunately have developed carcinoma due to the action of the asbestos, while on the other side, there are the untold millions of people who have been saved by the truly God-given characteristics of this material to prevent fire.

We have seen a reaction to fireproofing of babies' garments on the theoretical aspect that at some time in the future they will develop a cancer. However, if they burn to death when they are six months of age, they will never develop a cancer.

There are necessary regulations for radioactive materials. We use infinitesmal doses in the clinical laboratory but the regulations require us to monitor our radiation use. It is going to cost us three thousand dollars to buy a monitor sensitive enough to even get a reading! I do think we have to measure the benefits derived from various procedures and substances versus the possible hazards.

MR. SCOTT THACHER (Graduate Student in Biophysics, Harvard University): I have been a critic of the NIH and its development of the guidelines. However, I have a great deal of respect for them now. But I would like to disagree with Doctors Stetten and Davis over the lessons to be learned.

I think initially, after the Asilomar conference, the scientists took it upon themselves to write further guidelines. They used the elitism to which Dr. Davis alluded to exclude the public from any of the decision-making processes. That elitism was so strong that many people in occupational health and in epidemiology who would have had informative things to tell them were also excluded. Therefore, in many senses, the scientists writing these guidelines cut themselves off and had to deal with the public across a fence, so to speak. They had to deal with the fears of the public without consulting the public directly.

Recently, I think they made a very productive move and expanded the recombinant DNA advisory committee to include members of the public, some of their scientific critics, and people in occupational health. I think this is a good move. I wonder if Dr. Davis or Dr. Stetten have reflections on the nature of the specialization of people in recombinant DNA research that prevented them from realizing that outsiders could help them understand the dangers. I think this is the real lesson to be learned from the controversy: that scientists tend to be out of touch with the public. I wonder if they share my concerns.

DR. STETTEN: You are quite correct that the early picture, the early image of what was needed, was chiefly related to aspects of molecular biology. The original committee, which I think had twelve members, was predominantly made up of biologists, microbiologists, and molecular biologists, with one expert in bacteria and one expert in plasmids carefully selected for those specialties.

However, within the first twelve months, first one and then two nonscientists members were added to the committee and they sat with the committee consistently. One of these was in the field of government, the other was in the field of theology and ethics. They were, incidentally, contributory members.

One year after the Asilomar meeting, that is, February 1976, the matter was presented to the Directors Advisory Committee of NIH (which was specially expanded on that occasion) and approximately a hundred representatives of outside interested groups, including scientists groups, consumerist groups, labor union groups, law, the judiciary, and the body politic, were all included. The proceedings of that meeting were widely distributed and the public at large was invited to submit comments on the proposed guidelines.

I will admit that the *Federal Register* may not be the most popular journal in the world nor the easiest to read, but that was the channel of publication which was prescribed by the government's rule-making procedure.

PROF. DAVIS: The Asilomar conference, as described in subsequent publications by some other participants, illustrates the problem I raised of the need at certain stages of discussion to separate technical analysis from moral, legal, and social discussion. The speaker questioned whether the scientists weren't being too elitist.

Well, I spoke of the inherent elitism of science, using the word *elitism* to be provocative, perhaps; but let's substitute for the words *inherent elitism* the word *knowledgeability*.

Now, the people who set up the Asilomar conference, recognizing the temper of the times, made it their business to have not only a hundred and more scientists but also a certain number of lawyers, theologians, a dozen members of the press, et cetera. And one of the lawyers, Harold Green, wrote an essay in a book on recombinant DNA edited by John Richards in which he complained that he was simply lost, and couldn't understand the technical discussion that was going on at all. Similarly, the president of the Hastings Center, from which Dr. Callahan comes, wrote a very nice editorial in the *New England Journal of Medicine* on the effort of the scientists to be thoroughly responsible in handling the recombinant DNA issue, but he said there was one problem at Asilomar. It ended up "we" and "they," for "They talked over our heads."

Well, I ask, what was the purpose of the conference? If its purpose was to answer the questions of how we best assess the hazards and what can we do about them, that is a technical conference. And if you're going to have to talk throughout a long technical conference at the level of a press release or at the level of a general discussion of social issues, then you lose the purpose of the conference.

The great thing that came out of the Asilomar conference was the idea of seeking deliberately weakened strains as hosts for these recombinants. That idea could not come out except on the basis of highly technical discussion.

3

ETHICAL ISSUES IN THE CONTROL OF SCIENCE

DANIEL CALLAHAN
The Hastings Center
Institute of Society, Ethics and the Life Sciences
Hastings-on-Hudson, New York 10706

There are many ways of understanding the term *science*. Science can and has been understood as the search for the truth about the natural world. Science can be and also has been understood as the application of a particular methodology—the scientific method—for the sake of developing valid propositions, generalizations, and predictions about empirical reality. Science can be and also has been understood as a human enterprise devoted to the betterment of humankind. All of these understandings have a basis in the history of science, and an equally strong basis in the stated ideals of science. For many, science means all three of those understandings together. For others, science may be seen as only one of the three. My own view is that all three are pertinent to and descriptive of the full and rich reality of science.

One may ask: What ethical issues does the pursuit of science raise? It seems clear that some understandings set the stage for the generation of more ethical problems and dilemmas than others. While some have worried that an excessive emphasis upon the scientific method as the preeminent, or only, form for the pursuit of knowledge can result in a sterile and narrow positivism, the scientific method itself has generated few genuine ethical problems. It certainly generates problems in epistemology and in the philosophy of science, but those problems belong within the domain of science rather narrowly construed and rarely occasion much ethical dispute. The same might be said for the concept of science as a search for truth about the natural world. There are very few who would deny the value of such an enterprise. The only questions that, on occasion, arise are those turning on the question of whether there are some forms of truth that ought not be pursued. Thus we have seen recent debates on the question of whether research on IQ should be carried out at all, whether the enterprise of sociobiology is inherently hazardous, and whether there are certain forms of biological research—recombinant DNA research, for example—that should not be pursued because of some inherent hazards. By and large, however, those debates have taken place at the fringes of science, in those gray areas where some, at any rate, are concerned that the very quest for the knowledge may be in itself corrupting.

In this chapter I want to put aside ethical issues of that kind. I want to focus on science as a human enterprise, devoted to human welfare. For it is essentially in that context that a number of recent issues concerning the freedom and control of science have arisen, occasioning a number of difficult ethical questions. At the heart of the concept of science as an enterprise devoted to human welfare is an inherent tension between the freedom necessary to carry out scientific research at all and the good of the society within which that freedom is exercised. If at least one understanding of science is that of the pursuit of knowledge for human welfare, then it is quite legitimate to ask: At what point can that pursuit itself threaten human welfare? For the purpose of this discussion, I want to take as a given the widespread and to me quite valid belief that the best science is that which is unfettered by social constraints—where the only limitations upon science are the limitations of knowledge and methodology: internal, inherent constraints. A widely accepted moral principle stands behind that belief: that the principle of freedom of scientific inquiry is a fundamentally valid and valuable principle, which should be preserved, continually strengthened, and protected from all external incursions.

At the same time, the fact that science is now a public and social enterprise, directed at human welfare, means that we must now reconsider what that freedom should consist of and, specifically, whether and in what circumstances that freedom should be limited. I want to develop two basic lines of argument. The first is that no human freedom, including the freedom of scientific inquiry, can be absolute and unlimited. All freedoms are subject to other ethical claims, and all must be understood in the context of general human welfare. The second point to be developed is that it is insufficient to leave the question of scientific freedom, and any limitations on research, wholly in the hands of scientists themselves. When there is a potentially decisive impact on the lives and welfare of others in the society, it is inadequate to allow to scientists the privilege of internal regulation only. Let me now turn to the first issue.

It is important to realize that there are now, and have been for some decades, a number of practical constraints on totally unfettered research. These include relative limitations on funding from governmental and private sources, shifts in national interests and priorities, disagreements within the scientific community about the value of some lines of research (which can significantly affect some forms of research, which can then gain or fail to gain the benefit, both social and financial, of peer support), and various forms of governmental regulations that can alter the conditions under which research is done (e.g., present regulations concerning informed consent for human experimentation). In short, there are already a number of *de facto* social and political constraints at work. In that sense, science has never been totally free and no doubt could never be totally free as long as the scientific community is made up of human beings and scientific work carried out within a social context. Nonetheless, I do not believe those are the fundamental problems concerning the control of science, however much they may prove annoying or bothersome in the day-to-day life of the scientist.

Why, in our particular era, in our time and place, has freedom of research once again become a significant public issue? At least in the West, did we not solve that problem long ago? Did we not as a culture decide that the search for truth must be totally free, that neither church nor state should have the right to interpose themselves between the scientist and his quest for truth, however unsettling that truth might turn out to be? The name of Galileo is constantly invoked to remind us of the long struggle on the part of science for freedom.

Yes, in one sense as a culture we did decide those questions—but in another time and era. A number of developments in recent decades force us to take a fresh look at the problem. First, a significant portion of scientific research is now supported by public funds. This fact has an important implication: when the public supports any enterprise, whether it is a school system, a network of highways, or a welfare program, the public demands the right to have a say in how its money is spent. That is a well-established rule in our society, and one that seems in general quite legitimate. Only recently, however, has the public insisted on a role in science analogous to the kind of role it plays when other public funds are spent. This may be unsettling to many in the scientific community, but it would be asking very much to expect the public to establish different rules for science from those for other publicly supported enterprises.

Second, science is now a public and social enterprise. Science is now a public enterprise not only because much of its financial support comes from the public but just as importantly because the implications and impact of science are public and social. Not only basic research, but obviously applied technology as well, affects the lives and welfare of all citizens. It is no longer possible, if it ever was, to make a sharp distinction between what is private and limited in science and what is public and potentially unlimited in its impact.

Third, not everything that results from scientific research automatically or necessarily fosters human welfare. It is no longer possible to hold the belief that the fruits of science are inevitably and always beneficial. Science can produce both good and evil; it is a two-edged sword. That knowledge, borne in on us by the development of nuclear weapons, technological hazards to the environment, irreversibly comatose patients kept unnecessarily alive by artificial means, forces us into a new wariness, and to a basic question: How will the freedom we as a society grant to science affect our lives and welfare?

These three new realities of scientific research force a reconsideration of the question of scientific freedom and, most pointedly, force us to ask whether and how some limits may or must be set to that inquiry. If I read the culture correctly, there seems to be very little significant opposition to the limits to inquiry when the issue is clearly one of applied scientific technology. The technology assessment movement and the environmental impact movement both resulted in federal agencies. There are now very few who would argue that we should not look before we leap in the technological applications of scientific research, and some who would argue that we should not leap at all. While there is considerable and often justified complaint in the scientific community about some of the bureaucratic restrictions imposed by new governmental regulations, there are few who would be prepared any longer to argue that the technological applications of science should be under no restrictions whatever.

The most difficult issue is, instead, the freedom of basic scientific research, which I will define as that research where the primary motive is knowledge for its own sake. Much recombinant DNA research, for example, would fall into that category.

I want to argue that basic research is as legitimately subject to the same sociopolitical understanding as applied technology, at least to the extent that it is publicly funded and has, in principle if not always in practice, social implications. Put another way, the three points noted above, which justify a public role in science, apply as well to basic research as to technological applications of research.

There is often alleged to be a basic distinction between basic research and applied technology. The difference is taken to be that basic research seeks only knowledge and understanding, while technology seeks the application of knowledge to change, modify, or manipulate the world.

The contemporary validity of this distinction can easily be challenged. Basic knowledge is gained in many if not all cases, not by mere thought, or by the development of ideas and concepts alone, but by action in and on the world. Genes are understood by manipulating them, bodies by experimenting on them, and nature by altering it. Only astronomy and pure mathematics deal with ideas and theories as such. The other sciences now must almost always intervene in nature to understand it, and must, moreover, make use of technology as a mode of understanding. Basic research now conflates thought and action; it uses action as a means to basic understanding. But when action is at stake, then the usual moral rules of responsibility for action come into play: one is obliged to take responsibilities for one's actions, and one can be held accountable for those actions.

While we may say there is a right to free scientific inquiry and that freedom to carry out scientific research should exist, it is important to realize that no right or freedom can be unlimited or unrestricted (save the right to think what we please in the privacy of our minds). If even basic research involves action and not just thought, then in principle it seems that we must conclude that the moral rules applying to action in general will have a bearing here as elsewhere. Moreover, unless one is prepared to argue that the freedom to do scientific research is the highest of all human freedoms, and the right to do it the highest of all human rights, one is forced to recognize that other kinds of freedoms and rights may, on occasion, have some legitimate competing claims. In this country we affirm the freedom of speech—but not to the point of allowing someone to falsely cry "fire" in a crowded theater; and we affirm freedom of religion—but not the right of a parent to deny a blood transfusion to a child in the name of the parent's religious convictions; and we affirm freedom of the press—but not the freedom of the press to libel people.

In sum, while a number of vital freedoms and rights are affirmed in this country, all have some established limits. The most obvious and common reason for establishing a limit is a simple one—people are not allowed to use their right to freedom to jeopardize the right to freedom, or the welfare, of others. Freedom can, paradoxically, only exist in a meaningful way if limits to freedom are recognized. Hence there is an obvious implication for the principle of freedom of scientific research: it can only be affirmed up to that point at which such a freedom does no harm to the freedom or welfare of others. If the freedom the scientist claims does not promise to harm the freedom or welfare enjoyed by others, it can readily be accepted. But it cannot claim absolute hegemony over all other freedoms or human goods. Were that the case, then literally anything and everything could be done in the name of scientific freedom, a point that few either would or should accept.

Like all other cherished freedoms, then, scientific freedom must be balanced against other rights and other freedoms; science does not have a sole monopoly on freedom. The next question, thought, is a very hard one: How should we go about setting limits to the freedom of basic research? The question is all the harder since, usually by definition, we cannot know whether the outcome will be good or evil, or some mixture of both. Given our knowledge of evils done in the name of putting social good ahead of scientific freedom (Lysenkoism in the Soviet Union, persecution of those working on the IQ or the XYY problem, and so on), and given also our historical knowledge that good science requires freedom as a condition for its very possibility, I would conclude: (1) It is as legitimate to limit the freedom of scientific research as any other human freedom, but it should be limited only when it comes into a clear and present conflict with other equally cherished human freedoms and rights. (2) Although it is in principle legitimate to limit freedom of research and the right to do such research, there should always exist a *prima facie* right to do such research. By that I mean that right will normally obtain unless it can be shown that an honoring of that right will do significant harm to other rights and other freedoms; thus the burden of proof should lie with those who would restrict the right of free inquiry. Put another way, the importance of free scientific inquiry is of such significance that it should be set aside only for the most serious and immediate potential threats to human welfare, and the burden of demonstrating those threats should lie with those who would propose restrictions on freedom. That would rule out arbitrary restrictions, a demand that scientists somehow prove to others that their research will not do harm, and any attitude that automatically casts doubt on the right of scientists to pursue their research.

More specifically, let me suggest that any limitation upon the freedom of scientific inquiry requires that a number of conditions be met, in addition to a general showing that significant harm may result if the freedom of research is honored in an unlimited way: (1) that the principle of "due process" be honored, that is, no arbitrary or capricious limitations on freedom; (2) that freedom be limited only by public agreement; (3) that any public agreement to limit freedom be, in principle and in practice, open to later reversal (e.g., the use of moratoriums rather than permanent bannings); and (4) that public agreement involve both scientists and the public (that is, scientists are not hindered from making their case to the public, or excluded from any final decision).

I believe this kind of a general process would protect the legitimate interests of science in freedom, yet at the same time protect the public as well. It would establish a good balance between the rights of the public and the right to freedom of scientific inquiry.

While one might well agree that, in principle, scientific inquiry might be limited or controlled under certain circumstances, it could also be argued that any limitations and controls should be imposed exclusively by scientists themselves. Why not self-regulation rather than public regulation? Quite apart from the personal virtues or vices of individual scientists, or of the scientific community as a whole, there seems no special reason to believe that either science, or any other discipline or profession, can or will adequately and fully regulate itself. That has not worked well in other fields and there is no reason to believe it would work any better in science. This is in no way to cast doubt upon the integrity of scientists. It is only to say that in cases where self-interest is involved, it is never wise to wholly trust that those with a stake in the outcome will be the best judges. The point here, then, is not to single out science for any special distrust. It is the principle of self-regulation itself that is to be doubted, in all disciplines and professions.

Quite apart from personal integrity, it should at least be recognized that the interests of the public and of scientists will not necessarily be identical. Scientists cannot fully regulate themselves in the public interest because they are not necessarily representative of the public. There is no special reason to believe that scientists are a representative group of human beings, or to suppose that they can so fully detach themselves from their own special inclinations that they can act fully on behalf of the public.

It can also be pointed out that there is nothing in the training of the scientist *qua* scientist that gives him or her any special expertise or wisdom about what will serve the public interest. A scientist is certainly an expert on scientific methodology, on scientific data, and the like, but none of that expertise automatically entails any special wisdom about the public interest or about the welfare of the society that has that kind of wisdom, and thus all the more reason to leave the issue to open public debate and the democratic process.

DISCUSSION

PROF. CHARLES FRIED: I would like to suggest a further distinction in that valuable set of distinctions.

I'm sure you don't mean to suggest that just any effect on society is a sufficient predicate for regulation. Also, one has to consider what the mechanism of that effect is. If the effect is through ideas, I doubt you would want to allow that as a predicate for regulation. Because otherwise, Newton, Darwin, Kerouac, and Sylvia Plath, I suppose, would have had to file environmental impact statements before publishing their works, which contained many dangerous ideas with very important effects on society.

DR. DANIEL CALLAHAN: I accept that further distinction.

PROF. BERNARD D. DAVIS: There clearly is a significant difference of opinion between Dr. Callahan and me about the question of self-policing.

What he presents is indeed, philosophically, a very neat picture: No group can be trusted to police itself. History tells us that. And why should we be asked to trust scientists any more than any other group?

I think it is very neat, but I don't think it is pragmatically right. Because history has shown us centuries in which scientists pretty much policed themselves and decided what they were going to do and how they were going to do it, and it did not work out badly.

I do not mean that there are not areas where we discover that a traditional pattern is no longer satisfactory. The most conspicuous one will come up for a lot of discussion in this meeting: the tradition in which the paternalistic physician doing some research could decide who was going to be the subject and what was going to be done to them without any informed consent.

We have gone beyond that. But it is another matter to accept the very broad principle that somehow we cannot trust bodies of scientists to set up adequate regulations within the area that Dr. Callahan defined as needing regulation, and on which I think all of us in science would agree. Nobody asks for absolute freedom in science. We all realize if you want to study burns, you can't take a Bunsen burner and go around and burn subjects, and if you want to use inflammable solvents, you have to have rules for how you store them in chemical laboratories; there are any number of such areas.

So there are no absolute freedoms to do whatever you damn please because you are labeled a scientist. But in the very restricted area where the activities in the research lab could be dangerous and where regulations are necessary, the simple assertion that somehow scientists cannot adequately set them up for themselves doesn't seem to me to fit the historical fact that for a long time they have done it and there hasn't been any trouble.

The troubles have arisen or the uneasiness has arisen largely in the areas where we are not dealing with the well-defined hazards that the scientists themselves want to regulate because they are the ones most exposed. Our problems arise in the areas where we are dealing with hypothetical, vague, conjectural hazards of the sort that I described and there, by and large, I think regulations are not in order at all.

DR. CALLAHAN: Professor Davis, you have really neglected the second point I made about regulation, namely, that the scientific community may have very different standards and sets of values than the public at large. Therefore what a scientist might accept as a risk or acceptable risk might be very different than what the public would, and therefore you may have a very significant discrepancy of values there.

I think equally important, perhaps, is that in one sense science surely has policed its ranks very effectively. I

25

would think in a time when the impact of science is increasingly great, how much trust ought one to have? And it seems to me there is room here for a certain wariness of absolute trust.

I would simply use as an example your stressing the need for technical discussion before issues go public. Well, how is one to know that one can trust the process of that technical discussion, first of all, and, secondly, how can one be certain that the value standards employed for that technical discussion are standards that the public would find acceptable? I'm not that confident in the private discussion of scientists. But, again, I wouldn't be confident about the private discussions of any groups.

DR. DANIEL STOWENS (Utica, New York): The problem here is the use of the words *science* and *scientist* and self-regulation by scientists. There are people in this room with diverse interests. Within science there are innumerable specialized groups, none having complete knowledge of what the others are doing, but they share a great deal.

Self-regulation comes from the fact that a physicist can sit down and work out a problem and his mathematician–scientist brother sits down and says, "Well, look, you have made a fundamental error here." Similarly, the human geneticist when he begins to talk to the theoretical geneticist may find that he has some awful problem.

The regulation of science is within science itself by its very nature; and trying to regulate the whole body science, I think, is just an impossible thought and certainly is based on the erroneous idea that all scientists share everything.

MR. RIHITO KIMURA (Tokyo Lawyer; Visiting Scholar, Harvard University): My question is to Dr. Callahan about the limit of regulation within the scientific community. My question is focused on the issues of the Hiroshima and Nagasaki experience in Japan. As you know, a hundred thousand people disappeared almost immediately and still almost one thousand people are dying every year because of this tragedy.

I think scientists must have consent or know the results beforehand. But I don't know about the responsibility of scientists at this moment, thirty years after the Hiroshima and Nagasaki experience. How do you see the ethical responsibility of the scientists in the field of national and international politics? This is a very big question. But, as a Japanese, I really feel the big tragedy and the responsibility of the scientists to see these issues beforehand.

DR. CALLAHAN: You asked a very large question and I have to give a very large answer.

It seems to me that it is absolutely imperative that scientists do try to foresee the consequences not only, obviously, of their applied research but also of basic and theoretical research as well. The scientist has, I believe, a major task to visualize what the likely consequences of the scientific work are going to be. Once there is any notion whatever of some possible harm to the public, that information should be immediately communicated to the public.

In short, I see the scientist as a fundamental moral agent but one who must also be in very close relationship to the public. I think those of us who are not scientists are very dependent upon the moral responsibility of the scientists in calling issues to our attention. We are not in the laboratories; we are not the ones who are going to see the first glimmerings of new ideas, whether good or bad. Hence, the scientist is in that sense the first person who has a moral responsibility.

DR. RICHARD RATZAN (University of Connecticut): A question to Dr. Callahan, who has written about this before. The same question is whether scientists should, or how they should, go about trying to figure out if forty-three years from now something is going to turn out very detrimental to the public welfare. Although that is, I think, a real issue that scientists should face up to for themselves and although it is extremely hard to figure out whether what you are going to do could be used by others negatively, isn't that in the last analysis passing the ethical buck upon yourself, when it really rests with someone else?

If you designed the atomic possibilities for the atomic bomb, should you not do so simply because someone else may use it for that purpose? If *everyone* really followed their own ethical principles of not harming others no matter what ethical or religious system is followed, wouldn't it be acceptable to devise atomic energy or asbestos?

In other words, is the responsibility really on that scientist for having constructed something which can be used harmfully? If someone else did not use it harmfully, then it would never have been a problem for the scientist who devised it. Similarly, if the scientists at Asilomar had raised the issue, the public might have said it's too early for their involvement in this technical discussion; it is a problem and we're glad you brought it to our attention, but now the next process is so and so. In other words, I don't think it is fair to assume the responsibility of other people who use your information wrongly.

Dr. CALLAHAN: One surely can't be responsible for somebody else's misuse of the knowledge or device you develop. But it seems to me one ought at least to foresee that as a possibility. In some cases one can see that there is almost a likelihood that something will be misused. If one can foresee the likelihood of misuse, then you do, it seems to me, bear a moral responsibility for something.

Granted, in most situations we are going to be relatively ignorant. But I would simply argue that any moral agent at least has to take account of what may happen when the good idea goes awry or the good idea is put in the hands of somebody who will not use it properly. If it is a case that the idea or the device is almost certain to be misused, then it seems to me one might have a real obligation to refrain from doing it in the first place. That gets very tricky, as to how high the likelihood would be.

4

The Scientist's Right to Research and the Legitimacy of Governmental Regulation

John A. Robertson
Law School and Medical School
University of Wisconsin
Madison, Wisconsin 53706

The recent controversy over recombinant DNA research has raised important questions about the legitimacy of government regulation of scientific research. Many scientists claim that they have a right of free inquiry—a right to research—that governmental constraints on a scientist's choice of research topics and methods violate. The purpose of this chapter is to examine the constitutional status of this claim and to discuss its implications for state regulation of research.

In my view, the claim for a constitutionally based right to research has substantial support in First Amendment doctrine.[1] While most First Amendment cases involve state restrictions on dissemination or publication of already acquired information, there is growing recognition of a right to receive or acquire information.[2] Indeed, protection for the acquisition of information would seem to be required by the First Amendment if dissemination is to be fully protected. Otherwise the state could limit information flow by focusing restrictions unacceptable at the dissemination stage on essential predissemination activities. Although the Supreme Court has not yet faced this question, there is substantial reason for thinking that the right to acquire information from willing sources would be recognized.[3] On this view of the First Amendment, the right to research, at least with willing subjects and materials under a scientist's lawful control, is constitutionally protected.

The meaning of a constitutional right, however, needs to be clarified. Rights may be either positive or negative, depending on whether they give the holder a claim (and thus impose a duty) on another to provide resources or merely impose a duty not to interfere with the holder's exercise of the right.[4] Most constitutional rights, including rights of free speech and research, are negative rights. They protect the holder against (or negate) governmental interference with an individual's decision to exercise the right. But they

TABLE I.   Types of Government Regulation of Research

Sponsor	Source of harm	
	Research methods	Research products
Public	Always legitimate	Always legitimate
Private	Possibly legitimate	Almost never legitimate

create no entitlement in the holder to have the state provide the resources necessary for its exercise. For example, the state may not stop a citizen from going to Washington to protest governmental policies, but it is not obligated to give him a plane ticket to do so.[5] Similarly, a constitutional right to research would create a duty in the government not to interfere except in very compelling circumstances with a scientist's choice of research topics or methods. It would not impose a duty on the state to provide the resources essential to conduct the research.

Also important in understanding the scope of a right to research are distinctions about the focus of government restrictions on research. The state may restrict research because of harms that flow from the research process itself, or because of harms that arise from the uses and consequences of the research product. Second, restrictions on research methods or products can be applied to publicly or privately funded research. Assuming a First Amendment right to research, the legitimacy of research regulation will vary with whether the research is publicly or privately funded and, within each category, with whether the regulation is aimed at harms arising from the research process or the research product. To avoid confusion, one must be clear about the type of regulation in question. Table I illustrates four possible variations in the legitimacy of governmental regulation.

GOVERNMENT RESTRICTIONS ON PRIVATELY FUNDED RESEARCH

In the remainder of this chapter I will explore the legitimacy of governmental regulation of research on the assumption that research with willing subjects or materials under a researcher's lawful control is constitutionally protected by the First Amendment. Looking first at governmental restrictions on privately funded research, we see that the state's power to restrict such research turns in the first instance on whether regulation is aimed at harms from the research product (the topic or content of research) or harms from the research process (methods of research).

Restrictions on Privately Funded Research Product

If research is constitutionally protected, the state would have little power to restrict research because of the ideas being explored or the applications to which they might lead. State suppression of knowledge because of fears about potential uses quite clearly conflicts with the Constitution. The First Amendment prohibits government from restricting the flow of information because of an assessment of the worth, utility, or potential impact of the information. It follows that the state has no more authority to restrict the acquisition or production of knowledge on the basis of an evaluation of its worth or content than it does to prevent the dissemination of knowledge and ideas that have already been developed. Since the state could not ordinarily prohibit publication of research results, it could not prohibit production or acquisition of the knowledge that would be disseminated. On the other hand, in the extreme case when publication could be prohibited because of its content, then it is likely that the research necessary to produce the information could also be prohibited.

When might the government restrict publication of research findings and hence of the research precedent to publication? The answer is rarely, for a very rigorous scrutiny would be applied and the government would have the burden of establishing that significant harm would imminently occur from the publication. Under the evolving permutations of the clear-and-present-danger test, as most recently articulated in *Brandenburg v. Ohio*,[6] the state could ban publication only to prevent imminent lawless action or some other substantial harm from occurring.

On this standard only clear cases of direct, imminent harm to others would support a ban on publication. What might that be? Perhaps a formula for deadly nerve gas constructed from potatoes and onions and thus within the reach of everyone would qualify. Perhaps national defense secrets that would clearly give an advantage to enemies would also qualify (although even the engineering principles of the H bomb might not fall into this category).[7] But information and research findings deemed morally offensive or subject to misuse would not. None of the research that some have recently tried to stop would seem to meet this rigorous standard for banning publication of research findings. Information about cloning, genetic engineering, sex selection, the genetic basis of intelligence, aging, and interstellar communication[8] may be undesirable because of potential social disruption or oppressive use, but it does not pose such a substantial or imminent harm that the government's heavy burden of justification for restricting publication on content grounds would be met.

Since publication of research findings can be stopped only in extreme situations where there is a high probability of serious risk to others uncontrollable by less restrictive means, state interference with privately funded research because of the ideas to which it might lead would also be unconstitutional. Research could be banned because of the type of knowledge sought only in circumstances where publication of the results of the research could also be penalized. If research is constitutionally protected, a lesser showing of imminent harm would not be acceptable simply because the government focused its restriction on the research rather than on the publication stage of communication. In either case the government would have to meet a very heavy burden of justification that may be formulated as follows: a high probability, approaching imminence, that publication of research findings would cause substantial harms within the power of government to prevent that no other alternative would as effectively prevent.

Restrictions on Privately Funded Research Methods

If research were constitutionally protected, the state would have wider leeway in restricting privately funded research methods than it would have in restricting the content of such research. Just as the state may require a speaker to change the time, place, or manner of disseminating his message in a public forum in order to accommodate community concerns with privacy, peace, and competing uses, it may require researchers to alter the methods employed in acquiring knowledge to protect the rights of research subjects, the safety of the community, or other interests that might be infringed by unrestricted choice of research methods.[9] If the method restriction "further[s] an important or substantial government interest . . . unrelated to the suppression of free expression" and the restriction only incidentally restricts content with the least restrictive means possible, it is constitutionally valid.[10]

This standard means that government regulation of research methods to promote the health, order, safety, and welfare of subjects and nonparticipants is valid if there is a reasonable basis for the concern, if it is independent of the research product, and if less onerous means of protecting the same interests are not feasible. Under this standard researchers would not have a First Amendment right to violate criminal statutes or harm others just because scientific knowledge results. An experiment to test poisons may yield useful scientific knowledge, but the state's interest in protecting human life—an interest unrelated to suppression of the knowledge sought—allows the state to punish such an experiment. Similarly, the state may punish the researcher who invades privacy to advance knowledge, just as it might punish the journalist who trespasses[11] or invades the privacy of a sick hospital patient to gather news.[12] Thus laws requiring informed consent from research subjects are valid, for they promote interests in individual autonomy and welfare independent of the research product, even if they sometimes prevent recruitment of subjects in scientifically important research. Similarly, laws restricting research with fetuses, children, or incompetent persons promote state interests in the protection of subject's welfare, and are valid despite a dampening effect on research with those populations. Laws restricting DNA research to certain laboratories or containment procedures would be valid if they protected noncontent interests actually threatened by the manner in which research is conducted.

One consequence of the less rigorous standard for non-content-related regulation of speech (and hence research) is that speech (research) could effectively be stopped if alternative methods for conducting it are not feasible. In such a case, however, the government's aim is not to diminish ideas but to protect non-content-related interests. The reduction of information flow is an unavoidable, incidental effect of protecting those interests. The First Amendment allows the state to choose to avoid these costs at the cost of the lost benefits of the research, for the invidious discrimination among ideas prohibited by the First Amendment contained in content restrictions is not present. In most cases the research can usually be done in some acceptable way, so that the loss in benefits from knowledge not obtained is generally small and less than the losses incurred if the research methods are unrestricted.

Scientists, however, may be skeptical about such legal distinctions. Since method restrictions have a lesser burden of justification to meet, scientists may fear that the government will use restrictions on methods as a subterfuge to control content. This problem arises with any test of legislative motive and exists in many other constitutional contexts.[13] The main safeguard here is judicial scrutiny of the government's claim, to make sure that the method restriction is reasonably related to noncontent interests and narrowly drawn to restrict research as little as possible. Under this standard the courts might invalidate as overbroad a ban on all DNA research that purports to protect health and safety. Given the possibility of conducting some DNA research safely through physical and biological containment, such regulation may be an attempt to limit knowledge in the guise of regulating method. However, restricting DNA research to physically safe facilities would be constitutional if there were a reasonable basis to think there were health and safety hazards from unconfined DNA research, and if the restrictions were precise, narrowly drawn, and applied only to DNA research that posed risks to health and safety.[14] Under the constitution the legislature is free to prefer the interests of nonconsenting subjects or persons likely to be harmed by the research over the benefits of increased knowledge.

GOVERNMENT REGULATION OF PUBLICLY FUNDED RESEARCH

While the First Amendment provides wide protection against government interference with privately funded research—indeed, at least the same protection extended to publication of the research results—it provides much less protection against governmental regulation of research content and methods in funded research. Government decisions to allocate research funds to one category of research over another and to attach conditions regulating research to such allocations undoubtedly exert a powerful influence on the research activities of scientists. However, funding decisions are prudential or policy judgments within the constitutional authority of government to spend for the general welfare. Ordinarily, funding decisions do not violate rights against government interference with individual choice: Constitutional rights do not protect against every conceivable influence and do not create positive entitlements to a share of state resources even when resources are necessary to exercise the right. The right to travel to Washington, have an abortion, or send one's children to private school implies[15] no duty in the government to pay the bill for exercising the right.

As a result, the state may constitutionally influence the content and direction of research through its funding power. It may through grants and subsidies promote one line of scientific inquiry and starve another. For example, a decision to divert funds from recombinant DNA or laser research to support drilling into the earth's mantle raises no constitutional problem, however unwise or unproductive in terms of science or social utility the decision seems.[16] Similarly, there is no question that the unsuccessful Bauman amendment to the 1976 National Science Foundation appropriations bill, which would have given Congress veto power over any National Science Foundation funding decision,[17] would have been a constitutional exercise of Congress's funding power. Under our constitutional scheme of government, the remedy for objections to funding policies is the ballot box and not the courts.

An important implication of the power to fund is that the state may condition research grants on content and method restrictions that would be beyond the power of the state to impose directly on nonfunded research. Thus a government funding agency may give grants only to those who conduct certain kinds of research, and then only when they agree to use methods that protect the health and safety of subjects and the surrounding community. Just as the funding agency is free to make content assessments in its allocation decisions by choosing to fund only research that it thinks is scientifically more valid, socially more useful, or morally more desirable, it may likewise choose to fund only those methods of producing the desired knowledge that protect third-party interests threatened by the research. Because the researcher has no right to receive funds for particular research ends, *a fortiori* he has no right to receive such funds free of restrictions on the manner of conducting the research. Thus federal regulations requiring institutions that receive federal research funds to create institutional review boards to review funded research with human subjects, and biohazard committees to review recombinant DNA research, are constitutional exercises of the state's power to fund.[18]

Broad authority over funding decisions also exists at the state level. State institutions may allocate research funds and positions according to content and method criteria that they choose. A state institution may legally staff research positions with those persons who agree to research certain topics, to conduct research in a certain manner, or to abide by university rules for the conduct of research. Thus it may condition hiring faculty on their agreeing to submit research to human subject and biohazard committees, even though these involve prior restraints that would raise serious constitutional problems if imposed on unfunded research directly by the state.

The Conditional Spending Power as a Limit on Regulation of Federally Funded Research

There, are, however, two important constraints on the government's power to restrict topics and methods in publicly funded research. One concerns the power of government to regulate nonfunded research as a condition of the research that it does fund. Congress currently uses its funding power to regulate nonfunded research that probably could not be regulated directly.[19] The National Research Act requires the secretary of HEW to require institutions receiving federal funds for research with human subjects to create IRBs to review all research with human subjects conducted or sponsored by the institution, whatever the funding source.[20] A similar requirement has been imposed on institutions doing federally funded recombinant DNA research.[21] Regulation through the funding power thus extends the government's hand over nonfunded research considerably further than occurs with direct government regulation of nonfunded research. The state's power to regulate nonfunded research through the funding power has not been clearly established. The validity of such regulation hinges on the scope of the funding power—on the degree to which it independently permits Congress to regulate, as a condition of funding, nonfunded activities otherwise beyond its power. If the Supreme Court ever confronted the issue, there is a reasonable probability that it would limit the conditional spending power to conditions on the nonfunded activity that are reasonably related to the funded activity. On this standard one could argue that subjecting all research—funded and nonfunded—to the same scrutiny is rationally related to the success of the funded project and thus is constitutional, for it facilitates the ethical review and conduct of the funded research. However, that conclusion is not obvious. Closer scrutiny may show very little actual effect on the funded research, and considerable costs, as many scientists, institutional administrators, and IRB members can confirm.[22] Scientists and universities jealous of their rights might consider litigation challenging regulation of nonfunded research through the funding power. Such a suit would not be frivolous and would clarify the state's power to regulate nonfunded research through the funding power.

The Doctrine of Unconstitutional Conditions

A second constraint on the power to regulate research through funding is the doctrine of unconstitutional conditions. Funding conditions must be within the state's funding power and must also avoid penalizing a person for exercising constitutional rights in areas not funded or supported by the government. Conditioning a government research position or grant on, for example, voting Democratic in the next election would be an unconstitutional condition, which could not be the basis for denying a person the job or grant, because it uses the incentive of public support "to achieve what it may not command directly."[23]

The doctrine has applicability to research funding decisions that are conditioned on the recipient waiving constitutional rights in nonfunded areas. Thus state institutions in hiring staff may legitimately require that they conduct certain types of research according to certain methods on institutional time. A scientist who does brilliant basic research in physics may legitimately be fired if he does not do the biochemical research he was hired to do. However, the doctrine of unconstitutional conditions means that a condition of employment that requires a scientist to refrain from certain kinds of constitutionally protected activity outside institutionally funded time would be open to serious question. If research is protected by the First Amendment, then restricting the content or methods of one's nonfunded research, such as by requiring prior IRB approval, could constitute an unconstitutional condition on employment.[24] A state institution can prevent its resources from being used in genetic engineering, but it cannot deny a researcher a job or funds just because he conducts such research in the proverbial basement laboratory on his own time. Similarly, the state could not deny research funds to a researcher because of his political views.

CONCLUSIONS

As the boundary between science and society shifts toward greater public control of scientific research, the restrictions on research methods developed through the federal funding power are likely to be expanded. Since excessive government regulation can hurt science and may continue to demoralize scientists, the constitutional limits on state control of research become significant. Because research is an essential precondition for publication, state regulation must satisfy the exacting scrutiny required by the First Amendment for limitations on information flow. This provides considerable but not absolute protection for scientists' decisions about privately funded research. When the state restricts private research because of fears about the research product, the Constitution requires the state to demonstrate a very compelling need that could rarely be met. Restriction of research methods to protect subjects and the surrounding community is more easily done, but even then the state must show a health and safety interest independent of the uses of the resulting knowledge that outweighs the burdens on research from restricting research methods.

Although scientists might prefer even greater protection, the importance of these limitations should not be minimized. Constitutionally, there is no forbidden knowledge, but only forbidden ways of seeking knowledge. Scientists cannot be punished because of the ideas they research or the knowledge they create, but they can be for failing to get consent or unleashing dangerous organisms on the community. Galileo is protected; Dr. Frankenstein is not.

In contrast, the government's powerful influence on research through the funding power largely escapes the constitutional net. The Constitution protects negatively, and it gives few positive entitlements to resources. If the government is not directly interfering with unfunded research, it has much room to encourage or influence science to take the paths that maximize values preferred by government, subject only to the constraints of electoral politics. The Constitution barely touches the state's power to condition research expenditures on the pursuit of certain ends by methods that maximize interests other than pure science.

Scientists who disagree with funding decisions, then, must leave off claims of principle and pleas for judicial relief and resort to pragmatics and politics. Given their influential role in the funding process, they are well situated to have their voices heard on the merits. If the past is a guide, those voices will be loud and not easily ignored. In these battles, however, one should not forget that funding decisions ultimately concern trade-offs among competing values, and conflicting views of the desirability of the distributions of burdens and benefits that the trade-offs entail. The question of which trade-offs and distributions to prefer is paradigmatically an unscientific question. On this question scientists have no more authority or insight than other citizens.

REFERENCES AND NOTES

1. See Robertson, J., The scientist's right to research: A constitutional analysis, *So. Calif. L. Rev.* 51:1203 (1970). Hereinafter "Robertson."
2. *Martin v. Struthers*, 319 U.S. 141 (1973); *Virginia State Board of Pharmacy v. Virginia Citizens Consumers Council, Inc.*, 425 U.S. 748 (1976); *Procunier v. Martinez*, 416 U.S. 396 (1974); *Lamont v. Postmaster General*, 381 U.S. 301 (1965); *Stanley v. Georgia*, 394 U.S. 577 (1969).

3. Robertson, p. 1215.

4. For a philosophical discussion of the distinction between negative and positive rights, see Macallum, G., Negative and positive freedom, *Philo. Rev.* **76**:312 (July 1967).

5. This is true even if the citizen is indigent and would not otherwise be able to exercise his right.

6. 395 U.S. 444, 447 (1969).

7. See *United States v. The Progressive, Law Week* **47**:2636 (April 10, 1979). For an account of the dismissal of the case by the government see *New York Times*, p.1, col. 1, Sept. 18 (1979).

8. See Sinsenheimer, R., Inquiring into inquiry: Two opposing views, *Hastings Cent. Rep.* **Aug:**18 (1976); *Daedalus* **Spring** (1978), entitled *The Limits of Scientific Inquiry*, also contains many examples of research that some have tried to stop.

9. Robertson, p. 1226.

10. *United States v. O'Brien*, 391 U.S. 367, 376 (1968); *Procunier v. Martinez*, 416 U.S. 396 (1974).

11. *LeMistral, Inc. v. CBS*, 61 App. Div. 2d 491, 0402 N.Y.S. 2d 815 (1978) (damages awarded against television station for entering restaurant with cameras rolling to record health code violations).

12. Such a violation is graphically presented in the German film of Heinrich Boll's novel *The Lost Honor of Katerina Blum*.

13. See, for example, the problems of determining whether there is a discriminatory motive in situations where the validity of state action that has a disproportionate racial impact turns on the motive for the action. *Washington v. Davis*, 426 U.S. 229 (1976); *Arlington Heights v. Metropoliton Housing Corp.*, 429 U.S. 252 (1977).

14. Indeed, if the claim of many scientists that recombinant DNA research poses no realistic threat of injury to anyone is true, then such legislation would be unconstitutional as not rationally related to a valid state purpose.

15. *Maher v. Roe*, 432 U.S. 464 (1977).

16. For an account of a project attempting to drill to the earth's mantle, see Greenberg, D., *The Politics of Pure Science*, p. 171, New American Library, New York (1967).

17. Bauman amendment's chances down, *Science* **189**:27 (July 4, 1975).

18. 42 U.S.C. 289 1 (a) (1976). 43 Fed. Reg. 60123, Dec. 22, 1978.

19. See discussion in Robertson, J., The law of institutional review boards, *UCLA L. Rev.* **26**:484, 498 (1979).

20. 42 U.S.C. 2891 (a) (1976).

21. 43 Fed. Reg. 60123, Dec. 22, 1978.

22. See Robertson, op. cit. in note 19 *supra*, at 500.

23. *Elrod v. Burns*, 427 U.S. 347, 360 (1976), Van Alstyne, W., The demise of the right-privilege distinction in constitutional law, *Harv. L. Rev.* **81**:1439, 1445 (1968).

24. In this context "nonfunded" means research that receives no state support at all. It would not cover research conducted by a salaried employee of a state institution as part of his duties, even though not funded by a specific extramural government grant.

DISCUSSION

PROF. CHARLES FRIED: I would like to invite Professor Robertson to consider some further issues. It really is cold comfort to be told that the government cannot make the publication of or the engaging in research a crime so long as there is no federal funding.

I suppose the idea that a pencil-and-paper mathematician cannot be prohibited from publishing his research findings is not all that encouraging. As we know, the federal government in one way or another is involved in the funding of practically everything in science, with the exception of industrial research. And it would be a most ironic result if the notion of academic freedom were to receive its only constitutional protection within industry-funded research. That would be a very strange result indeed.

I suggest that a little more constitutional ingenuity is required to prevent that result. After all, it would be strange if the government through its funding power could use that power to withhold grants from universities which, let us say, insisted on an appointments process or a promotion process which did not use a lottery or which did not insist on promoting certain percentages of particular groups in the population but, rather, rigorously appointed or promoted scientists simply on the basis of demonstrated competence.

The idea that because the federal government is supporting those institutions it could constitutionally threaten to withdraw its support on that basis is an idea which is bad news indeed. And I wonder if Professor Robertson could tell me if that is the bad news he intended to deliver. The idea that the government could withdraw its support of a university because it did not fire a professor who published a particular research finding or indeed who participated in a conference at which the heavy-handedness of government regulation was criticized is bad news indeed. And I wonder if that is the bad news he really meant to deliver to us.

The former president of Yale once said that the federal government, because it has bought the buttons, thinks that it can determine the cut of the coat. Well, I hear Professor Robertson saying, yes, that's indeed true. Or at least if it can't determine the cut of the coat, it can threten to rip off the buttons if the private or university groups will not tuck in here and remove a pocket there and sew on a carnation in the other place. I wonder if that really is the bad news that was intended to be delivered.

PROF. JOHN A. ROBERTSON: With regard to the buttons on the coat, the bad news I have is that the state is free not to buy the buttons on the coat because it dislikes the cut of the coat for which the buttons are going to be used.

Now, there are a few limitations on the funding power of government—several constitutional limitations. Let me just mention two of them that, I think, respond to some of Professor Fried's points.

One is that while the government is free to set restrictions on funded research, there is a serious question of whether it may use the funding vehicle to also set conditions on unfunded research. As many of you are aware, the government now makes it a condition of receiving research funds that institutional review boards or biohazard committees also review nonfunded research going on in the same institution.

I must say that I am surprised that the academic and scientific communities have not litigated this issue. Because the power of the state through the funding power to get at unfunded research is open to serious constitutional doubt, even though no one has challenged it.

A second area of possible constitutional limitations which I'm sure Professor Fried has in mind is what is known as the doctrine of unconstitutional conditions. There is a limit on the conditions which can be set on unfunded research. Essentially, they cannot penalize, they cannot use the research funds as a penalty for engaging in some protected constitutional activity elsewhere. And the only import of that would be that they cannot deny me research funds solely because you say you voted Democratic in the last election or chose to engage in a certain kind of research in some unfunded area. That provides some protection, but it still leaves a lot of bad news here for scientists who want to assert claims over the power of government to restrict research through the funding power.

There are some limits here but they are not nearly as great as the requirement which the Constitution provides for unfunded research and this leaves a great deal of room in the funding agencies to control both the content and methods of the research which they have funded.

PROF. FRIED: I feel greatly reassured.

MR. RICHARD VODRA (Cystic Fibrosis Foundation, Brockhill, Maryland): I must say that I am somewhat less sanguine about the limitations on the government's power to regulate research than Dr. Robertson. When I look at the cases involving journalists' power to gather information as opposed to their power to disseminate information, I see the Supreme Court ruling that they cannot go into jails unless the warden lets them into the jail. They cannot go into meetings unless the government agency opens the meeting. I would give as an example last week's mark-up of the labor HEW appropriations bill by the House subcommittee, which has been closed to the public and even to the staffs of the congressmen.

So using the example of this in the research field, they may not be able to say that a journalist can't write about the mark-up. They simply forbid him from learning about the mark-up and conducting the research that is necessary. And I would suggest that it would be an effective way to eliminate research in an area to prescribe such rigorous regulations on the conduct of that research as we were discussing earlier today, so that the research itself becomes impossible.

So you don't say people can't look at genetic issues. You simply make the conduct of genetic research so burdensome and then cut off the money, which the government also has the power to do, through the appropriations process. It seems to me that the power of the government to shape the content of research is much greater than your remarks would have suggested.

PROF. ROBERTSON: I talked about the right of journalists to gather news from willing sources. But the governmental bodies you are talking about in these cases are unwilling sources. As a matter of fact, at least four members of the Supreme Court are willing to extend this right to unwilling governmental sources in certain cases. I think if you look more closely at those cases, you will see that the court is at least implicitly recognizing a right of journalists to gather news from willing sources. That's precisely what the scientist is doing. He or she is gathering scientific information and experimentation from willing sources or materials under his or her lawful control.

It is true that setting restrictions on methods can ultimately make it impossible to do the research altogether. But the state would have to meet a very heavy burden there of showing that the reason for the restriction was not the content or ideas being investigated but the noncontent interests that are affected by the research methods themselves. If there was some other method of doing the research without infringing those noncontent interests, I think the scientist would have a right to pursue it.

Obviously, the funding power gives the government a great deal of authority here to influence research; and the Constitution, as I said before, seems to provide very little protection against that.

MS. TABITHA POWLEDGE (The Hastings Center, New York): I share Professor Fried's dismay at your tying arguments so closely to the funding mechanism, although my concerns are quite different from his. I would be interested to know what, if any, kind of constitutional basis there might be for government regulation of private research? I suppose the best known example at the moment would be recombinant DNA research in industry, particularly the drug companies, which are already heavily regulated.

PROF. ROBERTSON: This matter of the government funding research is something that bothers us a great deal because it gives the government an enormous amount of power. The structure of rights in the Constitution is one that seems to hinge a great deal on the funding power.

I recall Justice Black's comments in *Epperson v. Arkansas*, the teaching-of-evolution case in Arkansas, where he made a comment that no one would claim that if a school board chooses not to fund a program in mathematics, that anyone's constitutional rights would be violated. The school board would be free to fund one area rather than another if it so chose to. The problem in that case was that there was a religious motivation in banning the teaching of evolution, and that is what was wrong with that case.

Now, on your point about private research, the government's constitutional power, at least on the federal level, to pass laws restricting recombinant DNA research, comes probably from the commerce power.

However, the basis the states could use to restrict the research is to protect health and safety. It would probably be impossible for the government to restrict recombinant DNA research because of a dislike or feared effects of the ideas being investigated and the uses or applications to which they might be put at some point in the future. The only basis for restricting that kind of research is that in carrying out the research itself, there are definite tangible harms to subjects or members of the surrounding community that are independent of the nature of the ideas being investigated or the applications to which they might lead. And it is that which would enable them to regulate.

Ms. POWLEDGE: Then you tie it exclusively to rather narrow questions of immediate health and safety?

PROF. ROBERTSON: Exactly. Now, in an extreme case you could prevent research into certain ideas just as you might be able to, for example, stop the *Progressive* magazine from giving out information about how to construct an H bomb. But that would only be an extreme case where there is an imminent danger of great harm to others. Aside from that limiting case, restrictions based on the nature of the ideas would be unconstitutional.

5

Public Control of Genetic Research

Seymour Lederberg
Division of Biology and Medicine
Brown University
Providence, Rhode Island 02912

The Biological Sciences Coming Under Public Scrutiny

This decade has seen enormous increases in our understanding of biological processes dependent on genetic mechanisms. Some of the most noteworthy research findings have come from the areas of prenatal diagnosis, *in vitro* fertilization, DNA behavior during differentiation, DNA analysis and recombinant DNA synthesis, and the induction of mutation and cancer by chemical agents.

Prenatal Diagnosis

The biological qualities of a fetus have been estimated by using the loose fetal cells found in samples of amniotic fluid. This has allowed determination of chromosome sex and gross qualities of chromosome number and type, and also has permitted analysis of those selected metabolic properties expressed by fetal cells. However, the process was not easily applied to specialized tissue products such as hemoglobin. Recently, techniques based on restriction enzymes have allowed the DNA sequences in the fetus's genes for hemoglobin to be compared with DNA from reference hemoglobin genes, permitting prenatal diagnosis of thalassemias and sickle cell anemia.[1] In principle, the method is applicable to all human fetal genes for which adult human DNA references can be obtained.

In Vitro Fertilization

Although experimental fertilization of human ova had been barred in this country, programs of this type had been allowed abroad. An increased understanding of the

41

physiological transformations of the ovum following fertilization and of the womb receiving the implant has now allowed for a successful maturation and normal birth.[2] Surrogate roles for donors of sperm, ova, and wombs are obvious so that specific causes of infertility can be bypassed by a medical facility with the appropriate experience.

Gene Expression

Mechanisms of genetic control in higher cells now appear to be much more complicated than the simple patterns used by bacteria. Regions of DNA for genes and for elements controlling those genes rearrange their relative positions during differentiation as part of the regulation of that gene.[3] Some of these rearrangements may be reversible. Others, like the genes for antibodies, may experience stochastic processes at the boundaries of the rearrangements or otherwise be irreversible in their commitment. These mechanisms give an added direction and individuality to everyone's biological makeup. If the genes irreversibly fixed are needed in the general state for early stages of development, then it may be that the scenario of "cloning" mature cells will not be possible for higher animals, and the apparent success with amphibia may have depended on the use of uncommitted cells. In any event, it strongly suggests that monozygotic twins and cloned forms of life will be found to be imperfect in their genetic likeness as a result of random changes at the boundaries of rearranged genes.

Recombinant DNA

Analysis of DNA by restriction enzymes and the recombination of alien fragments of DNA to make operative hybrid DNA molecules symbolize the acme of reductionist approaches to biology.[4] The fact that restriction enzymes make identical cuts in each copy of a given DNA molecule, and the ability of the cut fragments to be recombined with other fragments have allowed a detailed analysis of regions affecting gene function. These techniques have provided the means for the refined analyses for prenatal diagnosis and for studies on changes of DNA during differentiation. They have been successfully used in studies of sites of DNA at which viruses attach and integrate, sites at which controlling enzymes act on DNA, and sites at which hormonal, metabolic, and environmental signals interact. The production of animal proteins and hormones by bacteria carrying animal genes heralds a large-scale development for these and other useful, scarce biological products of medical and industrial value.

Environmental Mutagenesis and Carcinogenesis

Studies on bacterial strains in which damage or mutation of DNA could easily be measured have yielded an extraordinary correlation of mutagenic potency for agents that are known to be carcinogenic for animals or man.[5] Conversely, most agents that are not carcinogenic for man correlate strongly as not mutagenic. The few exceptions of either class have reasonable interpretations that still allow the major generalizations to hold: (1) chemical carcinogens are mutagens, and (2) chemicals found to be mutagenic in bacteria or other life forms have a very high probability of being carcinogenic to man. Mutagenicity in bacteria can be readily determined in only a few days. Other types of chromosome damage, as well as mutation, can be detected by yeasts and other fungi in a similar rapid manner. A reasonable early-warning monitor of genetic hazards is therefore available as an index of cancer risk in materials that we breathe, eat, drink, or contact.

A separate analysis of the origins of atherosclerotic lesions that are associated with cardiovascular diseases has indicated that the plaques arise from the proliferation of single cells.[6] These are correlated with a diet rich in fat and therefore rich in mutagens whose chemical structure renders them fat-soluble. Other mutagens abound in cigarette smoke, which also is known to increase the risk of arteriosclerosis and cancer of the lungs and bladder. Therefore, complex chemicals contributing to three of our main health scourges—cancer, heart attack, and stroke—can now be quickly approximated by virtue of their genetic activity as mutagens.

THE BASIS FOR PUBLIC CONTROL

These findings in genetics and in interacting areas of biology and medicine are representative of an accelerating growth in our knowledge. We have the obligation of using that knowledge to do no harm, even as we have the opportunity to reduce major portions of our health burden. This opportunity did not come lightly, and the scientific power it represents has raised concern, calling for regulation at several general levels:

1. The genetic experiments on which this knowledge is based may involve or create hazards to our health and environment—recombinant DNA work has been considered mainly in this context, and environmental mutagenesis may also fit here.
2. The research may be controversial to some political or moral persuasions—research on products of fertilization is an example.
3. This research may allow for applications that may be controversial for certain views of social stability—prenatal diagnosis for fetal sex, screening for XYY chromosomes, and research on aging are appropriate examples.
4. This research competes for limited public resources with other public needs, and the allocation of these resources calls for greater selectivity and regulation.

Although the self-regulation track record of many scientists in a specific field may be faultless, the Declaration of Helsinki has not prevented experimentation on classes of people to their detriment and without their consent, or even knowledge. In this light, who does not groan on hearing that someone under the banner of science or medicine proposes to test a viral vaccine on retarded children? There is also no reason to expect a scientist to balance better the competing social values attendant upon controversial work. Finally, the dramatic failure of technical myths generated about a regulated nuclear power industry makes it even more unlikely that the disputed pronouncements of biological scientists on the safety of their own experimental activities will go without public examination and regulation.

Given a demand for some public voice, what part of the scientific processes in genetic research can be regulated, and what are the limits of this regulation? We can distinguish elements such as reading, formulating hypotheses, planning of experiments, and communicating and receiving these ideas and expressions. These elements are sufficiently akin to speech protected by the First Amendment that it is reasonable to believe they *per se* would be unregulated by prior restraints. Justice Douglas, speaking for the Supreme Court in *Griswold v. Connecticut*,[7] held that "in reaffirmation of the Pierce and Meyer cases, the State may not, consistently with the spirit of the First Amendment, contract the spectrum of available knowledge. The right of freedom of speech and press includes not only the right to utter or to print, but the right to distribute, the right to receive, the right to read ... and freedom of inquiry, freedom of thought and freedom to teach. ..."[7] Therefore, although not specifically mentioned in the First Amendment, the cerebral processes of science would probably qualify for protection.

What of the elements of actual experimentation? What of the receipt, storage, handling, transfer, and disposal of materials, and the interactions of experimental materials with other people and with the environment? What accountability is there for such actions? In other contexts, some or all of these activities appear to be covered by federal legislation such as the Public Health Services Act, the Occupational Safety and Health Act, the Toxic Substances Control Act, the Clean Air Act, the Clean Water Act, and the Solid Waste Act.[8] If these elements of technical activity are regulatable by the federal government, we might expect to find the Court decision in *United States v. O'Brien* to apply.[9] In that decision, Chief Justice Warren held: "When 'speech' and 'nonspeech' elements are combined in the same course of conduct, a sufficiently important governmental interest in regulating the nonspeech element can justify incidental limitations on First Amendment freedoms. ... [W]e think it clear that government regulation is sufficiently justified if it is written in the constitutional power of the Government; if it furthers an important or substantial governmental interest; if the governmental interest is unrelated to the suppression of free expression; and if the incidental restriction on alleged First Amendment freedoms is no greater than is essential to the furtherance of that interest."[9]

By example, this might mean that a scientist could not be restricted in lecturing on a regulatable procedure to a class that had no connected intent to carry out that procedure. But if a scientist prepares an outline of regulatable experiments that he plans to perform, then regulations governing those experiments might require prior approval of those plans by an agency delegated that responsibility before those experiments could proceed. This is now done for work on recombinant DNA and for research on human subjects.

If federal legislation regulating genetic research is enacted, can there also be state or local regulation? This would depend on congressional intent, since the doctrine of preemption sets federal law above state law when the latter conflicts with constitutionally valid acts of Congress.[10] The federal preemption can be explicitly stated by Congress or be implied in the legislative history of the act or arise from irreconcilable conflict between compliance with federal and state laws. Federal approaches to legislation on recombinant DNA, the *Rogers Bill*[11] and the *Nelson Bill*,[12] provided for federal preemption but allowed local laws when these were demonstrated necessary to protect health or the environment. Concern over regressive local harassment has motivated some scientists to welcome federal controls with strong preemptive content.

In the absence of preemptive federal regulatory legislation, states have the police powers to enact "reasonable regulations to protect the public health and safety."[13] If a state or locale tried to ban totally a form of research that is federally funded, the question of conflict with the federal research funding authorization would arise and presumably the issue would move through the federal district court for adjudication. If instead, for example, a state or locale were to set standards for laboratory facilities and workers that were not in conflict with the federal Occupational Safety and Health Act,[14] such regulations might stand the challenge of a constitutionality test. Their defense would be heightened by having such regulations cover comparable biological hazards in hospitals and clinics dealing with infectious materials and patients suffering communicable diseases. This might create a reasonable balance between caution and efficiency at the local level. Few communities would want to banish a medical facility so that it became inaccessible to their own membership, and few hospitals would not be improved by reducing their contribution to iatrogenic health problems such as antibiotic-resistant pathogenic bacteria and carcinogenic ethylene oxide residues.

The regulation of genetic research whose goals are socially, politically, or morally controversial and whose procedures pose no harm has frequently been urged. There surely will be social dislocations as research identifies the factors that contribute to aging and makes them susceptible to control. But it is equally certain that it would be impossible to prevent totally the acquisition of this knowledge by restricting projects targeted for this goal. This impossibility arises because the factors involved in aging are common to all life and are part of all biological processes. To restrict the catalog of research that bears on aging would displace most of health research. Furthermore, the very complexity of aging means that the ability to control it will come in small quanta from many sources, and social adjustments can be planned and made incrementally. A policy of preventing people from using available knowledge to live longer when this is desired is unacceptable to our culture. The need for opportunities for younger people need not be limited to those opened up by death and retirement of the elderly. Science and the pursuit of understanding are frontiers that cannot be emptied of opportunities. It may be in that value system that we can accommodate longer lives and new births.

THE REAL MEANS OF PUBLIC CONTROL

Public regulation of science has succeeded through control of the purse with fewer constitutional questions than when attempted through statute. As scientists moved from questions of gene transmission to complex projects on gene function, sophisticated analytical equipment and supporting laboratory facilities and materials became increasingly important. The cost of such research could not be borne personally or by institutional funds, and by necessity shifted to external agencies. With the growing dependence on public funds, the priorities in the allocation of such funds became an effective means of public direction of science. A reallocation of fewer dollars in one area and more in another will, in time, result in a corresponding channeling of researchers. A scientist with a proposal approved by a granting agency, but for which there are no public funds, may change to a problem for which funds are available. This redirection is achieved more subtly when a scientist submits proposals on several problems in order to be able to follow even one that becomes the bread-and-butter project for his laboratory.

A more chilling effect has occurred with the withholding of federal funds for specific projects such as research on products of fertilization *in vitro* of human ova. Research in this area was exiled abroad until the American public learned of the first successful human birth arising from fertilization *in vitro*. One may ask whether this instance will repeat itself into a pattern of export of controversial scientific problems to host countries where they are supported. Dr. Stetten[15] has reminded us of Bronowski's belief that science shifted out of Italy as a result of this inquisitorial process.

In practice, therefore, the immediate issues are not whether modern Galileos will be regulated in their ability to disturb society with new observations, or whether they should be allowed to use telescopes in making their disturbing observations, or whether they are subject to strict liability if they drop their telescope on a nonconsenting person. The foremost barrier to creative scientific inquiry today is none of the above. It is the limited allocation of public support that throttles the majority of all approved research proposals by established scientists in this country and sends graduate students and postdoctoral associates in science into substitute careers. A much-needed corollary to the First Amendment protections of inquiry and knowledge is the obligation of the public through Congress to ensure that scientific inquiry receives greater support.

REFERENCES

1. Orkin, S. H., Alter, B. P., Altay, C., *et al.*, Application of endonuclease mapping to the analysis and prenatal diagnosis of thalassemias caused by globin-gene deletion, *N. Engl. J. Med.* **299**:166 (1978).
2. Reproductive technology: Whose baby? (Editorial), *Nature* **274**:409 (1978).
3. Dryer, W. J., and Bennett, J. C., The molecular basis of antibody formation: A paradox, *Proc. Natl. Acad. Sci. USA* **54**:864 (1965); Wiegert, M., Gatmaitan, L., Loh, E., *et al.*, Rearrangement of genetic information may produce immunoglobulin diversity, *Nature* **276**:785 (1978); Seidman, J. G., and Leder, P., The rearrangement of antibody genes, *Nature* **276**:790 (1978).
4. Beers, R. F., Jr., and Bassett, E. G. (eds.), *Recombinant Molecules: Impact on Science and Society*, Raven Press, New York (1977).
5. McCann, J., Choi, E., Yamasaki, E., *et al.*, Detection of carcinogens as mutagens in the *Salmonella/ microsome* test: Assay of 300 chemicals, *Proc. Natl. Acad. Sci. USA* **79**:5135 (1975).
6. Benditt, E. P., and Benditt, J. M., Evidence for a monoclonal origin of human atherosclerotic plaques, *Proc. Natl. Acad. Sci. USA* **70**:1753 (1973); Juchau, M. R., Bond, J. A., and Benditt, E. P., Aryl 4-monooxygenase and cytochrome P-450 in the aorta: Possible role in atherosclerosis, *Proc. Natl. Acad. Sci. USA* **73**:3723 (1976).
7. *Griswold v. Connecticut*, 381 U.S. 479 (1965).

8. Public Health Services Act 42 U.S.C. sections 262, 264 (1970); Occupational Safety and Health Act 29 U.S.C. sections 651 *et seq.* (1970); Toxic Substances Control Act 15 U.S.C. section 2601 *et seq.* (1976); Clean Air Act 42 U.S.C. section 1857 (1970); Clean Water Act 33 U.S.C. section 1251 *et seq.* (1977); Solid Waste Disposal Act 42 U.S.C. section 6901 *et seq.* (1976).
9. *United States v. O'Brien*, 391 U.S. 367 (1968).
10. U.S. Constitution, article VI, clause 2.
11. H.R. 11192, 95th Congress, 2nd Session ¶106 (1978).
12. S. 1217, Amend. No. 754. 95th Congress, 1st Session, ¶484 (1977).
13. *Jacobson v. Massachusetts*, 197 U.S. 11 (1905).
14. 29 U.S.C. section 651 *et seq.* (1970).
15. Stetten, D., Jr., Freedom of inquiry, *Genetics* **81**:415 (1975).

DISCUSSION

PROF. BERNARD D. DAVIS: I believe the preamble to the Constitution contains some general phrase about the obligation of the government to promote the public welfare. Since the concept of what promotes the public welfare has expanded a great deal since 1776, I wonder whether that part of the Constitution could conceivably be invoked by an investigator in a field that was politically unpopular and not being funded.

PROF. JOHN A. ROBERTSON: No. Because the allocation of authority there among branches of government for determinations of the general welfare is to the legislature to determine what the general welfare is. And it has a great deal of leeway in determining what the welfare of the community is, subject to some precise limitations in the Bill of Rights and other constitutional provisions.

It would be a very desirable thing as a matter of general welfare to give scientists an entitlement to research funds as it would be to give poor people an entitlement to health care resources and other things, but we would have to amend the Constitution to do that.

DR. DANIEL STOWENS: It seems that this limitation on research has something else involved, and that's a rare character known as money. We have heard about *in vitro* fertilization and the hue and cry that has been raised. But at the same time, this is a commercial enterprise in cattle. This is the way the very profitable Siementhaler cows are bred today in Texas. They use Clomid and they flush out the fertilized eggs and implant them in the host cows and right now one of the ranches is making five million dollars a year. Nobody says anything about that. I have nothing against cows but I have nothing against people, either. But what is a perfectly acceptable procedure commercially, if it is thought of in purely human scientific terms, all of a sudden becomes anathema.

I would also like to say that your restrictions on what the government can and cannot do didn't seem to apply several years ago when Linus Pauling, a rather vociferous advocate of peace, was suddenly deprived of all his research funds. That really happened; and for what reasons, we know.

PROF. ROBERTSON: Well, just on that last point, you know, there is one other limitation on governmental funding that I haven't alluded to in referring to the *Epperson* case in Arkansas. And that is if you could show an impermissible motivation on the part of the governmental funding agency, that could perhaps nullify a grant which it made. For example, if it gave a grant in order to establish the racial inferiority of blacks, if that was the precise purpose of it, that perhaps could lead to a nullification of that grant. However, it would not create an entitlement in someone who was not funded with those funds to receive those funds. So we really don't have very much protection there.

PROF. FRIED: Wouldn't you say, though, that in a case like the Pauling case that if you could demonstrate that Pauling's access to funds had been cut off, if you could demonstrate that there had been directives that his applications not be considered, that you could have some kind of a constitutional handle there to require that that directive be rescinded and that his application be considered on whatever basis other people's applications were considered?

PROF. ROBERTSON: Yes, I would agree with that. But that would be an example again of an impermissible invidious governmental motivation for a certain action: no funds for Linus Pauling. It could nullify that. That could then lead to further funds for him, but it still would not create a general entitlement to research funds just because scientists or the community thinks it would be a good and desirable thing to have those funds spent.

MR. MITCHEL KLING (Student, Harvard Medical School): I have a question for Professor Robertson on the issue of regulation of private research, since you had stated that the government might have the right to control nongovernmental funded research if there was an imminent danger to public welfare. The NIH guidelines, although they are not law, were at least partially, if not primarily, designed to protect the public from the dangers of research. Could this not conceivably be applied toward private industries doing this kind of work?

PROF. ROBERTSON: No. Because the NIH guidelines on recombinant DNA research were restrictions on research funded by NIH. Therefore there was much more leeway there for the government to restrict research under its power and to set the conditions on how to spend money which it allocates. It would need some independent basis to get at privately funded DNA research, and I think the basis for that would be something like the commerce power; or states could get at it through their general police power. But, again, only because of harm due to the medical research methods themselves and not because of the nature of the ideas being investigated.

PROF. SEYMOUR LEDERBERG: Actually, the NIH guidelines could come into play in the private area, but only after the fact. They could be used as examples of what reasonable procedures are, have been accepted to be in that particular science, and therefore any violations of those procedures which led to a tort action would presumably put that private sector in a poorer light.

PROF. ROBERTSON: NIH now, with the guidelines on recombinant DNA research, applies them not only to the DNA research funded by NIH but also to other DNA research that is unfunded going on in the same institution. And it is that attempt to get at unfunded research through the funding power, about which there is a serious constitutional issue, that has not yet been raised by any institution.

PROF. FRIED: Well, maybe you have provided a cause for some underemployed public interest legal organization.

PROF. ROBERTSON: Yes, or some underemployed university counsel who wants some suit to bring against the increasing governmental intervention into academic affairs. As I say in my paper, I can't guarantee the success of such a suit and I present analysis for the argument which he or she could apply. But it would not be a frivolous suit and it would go a long way to clarifying the power of the government to get at unfunded research through the funding power.

DR. DEWITT STETTEN: In regard to participation in the guidelines by private industry, we at NIH made an effort to secure a voluntary concurrence on the part of the Pharmaceutical Manufacturers Association. This was forthcoming, with certain exceptions, which were not without interest.

For instance, in the guidelines we absolutely prohibited experiments involving volumes larger than 10 liters of solutions containing infectious material. This, because large spills are more difficult to clean up than small spills. Industry said that they would have to break this regulation and, in effect, there was nothing we could do to stop them.

I would, however, like to ask Dr. Robertson, perhaps, or whoever may feel qualified, concerning recent action taken by the Food and Drug Administration. My understanding is that the Food and Drug Administration approves or disapproves intervention of drugs on the basis of the final product, its effectiveness, and its safety, and up to this time has never concerned itself with the method whereby that final product was prepared.

There is now, I believe, an order extant instructing the Food and Drug Administration not to approve products regardless of their safety or efficacy if in the process of manufacturing, recombinant DNA techniques were used in violation of the guidelines. I would be interested to learn from the attorneys whether this is likely to be contested in court and what they think might come of it.

PROF. ROBERTSON: The point that this is the first time the FDA has done something of that sort I'm not sure is accurate. I believe the 1962 Kefauver amendments requiring the safety and efficacy standard also required that the research done to establish safety and efficacy be done with informed consent of the subjects involved. If informed consent was lacking in those cases, the actual data generated in those studies could not be considered by the FDA concerning safety and efficacy. And I think it is that kind of parallel which you are talking about.

PROF. FRIED: Sort of an exclusionary rule.

PROF. ROBERTSON: An exclusionary rule for violating method restrictions or the ethics of doing the research. It would seem to me it would be just as valid in the case of violating the rule about recombinant DNA research, as long as there were some reasonable basis for concern for the health and safety of the community.

MR. RICHARD VODRA (Cystic Fibrosis Foundation): The FDA does regulate drug manufacturing processes, so that it would not regulate the process by which a drug is discovered; it would regulate the process by which a drug is tested and the process by which a drug is manufactured. So if indeed recombinant DNA is used in the manufacture of the drug, that would come under the FDA's jurisdiction.

6

On Mandating Diagnostic Practices

STANLEY JOEL REISER
Harvard Medical School
Boston, Massachusetts 02115

As American society faces the perplexing task of harnessing medical knowledge for maximum benefit and minimum harm to person and purse, the temptation grows to take medical decision making out of the hands of the private sector and embed it within a structure of laws and rules. The large question of whether a pluralist health care system that encourages independent choices by individuals and agencies working within a competitive framework is superior to a corporate system based on central direction, comprehensive planning and regulation, and bureaucratic control is beyond the scope of this chapter. But I wish to comment upon one aspect of medical practice significantly harmed by efforts to take decisions away from individuals—the diagnosis of illness.

During the 20th century, and particularly in the last two decades, procedures and rules have been developed by institutions, private and public, that require physicians to gather certain kinds of evidence to detect disease. I offer several examples.

1. Two outstanding features of modern medicine are its harnessing of techniques of biochemistry and microscopy in the service of diagnosis, and its use of X rays to gain knowledge of human pathology. The discovery of X rays by Wilhelm Roentgen in 1895 and the integration of techniques of chemical and microscopical analysis through the diagnostic laboratory, which was created at the turn of the 20th century, were events at the center of the dramatic increase in the power of medical care in modern times. Physicians, exhilarated by the recognition that through these procedures they might at last practice "scientific" medicine, impressed that these procedures might detect disorders hidden from their own and the patient's senses, and encouraged by a popular view that more tests meant better medicine, began to order laboratory tests and X rays with increasingly greater frequency and in larger and larger numbers.[1]

Gradually, private decisions by doctors to use these tests were supplemented by institutional decisions to require them. As the 20th century progressed, growing numbers of hospitals established rules that made each patient admitted undergo a number of technological procedures (usually chest X rays, urine analysis, and a variety of blood tests) whether or not the attending physician thought them necessary. Only in 1979, in the first major challenge to this practice, have Blue Cross/Blue Shield Associations declared they will not pay for such routine tests given to patients entering a hospital for nonsurgical complaints.

2. Another instance in which the doctor is required to use a given diagnostic procedure involves a urine test for a genetic disorder that may cause severe mental retardation—phenylketonuria (PKU). Beginning in 1963, under the impetus of efforts by individuals and organizations devoted to reducing the incidence of retardation in the population, state legislatures in this country began to pass laws that compelled physicians to order a PKU test for all newborns. By the end of the 1960s, almost all states had such a statute in effect.[2]

3. The final example I offer comes from the courts. In 1974 the Supreme Court of the state of Washington, in the case of *Helling v. Carey and Laughlin,* reversed a lower court decision and declared that two physicians, ophthalmologists and partners in practice, were negligent for failing to administer a pressure test for glaucoma to a patient, now 32. She had been coming to them with visual complaints from 1963 to 1968. The doctors testified that they had attributed her symptoms to complications associated with contact lenses prescribed at her first visit to them in 1959. Only on her last consultation with the physicians in 1968 did they do the pressure test; it was positive. By then the patient's vision was seriously damaged, the result, as none disagreed, of having had glaucoma for at least the 10-year period that encompassed her treatment by the doctors. The standard practice in the medical profession at the time of the therapy was not to routinely do the pressure test on patients under 40 unless symptoms indicative of glaucoma were present. Nevertheless, the court wrote its opinion, as legal expert William Curran comments, "virtually holding the entire specialty of ophthalmology to be negligent in not administering pressure tests routinely with all patients regardless of age."[3]

The search for hidden disease in unsuspecting victims, which all of the above examples represent, deserves a high priority in medical practice. I support it. But I want to examine some of the issues raised by making such procedures legal or quasi-legal requirements.

First, let me emphasize the exclusion from this discussion of policy involving contagious disorders or national emergencies. Such circumstances may, properly, require standard and compulsory diagnostic actions by physicians such as the testing by X ray of all army recruits during World War II to find tuberculosis and prevent its spreading through the ranks. This discussion deals solely with disorders that threaten individual health alone.

By compelling physicians to use a given diagnostic procedure, we imply that at least three circumstances prevail: that it accurately identifies the disorder in question, that the disorder is serious and warrants preventive measures, and that misgivings about the professional judgment and knowledge of physicians as a group cause us to worry they will not effectively use the procedure, given free choice. Let us hypothetically assume that these circumstances hold true for the tests under discussion (controversy on some of these points does in fact exist). On what basis, then, can tests for other disorders be included or excluded from compulsory performance? If we worry about PKU or glaucoma, should we not be concerned also with finding genetic and acquired diseases that are serious, and for which equally good tests exist? Our dilemma is intensified by the absence of a reliable set of rules or a principle to decide this question. The number of possible procedures we could use to check for unapparent disease is large, their cost great in terms of physical and financial risks to patients. We thus cannot do everything diagnostic medicine is capable of doing. In the absence of such a precept, we allow combinations of random social, scientific, and political events to determine which tests become compulsory. Yet we are left with a gnawing problem. If mistrust of professional judgment has caused us to initiate actions to require doctors to check for certain significant disorders, how can we feel comfortable leaving use of other equally important diagnostic tests to their personal judgment?

It is not possible to practice good medicine by formulas. Wise choice from among a large number of alternative actions remains the heart of doing superior clinical medicine. The development of this crucial ability in physicians requires a sense of ultimate responsibility for the choices involved. Creating—cookbook style—a list of required diagnostic actions is the best formula I can think of to erode this sense of responsibility and the excellence it helps develop and sustain. As such a list grew, excellence of clinical judgment would surely diminish. Medicine has not reached a stage (nor do I believe it will) when canons of care can be epitomized in checklists and physicians, like actors, successfully perform their tasks following a script.

A second general problem of regulation-determined testing is its lack of quick response to scientific change. Practicing physicians, reading a convincing argument in a respected medical journal against a standard procedure, should be free to change their practice that very day. Yet if the procedure is written into rules, doctors face a difficult choice: to follow their own judgment and, by so doing, also remain faithful to the ancient precept of not harming patients, or to follow the rule until officials change it. A new finding that challenges an accepted dogma is a commonplace event and a prevailing feature of modern medicine. This climate of rapid change is antithetical to clinical practices determined by regulation.

I thus think unwise those legal and quasi-legal rules that make physicians follow certain patterns of diagnosis. I support instead other routes to secure high standards in such medical decisions: among them, increasing requirements for the continuing education of doctors, more frequent and rigorous testing of doctors to maintain their licenses, just and expeditious professional and legal censure and punishment for wrongs committed in practice. Requiring physicians to perform a given test on a given patient diminishes their application of, and endangers the sensitivity of, the most important instrument physicians possess in providing medical care—keen clinical judgment.

References and Notes

1. The evolution of these events is examined in Reiser, S. J., *Medicine and the Reign of Technology*, Cambridge University Press, New York (1978).

2. A summary of these laws and the events leading to them is found in Swazey, J. P., and Bessman, S., Phenylketonuria: A study of biomedical legislation, in: *Human Aspects of Biomedical Innovation* (E. I. Mendelsohn, J. P. Swazey, and I. Taviss, eds.), p. 49, Harvard University Press, Cambridge (1971).
3. Curran, W. J., Glaucoma and streptococcal pharyngitis: Diagnostic practices and malpractice liability, *N. Engl. J. Med.* **291**:508 (1974).

Discussion

PROF. CHARLES FRIED: I had hoped that I might provoke the discussion in this final segment of the morning by noting an irony, and that is that we all applauded as Dr. Reiser called for less regulation of physicians and indeed gave examples of the awkwardness, ham-fistedness, and nonsense of all of these regulations of physicians. And yet the earlier part of our program was devoted to nodding approval to the similar regulation of *scientists*. I think that is indeed ironical. Because it would seem to me that physicians are in a much more direct position to do harm to people than are scientists, whose ability to do harm to people is, after all, somewhat remote and speculative. It often requires considerable imagination to see harm in the work of pure scientists, relative to the harm that doctors can do. And yet here we have heard a passionate plea for self-regulation and for keeping the regulators and the judges and all of those troublesome and relatively ignorant folk out of the matter.

Now, I hope other people have noted that irony and are going to make some remarks which will help resolve it.

PROF. MIRIAM LARKIN (Mount Saint Mary's College, Los Angeles): I think some of us were not applauding Dr. Reiser's conclusions; I think we were applauding the thought that he brought up the issue. I think that will take care of that one point.

Second, with respect to the first point made by Dr. Reiser, I believe he is assuming a particular image of the doctor–patient relationship in which the doctor is the one who makes the decision and is the only one involved in making the decision. I believe part of the regulation by the state government may have developed because someone has to stand on behalf of the patient and perhaps there is more than one way in which the medical profession should have been standing on behalf of the patient. The patient is in a state of being deprived of a certain expert knowledge, and it is possible that the patient should know the kind of thing to ask for in a given instance, the kind of test to request. It seems to me that the medical profession bears some responsibility there.

With respect to Dr. Reiser's third point, the technological procedure is not sufficient for determining what might be wrong, but in many cases it is necessary. Without it, what is wrong will not be determined.

DR. STANLEY JOEL REISER: In regard to your last point, I was not saying technology was unnecessary. I was saying that too often it is thought of as all-sufficient, and it is not.

On the first point. I could have added to my list at the end that people should be told much more about medicine so they can bring these requests to doctors. But I was trying to argue that the best protection for any patient in a doctor–patient encounter is a doctor who is forced to use his clinical judgment and develops it as a high art and skill. One of the ways that one can diminish that kind of art is by imposing on him certain routines and regulations which make it routine. The minute practice becomes reduced to a checklist where you have to do all these things on a list, then you get very bad diagnoses.

It is very, very evident that this happens when you look at videotapes of medical students who are in their second and third year. When they interview patients and look at a diagnosis, they think if they ask all the questions on the list they are going to get the diagnosis. Sometimes patients say things that they miss out on because it is not on their list.

I think this frame of mind can be very pernicious to the practice of good medicine.

DR. HARRY KAY (Physician, Huntington Beach, California): I understood Dr. Reiser to offer another form of regulation at a more basic level—to order that the doctor is regulated by requiring him to prove periodically and frequently that he is still qualified. That was my understanding of it. He also added the supplement in answer to the previous question about teaching people what to expect. I come into it every day, especially in the older patients. They are much more pleased if I give them answers before I even ask them the first question and many object to disrobing at all. That's the way they have been taught. That's what they have been exposed to, and they think they are getting treatment. We have to educate them what to demand or what to expect of their physicians.

DR. MARK DINE (Pediatrician, Cincinnati, Ohio): The law is fairly responsive to medical change. When the smallpox vaccination became unnecessary, there was a very rapid change in the state law so that we need not give it. The state law requiring immunizations might be interpreted as a means of telling the physician what immunizations a child should get, but it also had the effect of forcing the parents to bring their children to a physician or a clinic to get that which society and medicine together felt was very important.

So that what we are saying, what I hear Dr. Callahan saying, for example, was that the experts have to tell the public what is necessary, and therein lies some of the problem. I see our discussion this whole morning as being related to problems of paternalism. I am intrigued to see Professor Fried not be quite as angry with the paternalism of the scientist as he is with the paternalism of the physician.

PROF. FRIED: I wasn't angry at anybody. I was simply pointing out a certain contradiction in the various approaches, hoping it would stimulate discussion.

DR. DINE: Dr. Stetten brought out a very interesting point. I attended sessions at the Hastings and Kennedy Centers, and we talked about health care and the fact that public health care measures where people don't even see a doctor are far more effective in saving lives and improving the health of the country. Therefore funds devoted to decreasing the speed limit to fifty-five miles per hour or funds devoted to poverty do a lot more than money expended, for example, on PKU screening. And then we hear him say, why are we spending so much time on being concerned about recombinant DNA? Shouldn't we be spending our time elsewhere? There seem to be some curious parallels to the ethics of teaching medicine that are going on this morning.

DR. VINCENT AVERY (Ethicist, Philip Andover Academy): Dr. Reiser mentioned education, and I am concerned with my students about a very prevalent attitude that I see toward ethics as not being a very viable area of human discourse. I wonder if there is a role for education either alongside or in place of regulation in the area of bioethics.

I think my statistics are a little old, but I believe in 1974 about ninety percent of medical schools in the United States had ethics dealt with in some form but only six percent of those schools required it in any form.

I wanted to ask Dr. Reiser if he thought that education had any role alongside of or in place of regulation and also I would like to ask about other sciences, not just medicine.

DR. REISER: Clearly, education is one of the key alternatives to regulation. I think that that has been stressed all morning.

As far as your statistic about ethics courses is concerned, my conclusion in teaching medical students is that requiring them to do things isn't necessarily the best way of transmitting good knowledge. Because they have so many things that can be required of them, they are stressed a great deal. It strikes me that ethics should stand on its own and if the courses in ethics are good and there is time in the curriculum that is allowed, students will take them. So I don't think requirement is a precondition for a lot of people learning ethics.

DR. ALLEN STOOLMILLER (Health Science Administrator, NIH): I was interested in the fact that Professor Robertson's lucid and objective analysis of the realistic surroundings for the kind of science in our own participatory democracy caused such shock waves to go through this audience. In my opinion, if there is a message for this group from Professor Robertson's comments, however, it is a reminder that scientists have both a professional obligation and certainly a personal interest in a continuing dialogue with the public and with various legislative bodies, be they federal, state, or local, as in the case of Cambridge politics here in Massachusetts.

I might add that it raises another interesting specter or analysis in this debate and that is that many scientists are reluctant to engage in conversations with the press, because of difficulties of communication or difficulties in translation between their view and what ultimately appears for public consumption. I don't know how the chair would reflect or the panel would reflect on this aspect of communication, yet is is one that seems to create difficulties for scientists.

PROF. JOHN A. ROBERTSON: Yes, more communication is in order and on the whole, scientists seem to have done very well in communicating their needs in the legislative process. The point is that the amount of funding is a political question, a question of how much resources we want to go into a certain area. That will be dependent upon political activity and communication by scientists.

DR. SUE PAUWKER (Medical Geneticist, Harvard Community Health Plan): There is a growing use of the technique called decision analysis, which allows the physician to blend the patient's values with the estimated risks of certain tests or treatments in order to arrive at the best decision for each individual patient. This is a growing use of a technique which takes a lot of time, but which is one answer to the conflict of how to put together the patient's values with the available treatments or tests. This technique can particularly be used in the case of tests of very high cost, either economically or in terms of patient risks, or in treatments which may or may not be effective in decreasing pain, such as coronary surgery.

DR. RICHARD RATZAN (University of Connecticut): I agree with Dr. Reiser wholeheartedly. I think there wasn't any irony there. He did talk about regulation and juridical punishment. I agree with Dr. Reiser because I think he has approached it at least from the standpoint of medicine on a rational basis. I think the problem is that bad medical care arises from their becoming bad physicians. You can be a bad physician easier than you can be a "bad" scientist. The scientist will certainly not act upon a person directly that afternoon in his office. Whereas the action of a physician is more tangible.

I think the root of the problem is to try and turn out only well-educated and good physicians. Another aspect which he didn't mention is to try and not just punish but actually remove the license of bad or incompetent physicians, the addicted, the alcoholic, or the senile, who may be able to pass a written test on a given day but who are incompetent physicians.

PROF. FRIED: The point, of course, is that Dr. Reiser was arguing for self-regulation in respect to doctors, who, as you have pointed out, are in a much better position to do harm; and the great movement is for all kind of external regulation of scientists, who, as you pointed out, are in a lesser position to do harm. That I consider an irony.

Also, one has to recognize the fact that doctors are not as subject to peer review as are scientists. A scientist before he publishes his work usually has to go to a refereed journal; he is usually working with colleagues. There are departmental heads and so forth. Doctors very often do their thing in the privacy of their offices. So I think the irony remains.

There was another question. And may I say, the last; so you bear a heavy responsibility.

MR. LEONARD RISKIN (University of Houston Law School): I wanted simply to take issue very mildly with one of Dr. Reiser's suggestions. The tort law insofar as it regulates the practices of physicians is not responsive to change.

I think the example you cited of the *Helling v. Carey* case from the State of Washington does not illustrate a resistance to change. It accommodates change. Because one of the fundamental assumptions that the Washington Supreme Court made in that case was that the glaucoma test was simple, inexpensive, and safe. I think that if it turned out that this test was not safe, that a physician would not, in any reasonable understanding of the law, be required to give the test.

The problem, of course, is interpreting this ruling to physicians, and I'm sure it is the case that ophthalmologists will be giving that test for a long, long time no matter what turns out to be the case.

DR. REISER: Of course, there is sometimes a gray area when any given doctor might decide, "Well, I would really rather not use that test." Certainly the law is not going to change in time for him to make that decision for the next patient.

GENETIC COUNSELING AND SCREENING

MODERATOR: MARGERY W. SHAW

7

Prenatal Genetic Diagnosis and the Law

Aubrey Milunsky
Harvard Medical School
Boston, Massachusetts 02115

Genetics Division
Eunice Kennedy Shriver Center
Waltham, Massachusetts 02154

Massachusetts General Hospital
Boston, Massachusetts 02114

The "right" to be born of sound mind and body has recently been given impetus by at least two recent court decisions.[1,2] Technological advances. especially during the past two decades, have enabled the development of procedures that allow for the identification in early pregnancy of specific genetic disorders of the fetus. These advances have recently been reviewed elsewhere[3] and only the briefest summation is provided here.

It is possible today to diagnose all recognizable chromosomal disorders *in utero.* The commonest specific indications for these studies are advanced maternal age (≥ 35 years of age), a previous child with trisomy 21, and a parental translocation or other chromosomal anomaly. Fetal sex determination can be accurately provided for the express purpose of determining the presence of a male fetus when the risk is 50% that such a fetus would be affected by an X-linked genetic disorder. Among the approximately 200 sex-linked disorders in this category, only 6 can specifically be identified in the male fetuses. Over 100 different biochemical genetic disorders of metabolism can now be diagnosed prenatally based almost invariably upon the demonstration of a deficient enzyme in cultivated amniotic fluid cells.[3] In the last few years, it has become possible to accurately make the prenatal diagnosis of open neural tube lesions with a 90–95% efficiency. The use of additional ancillary tools including ultrasound, amniography. and fetoscopy is providing important new dimensions that are expanding the scope of prenatal diagnosis. Continued progress using new techniques including restriction endonucleases[4] and somatic cell hybridization[5] is assured.

Notwithstanding the fact that prenatal genetic diagnosis will never be able to approach all the recognizable genetic disorders known (over 2800 at the present time), the steadily growing list of conditions that are diagnosable will undoubtedly increase the public awareness of the opportunities for the prevention of fatal genetic disease or serious mental retardation. The dawning public realization of opportunities for prevention will inevitably, I believe, encourage the philosophy concerning the "right" to be born of sound mind and body.

GENETIC COUNSELING

Professor Joseph M. Healey, Jr., addresses questions concerning genetic counseling in Chapter 8, as Hirschhorn did in the First National Symposium proceedings.[6] While a failure to counsel about the risks of bearing defective children has already had consequences in litigation, information leading to unnecessary abortion, sterilization, contraception, artificial insemination, or adoption may also incur liability. I have discussed genetic counseling as a prelude to prenatal diagnosis elsewhere.[3]

UTILIZATION OF PRENATAL DIAGNOSTIC FACILITIES

Routine prenatal genetic studies have been available for over a decade.[7,8] Nevertheless, utilization of this new technology has been remarkably tardy. I estimate that in 1978 fewer than 10% of women 35 years and over in the majority of states actually had amniocentesis for prenatal chromosomal analyses. This figure is closer to 20% in Massachusetts, where much effort has been expended in professional and public education about these new advances.[9] The primary reasons for this underutilization are probably multifactorial, including a lack of physician and public awareness of the indications for prenatal genetic studies, concern about the risk of amniocentesis during the second trimester (which should have been assuaged by the National Institutes of Child Health and Human Development and Canadian reports,[10,11] the antipathy toward abortion, and the considerable costs of the required tests (probably over $500 per patient). Broadly based public and professional education, realization of the low risks of amniocentesis (despite the United Kingdom's Medical Research Council report[12] with contrary but disputable evidence[13,14]), and the increasing third-party coverage for these studies will undoubtedly lead to an increased utilization of prenatal diagnosis. Certainly, any claims in North America that amniocentesis during the second trimester is a high-risk procedure will be negated by the published U.S. and Canadian studies referred to previously.

Problems and Pitfalls

There are two main procedural variations in how patients undergo amniocentesis and prenatal diagnosis. Ideally, and most expensively, patients are seen for counseling in a university medical center and then have ultrasound prior to amniocentesis. Often, however, patients are briefly counseled by their obstetrician, who performs an amniocentesis possibly without prior ultrasound and dispatches the amniotic fluid with the patient (or by messenger) to a prenatal diagnostic laboratory. Informed consent would be uniformly obtained in the university setting prior to amniocentesis, the procedure including a discussion and signing about purely laboratory procedures. Physicians in private offices obtain consent for a minor surgical procedure (amniocentesis) but often leave further discussion about laboratory aspects to the facility to which the fluid is brought. Here a patient may discover for the first time that no result is guaranteed, that cell cultivation may fail, that the cells may become contaminated, that a urine sample might have been obtained inadvertently, that maternal cells may be admixed in the sample, that uninterpretable mosaicism may be found, or that slides or samples could be involved in a mix-up. In the absence of ultrasound, twins or multiple fetuses may have been missed, the sample provided reflecting the genotype of only one fetus.

After full verbal disclosure, signed consent is also obtained.

Errors in Prenatal Diagnosis

Errors in prenatal diagnosis have generally occurred with a frequency of less than 1%[3] (0.6% in a U.S. study, 0.4% in the Canadian MRC study, and 0.2% in a worldwide survey of over 32,000 cases). These data indicate that inadvertent maternal cell admixture with the amniotic fluid remains the commonest source of error.

Errors in both omission and commission have occurred. Errors of omission mainly abound in the fact situation where obstetricians fail to offer prenatal genetic studies. A countrywide rash of lawsuits has arisen on this basis mainly related to a failure to offer amniocentesis on account of advanced maternal age in situations where parents at risk have had a child with Down syndrome. I will predict that many of these suits will ultimately end up with settlements out of court, especially those dated after the fall of 1976 (after the issue of the U.S. and Canadian amniocentesis reports). Meanwhile, in numerous cities across the United States, major battles have ensued, with the physicians claiming that they informed their patients, who counterclaim no such discussion.

Park v. Chessin represents the first claim for damages as a result of failure to provide proper *preconception* genetic counseling.[15] The parents had had a previous child die with polycystic kidney disease. The defendant obstetricians. who had been sued for wanton and reckless negligence, counseled these parents that the risk of recurrence for this disorder was "practically nil." Based on this advice, they conceived their next child, who was afflicted with the same disease. The risk of recurrence was in fact 25% for this autosomal recessive disorder and not "practically nil." The plaintiff's claim on behalf of their decased child was therefore held to be legally enforceable.

The fundamental importance of ethnicity relative to genetic disease has become strongly apparent in *Howard v. Lecher*.[16] The plaintiffs alleged that their obstetricians were aware of their Ashkenazic Jewish ancestry and that they were therefore at special risk for having a child with Tay–Sachs disease. They claimed that no effort had been made to use tests to determine their carrier status for Tay–Sachs disease (about 1 in 30 Ashkenazic Jews are carriers of this disorder), to analyze their family pedigree, to advise them of the risks of having a child with Tay–Sachs disease, or to recommend or offer amniocentesis or prenatal studies. Their daughter, born in 1972, died about 2 years later of this lipid-storage disease. The parents claimed that they would have terminated the pregnancy if the diagnosis of Tay–Sachs disease had been made *in utero*. They sought damages for their own suffering. The New York Court of Appeals did not allow the parents of this child to recover damages for the mental suffering endured watching their child suffer, then die. The court, however, did not exclude the possibility of the plaintiff's bringing a cause of action for the mental suffering endured by the child.

In yet another ongoing case, in Michigan, action has been brought against an obstetrician for not offering amniotic fluid alpha-fetoprotein studies for a couple who each had a sibling with a neural tube defect. They subsequently had a child with a catastrophic spina bifida defect. When it became apparent early in the proceedings that the suit was not likely to be won, since reagents for the alpha-fetoprotein assay have not even today been released for commercial sale and distribution, the claim was focused on the alleged failure of the physician to provide counseling related to their risks of having a child with a neural tube defect.

An error of commission is best exemplified by a Philadelphia case in which the prenatal diagnosis of Tay–Sachs disease was not made due to problems in the interpretation of the assay used.[17] A further case in Philadelphia, still to be fully adjudicated, developed as a consequence of an alleged failure by the laboratory to communicate a failure of cell culture early enough for the couple either to have a second amniocentesis or to electively terminate the pregnancy simply on the risk of advanced maternal age. The couple allegedly were appraised of the cell culture failure only after 24 weeks, when it was regarded as too late to perform an elective abortion. The plaintiffs subsequently had a child with Down syndrome. The couple claimed that since they had previously had an elective abortion in a previous pregnancy without increased genetic risks, they would clearly have taken a similar action because of their real concern about the outcome of the pregnancy in question.

Not surprisingly, obstetricians increasingly feel pressured by their patients about the risks of birth defects and genetic disease and the increasing opportunities for prevention. The age of the patient, the ethnic background, the possibility of carrier tests, the varying risk figures for a large number of genetic diseases, and the rapid advance in somatic cell genetics are all factors that have led some obstetricians to practice defensive medicine. In fact, one group in Michigan, when asked any question about genetic disease, provide no answers but immediately produce a form in which they advise the patient to seek genetic counseling from a university medical center!

Prenatal Diagnostic Laboratories

The majority of laboratories in the United States now performing prenatal diagnostic studies have evolved from subjacent research activities. Hence many of these laboratories have not in the past and do not now fall under the jurisdiction or aegis of licensing or regulatory agencies. Therefore, in the absence of regular monitoring and surveillance, it is not unexpected that a wide range of standards now exists. The matter is further compounded by the entry of commercial laboratories, where again the normal built-in constraints of a university medical center are lacking and where peer scrutiny is absent. Knowledge about the failure rate for cell cultures, the adequacy and confidentiality of record keeping, the time taken for laboratories to respond with results or requests for repeat samples, the frequency of their false positive or negative results for biochemical studies, and the frequency with which efforts are made to confirm the prenatal diagnosis of abnormality are simply some of the factors that currently receive little or no attention from any specific agency. The fact that so few problems have emerged is undoubtedly a reflection of the location of most of these centers in medical school/hospital settings.

It is unnecessary for small inexperienced laboratories to perform prenatal diagnostic tests for rare biochemical genetic diseases. Optimally, such responsibility is best lodged in supraregional medical school/hospital facilities where multidisciplinary expertise (e.g., biochemical genetics, ultrasound) is available. Such facilities are more likely to respond sensitively to questions regarding inter- or intraassay variability and reproducibility and have satisfactory studies about the quantity and quality of controls required for some of these assays. The academic center philosophy makes it unwise to lodge some of these studies in commercial laboratories without a monitoring or surveillance program. Commercial laboratories are in the business of measuring substances but not of interpreting data (dozens of examples exist where alpha-fetoprotein assays have been performed on serum and the result in ng/ml provided without any note on normal values or the meaning of the value obtained).

I have long maintained that physicians have a responsibility to their patients in being fully aware of the competence of laboratories to which their patients' samples are being sent. In general, if some kind of harm results from a physician relying on tests negligently conducted by an independent laboratory, liability would not usually accrue.[18] However, if the physician was negligent in the choice of such a laboratory or he failed to detect the error, liability could attach.

Responsibility of the Physician

The physician's duty vis-à-vis prenatal genetic diagnosis is largely governed by normal standards of medical practice.[3] Some specific guidelines might be useful.

GENETIC COUNSELING.　The physician clearly has the responsibility of correctly advising the couple about the risks of occurrence or recurrence of the specific genetic disorder in question.[3,6,7] Obviously, it is not possible for an obstetrician or, in fact, any physician to be aware of the specific risks of all recognizable genetic disorders, nor is it even reasonable to expect an extremely extensive family pedigree analysis. Minimal expectations, however, are that the physician not communicate risks that are significantly lower or higher than are actually known, and that the ethnic background of the patient be recognized [especially relevant to disorders such as sickle cell anemia (blacks), Tay–Sachs disease (Ashkenazic Jews) and thalassemia (Italians)], as well as other specific risk situations. It could also be expected that the physician communicate known options, which include carrier detection tests, vasectomy, tubal ligation, prenatal diagnosis, artificial insemination by donor, and adoption.[7]

COMMUNICATION IN TIME.   There is a need for both the physician and the laboratory to respond to women being studied during the second trimester in order that results be provided before it is considered too late for an elective abortion. Should results not be available in time, the parents need to be advised about their option to terminate the pregnancy on the basic risks of the disorder in question. A failure to communicate in time in association with a pregnancy that ends with a defective child in which the defect could have been diagnosed *in utero* clearly invites litigation.

INFORMING BOTH PARENTS.   While it would be optimal practice for the physician to inform *both* parents about special inherent risks in a particular pregnancy, The U.S. Supreme Court has recently emphasized the primacy of the woman's interest.[19] On rare occasions, there may be a lack of agreement between the parents regarding the decision for elective abortion in the face of results about a defective fetus. Again, it would appear that the husband has only a secondary right in this situation.

AMNIOCENTESIS AND THE UNMARRIED PREGNANT MINOR.   On rare occasions, the physician may be confronted by an unmarried pregnant minor requesting amniocentesis because of a family history of genetic disorder (example—a sibling with a neural tube defect). Where possible, it is obviously preferable to obtain parental permission. However, based upon the recent Supreme Court decision of *Planned Parenthood of Central Missouri v. Danforth*,[19] there does not appear to be an actual legal duty for the physician to seek such permission.

MISSED TWINS.   Routine ultrasound prior to amniocentesis would preclude most but not all occurrences of missed multiple pregnancies. In my view, ultrasound prior to amniocentesis is mandatory in the face of high genetic risk (10% or more). Today, if an affected twin in such circumstances is missed because of a failure to use ultrasound, then I believe the physician is open for suit.

FAILURE TO INFORM.   The physician is expected to communicate all the results of prenatal genetic studies to the parents. Should the doctor decide to withhold information about an unusual or perhaps difficult-to-interpret abnormality (for example, the XYY syndrome) legal action could ensue. The principle of informed consent requires that the information be made available and that every effort be made to interpret its implications.

INFORMATION ABOUT FETAL SEX.   Fetal sex determination is routine for the management of X-linked genetic disorders. There are patients, however, who even in the second trimester of pregnancy will seek to determine the fetal sex and aim to abort a fetus with the unwanted sex. While we adhere to the philosophy that the parents have a right to know all information about their fetus, we do not ourselves provide information on fetal sex simply for family-planning purposes during the second trimester of pregnancy (see Chapter 22). Physicians have no legal duty to provide information on fetal sex on demand and patients might therefore have to seek laboratories elsewhere.

LEGISLATION AND PRENATAL DIAGNOSIS

Philosophically, genetic disease is clearly not a public health hazard. There is no need to worry about any contagious phenomenon. Legislative actions, however, may develop when public policy concerns reach into the fiscal arena and spawn taxpayer burdens. Predicted lifetime institutional and medical care for a single child with Down syndrome now far exceeds half a million dollars. It is known that about 20,000 infants with some kind of chromosomal anomaly are born each year. If we assume that 5000 have Down syndrome or similar chromosomal anomalies, the need can be anticipated that *each year* society will commit itself to the expenditure of over 5 billion dollars expecting a 50-year survival (today, patients with Down syndrome aged 60 or 70 are known).

Society has already been seen to have developed an interest in both the number and the quality of children born—witness the family-planning clinics, court decisions ordering medical care of children against the wishes of their parents, laws on child abuse, etc. It is also an established public policy posture to minimize illness in the community—for example, compulsory vaccination for rubella or smallpox, fluoridation of water supplies, and, most recently, mandatory genetic screening (but strangely not mandatory treatment for the inborn errors detected).

Hence, rather than anticipate restrictive or constraining legislation, one would have hoped for enabling legislation, the essential thrust of which would be to provide diagnostic services and to educate both the profession and the public alike about the "new genetics." Happily, the states of both Oregon and Alabama have enacted such legislation in which the costs of amniocentesis and prenatal genetic studies are provided for those shown to be unable to obtain the necessary medical coverage. There truly are considerable fiscal concerns about the ability of low- to middle-income individuals being able to afford the studies. Recent records of most prenatal diagnostic centers will reflect that these tests have been mostly utilized by middle- to high-income families.

Despite the enactment of the National Genetic Disease Act in 1975, the abysmal appropriations lend no encouragement that low-income families will soon or ever benefit from the new genetic technologies.

SOCIAL JUSTICE (AND THE POOR SHALL HAVE RETARDED CHILDREN)

Low-income and poorly educated people are at risk for suffering a cascade of complications in relationship to pregnancy. Unwanted or unplanned pregnancies; poor nutrition; poor, irregular, or late antenatal care; multiparity; a higher frequency of low-birth-weight infants—all have the net effect of resulting in higher perinatal morbidity and mortality. The increased frequency of low-birth-weight infants among the economically disadvantaged has further consequences in their association with a higher frequency of mental retardation, congenital malformation, neurological defects, learning disorders, deafness, and blindness. Even a relative increase in the prevalence of genetic disease can be anticipated because of the failure of utilization of carrier detection tests and prenatal genetic studies. Genetic counseling is also still largely used by middle to upper classes only.

More recently, an ominous pall has fallen across these services for low-income families. The present inability to use federal funds for abortion, even when the fetus is genetically defective, makes for a major problem for the economically disadvantaged. Until recently, it was possible for women on Medicaid to obtain an elective aborion for medical reasons if certified by one physician. As of May, 1979, the U.S. Supreme Court ruling in the case of *Preterm v. Dukakis* means that elective abortion is available to a mother on Medicaid only if her life is threatened or following rape or incest. Further, two physicians will still need to certify that her *physical* health would be damaged. Hence, if a poor woman was found to be carrying a fetus with Down syndrome or another major genetic defect, she would only be able to abort the pregnancy if she were able to pay for the procedure. Since the Supreme Court decision in *Roe v. Wade*,[20] the state cannot restrict abortion during the second trimester, but at the same time it does not have to pay for it. The only way a poor woman *might* ultimately succeed in having the procedure covered is for two physicians to certify that severe psychological damage would occur and lead to physical damage thereafter. Exclusion of these services described in the Hyde amendment were declared unconstitutional in a federal court in Illinois (see Chapter 18 by Professor Glantz). One can only hope that similar action eventually occurs through the U.S. Supreme Court. Meanwhile. in our society, any judicial philosophy relative to being born of sound mind and body is countered by legislative acts that serve to punish the poor!

REFERENCES

1. *Smith v. Brennan*, 31 N.J. 353, 364, 157A.2d, 497, 503 (1960).
2. *Sylvia v. Gobeille*, 101 R.I. 76, 78, 220A.2d, 222, 223 (1966).
3. Milunsky, A. (ed.), *Genetic Disorders and the Fetus: Diagnosis, Prevention, and Treatment*, Plenum Press, New York (1979).
4. Orkin, S. H., Alter, B. P., Altay, C., *et al.*, Application of endonuclease mapping to the analysis and prenatal diagnosis of thalassemias caused by globin-gene deletion, *N. Engl. J. Med.* 299:166 (1978).
5. Kan, Y. K., and Dozy, A. M., Antenatal diagnosis of sickle-cell anaemia by D.N.A. analysis of amniotic-fluid cells, *Lancet* 1:910 (1978).
6. Hirschhorn, K., Medicolegal aspects of genetic counseling, in: *Genetics and the Law* (A. Milunsky and G. J. Annas, eds.), p. 105, Plenum Press, New York (1976).
7. Milunsky, A. (ed.), *The Prevention of Genetic Disease and Mental Retardation*, W. B. Saunders, Philadelphia (1975).
8. Milunsky, A., Current concepts in genetics: Prenatal diagnosis of genetic disorders, *N. Engl. J. Med.* 295:377 (1976).
9. Milunsky, A., *Know Your Genes*, Houghton Mifflin, Boston (1977).
10. The NICHD National Registry for Amniocentesis Study Group, Midtrimester amniocentesis for prenatal diagnosis. Safety and accuracy, *J. Am. Med. Assoc.* 236:1471 (1976).
11. Simpson, N. E., Dallaire, L., Miller, J. R., *et al.*, Prenatal diagnosis of genetic disease in Canada: Report of a collaborative study, *Can. Med. Assoc. J.* 115:739 (1976).
12. Medical Research Council Working Party on Amniocentesis, An assessment of the hazards of amniocentesis, *J. Obstet. Gynaecol. Br. Emp.* 85: Suppl. No. 2 (1978).
13. Milunsky, A., Hazards of amniocentesis, *Lancet* 1:546 (1978).
14. Alexander, D., Lowe, C. U., Seigel, D., *et al.*, Risks of amniocentesis, *Lancet* 2:577, 1979.
15. *Park v. Chessin*, Sup. Ct., Queens Co., N.Y.L.J., Sept 16 (1976), p. 10; 400 N.Y.S.2d 110 (App. Div. 2d
16. *Howard v. Lecher*, 42 N.Y.2d 109, 397 N.Y.S.2d 363, 366 N.E.2d64 (1977).
17. Curran, W., Tay–Sachs disease, wrongful life and preventative malpractice, *Am. J. Publ. Health* 67:568 (1977).
18. Reilly, P., Genetic counseling and the law, *Houston L. Rev.* 12:640, 656, (1975).
19. *Planned Parenthood of Central Missouri v. Danforth*, 98S.Ct., 2831 (1976).
20. *Roe v. Wade*, 410 U.S., 113 (1973).

8

THE LEGAL OBLIGATIONS OF GENETIC COUNSELORS

JOSEPH M. HEALEY, JR.
Department of Community Medicine and Health Care
University of Connecticut Medical School
Farmington, Connecticut 06032

LEGAL LIABILITY

Four years ago, at the First National Symposium on Genetics and the Law, Dr. Kurt Hirschhorn, asked to comment on a topic similar to mine, pointed out his discomfort in complying with the request:

> When I was asked to prepare a paper on potential situations in genetic counseling which may result in questions of malpractice, I felt like a little like the unarmed Christian Martyr who was told to enter the Roman Arena in order to fight a group of lions with large teeth. The current attitudes in the United States encouraging such suits while reducing protection make me quite hesitant to engage in an exercise designed to sharpen those teeth.[1]

Dr. Hirschhorn was reflecting the sense of frustration, bewilderment, and anger that many genetic counselors feel about their vulnerability to legal liability and their uncertainty about what society expects of them.

This confusion is created in part by litigation like the recent New York cases of *Park v. Chessin* and *Becker v. Schwartz*.[2] Those who regularly encounter patients at increased risk for Down syndrome and who encounter, though less frequently, patients at risk for polycystic kidney disease (the contexts of these two cases) have difficulty identifying the precise nature of their responsibilities. It is of little consolation to genetic counselors that the primary issue in these cases was not the guilt or innocence of the particular parties but a more general question of whether there existed a legally cognizable cause of action allowing recovery for the damages alleged. Neither the fact that the New York Court of Appeals, the highest court in the state of New York, refused to recognize a cause of action for "wrongful life" nor the fact that in one of the cases a subsequent jury trial found the physicians not guilty reduces the bitterness or eliminates the confusion.

In all honesty, it may never be possible to eliminate entirely this bitterness and confusion. I believe it is possible, however, to restore them to proper perspective and to encourage a spirit of cooperation between lawyers and genetic counselors. In the course of this chapter I would like to clarify the legal obligations of genetic counselors by identifying forces that have shaped the contemporary context of genetic counseling, by examining three starting points for understanding legal regulation of genetic counseling, by describing several key legal obligations that counselors possess, and finally, by suggesting strategies for implementing and satisfying those legal obligations. Out of this will emerge, I hope, some clarity and a sense of perspective.

A discussion of the legal obligations of genetic counselors must begin with a conceptual framework for genetic counseling within which legal obligations can be identified. There is available a rapidly growing body of literature discussing the nature of genetic counseling, its goals, and its role as a health care service.[3] I shall devote attention to these issues in subsequent portions of this chapter. For the present, let me simply say I will be using the terms *genetic counseling* and *genetic counselor* in the broadest sense, including a wide range of services provided by a wide range of health care professionals.

TRENDS AND FORCES

Our discussion of the legal obligations of genetic counselors takes place in a context shaped by several trends and several prominent forces. They are familiar to us and need to be understood as contributing to the difficulties experienced by genetic counselors about their legal obligations.

1. There is a widespread absence of consensus about critical issues in health care. This absence of consensus ranges from one end of the life cycle to the other, from questions of reproductive decision making to questions of death and dying.[4]
2. There is a general absence of a clear societal mandate for health care professionals that could be the basis for a formulation of their rights and duties.[5]
3. We are currently in the midst of the period of major knowledge expansion in medicine generally, and in genetics particularly. This period is intimately linked to the period of the history of medicine associated with increased success in managing complex medical problems that had baffled previous generations of scientists and physicians.[6]
4. Success on such a grand scale has generated expectations of health care consumers, impressed with medical success and less willing to accept medical "failure."[7]
5. Over the past two generations there has been a dramatic evolution of the legal, medical, and ethical context of reproductive decision making, influenced by advances in prenatal diagnosis, contraceptive technology, a general reconsideration of the family, and the rise of the patient rights movement.[8]

It is not unexpected that the genetic counselor is in the middle of these societal currents and crosscurrents. It is inevitable, unavoidable, and not, in every case, undesirable. The forces of change represented by these factors can stimulate counselors to search continually for better ways to serve their clients.

UNDERSTANDING THE LEGAL OBLIGATIONS

Three starting points for understanding the legal obligations of genetic counselors are (1) an understanding of the general role of law in regulating the health care process, (2) an understanding of the role that health providers play in defining the mission of genetic counseling and in defining the legal obligations of genetic counselors, and (3) an understanding of the influence of the patient rights movement in the reshaping of contemporary physician–patient, counselor–client relationships.

There is a general feeling in our society that we are in the midst of a period of undesirable excessive legal regulation. One particular example of this general trend is excessive legal regulation of the health care process.[9] Unquestionably, there has been a dramatic increase in legislation and litigation involving health care. It is tempting in such a situation to claim that there is no legitimate role for the law and to overlook traditional justifications for the law's involvement in the health care process. Two primary reasons for the law's involvement in health regulation have been the protection of the public health, safety, and welfare (the police power)[10] and the protection of the rights of citizens in general and of providers and patients in particular.[11] It should be apparent that both of these are of concern in the genetic counseling context and that the primary question remaining is not whether any regulation is justified but what form the regulation should take. Answering that question and developing appropriate regulatory mechanisms are tasks not simply for lawyers but for us all.

One predominant characteristic of legal regulation of the health care process in the American legal system has been the major role played by health care professionals in determining their legal obligations. In developing a variety of legal mechanisms, including licensure and malpractice, legislatures and courts have acknowledged their lack of expertise and have afforded physicians and other health professionals primary responsibility for the determination that acceptable levels of competence have been attained by health care providers. During recent years there has been a gradual expansion of this responsibility beyond the health professions and into the lay community. The inclusion of nonphysicians on medical licensure and discipline boards and the elimination of the requirement that physicians establish the standard for disclosure in informed consent cases are examples of this trend. Nonetheless, there remains a major role for health care providers who engage in genetic counseling to assist in the formulation of their legal obligations. Exercise of this role by genetic counselors is complicated since the members of several professions, rather than a single profession, are involved. It is important for all those who can to contribute to the formulation of a comprehensive conceptualization of the goals and processes of genetic counseling that might serve as the framework within which legal obligations can be understood. The final responsibility for determining legal responsibility rests with legal institutions. However, there will be ample opportunity for influencing the formulation and its application.

A third basic foundation is an understanding of the role of the patient rights movement in the contemporary physician–patient, counselor–client relationships. Barber has contrasted effectively the traditional relationship and the modern relationship.

In the traditional pattern:

> the long established pattern, to which most participants in medical care are deeply committed, has a number of consistent characteristics. The doctor is superordinate, the authority, sometimes a person to be venerated. The patient and others are subordinate, respectful, even deferential. The doctor is active, knowledgeable and secure in the system: the patient and others are passive, ill informed, frightened and dependent.[12]

In the emerging pattern:

> patients, relatives and others are not so much subordinate of the authoritative doctor but more nearly partners even 'quasi-colleagues,' as Talcott Parsons has said [footnotes omitted], all participants are obligated to be knowledgeable independent and active. Mutuality and reduction of inequality are preferred to one-sidedness and inequality.[13]

The patient rights movement has clearly become a major force in shaping the context in which major health care decisions are made. The decisions at stake for the person or couple seeking genetic counseling are critical decisions: to conceive or not to conceive, to deliver a child or to abort, to undergo sterilization rather than risk the transmission of disease or defect. Genetic counseling represents one significant way in which information can be communicated to those who must decide, making possible for them an intelligent exercise of choice. The relationship between counselor and client/patient must be grounded in mutual respect. A fundamental principle of the patient rights movement is that the patient is a person who does not surrender his/her rights by entering the health care process. The contemporary counselor–client/patient relationship has been shaped by the patient rights movement, just as the physician–patient relationship has been reshaped.[14] Genetic counselors need to appreciate the impact of this reality upon genetic counseling and upon their legal obligations.

AREAS OF LEGAL OBLIGATION

An attempt to delineate the legal obligations of genetic counselors in a clear and concise manner is complicated by several factors. There has been only a relatively small number of cases and statutes dealing with genetic counseling. Most of these concentrate on limited aspects of genetic counseling. As a result, genetic counseling has not received comprehensive treatment in the law. The absence of precedent makes much of what is said speculative and subject to revision as more statutes and more cases give shape to this area of the law. The existence of different conceptualizations of genetic counseling, its nature and goals, and the professional identity of those who engage in it have also complicated the development of clear guidelines addressing legal obligations. Nonetheless, it is possible to identify several key areas of legal obligation and to consider strategies for satisfying them: (1) the obligation to possess sufficient knowledge, training, and skill to engage in genetic counseling; (2) the obligation to apply this knowledge, training, and skill in the evaluation of the patient/client and in the proper exercise of clinical judgment, within the limits of the counselor's competence; (3) the obligation to communicate sufficient information to the patient/client in an effective and accurate manner; (4) the obligation to respect and protect the rights of the patient/client.

Genetic counseling is a generic term embracing at least three major health care services:[15] (1) the identification of parties at risk for the transmission of genetic disease or genetic defect; (2) the use of diagnostic procedures to determine whether a fetus is affected with or is at risk for genetic disease or defect; (3) the communication of information to the parties as the basis for their exercise of choice in reproduction to conceive, to avoid conception, to deliver a child, to abort a fetus, or to engage in an alternative form of reproduction.

Those who engage in genetic counseling hold themselves out as possessing sufficient knowledge, training, and skill to provide health care in a competent fashion. The first legal obligation for genetic counselors is to possess the necessary competence to provide those services. Although there is some question as to who is best qualified to engage in genetic counseling[16] and there are strong arguments to suggest that an interdisciplinary team provides the best context,[17] there has been no clear resolution of this issue in the law.[18] However, a fundamental principle in other areas of health care regulation has been the requirement that health care providers respect the limits of their competence and avoid acting beyond the scope of their ability.[19] Such a principle has special importance in the context of genetic counseling if responsibility is shared by various health professionals as members of a multidisciplinary team. Use of such a team is consistent with the general shift in our society away from an exclusively physician-centered health care delivery system to one that allows properly trained nonphysicians a major role in the health care process.[20] In such a situation, members of the team should possess the appropriate knowledge, training, and skill, and operate within the boundaries of their competence.

The second level of legal obligation involves the application of skills in the evaluation of the patient/client. Genetic counselors have a legal obligation to provide health care services in a manner consistent with the standard of reasonable care. Genetic counselors are potentially liable for harms that are the direct and proximate results of their violation of the duty to act in a reasonable manner, which they owe their patients/clients. A person who engages in genetic counseling has no legal obligation to provide services to every person seeking them. There is no general right to health care in the United States.[21] However, once a client–patient is accepted, the counselor is obliged to provide services as a reasonable counselor would in the same or similar circumstances.[22] Potential violations of these obligations might include failure to identify a party at risk for transmitting genetic disease, failure to take an accurate family history or to employ diagnostic procedures accurately, improper diagnosis or communication of insufficient or inaccurate information. Many of the "wrongful life" cases contain claims of this variety.[23] The genetic counselor is expected to provide appropriate levels of quality care and is potentially legally liable for failure to do so.

The primary focus of genetic counseling is on the communication of information about the occurrence and risk of recurrence of a genetic disorder within a family. The importance of this communication is demonstrated by the importance of the decisions that will be based upon it. The genetic counselor has a legal obligation to provide the patient/client with sufficient information of an accurate nature in an effective manner to allow the patient/client to make an informed choice.[24] This legal obligation requires the counselor to form an accurate diagnosis and prognosis, and to communicate that information in a clear fashion. The New York cases cited previously[25] both involved claims that this obligation has been violated. The satisfaction of this obligation is not always easy, but if the patient/client is to be in a position to exercise choice, he/she must have access to this information.

The fundamental principle that patients do not surrender rights by entering the health care process requires that genetic counselors respect and protect the rights of their patients/clients. Two rights of patients are frequently involved in genetic counseling. First, the informed consent of the patient/client must be obtained before diagnostic tests are employed. The essence of informed consent is the assent of the patient/client after the risks, benefits, alternatives, and other related information have been carefully explained.[26] The second involves the right of the patient/client to have confidentiality respected. The patient has a right to expect that no information obtained in genetic counseling should be disclosed to any third party.[27]

STRATEGIES FOR SATISFYING LEGAL OBLIGATIONS

It cannot be assumed that those who engage in genetic counseling have been adequately prepared to understand and properly discharge their legal obligations. Strategies must be developed to allow those goals to be achieved. Four key areas of concern are (1) training of genetic counselors, (2) patient education, (3) continuing education of genetic counselors, and (4) additional research.

Undoubtedly, genetic counselors need a strong foundation in the areas of genetics, the transmission of genetic disease, and the use of statistics in health care. However, it is not sufficient for the genetic counselor to be only a competent scientist. Genetic counselors need special training in communication skills and in legal and moral aspects of genetic counseling. Such perspectives have not been a traditional part of the training of health care providers.[28] The addition of these dimensions to the training of genetic counselors may require new educational components in graduate programs in medicine, genetics, counseling, social work, and nursing.

A major problem confronting the health care provider is the presence of an erroneous concept of health that views health as a passive entity bestowed upon patients by physicians. Health is viewed not as within the control and responsibility of the patient but as the responsibility of the health care provider. When undesirable health outcomes occur, it is assumed that they have been caused by the health provider. In reproductive decision making, there is a tendency to assume that the birth of a child with disease or defect is something that happens to others but not to the patient him/herself. Gradually, over the past decade, this exaggerated expectation has been confronted by greater public awareness of genetic diseases and defects and of the responsibilities of parents. There need to be additional efforts of this kind designed to place the expectations and the fears of parents into proper perspective.

In order to enable genetic counselors to better serve their clients/patients, there need to be effective networks of communication about developments in the scientific, legal, and moral aspects of genetics. The counselor ought not to stand isolated but should have available the resources and the experience of other counselors. Conferences such as the National Symposium on Genetics and the Law are an important part of that process.

Finally, genetic counseling, as a health service only recently available, needs to engage in healthy self-examination. There is great need for additional research both in the genetic disorders that are the subject of genetic counseling and in the process of communication designed to facilitate choice by the patient/client. Many of the questions that have arisen during the first stage of the development of genetic counseling need to be examined carefully and critically.

CONCLUSION

Further clarification of the legal obligations of genetic counselors will occur as additional cases unfold and as legislation is proposed and enacted into law. The foundations, the trends, the forces described in this chapter have set the stage for this next step. I have no doubt that if we gather for another sympsoium, we will have quite a bit more to review.

REFERENCES AND NOTES

1. Hirshhorn, K., Medicolegal aspects of genetic counseling, in: Genetics and the Law (A. Milunsky and G. Annas, eds.), p. 105, Plenum Press, New York (1976).
2. 46 N.Y. 2D 401, 413 N.Y.S.2d 895 (1978).
3. Childs, B., Genetic counseling: A critical review of the published literature, in: Genetic Issues in Public Health and Medicine (B. Cohen, A. Lilienfeld, and P. Huang, eds.), p. 329, Charles C Thomas, Springfield, Ill. (1978).
4. Ryan, K., The legitimacy of a diverse society, J. Med. Assoc. 233:781 (1975); MacIntyre, A., Patients as agents, in: Philosophical Medical Ethics: Its Nature and Significance (S. Spicker and H. T. Engelhardt, eds.), p. 213, Reidel, Dordrecht, Holland (1977).
5. See Healey, J. Medical law and the physician, Conn. Med. 42:603 (1978). See also the comments of Spicker, S., Hellegers, A., Bok, S., et al., Round table discussion: the physician as moral agent, in: Philosophical Medical Ethics: Its Nature and Significance (S. Spicker and H. T. Engelhardt, eds.), Reidel, Dordrecht, Holland (1977).
6. See Mahoney, M., Report to the National Commission for the Protection of Human Subjects on the Results of a Literature Search Examining Fetal Research (1975); Fraser, F. C., Genetics as a health care service, N. Engl. J. Med. 295:489 (1976); van den Berg, J., Medical Power and Medical Ethics, W. W. Norton, New York (1978).

7. See Thomas, L., On the science and technology of medicine, *Daedalus* **106**:35 (1977); Knowles, J., The responsibility of the individual, *Daedalus* **106**:35 (1977); and Healey, J., The patient viewpoint on malpractice, in: *Social Issues in Health Care* (D. Self, ed.), p. 42, Teague and Little, Norfolk (1977).

8. See Powledge, T., and Fletcher, J. Guidelines for the ethical, social and legal isues in prenatal diagnosis, *N. Engl. J. Med.* **300**:168 (1979); Milunsky, A., *The Prenatal Diagnosis of Hereditary Disorders*, p. 171, Charles C Thomas, Springfield, Ill. (1973).

9. See Ingelfinger, F., Legal hegemony in medicine, *N. Engl. J. Med.* **293**:825 (1975).

10. See Grad, F,m *Public Health Law Manual*, p. 5, American Public Health Association, Washington, D.C. (1973); Reilly, P., The role of law in the prevention of genetic disease, in: *The Prevention of Genetic Disease and Mental Retardation* (A. Milunsky, ed.), p. 422, W. B. Saunders, Philadelphia (1975).

11. See Healey, J., Legal regulation of medicine: An overview, in: *Ethics in Medicine* (S. Reiser, A. Dyck, and W. Curran, eds.), p. 145, MIT Press, Cambridge (1977).

12. Barber, B., Compassion in medicine: Toward new definitions and new institutions, *N. Engl. J. Med.* **295**:939 (1976).

13. Ibid.

14. See Annas, G., and Healey, J., The patient rights advocate: Redefining the doctor–patient relationship in the hospital context, *Vanderbilt L. Rev.* **27**:243 (1974).

15. Childs, B., *op. cit.*, p. 329.

16. See Peterson, M., Should generalists provide genetic counseling, *Birth Defects: Orig. Artic. Ser.* **13**:171 (1977); Epstein, C., Who should do genetic counseling and under what circumstances. *Birth Defects* **9**:39 (1973).

17. See Milunsky, A., Genetic Counseling: Principles and practice, in: *The Prevention of Genetic Disease and Mental Retardation*, p. 71, W. B. Saunders, Philadelphia (1975).

18. Reilly, P., *Genetics, Law and Social Play*, p. 175, Harvard University Press, Cambridge (1977).

19. See, generally, Holder, A., *Medical Malpractice Law* (2nd ed.), Wiley, New York (1978).

20. Bliss, A., and Cohen, E. (eds.), *The New Health Professionals: Nurse Practitioners and Physician's Assistants*, Aspen Systems, Germantown, Maryland (1977).

21. Sparer, E., The legal right to health care: Public policy and equal access, *Hastings Cent. Rep.* **6**:5, 39 (1976).

22. Reilly, P., *Genetics, Law and Social Policy*, Harvard University Press, Cambridge (1977).

23. See Father and mother know best: Defining the liability of physicians for inadequate genetic counseling, *Yale L. J.* **87**:1488 (1978).

24. See Capron, A., Legal rights and moral rights, in: *Ethical Issues in Human Genetics* (1973); Reilly, P., *Genetics, Law and Social Policy*, p. 167, Harvard University Press, Cambridge (1976).

25. 46 N.Y. 2D 401, 413 N.Y.S.2d 895 (1978).

26. See Annas, G., Problems of informed consent and confidentiality in genetic counseling in: *Genetics and the Law* (A. Milunsky and G. Annas, eds.), Plenum Press, New York (1976); Reilly, P., *Genetics, Law and Social Policy*, p. 167, Harvard University Press, Cambridge (1976).

27. Ibid.

28. Katz, J., The education of the physician-investigator, in: *Experimentation with Human Subjects* (P. A. Freund, ed.), p. 293, George Braziller, New York (1969).

29. For an excellent example of patient education material see Milunsky, A., *Know Your Genes*, Avon Press, New York (1977).

DISCUSSION

Ms. TABITHA POWLEDGE (The Hastings Center, New York): Dr. Milunsky, you made an argument I have heard you make before on the question of amniocentesis for the purpose of choosing the sex of children. It is an argument with which I am uncomfortable for many reasons, one of which Mr. Healey has alluded to: the patient's right. It is not clear to me that that ought to be the physician's decision. But I don't want to argue with you.

I keep hearing more and more reports of people requesting fetal sex determination even to the point of making up medical histories. I am under the impression that this test is in fact being done more and more.

Would you advocate some sort of legal restriction should the demand increase sufficiently and, if you would, what form do you think that might appropriately take?

DR. AUBREY MILUNSKY: I would not advocate any legislative posture relative to fetal sex determination. I think it would be declared unconstitutional anyhow if an effort was made to direct people which sex infant they might have. I too am uncomfortable with the posture that a laboratory director has any kind of right or role in telling couples that they can or cannot find out the fetal sex in that laboratory.

However, I regard fetal sexing for nongenetic purposes as an inappropriate use of a very scarce and expensive technology and that single laboratories, just like single physicians, are in a position to say that they will or will not operate, will or will not perform certain studies in their laboratories. I recognise that this view flies in the face of the fundamental premise that everyone has a right to know all the available information about their fetus.

A PARTICIPANT: What kind of response would you be inclined to make if the scarce resources argument no longer applied?

DR. MILUNSKY: I regard it as an inappropriate use of not only scarce technology but also an inappropriate use of this technology. In my laboratories, I would be unwilling to use it for that purpose. Perhaps others would be quite willing to use it for that purpose.

PROF. MARGERY W. SHAW: Do we have any laboratories that are doing fetal sex determination on a routine basis?

(There was no response.)

DR. GEORGE HENRY (Denver): As more information is disseminated, more inquiries will arise from individuals who don't fit recognized genetic criteria. For example, what would you advise a thirty-three-year-old nurse in the pediatric ward requesting prenatal studies?

DR. MILUNSKY: I'm sure the general guidelines to the practice of medicine will always supervene. Amniocentesis for the nurse is not recommended, but is available as an elective option. Whether a sample would be accepted by a particular facility or not would probably be a reflection of the capacity of that laboratory and possibly the philosophy of that particular laboratory director.

From our point of view, a very solid case, of course, could be made mathematically and otherwise for women, let's say of age thirty and over, to have an amniocentesis in very good hands with ultrasound control. We all know that there is a three to four percent background risk for having a baby with a birth defect of consequence, mental retardation or genetic disease. A percentage of that three to four percent can be approached prenatally. If it is the perception of the patient at any age that they would take all risks as known and have the test and request it, if the laboratory would be willing to offer it, I see no reason why not. But it would have to be against the background of fully informed consent that they understand the risks, that it is not indicated, and that it is something they are requesting.

PROF. ARTHUR BLOOM (New York): I would like to ask Professor Healey his thoughts on an issue that he alluded to in raising the question of the polycystic kidney case that Dr. Milunsky had brought up. It seems to me, and I may be wrong, that in that case the suit was brought against a practicing physician who was not in fact a genetic counselor.

PROF. JOSEPH M. HEALEY: The obstetrician, yes.

PROF. BLOOM: Which raises another question. Many of us doing genetic counseling are interested in trying to encourage obstetricians and general practitioners to learn genetics, to provide genetic information. What are the legal obligations of a practicing physician who is not in fact really trained to be a genetic counselor?

PROF. HEALEY: I think that is part of the second of the legal obligations that I mentioned. The individual has to know the limits of his or her ability to make appropriate judgments and, yet, I think the answer to that or the solution to that problem is not simply to say that everyone has to be an M.D. with a Ph.D. in genetics in order to engage in genetic counseling. In the law's eyes, there is a possibility that a different standard might be applied to the obstetrician. The standard might be what a reasonable obstetrician would do under the same or similar circumstances as opposed to what a reasonable genetic counselor would do with all the training and available technology. But I would want to discourage obstetricians and others from making judgments that they are not properly trained to make.

In that case, my source of information about what ultimately happened is the *New York Times*. My past experience leads me to be a little skeptical about whether they have stated all the complexities, but there appears to have been some question as to the nature of the disease that the first child had, therefore some question as to whether the obstetricians had really improperly communicated information to the patients in the case. A second point was, I think that they said the recurrence rate was practically nil. They challenged the factual assertion that was allegedly made.

I think the answer is that you should never act in a way that is beyond the limits of your competence. You should have a threshold, an appropriate threshold, for referral and yet you should not be afraid to make those judgments which come within your competence. A nice, simple solution, right?

Ms. BEVERLY ROLLNICK (University of Illinois): I too have a question regarding the legal obligations of a genetic counselor. At least three of the eight items that you listed, it seems to me, are rather subjective. For example, communication of information to the patients in an effective manner, encouraging the patients to make or exercise responsible choices, and respect for patients' rights.

Are there in fact now sanctions, when you say legal obligations, if these subjective criteria are violated? There is a great deal of disagreement about what constitutes effective communication. And if not, are you advocating this?

PROF. HEALEY: There is no magic formula that these things represent. I have read as many of the cases as I could find dealing with the issues, particularly communication of information involving or affecting reproductive decision making. In the published chapter, you will see the citation of the cases.

In my mind, it is not so much subjective as it is the need to be responsive to the specifics of the case. There are guidelines within which one operates rather than a pat solution or answer to the problem. A simple printed form does nothing to satisfy any of one's obligations. Indeed, this question of whether there has been adequate disclosure of information is so difficult to identify that there always is going to be room for evaluating the judgment.

I try to encourage the physicians and the graduate students that we teach to develop some sense of the issues that are material to the decision and what you as a physician ought to disclose so that person can make a reasonable decision. The consent also has to be voluntary. You cannot coerce. The whole question of whether it matters if you tell somebody that there is ninety-five percent, no risk, or five percent risk is to me a question of whether you are seeking to *help* the person make the decision or not.

It is always after the fact that we look back to see if indeed that occurred. That is one of the inherent limitations of the legal regulation of these decisions. But I encourage people to meet on an annual basis with an attorney, to work with them about the method in which they satisfy or attempt to satisfy each of these elements, and in doing so to discuss all the problems that have occurred during the year. In that way I think you can avoid falling into the situation where one is relying on a style or a mechanism rather than substance.

9

THE CONTINUING WRONG OF "WRONGFUL LIFE"

ALEXANDER MORGAN CAPRON
University of Pennsylvania Law School
Philadelphia, Pennsylvania 19104

My central task here is to explore the serious side of the humorous epitaph W. C. Fields is said to have chosen for his tombstone: "On the whole, I'd rather be in Philadelphia." Of course, that is only a part of my thesis. In its broadest terms, the argument is that the term *wrongful life* has added needless confusion to the law—the courts are to blame for creating and employing it and the commentators are at fault for repeating it (a sin I suppose I am guilty of even now). The judiciary's use of the terminology and the reasoning to which it gives rise in the decision of cases has taken a fairly straightforward injury and turned it into a metaphysical conundrum. This problem is going to assume increasing significance as human genetics becomes a more accepted part of medical practice, particularly as prenatal diagnosis for an ever-growing number of conditions becomes incorporated into obstetrics. With increased frequency, perhaps, will come clearer analysis, which will lead the courts to the view that recovery for malpractice leading to an affected birth should usually be possible for the parents, under standard theories of negligence and informed consent, and should also be available, upon proof of appropriate facts, for a child born with a severe anomaly that was "preventable." It is particularly in the latter context that it may be necessary to explore Fields's much-quoted dig at the town I call home.

ORIGINS OF "WRONGFUL LIFE"

It is not difficult to understand the seductive appeal of the term *wrongful life*, as much as one may lament the term's ability to cloud clear thinking about the issues. The phrase originated in the early 1960s in an Illinois case, *Zepeda v. Zepeda.* [1] The plaintiff, an illegitimate child, sued his father, who had had an adulterous relationship with the child's mother. The Illinois appeals court concluded that a tort had been committed because illegitimacy was a real and substantial burden for the child to bear and because the defendant had been acting wrongly in promising to marry the child's mother when he was already married. Yet the court declined to allow the child to recover, fearing that it would be flooded with what it termed "suits for wrongful life" brought by children born under conditions they regarded as adverse. The unwillingness of the Illinois court to allow recovery is understandable because the claim presents the same problem that all claims by children against parents face, namely, the fundamental limitation on the parents' duty to act in any particular fashion toward their children. To sum up that barrier to recovery—the denial of children ever being permitted legal redress for their complaints about their economic, social, environmental, or genetic legacies—the court coined the phrase "wrongful life," which was a play on words, since it mirrored the statutorily created action for what is termed "wrongful death," the right of certain enumerated beneficiaries (relatives of the deceased person) to recover for the loss they suffer from the death.

The problems with the notion of "wrongful life" first became apparent in cases brought later in the 1960s in which the defendant was someone other than the child s parent and the burden was physical rather than social. In 1967 the New Jersey Supreme Court decided the landmark case, *Gleitman v. Cosgrove*,[2] in which the parents sought recovery for themselves and their son from the physicians who they alleged had erroneously assured them that the rubella that the mother had contacted early in her pregnancy posed no risk to her fetus. In consequence, rather than terminate the pregnancy, she carried to term a child who was born deaf, mute, probably retarded, and nearly blind. The court concluded that "life with defects" is preferable to "no life at all," reasoning that it was impossible to weigh benefits and burdens as proposed by the plaintiffs; accordingly, the child was not permitted to recover. Further, even assuming that Mrs. Gleitman could have obtained a legal abortion, the court decided that it would go against public policy to redress parents' economic and emotional injury for "the denial of the opportunity to take an embryonic life." In some ways it is difficult to separate the abortion issue from the court's conclusion in *Gleitman*, as is especially apparent in the concurring opinion of Justice Francis. But the notion, picked up by judges from the commentary on *Zepeda* that there is some "logico-legal"[3] impediment to recovery by parents or child for "wrongful life," persisted even after the abortion issue had been settled by *Roe v. Wade*.

TYPES OF HARM ARISING FROM NONDISCLOSURE OF GENETIC INFORMATION

Injuries from rubella are similar to those caused by hereditary disorders in that (1) the harmful condition is not correctable prior to birth and is inseparable from the child's existence, (2) the defects themselves do not originate in the defendant-physician's conduct, and (3) termination of the pregnancy is the only way to avoid the manifestation of the defects. This similarity has led courts to apply the "wrongful life" reasoning to the cases of alleged failure in genetic counseling that have arisen recently.

I leave to other accounts,[4] including the descriptions provided by Professors Milunsky and Healey in Chapters 7 and 8, an analysis of the obligations of various health care professionals to provide genetic counseling. For purposes of this chapter, we may assume that counselees might not have information material to their decision on account of either (1) counselors' failure to perceive the risks accurately or (2) counselors' conscious decision to withhold information. Both situations will be described here as nondisclosure, whether resulting from negligence or from intentional decision. The issue presented is whether a health professional is liable if nondisclosure leads to the birth of a child affected with a genetic disorder, the risk of which could have been predicted by the exercise of reasonable care under the circumstances.[5] This issue, whether analyzed by courts in terms of duty, causation, or damages, has raised many problems. First, can there, as a matter of law, ever be a cognizable injury to parents in producing a child, however afflicted with congenital burdens? Second, if the physical injury is suffered solely by the child, have the parents any basis for recovery for their nonpecuniary loss? And third, if the only means of avoiding the injury would have been to prevent the child's birth, is the child barred for reasons of policy or logic from recovering damages from a counselor?

The Parents' Economic Injury

The harm that has given the courts the fewest problems is the economic burden suffered by the parents. Indeed, at the very time that some courts were developing and applying the concept of "wrongful life" in the cases involving children with birth defects, other courts were recognizing claims on behalf of parents against physicians and pharmacists for the birth of healthy but "unwanted" children. Although the latter cases are not uniform in their outcome, they have generally held that a mother who has attempted to avoid reproduction can collect for the medical expenses, loss of income, and pain and suffering that result from her pregnancy.[6] The courts are divided, however, in deciding whether the birth of a healthy child confers so substantial a benefit on the child's parents that damages can never be recovered for the expenses associated with rearing the child.[7] But the direction of the recent cases is toward the sensible view that the prospective benefits from a child's services and companionship are merely elements that a defendant may prove in order to reduce damages. The alleged uncertainty of such proof, which led courts like the New Jersey Supreme Court in *Gleitman* to deny damages, is now seen as a problem for defendants rather than for plaintiffs who have provided the necessary data for calculating the economic detriment consequent to their child's birth.[8]

The irony, and indeed the injustice, of allowing parents of healthy children to recover, while denying relief to parents of children with serious physical burdens because of the misguided notion of "wrongful life," has recently been removed by decisions in several states. The first decisions came in cases which, like *Gleitman*, involved rubella-related injuries. For example, the Texas Supreme Court in *Jacobs v. Theimer*[9] concluded that the parents of such a child could recover the costs of care and treatment necessitated by the child's impairment. And the Wisconsin Supreme Court, which had earlier taken the view that no damages could be collected from a physician who failed to detect pregnancy in time for an abortion because the birth of a physically healthy child is always worth more than its nonbirth,[10] recognized a cause of action for the extra medical expenses of caring for a physically burdened child.[11]

In the past two years the first cases involving genetic burdens have come before a state's highest court for decision. All of these cases have arisen in New York, which by now has the most extensive jurisprudence on the subject of the liability of health professionals for nondisclosure of reproductive information about genetic risks.

In a decision handed down at the end of 1978, the court allowed the claims of parents in two cases in which children had been born with genetic defects that would have been preventable only by nonconception or by abortion following conception. In *Becker v. Schwartz*[12] the parents sought damages for the costs of "the long-term institutional care of their retarded child," who they alleged would not have been carried to term had the physicians performed an amniocentesis on Mrs. Becker, who was 37 years old at the time the child was conceived, and discovered that the child had trisomy 21. The parents in *Park v. Chessin* had produced one child, who died of polycystic kidney disease shortly after birth. They alleged that in response to their inquiry, their physicians had told them that polycystic kidney disease was "not hereditary," so their chances of conceiving a second child with that disease were "practically nil." When a second child with the same disease was born, the parents sought to recover from the physicians the additional expenses of caring for the child until her death. The New York Court of Appeals held that both the Beckers and the Parks stated valid causes of action in seeking to recover the sums expended in the care of their children.

In an earlier opinion, *Howard v. Lecher*,[13] decided in June 1977, the same court had before it an appeal from the dismissal of parental claims for psychic injuries arising from the plaintiffs' daughter having Tay–Sachs disease, from which she died at the age of 2½ years. The defendant physicians had not, however, objected to the plaintiffs' attempt to recover the additional economic costs of caring for the child.

Perhaps the most remarkable facet of these cases is that they prompted as much notice as they have,[14] since they seem unexceptional according to standard principles of tort law. Why should it merit headlines in the lay and professional press when a court holds that a physician must pay for the costs of medical care necessary to treat the injuries that have resulted from the physician's negligence? The attention probably resulted in part from the recognition that the area of genetics is suddenly coming into its own as a matter for physician attention and patient service—and hence also for malpractice litigation. Since the average physician does not know much about genetics, he or she is worried by the prospect of making mistakes that could give rise to heavy liability. The concern over the *Becker* and *Park* decisions results too from a misunderstanding of what the court has held—those cases came up on plaintiffs' appeals from the trial courts' dismissal of their complaints. Thus, in ruling in the plaintiffs' favor, the Court of Appeals was merely holding that if the defendants were proven to have been negligent, they could be held for the resulting damages to the parental pocketbook. The court was not holding, as some apparently believed, that the physicians were negligent as a matter of law in failing to provide the information that the plaintiffs alleged they needed to reach decisions about reproduction; such negligence could only be determined by reference, as is typical in medical cases, to the standards of the profession as supplied by expert witnesses at the trial of the cases. Some of the anxiety may have been dispelled when the Parks, having won the opportunity to have their case tried, were unable to convince the jury that Dr. Chessin and his colleagues had been negligent. [15]

Nonetheless, the decisions of the New York Court of Appeals are significant for other reasons, which seem yet to have gone unnoticed. The allowance of damages for the straightforward economic injuries to the parents carries significant implications for several other more vigorously contested issues of the harm that may flow from nondisclosure in genetic counseling.

First, the New York Court of Appeals has plainly dispensed with the notion that a counselor defendant's conduct is "not the cause of the [child's] condition," [16] as the *Gleitman* court had maintained. The New York court reached this conclusion both for the easier cases, such as *Park*, where negligent counseling led to a child's being conceived, and for the cases like *Becker*, where the nondisclosure occurred during pregnancy and abortion was the only way to avoid the manifestation of harm. [17] The court thus implicitly accepted abortion as a procedure that neither breaks the causal chain nor so contaminates the parents that they cannot be allowed to recover damages.

Second, the court has knocked the legs out from under the argument that damages are rendered noncomputable by the intangible nature of the offsetting advantages to the parents in having a child, even one with severe problems. This attitude toward mitigating benefits should carry over to a child's suit for damages as well as to other elements in the parents' claim, both of which have also been denied in the past on the grounds of being incalculable.

Third, since a living child will be the real (if not the formal) beneficiary of any recovery for health care expenses, remaining objections to allowing claims directly on behalf of an affected child seem beside the point, especially if based on metaphysical refinements.

Finally, a willingness to allow the award of damages to the parents for one item of damage they have suffered appears to recognize that the defendants owe the parents a duty of care. The notion that the parents should be entitled to prove that indirect harm has flowed from the duty's breach seems a reasonable, indeed unexceptionable, conclusion. Let us turn now to the implications of this last conclusion for the recovery of noneconomic harm by parents.

The Parents' Emotional Harm

The issue of noneconomic harm—namely, whether parents may collect for emotional injury—was at the heart of the decision by the New York Court of Appeals in *Howard v. Lecher*. [18] The plaintiffs alleged that although Mrs. Howard's obstetrician-gynecologist knew that the couple was of Ashkenazic Jewish background and hence at special risk for Tay–Sachs disease, he neglected to employ the tests available or advise them of the risks and the means of avoiding them. The Howards sought damages for, among other things, their "mental distress and emotional disturbances" consequent to their daughter's having Tay–Sachs disease, from which she died at age 2½. The defendant argued that this cause of action should be dismissed because damages cannot be awarded for one person's distress in seeing someone else's suffering.

In a 4-3 decision, the Court of Appeals upheld dismissal of the action, reasoning that parents of a child with a congenital disorder are in the same position as those who witness their child being injured in an accident. New York, like most jurisdictions, permits recovery for injuries resulting from fear for a member of one's family only if the plaintiff was also in "the zone of physical danger."[19]

One response to this difficulty would be to turn to a bystander rule that takes greater account of the range of harm reasonably to be expected from one's negligent actions and hence expands the scope of duty owed. For example, the California Supreme Court permits recovery "if defendant should foresee fright or shock enough to cause substantial injury to a person normally constituted."[20] It is apparent, however, that even the broadest rules for bystander recovery for emotional suffering fit the genetic counseling cases very poorly. Nondisclosure leading to the birth of a genetically affected child exposes parents to their child's course of suffering, perhaps drawn out over a long period of time, but not to a sudden, traumatic injury. (It is not surprising to discover that in other, nongenetics examples of alleged malpractice, parents have been denied recovery when their child's illness was drawn out[21] but were able to recover when, for example, a child choked to death in their arms.[22] Even when the parents are present, it has been held that they must have an "understanding observation" of what they see.[23])

The difficulty in fitting the facts of the genetic counseling cases to the Procrustean bed of bystander law would be alleviated if the obstetrician's primary duty to the parents, in their own right and not merely as representatives of the unborn fetus, were recognized. The failure to conduct tests that would uncover a disorder of the fetus would in this view be a breach of the due care owed to the parents, although the fetus would be the entity that "has the disorder." The principal injury would be depriving the parents of their role as informed decision-makers concerning their reproductive choices; the measurement of the damages here would be the costs, including the emotional or psychological costs—net of the benefits of parenthood—of raising and caring for the child.

Such a "duty" analysis has been followed by several New York judges in recent cases. In *Karlson v. Guerinot*,[24] the Fourth Department of the Appellate Division declined to follow the Second Department's decision in *Howard v. Lecher*[25] and held that plaintiffs should be permitted to prove, if they could, that the defendants had breached their "duty to plaintiff wife to provide her with proper medical care during her pregnancy . . . by not properly diagnosing the condition of the child, thus precluding a decision to abort."[26] If the defendants breached a duty owed to the plaintiffs (rather than one owed to their child), under accepted tort theory they would be liable for those injuries flowing to the plaintiffs *directly* from the breach, including mental anguish and other emotional injuries.

Shortly thereafter, the *Howard* holding of no action for emotional injuries was affirmed by the Court of Appeals. Nevertheless, a suit brought by a couple seeking damages for their emotional upset and mental anguish over their daughter's drug-induced birth defects recently survived a motion to dismiss in the New York Supreme Court.[27]

The defendants in *Vaccaro v. Squibb Corp.* contended that *Howard* precluded parental recovery. But Justice Ascione ruled that since the injection of the drug was the direct and proximate cause of the deformed infant's birth, the parents would be permitted to prove whatever other injuries the injection may also have produced for them. Although Justice Ascione's opinion lacks the authority of the New York Court of Appeals' decisions, the theory it articulates could remove several obstacles to clarity if the judgment is upheld.

First, the opinion recognizes that any suggestion that the duty is limited to avoiding financial but not emotional injuries is so unrealistic as to be nonsensical. Surely, in seeking medical advice and care prospective parents, particularly those at risk for a hereditary disorder, desire not only to avoid the medical expenses of having a child with a severe disease that will prove fatal within a few years, but also—indeed very likely, primarily—to avoid the terrible anguish of producing such a child (especially when the birth could have been avoided), caring for it, and then waiting and watching (helplessly) as it dies. Since, in Judge Cardozo's classic statement, "the risk reasonably to be perceived defines the duty to be obeyed,"[28] the duty of physicians in such circumstances must include the giving of professionally competent advice such as would permit the parents to avoid the grave risk of psychological and emotional harm.[29]

Vaccaro may also have an impact on the measure of damages. Since the parents' emotional injuries are, like their economic expenses, "costs of parenthood," the benefits that flow to this same interest (that is, the "joys of parenthood") should be allowed as a setoff. Justice Ascione had allowed damages at least in part because the injuries flowed "directly" from the injection of the drug; this may have been intended as a contrast to the genetics situation in which the underlying disorder occurs in the fetus independent of the physician's conduct. But this is a distinction not grounded in any significant difference. For whether or not the doctor is the active agent of the child's defects, the plaintiffs suffer an identical fate: becoming the parents of a child with defects whose manifestation was preventable. The *extent* of the emotional upset may be different when the parents, in the genetics case, realize that the alternative was to have terminated the pregnancy rather than to have had a "normal" child, which is the supposed alternative when the harm alleged is that the defendant's conduct led to the child's problem.[30] But if parenthood itself were always sufficient reward, then parents could never recover for emotional upset, whether within the "zone of danger" or not, so long as their child's life were spared, however maimed. And, by a parity of reasoning, the plaintiff in *Johnson v. New York*,[31] who was held to have a right to recover because she had been erroneously notified by a state hospital that her mother had died, should have had sufficient joy in her mother's continued life that it would have overborne her emotional upset at the false tidings of her death. Instead, these matters are left to an individualized weighing of felt benefits against genuine upset.

Within two months of *Vaccaro*, the New York Court of Appeals took a step in the direction of recognizing liability but then demurred on the very matter of the difficulty of calculating damages. As we see in other contexts, the alleged "speculative" nature of harm remains the last refuge for a principle trying to hide from its own shadow.

The Court of Appeals, in deciding the *Park* appeal, concluded that the parents' emotional harm should better be analogized to the rule of direct duty recognized in *Johnson* than to the bystander rule. In so doing, as Judge Wachtler observed in dissent, the court eviscerated *Howard v. Lecher* without actually overruling it. Yet, having recognized the applicability of *Johnson*, the court proceeded to distinguish it. The distinction was not that the defendants there owed a duty directly to Mrs. Johnson's daughter, while Mr. and Mrs. Park were owed no duty by the physicians whose advice about genetic risks they had sought; the court apparently did not believe this course was open, since earlier in the same opinion it had recognized such a duty as laying a basis for the Parks' right to recover pecuniary losses. Rather, the court declared that damages under *Johnson* had been limited to "the proven harmful consequences,"[32] and although the Parks' pecuniary losses were sufficiently certain, "calculation of damages for plaintiffs' emotional injuries remains too speculative to permit recovery notwithstanding the breach of a duty flowing from defendants to themselves."[33]

This rationalization ought not to prevent the New York courts from following the principle of duty they have recognized to its logical conclusion, and other jurisdictions ought not be persuaded by the dubious psychologizing of the Court of Appeals. To explain why damages are recoverable in *Johnson* but not in *Park*, the court apparently forgot such lessons on filial love as *Electra* and *King Lear* and declared, "That a daughter might receive such notice [of her parent's death] with mixed emotions lacks any rational basis in human experience," while "the same cannot be confidently said with respect to the birth of a child" although "parents of a deformed child will suffer the anguish that only parents can experience upon the birth of a child in an impaired state." Depending on the facts of the particular case, "parents may yet experience a love that even an abnormality cannot fully dampen."[34] This would merely lead one to mitigate damages, as the court recognized. This then served to provide a (spurious) distinction; on the basis that "the element of mitigation was not involved" in *Johnson*, the *Park* court found damages too speculative. Yet, as we have seen, the possible existence of mitigating joys of parenthood did not render the parents' economic damages *prima facie* unascertainable.

The Child's Damages

The confusion engendered by the "wrongful life" label has been most pronounced in the courts' consideration of suits brought by the children themselves to recover damages from physicians whose nondisclosure led to birth with burdens. The holding of the New Jersey court in *Gleitman* is again a landmark on this issue. Although that opinion found that weighing the value of life with impairments against the nonexistence of life itself was "logically impossible," neither of the two apparent readings of that phrase is persuasive.

The court first suggested that since there is no known way for the trier of fact to comprehend nonexistence, the difference in value between that state and the pain the child plaintiff experiences in life cannot be calculated. Yet it seems a matter of policy rather than logic—and very unfair to the plaintiff—to rule out any recovery because the damages require both great care and an imaginative leap for their calculation. There may be some situations in which common understanding would lead to the conclusion that it would be better to be dead (or never to have existed)—and it is for the finder of fact to determine just *how much better* it would be.[35]

Similar subjective calculations about the "value" of lives cut short, of pain and suffering, and other intangibles, are made every day. Indeed, the jury in evaluating a child's damages would be asked to compare two conditions to each other that in other cases jurors are permitted to compare separately with "normal life"; that is, in wrongful death cases jurors compare the plaintiff's condition before the defendant acted (which need not have been an entirely good or pain-free one) with nonexistence, and in injury cases they compare the previous condition with an injured one, which may include injuries like those suffered by a "wrongful life" plaintiff. Why should a comparison of these two states, nonexistence and injured existence, be regarded as insuperable, when comparing them with another condition ("normal" existence) is not?

The second reading of the "logical impossibility" presents a difficulty more of philosophy than of practicality. The *Gleitman* court would not allow the child to recover because it could find no logic in permitting a party to complain of its "wrongful life," when that very life was the precondition for the party's ability to appear in court to assert his or her claim. Although attractive at first blush, this conclusion is nothing more than a restatement of the idea that any life must be better than nonlife. There is nothing *illogical* in a plaintiff saying, "I'd rather not be here suffering as I am, but since your wrongful conduct preserved my life I am going to take advantage of my regretable existence to sue you." It may seem *improbable* that life with burdens—having to live in Philadelphia, to return to W. C. Fields—is even worse than nonlife; even though Fields did not think so, a jury might so find on the facts of a particular case. The damages may be difficult to prove and the value of the offset will probably be even more imprecise, but that, by analogy to the rule now generally accepted for measuring parental benefits, is for the defendants to show.

To say, as courts[36] and commentators[37] have, that plaintiffs in the circumstance under discussion here are suing for their "wrongful lives," mistakes the real basis of the children's complaints. The wrong actually being complained of is the failure to give accurate advice on which a child's parents can make a decision whether not being born would be preferable to being born deformed. The deprivation of choice causes a harm to both parties who have an interest in the parents being treated as informed decision-makers: the child and the parents themselves.

> If one objects to awarding damages for the violation of this right, it would seem that the objection is directed either at the policy of allowing abortions . . . or at giving parents who may have conflicting motivations the authority to make this decision. The fact remains that the *Gleitman* court departed from the rule that the choice in this matter lies with the patient, not the physician.[38]

The fetal patient being unable to exercise choice, it falls to its parents to choose for it. There is nothing inherent in the choice the parents claim they would have made had proper genetic counseling been provided that on its face disqualifies the child from recovering when the defendant's wrongful conduct has kept the parents from exercising choice.

On the contrary, the judicial refusal to allow a defective child to recover is unsupportable not only analytically but functionally as well. A purpose of malpractice law is to encourage good care by penalizing bad care. It may be that once having allowed the parents to collect for their economic losses, sufficient deterrence of deficient counseling has been achieved, but the absence of an entire category of damages (those suffered by the child) suggests likely underdeterrence because there is no assurance that the jury in setting the amount to be collected by the other plaintiffs (here, the parents) will take the excluded category of harm into account. The problem would appear most dramatically with disorders that involve only minimal additional expenditures by parents (where, for example, the conditions are not medically or otherwise correctable) but great pain and suffering for afflicted children. The predictable result is too much negligent genetic counseling because its true costs have been understated.

Finally, once the right of parents to collect for their economic losses is accepted, as it was in the *Becker–Park* decision, it becomes difficult not to allow the child to recover for his own injuries directly. The underlying theory of the child's case is the same as that of the parents: "Reduced . . . to simplest terms, the basic premise [of the parents'] claims, as well as that of the child, is that the child should never have been born,"[39] or, somewhat more elaborately, that the parents should have been placed in a position to exercise informed choice about what course to follow with their unborn child, and that finding the prospect of life burdened with the defects that should have been disclosed worse than no life, the parents would have terminated the pregnancy.

Perhaps the philosophical, functional, and analytical weaknesses in the current view of "wrongful life" claims by the affected children will prompt more sensible results as genetic counseling cases come before more courts for decision. The New York Court of Appeals' combined decision in *Becker v. Schwartz* and *Park v. Chessin* was the first by any state's highest court to pass on the issue in the genetics context. The lower courts had, with little explanation, upheld the children's claims; the Court of Appeals had little difficulty in disposing of the one reason given, that the defendants had breached "the fundamental right of a child to be born as a whole, functional human being."[40] The Court of Appeals rightly rejected this principle as unprecedented and as likely to produce untenable results, such as suits against parents "for less than a perfect birth."[41] Had the Appellate Division chosen instead to rest its decision squarely on the right of the child to have his decision made by his parents, the Court of Appeals could not have so easily dismissed its conclusion.

CONCLUSION

Slowly, and without explicit recognition of the fact, the courts may finally be groping their way out of the maze into which that catchy but deceptive phrase "wrongful life" led them 15 years ago. Although the path out is itself full of pitfalls, a theory of liability based upon conventional tort notions is surely preferable to the rules now prevailing in most jurisdictions that deny recovery to both parents and child on the metaphysical ground that recovery cannot be allowed for "the life" of a child who would never have existed had its parents' wishes been followed.

The main pitfall to be avoided is the substitution of "a legal right to begin life with a sound mind and body"[42] as the touchstone in place of "wrongful life." Such a principle not only is conceptually inapplicable to children who are affected since conception with genetic disorders but also misstates the relative rights and obligations of physicians, parents, and progeny. Whatever parents' moral obligations to their children, at the beginning of life and thereafter, the principle of a right to healthy birth is too far-reaching to establish a legal rule. The enforcement of such a rule by the state, through the courts and other agencies of social control, would quickly lead to unprecedented eugenic totalitarianism. Even in its mildest application, the rule would lead courts into the impossible and inappropriate task of reviewing for "good faith" parental judgments on all the aspects of "quality of life" that a child would experience.

In any event, the principle on which I have elaborated here, which is addressed to the obligations of physicians and other health care providers in the guise of genetic counselors, does not implicate the question of parents' obligations. Rather, it recognizes that the basic duty is to disclose information, in a comprehensive and comprehensible fashion, that parents[43] will need to weigh the probabilities and risks for the potential child and to decide whether life with defects is better than abortion or whether in some cases life with defects is "a fate worse than death." In order to give force to this right, it is necessary to permit recovery by an affected child and its parents of the economic and psychological damages that result from wrongful nondisclosure. In the process of recognizing the right, the courts will, I hope, end the continuing wrong of "wrongful life" and rid the legal lexicon of that unhappy phrase.[44]

REFERENCES AND NOTES

1. 41 Ill.App.2d 240, 190 N.E.2d 849 (1963), *cert. denied*, 379 U.S. 945 (1964). See also *Williams v. New York*, 18 N.Y.2d 481, 276 N.Y.S.2d 885, 223 N.E.2d 343 (1966); *Slawek v. Stroh*, 62 Wisc.2d 295, 215 N.W.2d 9 (1974).
2. 49 N.J. 22, 227 A.2d 689 (1967).
3. The phrase is from Tedeschi, G., On tort liability for "wrongful life," *Israel L. Rev.* 1:513, 530 (1966), which is relied upon in Justice Proctor's opinion, *Gleitman v. Cosgrove*, 49 N.J. 22, 29, 227 A.2d 689, 692 (1967), and quoted as a "logico-legal difficulty," by Judge Keating in his concurring opinion in *Williams v. New York*, 18 N.Y.2d 481, 484, 276 N.Y.S.2d 885, 888, 223 N.E.2d 343, 345 (1966).
4. See, e.g., Reilly, P., *Genetics, Law and Social Policy*, Harvard University Press, Cambridge (1977); Capron, A. M., Autonomy, confidentiality, and quality care in genetic counseling, in: *Genetic Counseling: Facts, Values and Norms* (A. M. Capron *et al.*, eds.), *Birth Defects: Orig. Artic. Ser.* 15 (No. 2) (1979), Alan R. Liss, New York; Note, Father and mother know best: Defining the liability of physicians for inadequate genetic counseling, *Yale L. J.* 87:1488 (1978).
5. Not treated here is harm of the opposite sort that could occur if inaccurate advice is given leading to unnecessary protective steps, such as abortion or sterilization; liability may also be imposed if the couple is counseled into unnecessary contraception, artificial insemination, or adoption, although these steps would be less likely to give rise to substantial irreversible harm.
6. See, e.g., *Coleman v. Garrison*, 327 A.2d 757 (Del. Super. 1974); *Clegg v. Chase*, 89 Misc.2d 510, 391 N.Y.S.2d 966 (Sup. Ct. 1977); *contra, Christensen v. Thornby*, 192 Minn. 123, 255 N.W. 620 (1934).

7. See, e.g., *recovery permitted: Custodio v. Bauer,* 251 Cal. App.2d 303, 59 Cal. Rptr. 463 (1967); *Troppi v. Scarf,* 31 Mich. App. 240, 197 N.W.2d 511 (1971); *Betancourt v. Gaylor,* 136 N.J. Super. 69, 344 A.2d 336 (1975); *Ziemba v. Sternberg,* 45 App. Div.2d 230, 357 N.Y.S.2d 265 (1974); *recovery denied: Shaheen v. Knight,* 11 Pa. D.& C.2d 41 (Common Pleas 1957); *Ball v. Mudge,* 64 Wash.2d 247, 391 P.2d 201 (1964).

8. *Rivera v. State,* 404 N.Y.S.2d 950 (Ct. Cl. 1978).

9. 519 S.W.2d 846 (Tex. 1975).

10. *Rieck v. Medical Protective Soc.,* 64 Wisc.2d 514, 219 N.W.2d 242 (1974).

11. *Dumer v. St. Michael's Hosp.,* 69 Wisc.2d 766, 233 N.W.2d 372 (1975).

12. 46 N.Y.2d 401, 413 N.Y.S.2d 895 (1978).

13. 42 N.Y.2d 109, 397 N.Y.S.2d 363, 366 N.E.2d 64 (1977).

14. Doctor held liable in abnormal births, *New York Times,* Dec. 28 (1978); Doctors take care, *Newsweek,* Jan. 8 (1979); Lavine, Wrongful birth decision termed victory for patients, *Nat. L. J.* **Jan. 8:** 3, col. 1 (1979). "New York doctors see twin impacts of the decision. The ruling, say these observers, may boost malpractice insurance premiums and could also figure in driving more physicians from their obstetric practices." Chapman, S., What are your odds in the prenatal gamble? *Legal Aspects of Medical Practice* **March:**30 (1979).

15. Jury clears two doctors of birth-defect liability, *New York Times,* April 9 (1979).

16. 49 N.J. at 29, 227 A.2d at 692.

17. Clearly, the case would be still easier were the options available upon diagnosis of a fetus being at "genetic risk" to include some form of corrective medical or surgical intervention *in utero.*

18. 42 N.Y.2d 109, 397 N.Y.S.2d 363, 366 N.E.2d 64 (1977).

19. *Tobin v. Grossman,* 24 N.Y.2d 609, 301 N.Y.S.2d 554 (1969).

20. *Dillon v. Legg,* 68 Cal.2d 728, 740, 69 Cal. Rptr. 72, 80, 44 P.2d 912, 920 (1968).

21. *Hair v. County of Monterey,* 45 Cal.App.3d 538, 119 Cal. Rptr. 640 (1975); *Jansen v. Children's Hosp. Medical Center,* 31 Cal.App.3d 23, 106 Cal. Rptr. 884 (1973).

22. *Huber v. Aengst,* 28 Citation 88 (1974) (Cal. Super. Ct., Los Angeles Co., Docket No. NEC 12214, May 17, 1973).

23. *Justus v. Atchinson,* 139 Cal. Rptr. 97 (1977).

24. 57 App. Div.2d 73, 394 N.Y.S.2d 933 (1977).

25. 53 App. Div.2d 420, 386 N.Y.S.2d 460 (1976).

26. 57 App. Div.2d at 78, 394 N.Y.S.2d at 936 (failure to perform amniocentesis on 37-year-old; Down syndrome child born).

27. *Vaccaro v. Squibb Corp.,* 412 N.Y.S.2d 722 (Sup. Ct. 1978).

28. *Palsgraf v. Long Island R. Co.,* 248 N.Y. 339, 344, 162 N.E. 99, 100 (1928).

29. The customary tort standard of "reasonableness" should apply to the professional's conduct but not to the highly personal and subjective decision making of the clients. Causation is made out if it is established that the clients (again, not the "reasonable person" but the actual individuals, with personal values on which the finder of fact determines they would have acted) would by contraception, abortion, artificial insemination, or other means have avoided the injury (manifested in the birth of the affected child) had the counselor or physician not breached his or her duty.

30. This bland assumption about what would have happened if the defendant had not been negligent is probably seldom justified. "Normal" children bring their parents untold trials and sorrows, and there is no assurance that another child of the same parents, though not affected by the particular problems manifested by the child then before the court, would even be "normal." *Vaccaro* itself illustrates that problem, as the drug in question, a progestational hormone, had been administered to Mrs. Vaccaro to prevent miscarriage because she had not had live issue from two previous pregnancies. *Vaccaro v. Squibb Corp.,* 412 N.Y.S.2d 722, 723 (Sup. Ct. 1978). The infant Vaccaro's cause of action against the same defendants for having caused her deformities were not treated in Justice Ascione's ruling on the defendant's motions, which were addressed rather to 9 of the 10 causes brought by the child's parents. But it is possibly significant that neither in ruling on the parents' actions nor in briefly considering the child's does the Supreme Court opinion suggest that all damages are moot because without the defendant's acts the infant plantiff would not have been born, i.e., that there is some logical barrier to her collecting for the injuries that *did* accompany the birth that the defendants facilitated for her.

31. 37 N.Y.2d 378, 372 N.Y.S.2d 638, 334 N.E.2d 590 (1975).

32. Ibid. at 383, 372 N.Y.S.2d at 643, 334 N.E.2d at 593.

33. 46 N.Y.2d at 409, 413 N.Y.S.2d at 902.
34. Ibid.
35. One way to educate the jury about the unusual task it has before it would be by use of instructions along the following lines:

> In calculating the damages of the child, consider first whether the defendant has through negligence failed to do anything that would have cured or alleviated the child's condition. If you find such a failure, the defendant is responsible for the measurable difference between the child's present condition and the condition he would have been in had the defendant not been negligent.
>
> If, on the other hand, you find that no treatment was available to cure or alleviate the condition of the child during the pregnancy, then you must decide whether the defendant negligently or willfully failed to inform the child's parents of the risks and alternatives, and whether if the parents had been so informed they would have terminated the pregnancy.
>
> If you find affirmatively on both these questions, you may find the following steps helpful in calculating the damages the child is entitled to collect:
>
> First, determine a dollar amount for the harm that a person would experience if, at birth, he or she were to experience injuries that left him or her in the condition you now find the child to be in.
>
> Second, determine a dollar amount for the harm that a person would experience if he or she were to die prior to birth; in this figure you may include compensation for the loss of life's pleasures and for the loss of life itself.
>
> Third, deduct the second figure from the first. If the result is positive, this suggests the amount by which the child's present condition is worse than the nonexistence into which the child would have been placed had his [or her] parents known of the risk or fact of the child's uncorrectable condition and chosen to terminate the pregnancy.

36. See, e.g., *Stewart v. Long Is. Coll. Hospital*, 30 N.Y.2d 695, 332 N.Y.S.2d 640, 283 N.E.2d 616 (1972); *Dumer v. St. Michael's Hosp.*, 69 Wisc. 766, 233 N.W.2d 372 (1975).
37. See, e.g., Lord Kilbrandon, The comparative law of genetic counseling, in: *Ethical Issues in Human Genetics* (B. Hilton *et al.*, eds.), p. 245, Plenum Press, New York (1973); Note, Father and mother know best: Defining the liability of physicians for inadequate genetic counseling, *Yale L. J.* 87:1488, 1500 (1978).
38. Capron, A., Informed decisionmaking in genetic counseling, *Indiana L. J.* 48:581, 603 (1973).
39. *Park v. Chessin*, 60 App. Div. 80, 94, 400 N.Y.S.2d 110, 118 (1977) (Titone, J., dissenting).
40. *Park v. Chessin*, 60 App. Div. 80, 88, 400 N.Y.S.2d 110, 114 (1977).
41. *Becker v. Schwartz*, 46 N.Y. 401, 406, 413 N.Y.S.2d 895, 900 (1978).
42. *Gleitman v. Cosgrove*, 49 N.J. 22, 28, 227 A.2d 689, 692 (1967), *quoting Smith v. Brennan*, 31 N.J. 353, 364, 157 A.2d 497, 503 (1960).
43. Judge Cooke, dissenting in *Howard v. Lecher*, 42 N.Y.2d 109, 116, 397 N.Y.S.2d 363, 368, 366 N.E.2d 64, 68 (1977) from the allowance of any recovery for emotional harm, would have limited the cause of action to the prospective mother since he concluded that a physician caring for a woman owes no direct duty of care to her mate. This conclusion, although perhaps constitutionally mandated if the woman forbade the physician to take the man's interests into account, is not justified in the usual case. A *couple* is planning to have a child; as both are undertaking certain obligations thereby and both have certain expectations, the physician or other genetic counselor owes a duty to both since they are both exposed to risks by nondisclosure of information material to the decision whether to continue or terminate the pregnancy.
44. Since this chapter was sent to the printer, a number of important legal developments have occurred, most notably the partial overruling of the *Gleitman* decision by the New Jersey Supreme Court in *Berman v. Allan*, 80 N.J. 421, 404 A.2d 8 (1979). See Capron, A., Tort liability in genetic counseling, *Columbia L. Rev.* 79:618 (1979).

Discussion

Ms. Sylvia P. Rubin (New York): What are the legal responsibilities of a genetic counselor when a patient literally falsifies her age just to obtain prenatal diagnosis for sex determination only? In my own experience I know that two abortions resulted because of the "wrong" sex as far as the patient was concerned.

Prof. Joseph M. Healey: What is the concern that you have in terms of potential liability? Are you afraid that somebody is going to sue the physician for providing the services to the person who is not legally competent to give consent, or that the decision to abort may have been based on information fraudulently obtained? I'm not sure what your concern is.

Ms. Rubin: Well, there are several different things here. Some patients may not have been able to obtain a needed test because of insufficient laboratory facilities.

Prof. Healey: I understand your question a little bit more clearly then.

When we teach truth-telling to medical and dental students, one of the examples we use is one of the incidents in the Chad Green case in which the mother of Chad Green lied to the physicians about whether Chad Green was still on the chemotherapy. The mother had taken him off for a period of time when she told the physicians he was on and there was the return of the disease.

So we try to place the obligation and the rights and duties in proper context. Both parties in the relationship have obligations to communicate honestly and effectively with each other from their point of view. So therefore I would hope that the client or patient has the same obligation to tell the truth as does the physician or genetic counselor.

In terms of any potential liability for acting on that sort of misstatement, I do believe it necessary to have to verify the age of people who come for prenatal studies. If, on reliance of what appeared to be a good-faith offer of information by the patient, you made available a diagnostic test not generally available to other people, no criticism should arise. In effect, you were saying that it is as much a responsibility of the person as it is our responsibility and we cannot be responsible for enforcement. It would introduce another note into the relationship that I think ought not to be present.

I'm not sure you should be concerned about that, although I have heard that it existed in another area, and that is the whole question, obviously, of abortion. Until the Supreme Court decides what a mature minor is, I think that is going to continue. It may even continue after that.

A Participant: For a genetic counselor to counsel patients as correctly as possible, it has been stated to me before that he or she should approach it being omniscient, detached, disinterested, and consistent.

Because counselors are only human and patients are not always capable of immediately assimilating information given at a counseling session, what stance should a counselor take when asked by a patient, "What should I do?" or "What would you do in my situation?"

Prof. Healey: The law will go first and medicine will follow quickly behind.

I think again it is impossible for a health care provider to avoid the temptation to influence the patient or the client in that type of situation. Indeed, one of the histories of consent to treatment is that patients or physicians or other health care providers are capable of manipulating the communication of information to get the so-called desired result.

I would like to argue that you have an obligation to avoid that for a number of reasons, including the primary goal of the relationship, which is to make it possible for the person to assume responsibility for himself or herself. When that is refused, obviously, there is an invitation for you to help the person to make the decision, and it may be necessary for you in order to do that to say what you would do in that situation. But I would do that only as a last resort, because I think your primary obligation is to help the person to make the decision. If that person does not want to make the decision, then you have to fall back to a different level.

DR. AUBREY MILUNSKY: I believe that when the patient signals the question of "What would you do, Doctor?" he or she is signaling failure in that genetic counseling situation and is saying, "Please make my decision for me." Of course, they really don't care what you would do about that decision in your family. What they are saying is, "Tell me what to do." My experience has been that there is a need in that situation to sit them down again and to reexplore the priorities in their own lives so they can reorder or discover those priorities important to them in their decision-making situation. These priorities may be entirely different from those of the physician.

I don't believe that the physician has a right or a role in visiting upon the patient his or her religious, racial, eugenic, or other prejudices, and I view that response of the physician as essentially inappropriate, saying that "well, if I were you, I would do the following." I think the physician simply needs to understand that the patient is asking for further help in the recognition of priorities.

DR. DANIEL STOWENS: My question is somewhat similar but it does raise a legal point. Counseling sessions are usually fairly lengthy. What sort of documentation would you advise or do you use to have some sort of proof of what you have told the patient? Secondly, what do you do when you can't give any advice?

PROF. HEALEY: You used the right word when you referred to the form as more or less evidence or proof of what has been communicated. Because I think you all are aware—or if not, you ought to be—the actual consent is not the document, not the signature on the document. It is the assent, the agreement of the patient; and the form serves as proof should there be some question as to what exactly was communicated between the health care provider and the patient.

So if one uses a document, one should use it with that in mind. A document does not make up for any defect in the process preceding the signature on the document or the use of some sort of consent form to keep a history of what occurred. And indeed, the second key word is *process*, that this is a process of communication and not a single event.

Obviously, when one particularly has to communicate very difficult information involving sometimes the necessity for the individual to accept the realities or accept probabilities which are not those which the patient expected to hear or wants to hear, it is important that this be done over a period of time so there is sufficient opportunity for the information to be absorbed. Any mechanism that properly records what was communicated plays the role of again being evidence should there be any challenge. But none of them is protection in and of itself.

DR. MILUNSKY: In practice, we send the patient a letter which is a succinct summary of the counseling provided. The original of that letter goes to the referring physician and a copy of that identical letter to the patient.

PROF. ALEXANDER MORGAN CAPRON: We use the same practice, except the original goes to the patient.

I would comment that the role of that document is not primarily, however, for its use as a legal memorandum or evidence. Rather, it is part of that process of information and counseling, because one of the things that the letter says is, having read this, please now be back in touch with us. We offer to answer more questions that may arise later. So it is really seen as part of the process, although it would be a valuable memorandum of what took place were there ever to be questions about that. But I see it primarily as instrumental to counseling rather than as an instrument for avoiding liability.

A PARTICIPANT: The concept of expertise suggests discontinuous knowledge, and in most cases the referral of one physician to another is an indication of that discontinuity. However, in certain interfaces between knowledge, one is confronted with situations in which one has to either be expert in more than one field or make a decision based on limited knowledge.

In a situation where a genetic counselor advises a patient that he or she suffers from a late-onset disease and that subsequently the disease will be manifested, and the patient undergoes some psychological trauma as a result of the knowledge, is the counselor liable because of insufficient psychological knowledge and preparation in the field of psychology to have properly prepared the patient for this kind of news?

PROF. HEALEY: I am not aware of any case in which that has been raised. Part of the problem is, again, that many patients in reproductive decision making do not appreciate the frequency of genetic problems. I think this is one of the barriers that those who would promote responsible decision making have to face. Because of that, it is very possible for a situation to occur much like you are describing, where there is some complaint by the patient about the failure to be adequately prepared.

I would ask Dr. Milunsky if at the early stages of communication with the patient it is his practice to mention that this does not eliminate the possibility of there being any subsequent problems that we are not able to diagnose.

DR. MILUNSKY: I think those who counsel would agree that an open-ended kind of exercise allowing for the lack of guarantees, allowing for the lack of substantive and final statements, is important and that the range of responses go from simply "I don't know" to "Here are the exact risk figures." But, again, there are always other possibilities relating to other disorders which might obtain in that particular pregnancy.

PROF. CAPRON: Let me add just one comment to that. If, as now seems to be the case, the law is at least articulating a stronger, clearer obligation on the part of physicians generally to disclose information to their patients, and if that is being communicated to physicians in a way that actually alters behavior, part of that change will have to be a greater training of physicians generally, but particularly those who call themselves counselors, to learn techniques of information conveyance which serve a counseling and informing function. This should not lead back into the circle which we have already entered, in which the fear of an adverse psychological reaction on the part of the patient provides the excuse for the physician to say, "Well, I better not disclose that." This approach leads back into what is really the circle which the law has already provided by the term *therapeutic privilege*, and which is a much-repeated, poorly understood, and easily abused notion that in certain cases it may be appropriate to withhold information where the conveying of the information would be harmful to the well-being of the patient.

If that is not going to become a trap in which the newly perceived obligations of physicians to disclose are washed out, then there is going to have to be necessary training or physicians will justifiably say, "Gee, I don't know how to do it without hurting you so I won't do it at all." That shouldn't be the answer.

10

The Law and Operation of Genetic Screening Programs

Y. Edward Hsia
John A. Burns School of Medicine
University of Hawaii
Honolulu, Hawaii 96826

Introduction

My charge in this volume is to discuss legal issues related to genetic screening programs. As I have had no legal training and few run-ins with the law, of necessity my discussion will be based on my viewpoint as a practicing medical geneticist. I have sought to serve community needs by participating in the planning, initiating, operating, upgrading, and justifying of various genetic screening programs. My contribution will be the only one in this volume based on personal practical experiences in the health services.

Operation of genetic screening programs in the context of our legalistic society[1-11] was the topic of an entire session in the first symposium.[9] Here I will first establish a factual framework, followed by illustrative experiences, and then I will offer my opinions and judgments about how future policies might mitigate some current anomalies and shortcomings.

The legal issues differ markedly according to the type and objectives of various programs (see Table I). Legal interaction is in three major areas: legislation, regulation, and litigation. There are four entirely different types of genetic screening: symptomatic, presymptomatic, parental, and research[1-4] (see Tables II and III).

Facts about the Law and Genetic Screening

Legal Aspects

LEGISLATION.   Legislation is the democratic political process whereby statutes are formulated, debated, enacted, and approved with or without appropriation of funds.

TABLE I. Types of Genetic Screening

	Basic objective	At-risk population	Primary service obligation
Symptomatic	Diagnostic	Undiagnosed patients	To responsible physician
Presymptomatic	Preventive intervention	Apparently normal population	To individuals tested
Parental			
Antepartum	Genetic counseling	Prospective parents	To prospective parent
Intrapartum	Fetal diagnosis	Pregnant women	To expectant parent
Research	Data collection	Various	To protection of participant

Aspects of legislation are covered in this volume both in the first part on government control of science and in the fourth part on the law and control of genetic disease. For genetic screening programs, all these discussions are relevant, but the service component of these programs has unique legal implications not covered elsewhere (see Table I).

Representative democracy is not an ideal political system for developing centralized preventive health programs. Before legislation can be passed, elected officials have to be at least sympathetic and receptive—if not enlightened—to support health issues constructively; public opinion has to be strongly favorable and not just permissive; vested lobbying interests have to be placated, assuaged, or outmaneuvered. A political consensus is a fragile structure that can be shattered by personality conflicts, backroom machinations, and sudden shifts in political power or public opinion.[12-14] A law often ends up finally as a hastily drawn up compromise of questionable wisdom or practicality.[15-20]

Funding allocations, which should be well planned, long-term, and equitable, are even more vulnerable to arbitrary priorities at many levels of government because of shortsighted goals, budget restrictions, dispensing of favors, "benign neglect," extravagant stipulations, or unrealistic demands for cost containment.[21-26] Both general and categorical funds are disbursed under a piecemeal system that is subject to patronage and other pressures.

The legislative process can initiate fact-finding hearings, set up committees of enquiry, convene task forces, and call for recommendations. Preexisting broad-ranging health-related statutes can facilitate or restrict genetic screening proposals, even if they had been written without any consideration of genetic screening issues. Categorical project-related legislation can also help or hinder, depending on the care and wisdom with which they were formulated, because rigidity can be too confining, and flexibility can lead to misinterpretations or to loopholes subject to abuse.

REGULATION. Regulation and control are discussed in the context of laboratory activities in Chapters 1–6. For the operation of community screening programs, regulations may define specific administrative structures; demand quality control of laboratory procedures; stipulate particular techniques; safeguard the rights and interests of the public, of other interested parties, or of screening staff; and mandate accountability.[3, 8-10, 15-21, 27]

Grant applications, agency clearances, and progress reports all drain the time and resources of key personnel. The ever-proliferating paperwork and committee reviews that

TABLE II. Hazards to Participants in Screening Programs

Hazards	Symptomatic	Presymptomatic	Parental		Research
			Antepartum	Intrapartum	
Mistaken result					
Skipped case	+	+	±	+	−
Wrong diagnosis	++	++	+	++	±
Misinformed subject					
Psychological upset	(+)	+	+	+	++
Social stigmatization	−	++	++	±	++
Mishap to participant					
Physical injury	±	+	+	Mother ++ Fetus +	++
Untoward outcome	+	+	+	++	+

TABLE III.   Legal Issues in Genetic Screening

Legislation	Regulation	Litigation
Planning	Certification	Conflicting interests
Needs assessments	Quality control	Errors
Recommendations	Personnel	Missed cases
	Facilities	Misdiagnosis
	Test Results	Misinformation
	Safety	Treatment failure
Enabling: guidelines	Accountability	
Funding	Availability	Consequential harm
		Physical
		To parent (pregnant mother)
		To child (fetus)
		Psychological or social
		Invasion of privacy
		Stigmatization
		Discrimination
		Inadequate support services

seek to ensure efficacy all tend to inflate program costs, thereby undermining efficiency and frustrating cost containment.

LITIGATION.   This relates to criminal and civil issues about diverted funds or protection of rights and privileges, or to recompense for harmful consequences of action taken or action omitted.[28-34] Litigation can arise from conflicts of interest or claims of attributed fault for injuries incurred (see Table III).

Genetic Aspects

SYMPTOMATIC SCREENING.   Symptomatic or high-risk genetic screening tests are performed on individual patients to answer the diagnostic question: Is it possible this patient has X-type genetic disease?

The objectives of symptomatic tests are to narrow diagnostic possibilities. Examples are chromosome analyses for nonspecific malformations, and amino acid screening for indiagnosed neurologic or metabolic diseases. Occasionally these tests will yield definitive diagnoses, e.g., "trisomy 21," but generally serve only to suggest or exclude disease categories, e.g., "generalized amino-aciduria, possibly due to an underlying genetic abnormality." These tests can be performed in theory by any registered pathology laboratory, using published protocols or commercial kits, but often are undertaken by research-oriented medical centers. Test facilities should be backed up by resources for more sophisticated confirmatory investigations. Legal issues arising from these tests relate to certification, accuracy, interpretation, competing interests, clinical responsibilities, and patient rights. These issues do not differ in principle from those for any routine blood, urine, or X-ray examination in an office, clinic, or hospital ward.[30]

PRESYMPTOMATIC SCREENING.   Presymptomatic genetic screening is focused on a target population to answer the question: Do any of these apparently normal people have disease X?

The target population could be relatives of a person with disease X,[35-37] a subpopulation with a high risk for disease X,[38,39] or the entire population in a community.[1-4,40-42] The objectives of presymptomatic tests are to find affected individuals in time for effective preventive treatment. Examples are presymptomatic examination of relatives at risk for hypercholesterolemia,[7] polyposis coli,[37] or breast cancer; testing specific ethnic populations for cystic fibrosis, sickle cell disease, or glucose-6-phosphate dehydrogenase deficiencies[7]; and testing all newborn infants for congenital dislocation of the hip, phenylketonuria, or hypothyroidism.[1-4,7,11,43,44] Some of these tests can be done simply and economically by any trained health professional; others are best done on a regional basis by carefully designed efficient mass-testing facilities.[41,42] There should be a comprehensive clinically supervised health care program, which integrates testing with interpretation, consultation, management, counseling, and follow-up services. General legal issues of certification, quality control, and competing service responsibilities are analogous to those in nongenetic presymptomatic screening, such as screening children for dental, visual, hearing, or learning difficulties; premarital tests for syphilis; and population tests for diabetes, hypertension, or glaucoma.[43,44] Particular legal issues relate to cost-effectiveness, missed cases, and consequential harm.[2-11,15-18]

PARENTAL SCREENING.   Parental screening, or heterozygote screening offered to a target population of potential parents, answers the question: What risk do these prospective parents have for disease X in future children?

Generally, the subjects have no health risks for themselves, except for dominant conditions such as hypercholesterolemia or polycystic renal disease.[35] The objectives of parental screening are to warn parents about identified risks and to inform them about possible preventive measures. Parental screening is offered *during* pregnancy for fetal diagnosis of chromosome abnormalities or major neural tube deformities[45-47] (see also Milunsky, Chapter 7), but heterozygote testing, e.g., for hemophilia, Tay–Sachs, or sickle cell carrier status, is best conducted *before* pregnancy. (Screening for rhesus blood groups and antibodies is routinely done in early pregnancy by obstetric services, yet hemolytic disease of the newborn is seldom regarded as a genetically determined disease that warrants detection and genetic counseling prior to pregnancy.) Legal issues unique to parental screening relate to offering tests of considerable sophistication that have no relevance to the health of the subject submitting to the test, but only to the subject's progeny. Special considerations arise regarding informed participation,[45] procreative rights, freedom of choice, hazards of test procedures, attitudes toward abortion, psychological harm, and invasion of privacy.[46,47] These issues differ for antepartum and intrapartum parental screening programs from other types of screening programs[1-11,34,46,47] (see Tables I and II).

RESEARCH SCREENING.   Research screening projects seek to answer the question: What can we learn from this population of test subjects for the sake of medical and scientific knowledge?

The objective of research screening is data collection for analysis. Whether the screening is for nutritional surveys or genetic data, the principles are similar. There are obviously overlapping objectives between research screening and all the other types of screening, but the research element involves different issues. No promise of benefit can be made for research screening; hinting at implied benefits may misrepresent the objectives of research projects. Legal issues therefore emphasize voluntarism, confidentiality, protection from harm of participants, scientific merit of project design, and accountability to supporting institutions.[3,6,27,29,46]

Experiences with Genetic Screening and the Law

Although concepts and issues related to genetic screening appear to be well defined,[1-11] in the establishment and operation of screening activities many anomalous paradoxes arise, which often vitiate or frustrate intended goals. A few examples based on personal observations provide vivid illustrations of how well-intentioned screening activities can fail to fulfill any useful objectives.

Legislation

A screening test casts a diagnostic net over a target population to catch individuals who exceed defined limits of normality. Without carefully planned objectives and programs, such testing can be worse than useless.

A LAW WITHOUT PLANNING.   One of the smaller states, which mandated a presymptomatic newborn screening test for phenylketonuria (PKU) in 1968, has required each responsible physician to have all newborn infants tested for PKU.[17] The law made no provision for record keeping, confirmation of diagnosis, quality control, supportive services, or appraisal. No projections were made about PKU infants likely to be detected. In 12 years an unknown proportion of infants have been tested in several independent laboratories; only one patient is said to have been detected, at an outlying hospital. (There was no knowledge centrally of this patient's existence.) If a second patient were to be detected, there is no planned or budgeted provision for this patient's medical supervision.

A PLAN WITHOUT LEGAL SUPPORT.   A proposal for an upgraded, coordinated, centralized newborn screening program for a state was turned down by state and federal agencies. The federal agency judged it by criteria established for parental sickle cell screening (sickle cell disease had negligible prevalence in this state); the state agency judged it with no local experience except of a preexisting PKU screening mandate. It was too service-oriented to receive any research support and too novel to receive service support. In the absence of categorical statutes and allocations, this proposal to embark upon new screening programs was disallowed because it could not demonstrate past efficacy before the program had begun.

UNJUSTIFIABLE LEGAL SUPPORT.   At the same time that a statewide newborn screening proposal was withering on the vine from ill-informed, inadequate support, an extravagant "international" specialized parental screening reference laboratory was approved and funded by state officials for a remote region of a state, primarily because of local political considerations during an election year. (The projected annual budget for this laboratory was several times that needed for an upgraded, coordinated, comprehensive newborn screening program.) Patently overambitious, unrealistic plans were accepted because of their vote-getting attraction. Opposition by local leaders in public health, medical education, and health administration to allocation of limited funds for the "international" center was ignored or suppressed by elected officials.

Lobbying

In a democracy, elected officials are responsive to representations from constituents. The loudest and most effective lobbists, however, often come from the best organized, most vociferous, or most generous interests, rather than the most deserving advocates. Narrow interests may receive better attention than more balanced, broader needs.

SUCCESSFUL LOBBYING AGAINST UNREASONABLE LEGISLATIVE DEMANDS. In response to a constituent's request, a legislator drew up a resolution demanding that a state university conduct a two-year study of the feasibility of setting up carrier detection facilities for a specific genetic disease. At very short notice, the university was obliged to draft testimony to the effect that reliable testing could be done by only a handful of referral centers, that adequate facilities would cost several hundred thousand dollars, that the state had less than a few dozen relatives at risk, and that carrier detection was already being conducted in collaboration with a major referral center. Legislative power was abused in making extravagant demands that could have been forestalled by a brief preliminary telephone enquiry.

UNSUCCESSFUL LOBBYING OWING TO UNFORESEEABLE LEGISLATIVE OPPOSITION. Special funding was requested from a state legislature for renovation costs of a fetal diagnostic laboratory. One legislator said he would always support prevention of major handicaps; there was general sympathy and no opposition at the formal hearings. The appropriation was readily passed but was vetoed on a technicality.

Operation

If a screening program is soundly planned, well organized, adequately funded, and carefully supervised, external controls would be redundant, except for purposes of standardization and data pooling among regional programs. Without external controls, however, the quality or adequacy of a program could remain unquestioned until a catastrophic error occurs.

A TEST WITHOUT STANDARDS. Symptomatic screening for amino acid patterns in one hospital laboratory was being conducted by technicians who had never seen an abnormal result, and who were reporting all blank results as normal. The test was performed by an approved laboratory under a board-certified pathologist and so charges for the test were paid without challenge by patients and third-party insurance agencies.

A TEST WITHOUT MEANING. An abnormal symptomatic urine screening test report led to intense interest by the clinicians, who wished to confirm the abnormality. The pathologist would not or could not interpret the frequency of false positives, as no attempt had ever been made to correlate test results with clinical findings.

On several other occasions, a commercial laboratory report of "slight increase" on urine amino acid analyses caused clinicians to leap to unjustifiable diagnostic conclusions. When the result of a test is meaningless to the clinicians, any charge is excessive for such tests.[48]

A TEST WITHOUT QUALITY CONTROL. A fetal diagnostic laboratory serving an obstetric community made no efforts to compare or confirm test results with pregnancy outcome. This was justified by the arguments that laboratory responsibility served solely to report test results, and that in the absence of known complaints, the laboratory results were obviously satisfactory.

AN UNEXPECTED TEST RESULT. Newborn screening revealed that an infant had defective activity of the enzyme for galactosemia. After extensive testing of the family it was determined that she was unaffected but probably a compound heterozygote, being a carrier both for galactosemia and for the innocent Duarte variant. In this patient, screening caused parental anxiety, which was not fully allayed even after all available measures had been undertaken.

A TEST OMITTED. An adoptive couple went to great lengths to ensure that the biological endowment and health of the infant to be adopted was optimal. All seemed to be satisfactory; the infant was adopted but developed features of infantile autism and delayed mental milestones by the age of one year. Ultimately, the infant was found to have irreversible brain damage from PKU, which was missed because the infant was born in a state without a PKU screening program.

A DELAY IN TESTING. A pediatrician saw a one-week-old infant with vomiting, diarrhea, liver damage, and prostration. He astutely suspected galactosemia and upon enquiry found that the infant's blood specimen was being screened at the regional laboratory that day. This infant benefited from the diagnostic acumen of her physician, not from the screening program, which was too slow to be of any use to her.

A DELAY IN SHIPMENT. An infant found to have PKU on newborn screening was being treated by special diet. A blood specimen for monitoring treatment was sent to the nearest PKU clinic by priority air shipment. The specimen disappeared in transit; all efforts to trace it were fruitless for two weeks, then it was delivered, too late to be of any help in managing this infant's diet.

A DELAY IN REFERRAL. A 39-year-old expectant mother was referred for amniocentesis by a comprehensive health service because of her age-related risk for fetal chromosome abnormalities. Administrative delays resulted in her arrival at the referral center too far advanced in pregnancy to have the test. Her baby was born with the Down syndrome.

Regulation

Regulations are drawn up to set standards and to govern competing interests, ostensibly to protect the consumer and to contain costs. These regulations may or may not fulfill their purposes effectively.

OPERATING INEFFECTIVELY WITHIN LEGAL REGULATIONS. Because mothers and babies were being discharged on the second day after birth, an established PKU program in one state recommended that blood sampling on infants be delayed and be done by the community pediatrician after discharge. One infant was not tested, the mother was not instructed, the pediatrician was not informed, and this infant presented after one year with brain damage from untreated PKU. The next year another PKU infant was also missed. This program had no means to identify untested babies and did not notify primary physicians or parents of negative results. No mechanism existed to recall infants who had not been tested. All of these were also standard operating practices among other newborn screening programs elsewhere at that time.[49,50]

OPERATING EFFECTIVELY AROUND LEGAL RESTRICTIONS.   A newly started medical genetic service offered symptomatic amino acid screening but could not meet the statutory requirement of being supervised by a certified pathologist and so could not recover the costs of the test, even though it was cheaper and more meaningful than that offered by the approved laboratories. Therefore, it could only be offered in the guise of research, requiring formal prior approval by a statutory human experimentation committee, without reimbursement.

Planning

Demands for screening tests can arise from various quarters, each of which may have quite different concepts of what should be involved in screening.[17-19]

DEMAND WITHOUT AVAILABILITY.   In a sparsely populated region with a very small Jewish population, an expectant couple wanted testing for Tay–Sachs disease. The estimated risk for the fetus was approximately 1 in 2000.[38] Testing could have been done for less than $20 at the nearest Tay–Sachs screening center, but costs of specimen collection and transshipment were over $100. The parents chose to forego testing.

INDEPENDENCE WITHOUT ECONOMY.   Despite all published data about the efficiency of centralized mass screening for neonatal hypothyroidism,[41,51,51a] several independent health units in a community each purchased costly equipment for automated radioimmunoassay, although no single system could generate enough specimens to recoup its investment without raising patient charges.

INDEPENDENCE WITHOUT COORDINATION.   Physicians in a community were in disagreement about whether a specific newborn screening test was indicated for their patients. Major concerns were expressed about the small risk of this condition to newborn infants (although several patients in that community had had grave neonatal complications in the past few years), about creation of unnecessary anxiety in families, and about possible ineligibility of diagnosed patients for future health insurance coverage. These physicians agreed to disagree. Any physician could request the test for his or her patients on an individual basis, using a more costly nonautomated test procedure that was of doubtful reliability.

INDEPENDENCE WITHOUT COOPERATION.   In a competitive independent medical community, physicians requested newborn testing for hypothyroidism from the laboratories but were resistant to any central clinical supervision. The consensus was that any trained pediatrician could interpret a numerical value for thyroid blood tests.[contra 48,51a] There was no appreciation of the needs for quality control, experienced interpretation of borderline or anomalous results, and sharing information to evaluate test results.[3, 23, 41, 42, 51a, 52]

SCREENING WITHOUT FORESIGHT.   One small isolated community with 3000 births per annum agreed to participate in newborn screening for phenylketonuria. After two years, when no PKU patients had been found, this screening was abandoned. There was no realization that the expected yield would have been less than two patients per decade.

SCREENING WITHOUT PRIVACY.   A city health department endorsed a sickle cell parental screening program for its schools but insisted that City Hall be given a list of individual participants and test results for the city health records. It was only after vigorous protestations that this demand for subject identification was relinquished. There was strong denial that these public records might ever fall into unauthorized hands or be used in a discriminatory way.

Training

Learning is via instruction, observation, experience, and deduction. What has been learned should be shared. Genetic screening programs have been found to have many pitfalls; the challenge is for those who have learned by experience to tell all[48, 51-53] and for those who need experience to be willing to learn from others; furthermore, there is a need for effective mechanisms of information exchange.

TRAINING WITHOUT EXPERIENCE.   A newborn infant found to have phenylketonuria was transferred to a tertiary medical center. No one at this center had ever treated such a patient, but physicians there were resistant to consulting an experienced clinic because of the desire to learn by their own mistakes.

TREATMENT WITHOUT TRAINING.   A newborn infant who had severe congenital heart disease was also found on screening to have hypothyroidism. Treatment with full replacement doses of thyroid hormone precipitated heart failure. The physicians concerned had not been aware of this well-known complication and had not realized there was any need for caution or any merit in consulting a specialist.[54]

SCREENING WITHOUT EDUCATION.   In another country, a centralized newborn screening program was testing for seven inherited metabolic disorders. Physicians at one regional referral hospital had never had any experience with any of these disorders and had been given no plans for confirmation and management of any disorder found on screening their patients.

RESEARCH SCREENING WITHOUT LEARNING.   In one region, scientists conducted a pilot survey for metabolic disorders by urine testing. The results were never matched with patient records and the findings were never published.

Errors

Human errors frequently frustrate carefully laid plans. Even the best plans can only minimize the possibilities of errors but can never totally eliminate wrong assumptions or careless actions.

CARELESS COLLECTION OF INFORMATION.   A 36-year-old expectant mother was referred for amniocentesis to test for age-related risks of fetal chromosome abnormalities. No physician elicited or heeded the fact that a previous child had died of a heritable lysosomal storage disease (this could have been detectable by amniocentesis). The amniocentesis chromosome test (no other test was done) was normal. The infant, after birth, was found to have the lysosomal storage disease.

CARELESS ACCEPTANCE OF INFORMATION.   An infant was assumed to have mucopolysaccharidosis because of an abnormal urine screening test. After the physicians had been convinced that confirmatory testing was necessary, enzyme assays on a skin biopsy specimen revealed a rare disorder that could mimic mucopolysaccharidosis.

MISUNDERSTANDING OF INFORMATION.   A worried expectant mother told her obstetrician that an X ray of her son had revealed spina bifida occulta. Not realizing that this was a benign finding in 5 to 10% of the normal population, the obstetrician sent his patient at considerable expense for unjustifiable amniocentesis.

MISINTERPRETATION OF INFORMATION.   A couple had two children with remarkably similar malformations. Their doctor assured them that since he and his colleagues had no knowledge of such a malformation syndrome, the condition must be a rare coincidence and of no threat to future children. Subsequently another child had the same malformations.

MISUSE OF INFORMATION.   A week-old infant with acute gastrointestinal, hepatic, and metabolic disturbances was found to have sugar in the urine. The infant was diagnosed as having diabetes mellitus and given insulin. Fortunately, another physician suspected galactosemia, showed that the sugar in the urine was not glucose, and stopped the insulin before it had caused severe brain damage.

RELIANCE ON LIMITED INFORMATION.   The mother of an infant diagnosed as having PKU was assured with confidence that diet therapy would be fully successful in protecting her infant from brain damage. Despite adequate diet control, the infant developed seizures and other neurological abnormalities. He had one of the recently described more severe variants that give the same test abnormalities as classical PKU in the newborn screening tests.[55]

INSISTENCE ON UNNECESSARY TESTS.   A physician caring for an infant with a well-recognized malformation syndrome insisted on an expensive chromosome test, although chromosomes were not abnormal in this syndrome.

Compliance

Genetic screening programs can only offer test result information and related support services. Individuals in target populations may refuse to participate, reject information obtained, or act contrary to intended objectives, declining all proffered support.[1,52] Coercion is constitutionally unacceptable, and persuasion has limited success in altering patterns of health behavior.[56]

OUTCOME NOT DESIRED.   A father understood that his daughter's PKU could be treated "so that she would be bright." He took the physician aside, and confided that he did not want a bright daughter, as no one else in the family was bright.

RESULT NOT BELIEVED.   A father refused to accept that a blood test showed that his infant son would develop the same muscular dystrophy that had crippled his older son. Presymptomatic diagnosis was intended to help prepare this man for the inevitable, but he was neither educationally nor emotionally mature enough for the test result to be of any value to him.

TREATMENT NOT FOLLOWED.   The grandfather of a child with PKU was receiving extra welfare allowances from the state because of his grandson's condition. When special low-phenylalanine cookies were baked by the dietician, he ate most of them himself.

PROCRASTINATION.   An expectant mother, at risk for having another child with a neural tube abnormality, was urged to discuss the possibility of fetal diagnosis with her husband before her pregnancy was too advanced for the test. When she was recontacted later, she said she had not found the time to discuss this with her husband yet. She never had the test.

REFUSAL TO PARTICIPATE.   A young couple lost their only two children, both of whom had a lysosomal storage disease. Their religious beliefs against abortion precluded their acceptance of the option of fetal testing to screen for this disease in future pregnancies. They decided to have no more children.

Summary

The examples in this section are not unique. Anyone with experience in genetic screening will be able to match these with his or her own examples. The purpose of these illustrations is to emphasize the naiveté of planning genetic screening programs without recognizing the myriad possibilities of things going contrary to expectations. Many of the shortcomings cited have since been corrected, but the lessons they offer are felt to be worth sharing.

THOUGHTS ABOUT CONFLICTS AND HAZARDS IN SCREENING

In a complex society, any program designed to benefit one party may conflict with the interests of another party. Furthermore, any procedure that can lead to benefit could also lead to harm.

Conflicts

CONFLICTING FINANCIAL PRIORITIES.   Crisis care can always claim urgency over preventive measures. Denial of a crutch to the crippled seems more reprehensible than curtailing programs to avoid crippling.[26] Most screening programs suffer from low priority because visible benefits may consist only in the disappearance of visible morbidity and mortality, often not demonstrable for many years. This is particularly true of savings achieved by investments in the personnel, equipment, and operating expenses of screening programs, where vast numbers of tests may have to be done before any effective preventive outcome is measurable.[51-53] In following cost-conscious guidelines, the needy new program is less likely to be supported than an outmoded program approaching obsolescence.[21-23,26]

CONFLICTING PROFESSIONAL RESPONSIBILITIES. The different types of screening program owe primary responsibilities to different interests (Table I). Symptomatic tests are requested by a physician as part of a diagnostic work-up, so the doctor is the person served by symptomatic screening programs. Presymptomatic tests are designed to find treatable disorders before any outward manifestation has appeared, so the affected individual is the one served. Parental tests are intended to give genetic risk information for family planning, so primary responsibility is to the prospective parents. Research screening has a dual responsibility, for scientific merit and for harm-free informed participation by all subjects screened.

Effective fulfillment of these responsibilities can only be in harmonious cooperation with other providers of health care, particularly the primary physicians. If care providers are antagonistic, indifferent, or uninformed, the health services to screened populations would be fragmented. In a free enterprise system,[26] especially with a mobile population, integration of health services is a Herculean challenge that cannot be even attempted without restricting the autonomy of physicians[57] and infringing on the commercial interests, for example, of private laboratories. When laboratories have purchased expensive equipment—e.g., for radioimmunoassays of thyroid hormone levels in newborn infants—they are unlikely to welcome proposals for centralizing all such tests at a regional laboratory. Powerful political pressures can be applied to oppose screening proposals that infringe on the profitability of the private sector.

Another complication of our mobile society is the movement of individuals into and out of geographically separated screening programs, rupturing continuity of administrative and clinical responsibilities.

Obligations to trainees often conflict with the best interests of the patient. If every patient was treated only in privacy by the most experienced physician, there would be no one trained to back up that physician.

Physicians do not always act exclusively in the best interests of their patients. With increasing consciousness of litigation risks, caution often overrules acumen, resulting in redundant testing or unnecessary treatment.[27,29-32,58] This can increase the cost and decrease the efficiency of screening programs.[5-10]

CONFLICTING SOCIETAL NEEDS. A community would be wise indeed if it could keep all its needs in perspective. Facilities supported by a community to serve special segments can be provided for altruistic, artistic, spiritual, or practical reasons. Criteria for cost-beneficial support of the arts, entertainment, and recreation cannot be the same as those for education, health, and safety. Screening programs make claims on a community's time, interests, and resources. These often demand that a large number of people contribute to the direct benefit of only a few. Potentially, however, prudent preventive measures can drastically curtail the cost of avoidable handicaps and their needs for complex support services.

CONFLICTING INDIVIDUAL RIGHTS

Right of Access to Knowledge and Resources.   Individuals are recognized to have a right to information and to equality of opportunity. The converse of these are the paradoxical rights to remain ignorant and to refuse opportunities.[59,60] Ignorance is dispelled by publicity, depriving individuals within earshot of their right not to know. Invitations to participate in screening programs can be associated with many subtle pressures to comply. Regulations define and restrict coercion from program staff but cannot control coercive pressures from relatives and peers, which can be for either compliance or non-compliance.[27,29]

Equitable dispersal of knowledge is a far simpler task than equitable availability of resources. Access to genetic screening programs can be impeded by geographic, financial, or cultural barriers. All that legal measures can achieve is fair distribution of resources that meet minimum requirements. Improved public awareness and legislative responsiveness can lead to progressive upgrading and expansion of these requirements, thus enhancing access to screening programs.

Right of Privacy.   Constitutionally, individual rights supercede community rights except in certain stipulated situations. This applies particularly to rights of privacy.[61,62] The implied contract between a patient and a physician gives intimate personal information to the physician for health supervision and illness intervention. The physician is privy to this information as a professional privilege and has no right to divulge it without the patient's permission. The information belongs to the patient and not to the physician. Complications arise because medical care is administered by many persons, some of whom must have patient identity for diagnostic, therapeutic, and billing purposes. This includes clerical and administrative personnel as well as health professionals. Beyond those directly responsible for patient care, use of patient information for public health and research purposes should preserve confidentiality.[63,64]

Right to Safety.   Freedom from harm can never be guaranteed, but every consumer of health care services has a right to expect reasonable protection from predictable hazards. These include physical harm to the test subject or to others from a screening program; psychological harm from anxieties, fears, or other stresses; and social harm from stigmatization, loss of status, or loss of economic potential.[29,30]

Hazards

DIRECT PHYSICAL HAZARDS.   Test procedures using simple urine or blood samples are associated with minimal threat of direct harm. More invasive procedures, however, notably the sampling of amniotic fluid, are known to have a small but unavoidable hazard[45,46] (see Milunsky, Chapter 7). Decisions as to whether these hazards are worth taking must be made with informed consent.[58,65-67]

HARM FROM ERRORS OF OMISSION.   False expectations can be raised by errors in planning or education that lead to incomplete understanding by regulatory and funding agencies,[18] by program personnel, by the target population, or by the public. Failure to fulfill unrealistic expectations will certainly harm the image of the screening program, but by association it will tarnish the reputations of related health programs and by producing disillusionment may lose public support and alienate target populations.

HARM FROM ERRORS OF DIAGNOSIS

Clerical and Administrative. The best-organized record-keeping system will fail if there is sloppiness in its execution. This includes misidentified cases, mislabeled specimens, and misplaced results.

Technical and Statistical. Errors in technical design or technician performance may undermine the selectivity or reproducibility of test results, producing both false positives and false negatives. Inherent statistical variability will cause normal extremes to overlap with truly abnormal test results.[48]

Biological. Human individuality and fluctuation of biological parameters in any one person often produce broad variations in test values from normal subjects that can overlap with the abnormal values being sought by screening.[51a]

Medical. Disturbances in test measurements can arise from more than one pathological cause; e.g., in screening for phenylketonuria, new diseases have been discovered that do not respond to diet therapy.[55] Therefore, diagnostic interpretation by skilled experienced clinicians should be an integral part of every screening program.[51a, 68, 69]

PSYCHOLOGICAL HAZARDS. Concerns about health are inevitably anxiety-provoking. If there is too little concern, there would be no compulsion to comply with health regimens or to participate in screening programs. Excessive concern, however, is too stressful and constitutes a psychological hazard that often leads to subjects misunderstanding the objectives of screening programs and the meanings of test results.[70,71]

Waiting for test results is a time when imagination can conjure up fearful possibilities. One moral obligation of screening programs should be to provide prompt reporting of results to participants (see Table I).

Breach of privacy can arise whenever names and addresses are recorded and stored. Traditional concepts of confidentiality have focused on physical access to papers in filing cabinets. Former guidelines restricting access are obsolete because of the burgeoning computerization of data files.[61,62] These have positive and negative consequences. Unprotected or poorly designed computer memories can be readily breached, but well-designed memory files can be programmed to release selective data to different users, enabling freer use of patient-related data while preserving privacy. Furthermore, increasing use of limited-access dedicated minicomputers will eliminate the specter of remote spies tapping into secret files via the telephone.[63,64,72,73]

RECOMPENSE AND REDRESS

Recompense

Consequential ill effects of genetic screening programs—or absence of such programs—can lead to litigation.[4,5,8–10,28–34,47] Civil suits have already been brought relating to failure to diagnose[10,34] and failure to forewarn.[33,74]

Monetary recompense is a poor and unsatisfactory replacement for preventable genetic illnesses, regardless of the generosity of damages awarded. Yet the notoriety accorded such suits does exert a powerful effect on the behavior of fellow-physicians,[30] who may overreact in their anxiety to avoid malpractice claims.[58] Despite the publicity attracted by malpractice claims related to genetic screening, it is probable that many more unfortunate events occur than are recognized or reported. Tort law offers an unsatisfactory legal mechanism for improving genetic screening programs.

Financial restitution for economic losses arising from restraint of trade would be a more appropriate issue for the courts, but I am not aware of any legal activity related to this issue.

Control

Regulation by licensure and quality control of laboratory tests have been well established in other medical fields; some regulation is by public agencies and some by professional associations. The Clinical Chemistry Division of the Center for Disease Control, Department of Health, Education and Welfare, is developing standardization criteria for laboratory tests used in screening newborn infants for metabolic diseases and for fetal diagnostic test procedures. The American Society of Human Genetics is formulating requirements for certification of geneticists responsible for various professional activities related to genetic screening. The impact of these control measures is yet to be determined, but they should result in improved and more uniform performance standards.

THOUGHTS AND RECOMMENDATIONS

It is always easier to bemoan deficiencies than to find constructive solutions. Planning and operating effective screening programs require clearly defined goals, carefully plotted strategies, closely monitored procedures, and critically conducted evaluations.[23,49-53,75-78] Because hazards and errors are inevitable,[63,64,68,69] programs should have built-in safeguards and measures to minimize harmful effects.

Genetic services should be given their due and appropriate priority within the context of general health services.[79,80]

Planning Strategy

In development of a screening program, ascertainment of need is an important prerequisite[1,7,40,49,50,77,81] that will enable better definition of program objectives. To win popular legislative support, general and specific health education of the public and their political representatives are essential. A carefully planned coordinated campaign will help to awaken awareness, win support, and build a consensus in favor of the screening program. Without indefatiguable persistence, ordinances and statutes for a program can be sidetracked or distorted, die unfunded, or fail to be implemented. Strategies for implementing screening programs have to respect the vested interests of different groups[57,82] and should seek to unify these interests toward the common goal of improving community health.

Compliance of target populations with health objectives has not been impressive in many nongenetic areas, such as automotive safety, tobacco and alcohol consumption, and limited population growth. Genetic screening programs are unlikely to be much more successful in altering health behavior and health habits.[83,84] Realistic objectives of these programs should include consideration of noncompliance[60] with development of measures to minimize this problem by education, persuasion, and other incentives.

Defined objectives, with scheduled timetables, will help to avoid unrealistic expectations from regulatory and funding agencies.

Possible complications should be anticipated so that they can be effectively forestalled.[48-53,61-71,75]

Funding Support

Inevitably, there will be piecemeal support that can lead to fragmented, uncoordinated, inequitable, maldistributed services. The challenge is to find the vision, patience, and skill to join the separate patches into a solid quilt that can provide comprehensive support for screening programs. Recruitment of support from governments, voluntary agencies, academic institutions, health insurance companies, and perhaps even industry may all help to finance various components of genetic screening programs. Complete long-range support is unrealistic and probably unwise, as it is subject to misuse and abuse. Partial support, however frustrating and shortsighted it may seem, is all that can be expected. Evidence of positive returns from moneys invested in partial programs will help justify funding of broader programs.

Regulatory Controls

Local communities should be encouraged to cooperate in setting up uniform standards, sharing resources, and comparing results of screening programs. Central governments should be made aware of variations in local requirements. Controls have to be flexible enough to accommodate new technical developments and changing medical needs, yet should be specific and unambiguous enough to achieve their aims of ensuring quality. Regulations will never be adequate where there is no willingness or ability to conform, and too rapid expansion of genetic screening activities can outstrip the availability of clinical and technical expertise to maintain high professional standards. Free interchange of administrative technical and clinical data among different programs, facilitated, it is hoped, by computerized data processing, can help to sustain standards.

Appraisal

The most effective way to convince all concerned that genetic screening programs are worthwhile is by carefully prepared program appraisals.

Inevitably, honest appraisals will question the justification for continuing some screening programs.[78] Evidence from scrupulously fair assessments may uncover some embarrassing program shortcomings.[49-52,70,71] Lessons learned from these assessments, however, would be invaluable for improving program designs and program operations. Furthermore, positive data about cost-efficiency and cost-benefit will justify program activities,[53,75] encouraging their continuation and expansion.

Overzealous bureaucratic demands for accountability can multiply in response to examples of malfeasance such as those given (pp. 102–108) on experiences with genetic screening and the law. These can be counterproductive in forcing programs to be overcautious,[58,76] to adhere to overly rigid stipulations, or to expend too large a proportion of time and effort on preparation of reports.

Litigation

The best insurance against medical malpractice suits is high-standard medical care by responsible personnel who retain personal respect for each of their patients. Even so, an untoward outcome is bound to result from time to time, and if there are harmful consequences, compensation may be legitimately due to the injured parties.[28-34] Genetic screening programs should be designed to be of high standard rather than to be litigation-proof.

CONCLUSIONS

In conclusion, only partially with tongue in cheek and probably as a spokesman for all my colleagues in this volume, I make the following claims:

1. I know what I am doing. The law should leave me alone to do what I know is right.
2. I know what is best for others. There ought to be a law to make everyone else do what I think is right for him or her.
3. I know what needs to be done. There should be ample financial support for what I want to do.
4. I know that what I have done was worthwhile. Everyone should trust me and respect my accomplishments, without demanding that I take the time to make detailed reports.
5. If anything has gone wrong, it was not my fault. The law should certainly recompense the unfortunate victims but not bother me.

What I am doing is important—at least to myself.

REFERENCES

1. Childs, B., and Simopoulos, A. P. (eds.), *Genetic Screening: Programs, Principles and Research*, National Academy of Science, Washington, D. C. (1975).

2. Childs, B., Genetic screening, in: *Annual Review of Genetics* (Vol. 9) (H. L. Roman, A. Campbell, and L. M. Sandler, eds.), p. 67, Annual Reviews, Palo Alto (1975).

3. Cohen, B. H., Lilienfeld, A. M., and Huang, P. C. (eds.), *Genetic Issues in Public Health and Medicine*, Charles C Thomas, Springfield, Ill. (1978).

4. Lappé, M., Genetic screening, in: *Counseling in Genetics* (Y. E. Hsia, K. Hirschhorn, R. L. Silverberg, *et al.*, eds.), p. 295, A. R. Liss, New York (1979).

5. Waltz, J., and Thigpen, C., Genetic screening and counseling: The legal and ethical issues, *Northwestern U. L. Rev.* **68**:696 (1973).

6. Bergsma, D., Lappé, M., Roblin, R. O., *et al.* (eds.), Ethical, social, and legal dimensions of screening for human genetic disease, *Birth Defects Orig. Artic. Ser.* **10**:6, Grune and Stratton, New York (1974).

7. Milunsky, A. (ed.), *The Prevention of Genetic Disease and Mental Retardation*, W. B. Saunders, Philadelphia (1975).

8. Reilly, P. R., Genetic screening legislation, in: *Advances in Human Genetics* (Vol. 5) (H. Harris and K. Hirschhorn, eds.), p. 319, Plenum, New York (1975).

9. Milunsky, A., and Annas, G. J. (eds.), *Genetics and the Law*, Plenum, New York (1976).

10. Reilly, P. R., *Genetics, Law, and Social Policy*, Harvard University Press, Cambridge (1977).

11. Task Force on Genetic Screening, The pediatrician and genetic screening (every pediatrician a geneticist), *Pediatrics* **58**:757 (1976).

12. Greenberg, D. S., Carter and health care, *N. Engl. J. Med.* **297**:1299 (1977).

13. Greenberg, D. S., Health "reform": Massacre on Capitol Hill, *N. Engl. J. Med.* **299**:1199 (1978).

14. Greenberg, D. S., Health and science policies—A look ahead, *N. Engl. J. Med.* **299**:1475 (1978).

15. Bessman, S. P., Legislation and advances in medical knowledge: Acceleration or inhibition? *J. Pediatr.* **69**:334 (1966).

16. Curran, W. J., The questionable virtues of genetic screening laws, *Am. J. Public Health* **64**:1003 (1974).

17. Reilly, P. R., Recent developments in state supported genetic screening, *Am. J. Hum. Genet.* **27**:691 (1975).

18. Culliton, B. J., Genetic screening: States may be writing the wrong kind of laws, *Science* **191**:926 (1976).

19. Schmidt, R. M., and Curran, W. J., A national genetic-disease program: Some issues of implementation, *N. Engl. J. Med.* **295**:819 (1976).

20. Reilly, P. R., Government support of genetic services, *Soc. Biol.* **25**:23 (1978).

21. Greenberg, D. S., Cost containment: Another crusade begins, *N. Engl. J. Med.* **296**:699 (1977).

22. Ayres, J. C., Decelerating skyrocketing health-care costs, *N. Engl. J. Med.* **296**:391 (1977).

23. Weinstein, M. C., and Stason, W. B., Foundations of cost-effectiveness analysis for health and medical practices, *N. Engl. J. Med.* **296**:716 (1977).

24. Coulter, C. F., Research instrument sharing, *Science* **201**:415 (1978).

25. Schwarz, W. B., and Joskow, P. L., Medical efficacy versus economic efficiency: A conflict in values, *N. Engl. J. Med.* **299**:1462 (1978).

26. Newhouse, J. P., Medical costs and medical markets, *N. Engl. J. Med.* **300**:855 (1979).

27. Loftus, E. F., and Fries, J. F., Informed consent may be hazardous to health, *Science* **204**:11 (1979).

28. Parker, W. C., Some legal aspects of genetic counseling, in: *Progress in Medical Genetics* (Vol. 7) (A. G. Steinberg and A. G. Bearn, eds.), p. 217, Grune and Stratton, New York (1970).

29. Katz, J., *Experimentation with Human Beings*, Russell Sage Foundation, New York (1972).

30. Holder, A. R., *Medical Malpractice*, Wiley, New York (1974).

31. Carter, G., Legal responses and the right to compensation, *Br. Med. Bull.* **32**:89 (1976).

32. Lynch, P. M., and Lynch, H. T., Genetic counseling of high-cancer-risk patients: Jurisprudential considerations, *Semin. Oncol.* **5**:107 (1978).

33. Curran, W. J., Genetic counseling and wrongful life, *Am. J. Public Health* **68**:501 (1978).

34. Reilly, P. R., The law and genetic counseling, in: *Counseling in Genetics* (Y. E. Hsia, K. Hirschhorn, R. L. Silverberg, *et al.*, eds.), p. 311, A. R. Liss, New York (1979).

35. Wolf, B., Rosenfield, A. T., Taylor, K.J.W., *et al.*, Presymptomatic diagnosis of adult onset polycystic kidney disease by ultrasonography, *Clin. Genet.* **14**:1 (1978).

36. Lamiell, J. M., Salazar, F. G., Polk, N. O., *et al.*, Pre-symptomatic surveillance for von Hippel Lindau disease in a large kindred, *Am. J. Hum. Genet.* **30**:57a (1978).

37. Crane, R. J., The case of the illegitimate gene, *Postgrad. Med. J.* **51**:637 (1975).

38. Goodman, R. M., and Motulsky, A. G., *Genetic Diseases Among Ashkenazi Jews*, Raven Press, New York (1979).

39. American Academy of Pediatricians, Statement on compulsory testing of newborn infants for hereditary metabolic disorders, *J. Pediatr.* **39**:623 (1967).

40. Levy, H. L., Genetic screening, in: *Advances in Human Genetics* (Vol. 4) (H. Harris and K. Hirschhorn, eds.), pp. 1 and 389, Plenum, New York (1973).

41. American Academy of Pediatricians, Committee on Genetics, Screening for congenital metabolic disorders in the newborn infant: Congenital deficiency of thyroid hormone and hyperphenylalaninemia, *Pediatrics* **60**:389 (1977).

42. Humphrey, M., Infant screening for metabolic disorders, *Congressional Record* s2527 (Feb. 28, 1978).

43. Whitby, L. G. (ed.), Screening for disease, *Lancet* **2**:819, 880, 939, 996, 1057, 1125, 1189, 1245, 1305, 1364, 1434 (1974).

44. North, A. F. (ed.), Screening in child health care, *Pediatrics* **54**:608 (1974).

45. Consumer Reports, Cutting the risk of childbirth after 35: The reassurance amniocentesis can provide, *Consumer Reports*, p. 302 (1975).

46. Milunsky, A., and Reilly, P., The "new" genetics: Emerging medicolegal issues in the prenatal diagnosis of hereditary disorders, *Am. J. L. Med.* **1**:71 (1975).

47. Powledge, T. M., and Fletcher, J., Guidelines for the ethical social and legal issues in prenatal diagnosis, *N. Engl. J. Med.* **300**:168 (1979).

48. Casscells, W., Schoenberger, A., and Graboys, T. B., Interpretation of clinical laboratory results, *N. Engl. J. Med.* **299**:999 (1979).

49. Sepe, S. J., Levy, H. L., and Mount, F. W., An evaluation of routine follow-up blood screening of infants for phenylketonuria, *N. Engl. J. Med.* **300**:606 (1979).

50. Hsia, Y. E., Follow up screening for phenylketonuria, *N. Engl. J. Med.* **301**:553 (1979).

51. Fisher, D. A., Dussault, J. H., Foley, T. P., *et al.*, Screening for congenital hypothyroidism: Results of screening one million North American infants, *J. Pediatr.* **94**:700 (1979).

51a. Green, O. C., Winter, R. J., and Wise, J. E., Neonatal thyroid screening programs: Caution, *Pediatrics* **63**:822, 1979.

52. Grover, R., Wethers, D., Shahidi, S., *et al.*, Evaluation of the expanded newborn screening program in New York City, *Pediatrics* **61**:740 (1978).

53. Bush, J. W., Chen, M. M., and Patrick, D. L., Health status index in cost-effectiveness: Analysis of PKU program, in: *Health Status Indexes* (D. Coddington, ed.), p. 172, Hospital Research Education Trust, Chicago (1973).

54. Greenwald, R. A., The need for subspecialists (automotive division), *N. Engl. J. Med.* **300**:326 (1979).

55. Anonymous, New variants of P.K.U., *Lancet* **1**:304 (1979).

56. Hsia, Y. E., and Hirschhorn, K., Response to genetic counseling, in: *Counseling in Genetics* (Y. E. Hsia, K. Hirschhorn, R. L. Silverberg, *et al.*, eds.), p. 261, A. R. Liss, New York (1979).

57. Chapman, C. B., Doctors and their autonomy: Past events and future prospects, *Science* **200**:851 (1978).

58. Tancredi, L. R., and Barondess, J. A., The problem of defensive medicine, *Science* **200**:879 (1978).

59. Bender, H., The right to choose or to ignore, in: *Genetic Responsibility* (M. Lipkin and P. T. Rowley, eds.), p. 71, Plenum, New York (1974).

60. Slack, W. V., The patient's right to decide, *Lancet* **2**:240 (1977).

61. Amarose, A. P., and Burks, J. L. (Coord.), Disclosure of genetic information: An invitational symposium, *J. Reprod. Med.: Lying-In* **11**:211 (1969).

62. Curran, W. J., Laska, E. M., Kaplan, H., *et al.*, Protection of privacy and confidentiality, *Science* **182**:797 (1973).

63. Martin, M. E., Statisticians confidentiality and privacy, *Am. J. Public Health* **67**:165 (1977).

64. Riskin, L. R., and Reilly, P. R., Remedies for improper disclosure by genetic data banks, *Rutgers-Camdon L. Rev.* **8**(3):480 (1977).

65. Chamberlain, J., Human benefits and costs of a national screening programme for neural-tube defects, *Lancet* **2**:1293 (1978).

66. Bennett, M. J., Blau, K., Johnson, R. D., *et al.*, Some problems of alpha-fetoprotein screening, *Lancet* **2**:1296 (1978).

67. Paulker, S. P., and Paulker, S. G., Prenatal diagnosis: A directive approach to genetic counseling using decision analysis, *Yale J. Biol. Med.* **50**:275 (1977).

68. Buist, N.R.M., and Jhaveri, B. M., A guide to screening newborn infants for inborn errors of metabolism, *J. Pediatr.* **82**:511 (1973).

69. Hill, A. Casey, R., and Zaleski, W. A., Difficulties and pitfalls in the interpretation of screening tests for the detection of inborn errors of metabolism, *Clin. Chim. Acta* **72**:1 (1976).
70. Blumberg, B. D., Golbus, M. S., and Hanson, K. H., The psychological sequelae of abortion performed for a genetic indication, *Am. J. Obstet. Gynecol.* **122**:799 (1975).
71. Finley, S. C., Varner, P. D., Vinson, P. C., *et al.*, Participants' reaction to amniocentesis and prenatal genetic studies, *J. Am. Med. Assoc.* **238**:2377 (1977).
72. Emery, A. E. H., Brough, C., Crawfurd, M., *et al.*, A report on genetic registers, *J. Med. Genet.* **15**:435 (1978).
73. Anonymous, Genetic registers, *Lancet* **1**:253 (1979).
74. Culliton, B. J., Physicians sued for failing to give genetic counseling, *Science* **203**:251 (1979).
75. Donabedian, A., The quality of medical care, *Science* **200**:856 (1978).
76. Frazier, H. S., and Hiatt, H. H., Evaluation of medical practices, *Science* **200**:875 (1978).
77. Sullivan, P. R., Damsgaard, A., and Rodkey, G. V., A critique of the Massachusetts Determination of Need Program, *N. Engl. J. Med.* **300**:794 (1979).
78. Farber, M. E., and Finkelstein, S. N., A cost-benefit analysis of a mandatory pre-marital rubella-antibody screening program, *N. Engl. J. Med.* **300**:856 (1979).
79. Fraser, F. C., Genetics as a health-care service, *N. Engl. J. Med.* **295**:486 (1976).
80. Scriver, C. R., Laberge, C., Clow, C. L., *et al.*, Genetics and medicine: An evolving relationship, *Science* **200**:946 (1978).
81. Chan, J. L., Hsia, Y. E., Oglesby, A. C., *et al.*, *Genetic Needs Assessment for Hawaii, June 1978*, Special Report for Health Services Admin., DHEW.
82. Gifford, J. F., and Anlyan, W. G., The role of the private sector in an economy of limited health-care resources, *N. Engl. J. Med.* **300**:790 (1979).
83. Breslow, L., Risk factor intervention for health maintenance, *Science* **200**:908 (1978).
84. Rogers, E. M., *Communication Strategies for Family Planning*, Free Press, Macmillan, New York (1973).

DISCUSSION

MR. PHILIP REILLY (Attorney): Speaking as a lawyer, I will be the last one to recommend that we write more laws.

I have been asked to make a few comments in regard to Professor Hsia's paper. I want just to make one assertion and then try to defend it to you.

He commented about genetic screening programs; I want to talk about genetic screening laws. My assertion is that government is becoming steadily more involved in the business of what I call "modern eugenics," that is, trying, either directly or indirectly, to reduce the number of pregnancies at risk for severe genetic disease. Now, I would like to back that statement up a little bit. I would like to convince you of the great activity that has gone on in genetic screening legislation.

I will just read off some statistics. Between 1962 and 1968, forty-three states wrote laws on the subject of screening for phenylketonuria. Between 1970 and 1973, thirteen states wrote laws on the subject of sickle cell anemia. Between 1972 and 1978, three federal laws were written on the subject of genetic screening: the National Sickle Cell Anemia Control Act, the National Cooley's Anemia Control Act, and, most recently, the National Genetic Diseases Act. In the meantime and before that, dozens of laws have been written pertaining to hemophilia in various states, usually to provide some sort of funding to support people suffering from this disease.

Now, this is neither a good nor a bad thing. It does show that all sorts of legislation is written without many people knowing very much about it. I think that this should be some cause for concern.

This policy of government support of genetic screening programs which are intended to provide information to people so they can make enlightened decisions about reproduction is continuing. I want to bring to you the latest statistics from the Bureau of Community Health Services in Washington, D.C., from a meeting I was at very recently. A summary of their first year in operation: forty-one applications for statewide genetic services programs were received in 1978 from thirty-six states and from Puerto Rico. Twenty-one of these applications were funded at a total dollar level of $3,500,000. Two were approved for a one-year project and others were approved for a longer period.

I think this expenditure, while small, represents a significant advance from a time not so long ago when no federal money was available for genetic screening programs to be carried out under guidance of a law. I still see no evidence based on my own studies of genetic screening laws around the country that adequate attention is being given to some of the concerns that Professor Hsia enumerated, that is, protection of genetic information, protection of its privacy, adequate information for genetic counseling, and genetic education programs.

The one note of caution I want to sound here is to make sure that if we are going to go forward with legislative support of genetic screening programs, we make sure that we have an adequate allocation of funding to account for what would be a well-rounded and comprehensive program. I think that would have to include some enforcement provisions to check up on these programs and make sure that the data were being adequately safeguarded and that genetic counseling actually was being provided to the people who are being screened.

DR. HARVEY L. LEVY: Genetic screening usually means general population screening. So, in effect, it takes us out of the first area that Professor Hsia was suggesting—and that is screening somebody who comes to you basically for medical attention—into an area of screening individuals who never come to you at all; in fact, they may not even be aware of the screening program.

We impose ourselves, particularly by newborn genetic screening programs, upon the population at large, upon newborns, upon families of newborns, and so forth. Is this bad? I don't think it is bad necessarily but I do think it raises a number of questions.

First, it raises the question of voluntarism versus mandatory screening. Patients have not volunteered to be screened in the type of genetic screening I just mentioned. But should we be imposing ourselves upon families in the sense that we screen them for diseases that they as yet may not be aware of? Should they indeed not volunteer for such screening? But the converse to that particular argument is: Should not, for instance, newborns be protected by some sort of legal fabric from having a disability from which they could have been protected but develop simply because of poor physician input or, for that matter, unwary or even neglectful or abusive parents?

So we are into that area of trying to decide exactly what we want laws for or whether we should have laws; and if we should have laws, how far such laws should extend.

The next area in which we impose ourselves is the area involving the extended family who may not have any knowledge or awareness about a particular screening result. This refers to the situation wherein one detects a disease in a given newborn and one has the ability to detect this disease or actual carriers in as-yet-undetected other members of that family and the family of the screenee does not want the information about the disease disclosed. They may be ashamed of having this disease, or may not have good relationships socially and otherwise with other members of the family.

What are our obligations as individuals, as physicians, as other health care workers toward these other individuals, who are totally unaware of this information but who may very well benefit by knowing it? Where are these obligations legally, medically, morally, and socially?

The third issue is the potential for divulging information from newborn screening programs and other kinds of general population screening programs to agencies such as life insurance companies and credit bureaus.

I do not have definitive answers to these questions. We did not have such answers four years ago at this session and we do not have answers now. But the issues are of serious import and we are deeply into these questions now whether we realize it or not.

11

DISCLOSURE AND CONSENT
In Search of Their Roots

JAY KATZ
Yale Law School
New Haven, Connecticut 06520

In my admittedly limited reading on genetic counseling, I have been struck by genetic counselors' preoccupation with the question of how much authority they should exercise in the counseling process; similar concerns rarely surface in the literature of other professions. Most striking is the debate over whether genetic counselors should provide only "information" to their counselees, allowing the latter to decide what course of action to follow, or whether counselors should guide in a variety of ways the decision-making process. At least in theory, the prevailing climate of opinion among genetic counselors seems to favor the former position, influenced to a considerable extent by second thoughts over earlier eugenic practices of proselytizing for "improvements" in society's genetic stock. Such strictures on guidance-persuasion are quite revolutionary, particularly when contrasted with traditional medical beliefs that patients should trustingly follow doctors' orders.

Yet genetic counselors' commitment to impart only information and to let counselees decide is threatened by competing considerations. One of these countervailing forces is the concern over the capacity of clients to understand what is being communicated to them. Although the debate on this problem is more balanced in the genetic than in the medical literature, it too tends to emphasize the counselees' propensity to distort what they have heard and their incapacity to comprehend what they have been told. The ways in which counselors contribute to these problems, however, has not received the attention it deserves in either the medical or the genetic literature. Any focus on counselors has mostly touched on their nagging doubts over whether their counselees had accurately understood what they tried to convey to them. Although an important issue in its own right, it is a derivative, not the primary problem, and I shall have more to say about it later. For the moment, suffice it to say that an apt analogy can be found in my own field, psychoanalysis, where the distortions that surface in the interactions between psychoanalyst and analysand are much better understood in terms of patients' transferences than analysts' countertransferences. The human propensity to project blame does not stop at the doorstep of the professional.

In my current work on disclosure and consent, I am increasingly appreciating the pervasive and distorting impact of such countertransferences on the dialogue between professionals and their patient-clients. Let me hasten to add that I define countertransferences broadly, including in this rubric not only the eruptions resulting from occasional emotional blind spots but also the influence on the dialogue of the counselor's personality, idiosyncrasies, and value preferences as well as of his/her personal and professional identity, uncertainties, knowledge, and ignorance.

In reading further, I was intrigued by occasional statements that genetic counseling should be distinguished from traditional medical practice because in the latter context deference to the greater expertise of physicians makes more sense than in the former, where genetic counselors may not have "more certainty or insight than the average citizen."[1] Such comments still puzzle me because they cannot be entirely explained by the uncertainty inherent in genetic counseling. Uncertainties over diagnosis, therapy, and prognosis loom equally large in medicine; yet one rarely finds in that literature statements that put physicians and their patients on an equal plane. Whether genetic counselors' greater awareness of professional fallibility and the impact of value preferences, as revealed in their theoretical writings, extends to their practices is a question I cannot answer. Counselors' conversation, unless great care is taken, can be manipulative; that danger does not disappear by ultimately "allowing" the counselee "to decide." We know that physicians all too readily, in order to obtain patients' assent, render "medical" opinions that on inspection turn out to be a hodgepodge of personal, interpersonal, emotional, philosophical, social, aesthetic, religious, and medical judgments.

Finally, and by way of introduction, let me note that the legal doctrine of informed consent has cast a dark shadow over genetic counseling as it has over medical practice in general. Elsewhere, I have begun to unravel the confusion that this doctrine has created in the minds of judges, practitioners, and commentators.[2] Here I can only note that on inspection, the doctrine's appealing name—informed consent—promises much more than its construction in case law has delivered; for after all is said and done, the courts imposed primarily a duty-to-warn on physicians, an obligation that is generally further limited to risk disclosures about proposed interventions. This judicially imposed obligation must be distinguished from the *idea* of informed consent, namely, that patients have a decisive role to play in the medical decision-making process. The idea of informed consent, though alluded to also in case law, cannot be implemented, as courts have attempted, by only expanding the disclosure requirements. In doing so, courts reinforced, as might have been predicted, physicians' traditional monologue of talking *at* and not *with* patients; they did not encourage what was so urgently needed: a new and unaccustomed dialogue between physicians and their patients. Without a commitment to a sustained dialogue, the idea of informed consent had to wither on the vine.

To command physicians to expand their monologue by requiring them to warn patients of risks, which patients are expected to accept trustingly, changes little; to urge doctors, on the other hand, to engage in a dialogue with their patients in which both, appreciative of their respective inequalities, make a genuine effort to voice and clarify their uncertainties and then to arrive at a mutually satisfactory course of action is quite another matter. Consider for a moment what impact such obligations might have on hope, reassurance, the placebo effect, and "therapeutic effectiveness," prescriptions so important to traditional medical practice.[3] Clearly, the idea of informed consent can never guide medical decision making without bringing about at the same time well-nigh revolutionary changes in the existing physician–patient relationship.

Judicial opinions, medical commentary, and some of the literature on genetic counseling as well have focused all too prematurely on what and what not to say to patients. All such attempts had to fail; indeed, they counterproductively had to spawn, as they did, a great many articles that either despaired of ever adequately informing patients or ridiculed such efforts.[4] However, it is not the physicians' monologue—the process of disclosing—that requires expansion; rather it is the dialogue between physicians and their patients—the process of informing—that deserves our attention.

But there is more. I would like to suggest—and this explains the title of my chapter—that disclosure and consent are concepts without roots and that the ground on which physicians and patients can meaningfully interact still needs to be prepared. To accomplish this task, professionals must be much clearer about three fundamental tensions inherent in the physician–patient relationship: tensions that have shaped, at least since the days of Hippocrates, their all too embryonic dialogue. The first tension is one of authority: Who should be in charge? The second tension is one of autonomy: Who can be in charge? The third tension is one of uncertainty: Can anyone be in charge?

AUTHORITY—WHO SHOULD BE IN CHARGE?

The traditional medical position on authority insists that patients must trustingly place themselves in physicians' hands; the renowned Harvard sociologist, Talcott Parsons, agreed and unequivocally stated that patients "must take [physicians'] judgments and measures on 'authority.' . . . We often speak of 'doctor's orders.' "[5] Patients, so it goes, have neither the expertise nor, in their hour of need, the capacity to make intelligent judgments on their own behalf. Thus professionals, operating within the traditional model, view disclosure as an irrelevant and meaningless charade, and patients, aware of not being talked with, listen with only half an ear, resigned to following doctors' orders or to marching to the beat of their own drums. While I question the need for blind trust, except in rare circumstances, I do not need to pursue this issue at this time. More crucial for my purposes is to assert the undeniable truths that the traditional medical position rejects the idea of disclosure and consent, and that disclosure and consent are severely compromised if authority is so defined.

For genetic counselors to remain faithful to only informing their counselees and leaving decision making to them requires, first of all, an abiding conviction that they and their counselees are companions, equal companions, on a voyage into the unknown. If, on the other hand, counselors see themselves as possessing greater insights, greater wisdom, and greater appreciation of their counselees' best interests than the counselees themselves, then the disclosure process wittingly and unwittingly becomes distorted and consent close to meaningless. Implementing the idea of disclosure and consent requires mutual trust and mutual respect on the part of professionals *and* patient-counselees alike. What exists today is unilateral trust, unilateral respect.

The words of my prescription can be simply stated: shared decision making. I do not have to remind you that translating such a prescription into practice is an inordinately hard task—it is opposed by thousands of years of tradition; it demands unrelenting vigilance, for the propensity to view patient-counselees as children who need to be led, however gently, is ever-present; it calls for a commitment that takes patience and time. Perhaps such a prescription asks for too much, perhaps even patient-counselees may not wish to interact with professionals on that basis. Be that as it may, the point I wish to hammer home is that in the absence of shared authority, disclosure and consent are at best of limited value and at worst a deception. What passes today for "informed consent," embedded as it is in the traditional cement of "doctor knows best," is neither informed nor consent. Perhaps genetic counseling can divorce itself from the medical-professional model; if it does, all of us will learn much from such a bold venture.

AUTONOMY—WHO CAN BE IN CHARGE?

The feeble call for informed consent began in the courts in 1957 with Justice Bray's decision in *Salgo v. Leland Stanford, Jr. University Board of Trustees.*[6] It was supported three years later by Justice Schroeder in *Natanson v. Kline*[7] with a powerful appeal to jurisprudential principles, soon forgotten in the very same opinions: "Anglo-American law starts with the premise of thorough-going self-determination. It follows that each man is considered to be master of his own body. . . ."[8] Yet it is precisely this proposition of individual self-rule that professionals have questioned since time immemorial. Listen to Hippocrates: "Perform [these duties] calmly and adroitly, concealing most things from the patient while you are attending to him. Give necessary orders with cheerfulness and sincerity, turning his attention away from what is being done to him; sometimes reprove sharply and emphatically, and sometimes comfort with solicitude and attention, revealing nothing of the patient's future or present condition. . . ."[9] Hippocrates, I believe, recommended this posture to physicians because he doubted patients' capacity for self-determination. Again the implications are self-evident: the idea of disclosure and consent cannot survive such doubts. Hippocrates made this crystal clear.

Let me speak briefly not of self-determination—man's constitutional right to govern himself—but rather of autonomy—"man's power, as possessed of reason, to give law to himself."[10] One must, if disclosure and consent have any meaning, rescue autonomy from the confusion that has surrounded it. Autonomy can never be absolute, it can only be relative. Biological givens, unconscious motives, intellectual endowments, developmental events—many of which are also unconscious—and situational factors—particularly when persons are under stress—shape, expand, and limit individual autonomy. Antagonists of informed consent have called attention to these determinants and have spoken disparagingly of "informed (but uneducated) consent."[11] Yet if autonomy were to imply total awareness of all factors influencing a person's decision, nobody would be autonomous, including physicians. Law's presumption of human competence would have to give way to a presumption of incompetence, and no one would be in a position to decide.

Human beings bring to the decision-making process their own unique individuality. Illness or stress, to be sure, can undermine autonomy and physicians have pointed to this important fact in order to claim that patients are incompetent to decide. On the other hand, commentators, particularly from law, have argued that fidelity to the principle of "thoroughgoing self-determination" demands that patients' assents and refusals be honored. With respect to refusals, some commentators have also suggested that patients' level of knowledge is irrelevant since it is the prerogative of patients to decide how much or how little information they want.[12] Neither position is satisfactory. Human psychology is too complex and human autonomy all too fragile. Autonomy should never be taken for granted; it requires nurture and care, which communication and dialogue can provide by bringing into awareness all kinds of unacknowledged and unconsidered influences on the choices about to be made. Is it not strange how little attention has been paid to this self-evident proposition? In most other areas of life we appreciate the value of acquiring as much knowledge as possible before embarking on a new venture into the unknown.

Thus, in the dialogue between physician-counselor and patient-client, the professional, through listening and talking, should spare no effort to maximize autonomy. Particularly in the face of patient-clients' doubts or disagreements, the professional should make as certain as he can be that he understands their objections and that he has been understood by them. Therefore, a professional's insistence that he and his patient-clients converse and think together should be honored, even if the dialogue has to be forced. That much invasion of privacy must be permitted in order to safeguard patient-clients' autonomy through insight, and not to abandon it to corruption by unnecessary fears, blind misconceptions, and false certainties. Even John Stuart Mill appreciated this: "Considerations to aid his judgment, exhortations to strengthen his will may be offered to him, even obtruded on him by others; but he himself is the final judge."[13] I appreciate that my recommendation of a possibly forced dialogue creates new dangers of overreaching and compulsion. But in insisting on conversation I have in mind that the physician-counselor interact with the patient-counselee in the spirit of Leonardo da Vinci's distinctions between painting and sculpturing.[14] In sculpturing—*per via di levare*—the figure comes into being by chiseling away the "distortions" introduced by the material; in painting—*per via di porre*—the painter adds his own "distortions" to the canvas. Thus the counselor must take great care to work with the counselee's distortions, and without adding his own to them. There is much here to be explored and to be learned.

The prescription sounds simple again: to nurture the patient-counselees' autonomous, adult functioning through a persistent dialogue. This recommendation goes counter to current practices that exploit the natural regression resulting from illness and stress. Command, a few words delivered in a monologue, a pat on the back send patients back further on the regressive path to childhood and impotence. By treating patients in this fashion, physicians overlook that patients' childlikeness, so frequently noted in the literature, is powerfully reinforced by precisely this kind of behavior on their part.

Again, what I said about authority is equally true for autonomy. In the absence of respect for patient-counselee's autonomy as it *ultimately* emerges (not necessarily as it is *initially* verbalized), disclosure and consent are at best of limited value and at worst a deception. Let me mention only in passing that fidelity to dialogue and autonomy may resolve the controversy among genetic counselors over whether only to inform or also to guide. The latter may prove less of a problem—it may even complement the former—if the parties are committed to *shared* authority, to dialogue, and to *mutual* trust.

UNCERTAINTY—CAN ANYONE BE IN CHARGE?

Genetic counselors' awareness of the uncertainties inherent in their specialty may explain their greater appreciation that traditional medical disclosure practices need be modified. The words of Dr. Alvan Feinstein, professor of medicine at Yale Medical School may sound much less foreign to genetic counselors than to many physicians:

> ... although anticoagulants, antibiotics, hypotensive agents, insulin, steroids have been available for 15 to 40 years, many of their true effects on patients and diseases are unknown or equivocal. Clinicians are still uncertain about the best means of treatment for even such routine problems as a common cold, a sprained back, a fractured hip, a peptic ulcer, a stroke, a myocardial infarction, an obstetrical delivery, or an acute psychiatric depression. . . ."[15]

Physicians, for many reasons, some of which go to the heart of traditional medical practice, have blinded themselves to the impact of uncertainty on their professional activities. As best as I can make it out, the prevailing climate of professional conduct is first to pay lip service to uncertainty and then to proceed, while interacting with self and patients, as if uncertainty did not exist. The consequences of this defensive and adaptive stance have been far-reaching.

Since professionals have had so little experience of living more self-consciously with uncertainty, it is difficult to know how much uncertainty professionals and their patients can tolerate. Thus the question: What is maladaptive about the denial of uncertainty? Would it be maladaptive or would it lead to a higher level of adaptation if therapeutic effectiveness no longer depended so much on the faith and trust that patients have in physicians and therapy as well as on the faith and confidence that physicians have in themselves and their therapy? For it is the therapeutic impact of conveying a sense of faith, trust, and confidence that is so often invoked in support of certainty over uncertainty. Would communication of uncertainty drive patients into the clutches of "charlatans" or would it have the opposite effect of promoting good medical care because patients could finally distinguish more clearly between faith healers and physicians? We just do not know. To be sure, uncertainty can be a terrible burden, and if a recommendation of greater awareness and disclosure of uncertainty has merit, questions of how much and what kind need be addressed.

The consequences of the defenses against uncertainty are much clearer; thus we must ask, Are they worth the price? For example: (1) Repression of uncertainty can lead to a false sense of certainty, making the professional more certain of his course of action than he can and should be. The at times acrimonious debates over which therapeutic interventions are best, e.g., in the treatment of breast cancer, carried on with zeal and conviction,[16] testify to this dilemma. (2) Since physicians, in defending themselves against their uncertainties, often project them unto their patients, the latter appear even more uncertain than they are to begin with. Such "perceptions," in turn, reinforce the doctor's propensity to giving unexplained orders. (3) An incapacity to cope with uncertainty, and the disquiet it creates, may account in part for the divergent opinions among genetic counselors on the significance of the burdens that various genetic diseases impose on individuals; on estimates of recurrence of chromosomal conditions; and on the respective weight to be given to clinical manifestations of disease, to the desire to have children, to the impact of defective children on marriage, or to the societal burden of perpetuating genetic defects.[17] Since in the dialogue between counselors and counselees uncertainty is ever-present, its denial must result in finding certainty somehow, somewhere—and where better than in one's own value preferences? Finally, defensive maneuvers against uncertainty that lead to greater certainty than is called for may explain why truth-telling becomes so readily compromised in interactions between professionals and their patient-counselees; why apodictic, authoritarian pronouncements have remained part of the fabric of medical practice; and why the idea of disclosure and consent are perceived as such a threat.

The more I think about the problem of uncertainty and the ways it has traditionally been dealt with, the more I see it as the major impediment to providing meaningful disclosure and to obtaining meaningful consent. I can only mention in passing that uncertainty is not only a consequence of the uncertainty inherent in the art and science of professional practice. Uncertainty, for example, also enters the dialogue between professionals and their patient-counselees due to the ubiquitous transference and counter-transference reactions that create their own uncertainties because they remain so frequently unnoticed and ill understood by physicians as well as by their patients—or, to put it most generally, due to the enormous inter- and intrapersonal difficulties of sorting out what is fact and what is fiction, what is informed by past, present, fantasy, temperament, ignorance, knowledge, or judgment.

Again, the prescription sounds simple: greater consciousness of the uncertainties inherent in diagnosis, treatment, and prognosis, as well as of the uncertainties inherent in the dialogue between the participants in the decision-making process. In asking for greater consciousness, I use "conscious" in its root meaning—*con-scire* to know together—i.e., for counselors to have a dialogue first with themselves and then with their counselees.

Only time will tell whether this prescription can become the medicine of tomorrow. At a minimum, and of this I am more certain, professionals can do much to eradicate some of the most pernicious manifestations of the denial of uncertainty once they have recognized its importance. What I said about authority and autonomy is equally true for uncertainty. Denying its influence makes disclosure and consent, at best, of limited value and at worst, a deception.

EPILOGUE

I have tried to make only one point, though a crucial one: Implementation of the idea of disclosure and consent will entail a radical change in traditional professional attitudes toward authority, autonomy, and uncertainty. I have said nothing about my reasons for favoring such a reorientation; this I must leave to another day. I only tried to suggest that without a fundamental change in professional attitudes, disclosure and consent will remain largely a cruel hoax.

Eventually we shall have to decide whether we wish to embrace or spurn a meaningful informed consent doctrine. Should we favor the former course, I appreciate as well as you do that reform of existing disclosure and consent practices will create formidable problems about which I have said very little. I know as well as you do that such a reform goes counter to the entire history of professionalism. And I also am aware as well as you are that powerful arguments have been made for entirely different models of physician–patient interactions. Dostoyevsky's Grand Inquisitor spoke awesomely for a model most opposed to the one suggested by me: "We have corrected your great work and have based it on *miracle, mystery and authority.* And men rejoiced that they were once more led like sheep and that the terrible gift [of free choice] which had brought them so much suffering had at last been lifted from their heart. . . . Tell me. Did we not love mankind when we admitted so humbly its impotence and lovingly lightened its burden. . . ?"[18] When I worked out my three categories I did not know how closely, though diametrically opposed, they are to the Grand Inquisitor's. His authority "[to take] possession of man's freedom" versus mine of shared authority; his mystery "which they must blindly obey" versus my confrontation of mystery-uncertainty; his miracle about which he asks, can man "reject [it] at the most fearful moments of his life and stick to the free decision of his heart" versus my appeal to respect for personal autonomy rather than a silent submission to miracles procured by others. Who knows which model comports more with human capacities? History sides with the Grand Inquisitor; but it may be a history of physician–patient interactions that has remained stuck in ancient practices, uninfluenced by medicine's march from magic to science and society's from feudalism to contract. I only have the audacity to propose my model because I am hopeful that a greater awareness of the contributions that professionals' countertransferences make to patient-counselees' childlikeness will eliminate unnecessary inequalities between them, inequalities that, as Hegel saw so clearly, penalize not only the slave but the master as well.[19]

ACKNOWLEDGMENT

The financial support of the Kayden Foundation is gratefully acknowledged.

REFERENCES AND NOTES

1. Sorenson, J. R., Biomedical innovation, uncertainty,. and doctor–patient interaction, *J. Health Soc. Behav.* **15**:366 (1974).
2. Katz, J., Informed consent—A fairy tale? Law's vision, *University of Pittsburgh L. Rev.* **39**:137(1977).
3. Katz, J., Informed consent: The therapeutic relationship: Medical and legal aspects, in: *The Encyclopedia of Bioethics* (Vol. 2) (W. T. Reich, ed.), p. 770, Free Press, New York (1978).
4. See, for example, Burnham, P. J., Medical experimentation on humans, *Science* **152**:448 (1966); Irvin, W. P., Now, Mrs. Blare, about the (complications...), *Med. Econ.* **40**:102 (1963).
5. Parsons, T., *The Social Systems*, pp. 463,465, Free Press, Glencoe, Ill. (1951).
6. 317 P.2d 170, 181 (1957).
7. 350 P.2d 1093 (1960) and 354 P.2d 670 (1960).
8. 350 P.2d 1104 (1060).
9. Hippocrates, Decorum, in: *Hippocrates* (Vol. 2) (Trans. W. H. S. Jones), p. 297, Harvard University Press, Cambridge (1967).
10. Webster's *New International Dictionary of the English Language*, Second Unabridged Edition, p. 188, G.&C. Merriam, Springfield (1959).
11. Inglefinger, F., Informed (but uneducated) consent, *N. Engl. J. Med.* **287**:465 (1972).
12. Goldstein, J., For Harold Laswell: Some reflections on human dignity, entrapment, informed consent, and the plea bargain, *Yale L. J.* **84**:690 (1975).
13. Mill, J. S., *On liberty*, p. 68, Basil Blackwell, Oxford (1946).
14. For an extended discussion, see Loewald, H., On the therapeutic action of psychoanalysis, *Int. J. Psycho-Anal.* **41**:16 (1960).
15. Feinstein, A., *Clinical Judgment*, p. 23, Williams & Wilkins, Baltimore (1967).
16. Compare, for example, Crile, G., Jr., Management of breast cancer: Limited mastectomy, *J. Am. Med. Assoc.* **230**:95 (1974) with Anglem, T. J., Management of breast cancer: Radical masectomy, *J. Am. Med. Assoc.* **230**:99 (1974).
17. Sorenson, J. R., and Culbert, A. J., Genetic counseling orientations—Unexamined topics in evaluation, in: *Genetic Counseling* (H. A. Lubs and F. de la Cruz, eds.), p. 131, Raven Press, New York (1977).
18. Dostoyevsky, F., *The Brothers Karamazov* (Trans. D. Magarshack), p. 301, Penguin Books, Baltimore (1958).
19. Hegel, G. W. F., *The Phenomenology of Mind* (2nd ed., trans. J. B. Baillie), Humanities Press, New York (1964). See also, Davis, D. B., *The Problem of Slavery in the Age of Revolution—1770–1823*, p. 557, Cornell University Press, Ithaca (1975).

Discussion

A PARTICIPANT: I sympathize very deeply both intellectually and emotionally with the position you just articulated. However, I would like you to comment particularly as a psychiatrist and psychoanalyst about the hypothesis that there is in fact something in human nature that wants an authoritarian relationship of that sort. All through history there has been the "medicine man" and now it is the psychiatrist and physician who in fact are trying to get to the kind of position you have mentioned. You have to fight very deeply against something very deeply ingrained in human nature. What do we do about it?

PROF. JAY KATZ: I have been thinking and talking with my students about this very same question for years, and of course I do not know the answer. What is so unclear with respect to attitudes toward professionals is the question what came first, the chicken or the egg.

In another speech which I gave some years ago, I talked about this problem. "Who Is Afraid of Informed Consent?" was the title of it, and among the points I made then was that of course patients are also afraid of informed consent. One of their earliest memories must continue to influence them, when they accompanied their mother or father or both to the pediatrician's office and saw how the pediatrician made children out of their powerful parents. This must make a great impression on children and may initiate the brainwashing process.

How much of it is brainwashing, how much of it is something inherent in the nature of man, I don't know. Again, of one thing only I am certain; that there is a group of human beings who do want to know more, who do want to interact with professionals on the basis of greater equality, and they have a terribly hard time, as I have a hard time, being a physician myself, to engage in a dialogue with others whenever I am sick or a member of my family is sick. It is an effort to raise myself up to my full five feet eight inches and force myself to ask my physician questions and to try to force him to give me meaningful answers.

So what the answer to your question is, I don't know, but it's worth learning more about.

12

Knowledge, Risk, and the Right to Reproduce

A Limiting Principle

Joseph F. Fletcher
School of Medicine
University of Virginia
Charlottesville, Virginia 22903

I have a question to raise and an answer to offer. My answer has been anticipated by Lawrence Ulrich of the University of Dayton. In 1976 he said, in a paper that I regard as a superb model for the ethical position I myself take, "The thesis which I wish to adopt is this: Reproductive rights are not absolute and those who are at risk for passing on clearly identifiable, severely deleterious genes and debilitating genetic disease should not be allowed to exercise their reproductive prerogative."[1]

I propose to take a somewhat different line from Ulrich's, using compatible though not identical reasoning to reach the same conclusion. The thesis we defend appears only rarely in the literature but it is, nevertheless, by no means rare in the thinking of practical people.

Pro forma, the conventional wisdom is set out in Article 16(1) of the U.N. Declaration of Human Rights: "Men and women of full age, without any limitation due to race, nationality or religion, have the right to marry and found a family." Since the United Nations has designated 1979 as the Year of the Child, my thought is that if child abuse is part of its concern we ought to recognize that children are often abused preconceptively and prenatally—not only by mothers drinking alcohol, smoking, and using drugs nonmedicinally but also by their *knowingly* passing on or risking passing on genetic diseases. There are more Typhoid Marys carrying genetic diseases than infectious diseases.

Another phrase that needs reexamination is "communicable disease." Given what we now know, it is tunnel vision, nothing else, to go on thinking of communicable diseases solely in terms of the infectious and not the genetic diseases. Diseases are transmitted vertically, from one generation to another, as well as horizontally from one contemporary to another. This being so, if infectious diseases are sometimes grave enough to justify both ethical and legal restrictions on carriers, why not some genetic diseases too? Is intergenerational responsibility possible otherwise?

A legal analysis is beyond my competence but I think the ethical questions at stake are reducible to three: (1) Is it right to judge that sometimes the condition of a conceptus falls below, or would fall below, a reasonable standard of quality? (2) Are we morally obliged to avoid having children with serious genetic disorders, if and when we *can* avoid it? (3) Can we justify interference by society through law to uphold minimum standards of human reproduction—i.e., by mandatory as well as voluntary controls? It is this third question that is most controversial.

Fairly obviously, several ancillary ethical issues are involved. Among them are the quality-of-life concept behind the first question; the morality of contraception, abortion, and sterilization behind the second question; and, behind the third one, a debate as to whether the individual's claims or a broader interest should prevail when they conflict—a deep and searching issue in social ethics.

I have discussed these questions in considerable detail elsewhere.[2] For our purposes here I shall simply answer the first two questions in the affirmative. We ought in conscience to have a humane minimum standard of reproduction, not blindly accepting the outcome of every conception. And we ought to act on our genetic information to prevent the birth of children below that minimum. Thus far, I assume, I am in agreement with the majority, both of genetic counselors and of people in the secular community.

As we extend our genetic and fetologic knowledge we also increase the frequency of value judgments and moral choices. For example, adding endoscopy to amniocentesis adds to the list of predictable or detectable genetic diseases requiring decisions. For some patients, and even some counselors, the moral burden grows intolerable, and they try to evade responsibile decision making by hiding behind simplistic absolutes.

Let me, however, focus on the third question. Should preventive control in some cases be mandatory, not only voluntary? Entailed here are profound questions of social philosophy, especially the question of human rights, and specifically the right to reproduce.

As I have done before, I appeal to philosophers of ethics and the law to reexamine the relation between human rights or claims and human *needs*. This is a far more basic question than many of the normative problems with which biomedical ethics has concerned itself. My own conclusion is in the record but I urge others to make a critical study, to see whether my view or another one stands up better.[3]

On humanistic grounds, it seems to me, human need constitutes and validates rights. No right is valid ethically if, in its exercise, it results in more harm or unhappiness than human benefit. The ethical imperative is to optimize the good, and therefore every act that on balance deprives or hurts people is wrong. This is the case even if the act happens to have a *prima facie* acceptance as a "right." In short, *rights may not be right.*

The relativity of rights is as clear in jurisprudence as it is in ethics. Both work by the principles of equity and "general justification." If reproductive partners are informed they both carry a dread disease such as Tay–Sachs or cystic fibrosis, yet even so conceive with the intention of bringing every conceptus to birth, their supposed right to reproduce becomes ethically invalid. This is why there is no constitutional right in the law to have children. This is why the Supreme Court declared that "neither rights of parenthood nor rights of religion are beyond limitation."[4] Thus ethical reasoning and legal reasoning are inherent in each other.

There is another ground for being wary of the conventional rhetoric about rights. Rights often come into conflict with each other. For this reason jurists and ethicists try to distinguish between rights that are "fundamental" and rights that are not. The difference actually is only that the fundamental kind are not as easily set aside when they conflict with other rights. We are speaking here of *legal* rights, i.e., supposed human rights that have explicit support in the law. Conflicts between rights, both moral and legal, are almost as common as their assertion.

Martin Golding has tried to distinguish between "option rights" and "welfare rights."[5] Option rights appear to be claims we might or might not make, as in becoming parents, and welfare rights lie in possessions or conditions, such as personal property and citizenship. Yet this is of little help since in reproduction the parents' "option right" to have a child could conflict with the child's "welfare right" to a sound mind and a sound body.

Something has to give. As the British philosopher Hare remarked about an alleged right to be born, "Rights are the stamping ground of intuitionists, and it would be difficult to find any claim confidently asserted to a right which could not as confidently be countered by a claim to another right, such that both rights cannot be complied with."[6]

It may not be too much to say that the very term *rights* is a conscious maneuver to cover up perceived or suspected contradictions. I cannot forget Bentham's remark, that talk of rights is "nonsense on stilts."

In ethics, as in law, rights at best are relative. In moral philosophy and theology rights are "imperfect." There are no valid claims of unique interest, and whenever a right is justifiable it will be for extrinsic reasons, not because of any intrinsic validity. Neither rights nor obligations are deducible. They are contingent on situations. Contrary to a purple passage in the Declaration of Independence, rights are not "unalienable."

If human rights, including reproduction, are relative, and relative primarily to human need and well-being, what then of the priority of individual or social interests when they conflict? Surely it is acceptable to hold that the commonweal comes first. This is what the principle of eminent domain means in property law, what search and seizure means in police powers, what conscription means in the national defense, what a traffic light means down at the corner, what compulsory school attendance means, what laws prohibiting consanguineous marriage mean.

Coercive or compulsory control is justified when people refuse to abide by humane standards of reproduction—first, as a protection of society's well-being, and second, to prevent harm to potentially misbegotton children. When we limit the freedom of such reproducers, they can conceive again; conceptuses are fungible. When their gonads dysfunction, such persons can resort to alternative modes of parenting with donated gametes, as in artificial insemination and enovulation.

The so-called right to reproduce is, therefore, in fact a privilege. At the level of logic, those who resist a limiting principle will insist that "a right is a right." Here is the circularity, the *circulo probandi* or *petitio principi* of all attempts to absolutize the finite. Like everything else we do, baby making is subject to the requirements of justice.

The radical individualism of some human rights advocates falsifies the social nature of man. We are members one of another in fact and not only in theory. The stark truth is that our gonads and gametes are not private possessions. Biologically they are shared values, realized in sexual union. They have to be combined and merged with the genetic values of others, except—someday possible—in asexual options such as vegetative replication and parthenogenesis. By and of themselves our sexual faculties are nothing. Testes and ovaries are communal by nature, and ethically regarded they should be rationally controlled in the social interest.

Another question calls for analysis that may have been even more neglected. Do we have any obligation to nonexistent, potential but not actual persons? In the present context, what do we owe to children who would suffer grievously if conceived or born? Here is a crucial question for the law in relation to medical genetics. If we have *no* obligation, then all talk about control of reproduction in their interest, either voluntary or mandatory, falls to the ground. Professor Tedeschi has, in effect, argued for no responsibility, in a paper in the *Israel Law Review*.[7]

I would contend that to agree with Tedeschi would be tantamount to rejecting the whole notion of preventive medicine, sanitation, environmental protection law, and all the other ways in which we express our obligation to the unborn. Nevertheless, I would welcome a more rigorous investigation of the question than it has yet been given.

In the meantime, I believe that life can sometimes be a fate worse than death or nonbeing, as against the absolutist view that a life however handicapped or diseased is better than no life at all. One Catholic theologian, Bernard Haering, feels the weight of this position so much that he has even suggested that his church authorities should reconsider their prohibition of sterilization, at least as a preventive therapy for diseased individuals.[8] In my own crystal ball I foresee what Dr. Margery Shaw thinks is a possibility, that the courts will more and more entertain suits of prenatal torts or "wrongful life" brought by misbegotten children against their guilty parents.[9]

To sum up, the obligation to prevent the birth of genetically diseased or defective children is plain enough to all but a few case-hardened "right to life" absolutists. Nevertheless, the obligation to mandate prevention for those who will not do it voluntarily is still moot for many geneticists, lawyers, and ethicists. We need to explore the reasons for their irresolution. Why should this particular wrong—the wrong of knowingly transmitting grievous genetic disease—be an exception to the general principle that a sane society prohibits by law substantial injuries by any one member of society to innocent others?

References and Notes

1. Ulrich, L. P., Reproductive rights and genetic disease, in: *Biomedical Ethics and the Law* (J. H. Humber and R. F. Almeder, eds.), p. 351, Plenum Press, New York (1976).
2. See my two volumes, *The Ethics of Genetic Control*, Doubleday, New York (1974) and *Humanhood: Essays in Biomedical Ethics*, Prometheus Press, Buffalo (1979).

3. *Humanhood*, pp. 17, 87, 117.
4. *Prince v. Massachusetts*, 321 U.S. 158, 64 S. Ct. 438, 88 L. Ed. 645 (1944).
5. Golding, M. P., The concept of human rights: A historical sketch, in: *Bioethics and Human Rights* (E. Bandman and B. Bandman, eds.), p. 44, Little Brown, Boston (1978).
6. Hare, R. M., Abortion and the golden rule, *Philos. Public Affairs* 4:210 (1975).
7. Tedeschi, G., On tort liability for "wrongful life," *Israel L. Rev.* 1:513 (1966).
8. Haering, B., It's wrong to knowingly beget defective children, *U.S. Catholic* **February:**14 (1976).
9. Shaw, M. W., Genetically defective children: Emerging legal considerations, *Am. J. L. Med.* **3:**333 (1977).

DISCUSSION

Ms. MARJORIE GUTHRIE: I think in a way I am rather fortunate or maybe unfortunate: I am neither a lawyer nor a doctor nor a genetic counselor. I am somebody who had a husband who had a disorder, who perhaps—and we will not know for several years—may have passed it on to his children. We are living with it and we are learning how to live with it, with help or not with help, depending on the attitude of the people with whom we come in contact.

I think it is terribly important to say that you may disagree with me or agree with me, and I enjoy hearing your opinion. You have used the term *humanistic grounds*. I would like to use the term *realistic grounds*.

Realistic is not always humanistic. I have met many people who would love to agree with the concept that no parent wishes to give life to a child who will suffer. I don't believe anybody really wants to. But there is a realism about the affirmation of life itself that somehow is inborn. I don't know where it comes from. There are those who are willing to say (and I've seen them) that they will not have children and it is based on good information. But there are those who say, "I must reproduce; I am glad that I am here; I believe that I have a contribution to make."

I had a child who lived only four years and when that child died, I remember turning to Woody and saying, "Woody, if someone had told you I'm only going to give you this child for four years and you know you'll never see her again, would you have agreed to take her?" And he and I on the day she died kissed and hugged and said yes, four years of a good life, of a reasonably happy life, *is* worth taking a gamble on.

Now, not everybody can be me. And not everybody can be you. I do believe it is wonderful that you are sitting here and talking. I do believe that we are going to have an explosion of knowledge, not only genetic knowledge but ethical and legal knowledge, which is going to help us come to decisions. But I do know that you do not know where to draw the line. So, to try to make that line is good. To discuss it is good. But to come to a decision today is truly premature.

REV. ALBERT MORACZEWSKI (St. Louis, Missouri): I have had a number of discussions with Dr. Fletcher before and we always differ, so this won't be any different from the past. I just have one question which has to do with the question of the obligation we owe to the children who might be born with a serious genetic defect.

There are two parts to this. If the child could be anticipated to be below certain critera that Dr. Fletcher has indicated for humanhood or personhood (he once listed fifteen criteria), then the question whether we owe something or not doesn't make sense, because that is a nonperson and we would not owe a nonperson anything.

On the other hand, if the anticipated child would be above that cutoff point and would have some lesser genetic defect, that is, would still be a person by Dr. Fletcher's criteria, then we do owe something, because then we are dealing with a person. Then the question is, if we are making a decision for another person, whether or not that person should exist or not exist.

My big problem then is whether we have the wisdom to make such a decision for another person even though that person is still *in utero*.

DR. JOSEPH FLETCHER: I think the question just put is the one that lurks behind the very serious debate over whether or not abortion can be justified. When people decide to terminate pregnancies, they have decided that a life *in utero* should not be brought to term birth and personhood. My personal feeling about it is that the termination of pregnancy for any number of reasons, certainly for therapeutic reasons, is justifiable. I know there are those who disagree with that.

But the more important question that both of these comments, I think, have underlined for us is that it is all very well to say, as I did, that there are some inherited diseases which are so grievous that they ought not to be allowed to happen to anybody. But just which ones are they?

This raises all kinds of questions for which I as a philosopher have no competence. That is to say, the relativity of severity and cost and age of onset and treatability, and so forth. But I am quite sure that if we turned to the clinical wisdom and experience of people who are actually doing the business today of bringing babies into the world, they could give us some very realistic, and I should say some highly acceptable, humanistic limits.

What they are specifically I don't know. But I would say, just to start off a discussion getting down to cases, that Tay–Sachs, for example, is a disease sufficiently deadly and dreaded to be one to put in the no-no, always-to-be-prevented class, if we can.

DR. DANIEL STOWENS: We talked about the common weal. Some people must have these children so that everybody can benefit.

A PARTICIPANT: One of the things that we got from Mendel was the idea that the genes are independent and individual, when in fact the genes are part of the genetic system. This means that if we would legislate against the gene, including something like Tay–Sachs disease, are we sure we are getting rid of something which is detrimental under most circumstances or even many circumstances? I may be homozygous to Tay–Sachs disease but in my particular genetic background it may not express itself in the form that it expresses itself in those individuals who died.

I am concerned about legislating against genes simply because under a particular set of circumstances they express themselves in a particular way. Who do you suggest should not reproduce? Carriers of such diseases where their offspring may be carriers and therefore may be protected by heterozygote advantage against certain conditions, or homozygotes who have a susceptibility to a particular illness but may not necessarily express that illness, or homozygotes who could be maintained by some medical treatment such as insulin? Do we get rid of all of these children or do you draw the line somewhere?

A PARTICIPANT: If you legislate gonad responsibility, how do you propose to enforce such regulations?

DR. FLETCHER: I don't know, and it may well be that those who do know more about these things than I do will convince me that, as arguable as it is ethically, it is impractical in actual fact. I don't know that this is true; but if it were shown to me, of course I would have to accept it.

PROF. Y. EDWARD HSIA: In the same way I feel there ought to be a law to support me. I think what Dr. Fletcher has proposed is admirable. We should ban all people who ride motorcycles and drink alcohol because these are greater priority problems; and we should try to ban the criminal, the sociopath, the retarded, the poor, and the accident-prone, because I think these are all much more important to our society than the rare person who has a genetic disease!

ETHICS, EUGENICS, LAW, AND SOCIETY

MODERATOR: MARY ELLEN AVERY

13

EUTHANASIA DECISIONS IN THE COURTS
The Post-Saikewicz Experience

CHARLES H. BARON
Boston College Law School
Newton Center, Massachusetts 02159

One of the many ways in which we are discovering doctors to be just like other mortals is that they would like to have their cake and eat it too. When it comes to euthanasia decision making, medical personnel show increasing concern over the lack of clear guidelines for making such decisions, but they don't want to have to go through the anguish of developing such guidelines for themselves and they don't want to go through the discomfort of having anyone else develop guidelines for them.

A recent issue of *New York* magazine features an article written by a nurse troubled by her own compassion-motivated, long-term involvement in euthanasia activities. At the outset, she notes the fact that another nurse had recently been indicted for murder in Baltimore for just such a compassion-motivated act. "Oh, God," she had said to herself on hearing the news, "somebody got caught. What I felt was what most nurses I know felt. I know because I asked them. They were truthful, because I'm from the inside of medicine, from the same family, so they shared with me."[1] In the remainder of the pseudonymously written article, she reveals details of many of her own and other nurses' euthanasia experiences and raises a number of troubling questions regarding the handling of the cases involved. She concludes the article with a plea:

> If we could acknowledge that the people in medicine get tired and upset, sometimes have lousy judgment, get emotionally involved; if we could realize that the doctors and nurses are not intrinsically better than everyone else, then maybe we could set up a better backup system. Something more effective than what we have now.
>
> What they say that nurse did in Baltimore is being done in hospitals and homes now. I don't know if she's better or worse than the rest of us but I do know that a system which allows the kind of scattered individual judgments for life-and-death decisions, whether they be made by a doctor or a nurse, is not viable and needs change.
>
> I tried to handle this from inside medicine. Once, I went to a directress of nursing and laid all this on her. She was progressive and obviously aware of my dilemma, but

what she said was "My dear, you're not talking about medicine. You're talking about issues that greater minds than ours are trying to solve." And she dismissed me, obviously uncomfortable that I had brought the whole subject up, she said, "You're talking about morality." It sounded like an accusation, like I had stepped out of line, and all I thought was "I wasn't talking about anything but me and my patients."[1]

On November 28, 1977, the Supreme Judicial Court of Massachusetts took a long step toward responding to such pleas for a "better backup system." In its opinion in *Superintendent of Belchertown State School v. Saikewicz*,[2] the court suggested to the medical profession that it should allow itself the luxury of concerning itself with just itself and its patients as regards the medical questions involved in patient care. The moral questions were to be left to the courts, where many such moral questions have always been decided. Using their time-honored power to develop common-law principles that create new societal responses to new problems while basing those responses on long-standing societal values, the court would make euthanasia choices for incompetent patients rather than leaving those choices to doctors. Over time and through the decision of many such cases, the courts would begin to replace a system that allows for "scattered individual judgments for life-and-death decisions" with one that provides principles for uniform decision making and commits itself to a public effort to constantly refine those principles as part of the decision-making process.

In light of the awesome responsibility that was lifted from the shoulders of the medical community by the *Saikewicz* decision, one might have expected that doctors, nurses, and hospital administrators would have greeted it with enthusiasm. Instead, the reaction has been one of outrage at the presumption displayed by the court in deciding to step in and fill the vacuum as to orderly and principled decision making that had been left by the medical profession. Suddenly these euthanasia questions were no longer "moral" but were and always had been "medical"—at least in the sense that they were the turf of doctors and not of judges. In an editorial in the *New England Journal of Medicine*, Dr. Arnold Relman was very clear about what was wrong with the *Saikewicz* opinion:

> This astonishing opinion can only be viewed as a resounding vote of "no confidence" in the ability of physicians and families to act in the best interests of the incapable patient suffering from a terminal illness. The latter's constitutional right to refuse or accept medical therapy under such circumstances, a right that the Court clearly recognized, is to be entrusted to no one but the courts. What has heretofore always been left to physicians and families must henceforth receive judicial sanction in advance. The Court thus asserts in effect that its duty is not simply to remedy abuses and settle disagreements that arise in the practice of medicine but also to take routine responsibility for certain types of medical decisions frequently needed for terminally or hopelessly ill patients.[3]

However, Dr. Relman was less clear about what might be done within the medical community about doing away with the existing system of "scattered individual judgment." In essence, he endorsed the existing system in which physicians "used their own best judgment, aided by the advice of colleagues and frequently the opinions of other health professionals, ministers and lawyers; sometimes, they are also advised by special hospital committees organized for this purpose."[3] And in a later article, he recognizes that "closed, totally private decisions, unreviewed by any third party, are a cause of legitimate concern," but his only suggested solution is: "When physicians make life or death decisions for incompetent patients (assuming, always, consent of the family or legal guardian), concurrence by several colleagues who have no vested interest in the decision should be documented in the medical record."[4] The theory is, apparently, that doctors have, as part of their medical training, been made privy to certain principles of right reason as regards how these euthanasia cases should be handled and that it only takes a little cool deliberation among fellow members of the fraternity of the enlightened to assure that the right principles are being applied in any given case. In fact, as all of us who have worked with doctors and nurses on such ethical problems know, the answer one gets (if one gets an answer at all) depends upon whether you ask Dr. A or Dr. B and whether you have asked him on day one or day two. There are no clearly worked-out principles and there is little agreement among doctors as to most of the difficult questions that are encountered in euthanasia cases.

What is really going on here, it seems to me, is that doctors are simply resisting the effort to have anyone outside medicine interfere with their decision-making process. Even if they are troubled by the difficult decisions that they have to make and find themselves unequipped to make them, and even if they are unwilling to take the pains to develop publicly acceptable and systematic principles and procedures for decision making, they are unwilling to let anyone outside medicine intrude upon their turf by developing such principles for them.

It is largely because of this attitude that the post-*Saikewicz* experience in the courts of Massachusetts has not been as successful as it might have been. "Apres *Saikewicz*, le deluge" was the original threat of the medical community. "Attorneys I have spoken with do not believe the probate courts can possibly handle the number of cases that will be generated by the Saikewicz decision," said Dr. Relman immediately after the decision.[3] But a year later he noted the fact that very few such cases had in fact been brought.

> "No-treatment" or "withdrawal of treatment" decisions for incompetent patients are being made all the time throughout the hospitals of Massachusetts, but very few are being brought to judicial attention. The reasons are obvious enough, and are implicit in what has already been said. Neither relatives nor physicians want to go to the trouble and expense of obtaining a court judgment, particularly when they have no confidence that the judgment will be medically or ethically sound. If there is real doubt that the court will consider the quality of life involved, and if the medical recommendations of the physicians in charge of the case, as well as the wishes of the family, are to be examined in an adversarial courtroom proceeding, then most families and most physicians would prefer to stay away from the courts. At present, it commonly is believed that they take very little risk in doing so, but that view could of course be changed by future developments.[4]

This "stonewalling" attitude on the part of the medical community is dangerous and unfortunate for many reasons. Among them is the fact that it has deprived the courts of the clinical material they need for case-by-case common-law development of guidelines for decision making beyond the substantive foundation laid by *Saikewicz*.

Even among the medical community, it is generally agreed that the *Saikewicz* opinion represented a laudable beginning as regards the substantive guidelines that it laid down. Among other things, it was the first decision in Massachusetts recognizing the right of competent patients to authorize termination of life-prolonging care under certain circumstances. It also recognized the same right for incompetents to be exercised on the basis of "proxy consent." Beyond that, it made tentative steps to deal with a variety of other issues that have troubled the medical community—issues that could be dealt with only in inchoate form because of the special nature of the facts of the particular case before it.

The Relevance of Quality-of-Life Criteria

Because Joseph Saikewicz had an IQ of 10, the court had to face the question of whether his profoundly retarded state could be taken into consideration in deciding whether the possible gain of an additional year of his life through chemotherapy was in his best interests. Could society sanction a felicific calculus that weighed the relative worth of life of a person with an IQ of 10 against that of a person with an IQ of 160? The court made very clear that the answer to that question was no. "[T]he chance of a longer life carries the same weight for Saikewicz as for any other person," the court said, "the value of life under the law having no relation to intelligence or social position."[2] But later in the opinion, the court restated its answer in language that has created some problems because of vagueness.

> The sixth factor identified by the judge as weighing against chemotherapy was "the quality of life possible for him even if the treatment does bring about remission." To the extent that this formulation equates the value of life with any measure of the quality of life, we firmly reject it. A reading of the entire record clearly reveals, however, the judge's concern that special care be taken to respect the dignity and worth of Saikewicz's life precisely because of his vulnerable position. The judge, as well as all the parties, were keenly aware that the supposed ability of Saikewicz, by virtue of his mental retardation, to appreciate or experience life had no place in the decision before them. Rather than reading the judge's formulation in a manner which demeans the value of the life of one who is mentally retarded, the vague, and perhaps ill-chosen, term "quality of life" should be understood as a reference to the continuing state of pain and disorientation precipitated by the chemotherapy treatment.[2]

As Professor Allen Buchanan has pointed out, the court is clearly ruling out judgments as regards the quality of life in the "social worth" or utilitarian sense, i.e., the assigning of "some comparative value to the life of an individual—to rank the worth of that individual relative to the worth of others for the purpose of calculating the costs and benefits of expending resources upon him."[5] But it is less clear, I believe, that Professor Buchanan is right when he says that the court has not ruled out as well at least judgments "about the value or quality of an individual's life *to that individual*, regardless of how society or would-be calculators of social utility evaluate it."[5] After all, the court suggests that "the ability of Saikewicz, by virtue of his mental retardation, to appreciate or experience life" should also have no place in the decision whether to refuse chemotherapy to him. Does this mean that the law can countenance only the prospect of physical or mental pain in measuring the worth of an additional year of life? Does this mean that the court would have ruled against authorizing denial of treatment to someone like Karen Quinlan, who might be experiencing no pain, because the court could not take into consideration the fact that the life prolonged thereby would not be a sapient, cognitive existence? I think not. I believe that the court spoke as it did in *Saikewicz* because it was concerned with protecting the mentally retarded from possible prejudices of the unretarded regarding the low value that life must have to the retarded themselves. I would predict that the court would rule that the low value of life to Karen Quinlan herself could be taken into consideration in a Quinlan-type case once it was shown that there is a workable line that can be drawn between her sort of brain damage and even the lowest IQ. But in order to work out such a subtle distinction in principle, the court needs to be presented with such a Quinlan-type case.

The Weight to Be Given to the Countervailing State Interests

In *Saikewicz*, the court suggested that a decision of even a competent person to accept death rather than continued medical treatment was limited by the extent to which the person's right to privacy and right to be free from invasion of his bodily integrity might be outweighed by certain contervailing state interests. The list of these state interests, which was culled from a variety of sources, comprised: (1) the preservation of human life, (2) the protection of the interests of innocent third parties, (3) the prevention of suicide, (4) maintaining the ethical integrity of the medical profession. Because the court found each of these interests easily outweighed on the facts of *Saikewicz*, there is not a very full exploration of the nature and extent of any of them. As a result, the statement of each is problematic standing alone, and there are additional problems in understanding the relationships they are to bear to each other.

As regards the first state interest, the court seems to recognize, for one thing, an interest that the state has in keeping each of its citizens alive. This aspect of the first interest is explored no further. Presumably it is based in part on the fact that the size of a state's populace affects its ability to function economically, defensively, etc. As another part of this interest, the court recognizes a state concern with "seeing to it that individual decisions on the prolongation of life do not in any way tend to 'cheapen' the value which is placed on the concept of living. . . ."[2] As to this, the court states: "The constitutional right to privacy, as we conceive it, is an expression of the sanctity of individual free choice and self-determination as fundamental constituents of life. The value of life as so perceived is lessened not by a decision to refuse treatment, but by the failure to allow a competent human being the right to choice."[2] But how are these statements to be reconciled with the assertion of the third interest, that of the prevention of suicide? Of that interest the court says only:

> The interest in protecting against suicide seems to require little if any discussion. In the case of the competent adult's refusing medical treatment such an act does not necessarily constitute suicide since (1) in refusing treatment the patient may not have the

specific intent to die, and (2) even if he did, to the extent that the cause of death was from natural causes the patient did not set the death producing agent in motion with the intent of causing his own death. Furthermore, the underlying state interest in this area lies in the prevention of irrational self-destruction. What we consider here is a competent, rational decision to refuse treatment when death is inevitable and the treatment offers no hope of cure or preservation of life. There is no connection between the conduct here in issue and any state concern to prevent suicide.[2]

But suppose a patient refuses a relatively benign therapy on the ground that it would only prolong an existence made extremely painful and meaningless by an underlying disease. Doesn't he have a "specific intent to die?" And does that make his refusal a suicide that is not to be sanctioned by the state? Or suppose someone refused therapy precisely because he wished to die for reasons having nothing to do with any underlying medical therapy. Suppose we would agree that his reasons for wanting to die are as rational as those of a person with a physically painful disease. Would this be an unsanctioned suicide? If so, why doesn't such an attitude on the part of the state undermine the value of life "by the failure to allow a competent human being the right of choice?"

The second state interest is described by the court as one of "considerable magnitude." That is the interest "in protecting third parties, particularly minor children, from emotional and financial damage which may occur as a result of the decision of a competent adult to refuse life-saving or life-prolonging treatment."[2] But this state interest, like the others, has the potential for swallowing up the asserted right of privacy on the part of the patient. Almost any person's death is likely to impose financial and emotional impact on third parties—even where there are no dependent children involved. And restricting it to the case of dependent children does not solve all such problems. As an example of a clear case where a strong countervailing interest of the third kind was present, the court cites *Raleigh Fitkin–Paul Morgan Memorial Hospital v. Anderson,*[6] where a blood transfusion was forced upon a pregnant patient to save her life and that of the fetus she was carrying. But if a mother has a constitutional right to an abortion that would end the life of the fetus, how can she be prevented from determining to end her own life just because the decision would mean the termination of the life of the fetus as well?

Finally, the fourth state interest, that of maintaining the ethical integrity of the medical profession, gets minimal discussion in *Saikewicz* as a result of the fact that "[t]he probate judge's decision was in accord with the testimony of the attending physicians" in the case and "[t]he decision is in accord with the generally accepted views of the medical profession, as set forth in this opinion."[2] This interest, as all the others, begs fuller development as to its parameters. But that development must await consideration by the court of cases that present a meaningful challenge to the interest.

LIFE-PROLONGING VERSUS LIFE-SAVING TREATMENT

In *Saikewicz* the court makes clear that it is only sanctioning, at least on the facts of the case before it, a patient's right to refuse life-prolonging, as opposed to life-saving, treatment. In light of the fact that man's mortal state would seem to make all medical treatment life-prolonging at best, this distinction seems irretrievably doomed to the status of being one of difficult-to-determine degree. All that can be said on the basis of the holding in *Saikewicz* is, first, that chemotherapy that is likely to lengthen a life by only 13 months is only life-prolonging, and, second, that the policy that must be looked to in developing criteria for determining when life-prolonging becomes life-saving is the state's interest in preserving life. As to this, the court states:

> It is clear that the most significant of the asserted State interests is that of the preservation of human life. Recognition of such an interest, however, does not necessarily resolve the problem where the affliction or disease clearly indicates that life will soon, and inevitably, be extinguished. The interest of the State in prolonging a life must be reconciled with the interest of an individual to reject the traumatic cost of that prolongation. There is a substantial distinction in the State's insistence that human life be saved where the affliction is curable, as opposed to the State interest where, as here, the issue is not whether, but when, for how long, and at what cost to the individual that life may be extended.[2]

As a result, clearer development of this distinction is linked with further development of the concept of the state's interest in the preservation of life. Both require for their development consideration by the court of cases that will necessitate a balancing of the conflicting interests involved.

Active Euthanasia versus Passive Euthanasia

On the facts before it in *Saikewicz*, the court approved only the withholding of medical care that would prolong life where the patient gives actual or "proxy" consent to such an omission to act. Because the doctors in that case were requesting permission for a course of behavior that seemed easily classifiable as just "letting nature take its course," the court did not have to face what it called "the sometimes subtle distinction between those situations in which the withholding of extraordinary measures may be viewed as allowing the disease to take its natural course and those in which the same actions may be deemed to have been the cause of death."[2] But many cases that confront medical personnel on a daily basis do require an effort to make that sometimes subtle distinction. Is refusing to treat an easily cured infection because it is clear that a patient suffering from a painful, ultimately terminal disease is better off dead just letting nature take its course? Can such treatment possibly be described as "extraordinary?" Is pulling the plug on a respirator so that someone like Karen Quinlan will die an act of euthanasia or just letting nature take its course? And why should this distinction make a difference? If it was determined by her doctors, her parents, and the Supreme Court of New Jersey that Karen Quinlan would be better off dead at the time of the decision, would it not have been merciful to her to take some step to ensure her speedy death once it was clear the she would continue to linger on once her respirator was unplugged? Where we decide that someone is better off dead—would *want* to be dead—rather than to go on suffering, don't we owe it to that person to end suffering quickly? And if we are unwilling to take responsibility for an act ending suffering, does that fact perhaps show that we are really uncertain about whether the decision is the right decision? Indeed, may we not be encouraging lax standards for decision making by allowing the decision-makers to believe that they are not responsible for the death, that they are only allowing nature to take its course?

The court gives only one clue as to how it is likely to deal with such questions when they are presented in the future. At one point in the opinion, it suggests that the view of these distinctions taken by the medical profession will be highly relevant.

> The current state of medical ethics in this area is expressed by one commentator who states that: "We should not use *extraordinary* means of prolonging life or its semblance when, after careful consideration, consultation and the application of the most well-conceived therapy it becomes apparent that there is no hope for the recovery of the patient. Recovery should not be defined simply as the ability to remain alive; it should mean life without intolerable suffering."
> Our decision in this case is consistent with the current medical ethos in this area.[2]

But it is generally admitted that the medical ethos in this area is muddled and contradictory once one gets beyond the easy facts of a case like *Saikewicz*. As a result, the court will have to struggle with this very difficult distinction, as with the others mentioned, on the basis of the more difficult facts of other cases that will present themselves in the future.

The four issues I have mentioned are not, of course, issues that are of concern to the courts alone. They are of concern as well to the medical community and to society as a whole. They were not invented by the Supreme Judicial Court of Massachusetts, and they would not go away if euthanasia decisions were taken out of the courts and officially placed in the hands of the medical personnel in intensive care units. As a result, we should all feel saddened by the fact that, in the period since the decision in *Saikewicz*, the appellate courts of the Commonwealth have been provided with so little opportunity to continue the work of developing guidelines in this area that was begun in *Saikewicz*.[7] From this point of view, it is somewhat ironic that, of the three *Saikewicz*-type cases that have been decided by the appellate courts of the Commonwealth since *Saikewicz*, one was brought to court largely as an effort to cut off the flow of such cases into the court system thereafter. In that case, *In Re Dinnerstein*,[8] the underlying substantive issue was whether the law would sanction a "no-code" order as being in the best interests of a 67-year-old woman with Alzheimer's disease whose condition was found by the court to be as follows:

> At the present time she is confined to a hospital bed, in an essentially vegetative state, immobile, speechless, unable to swallow without choking, and barely able to cough. Her eyes occasionally open and from time to time appear to fix on or follow an object briefly; otherwise she appears unaware of her environment. She is fed through a naso-gastric tube, intravenous feeding having been abandoned because it came to cause her pain. It is probable that she is experiencing some discomfort from the naso-gastric tube, which can cause irritation, ulceration, and infection in her throat and esophageal tract, and which must be removed from time to time, and that procedure itself causes discomfort. She is catheterized and also, of course, requires bowel

care. Apart from her Alzheimer's disease and paralysis, she suffers from high blood
pressure which is difficult to control; there is risk in lowering it due to the constriction
in an artery leading to a kidney. She has a serious, life-threatening coronary artery
disease, due to arteriosclerosis. Her condition is hopeless, but it is difficult to predict
exactly when she will die. Her life expectancy is no more than a year, but she could go
into cardiac or respiratory arrest at any time. One of these, or another stroke, is most
likely to be the immediate cause of her death.[8]

On these facts, one would have expected the court to authorize a "no-code" order,
although in doing so it probably would have felt forced to face the issues of whether
Saikewicz would allow it to take into consideration the low value of the quality of Mrs.
Dinnerstein's life to her and whether cardiopulmonary resuscitation constituted "extraordinary care." However, the court never reached the substantive issues since it was persuaded by the hospital that brought the action that it should not. In an effort to undercut
the further shaping of euthanasia guidelines by the courts, the hospital had urged the
intermediate appellate court before which the case was argued to take the case out from
under the requirement of judicial authorization that had been laid down by the Supreme
Judicial Court in *Saikewicz* and rule that the "no-code" decision in that case and others
like it could be made by medical personnel and relatives without judicial oversight. The
court so ruled. In the process, it deprived us of the further development of euthanasia
guidelines that might have been provided by an opinion on the merits in the *Dinnerstein*
case itself and in the other cases like it that would have been brought to court in the
future.

In fact, the *Dinnerstein* decision has been responsible for cutting off the flow of
euthanasia cases into the court almost completely. This is because the rationale of the
decision is unclear, and the medical community has felt encouraged to consider it as
making *Saikewicz* a dead letter. Indeed, the decision seems to have been motivated in
large part by a desire to accede to the medical community's unwillingness to comply with
the *Saikewicz* requirement of judicial authorization. At one point the court says of the
Saikewicz decision:

> It... , implicitly, would appear to establish a rule of law that unless... a court
> determination has been obtained, it is the duty of a doctor attending an incompetent
> patient to employ whatever life-saving or life-prolonging treatments the current state of
> the art has put in his hands. *As it cannot be assumed that legal proceedings such as the
> present one will be initiated in respect of more than a small fraction of all terminally ill
> or dying elderly patients*, the *Saikewicz* case, if read to apply to the natural death of a
> terminally ill patient by cardiac or respiratory arrest, would require attempts to resusci-
> tate dying patients in most cases, without exercise of medical judgment, even when that
> course of action could aptly be characterized as a pointless, even cruel, prolongation of
> the act of dying.[8] [Emphasis added]

Hence, the real rationale of the decision seems to be: (1) Doctors refuse to honor legal
requirements to bring euthanasia cases to court in order to obtain legal authorization not
to treat. (2) Without legal authorization not to treat, doctors will automatically continue
to treat even where that is pointless and cruel to the patient. (3) Therefore, the best
interests of the patients require that we drop the legal requirement that such cases be

brought to court before doctors may consider themselves legally authorized not to treat. However, since the court that decided *Dinnerstein* has jurisdiction inferior to that of the Supreme Judicial Court and did not have the power to overrule *Saikewicz*, it was forced in the remainder of its opinion to try to distinguish the kind of case that was involved in *Saikewicz* from the class of cases that it was authorizing doctors to decide on their own without advance court authorization.

I have dealt elsewhere[9] with the problems that I find with the *Dinnerstein* court's attempts to define the class of cases that it purports to take out from under the holding in *Saikewicz*. The point I wish to make here is that I think there is a way in which the *Dinnerstein* facts might be significantly different from those of *Saikewicz* that could be used to justify eliminating certain classes of euthanasia cases from the need for advance court approval. One must remember that the *Saikewicz* requirement of advance court approval was laid down in a case where the question of what the incompetent patient would have wanted for himself under the circumstances was a very close one. Indeed, there was evidence in the record that most competent people, when faced with Joseph Saikewicz's predicament, would opt for chemotherapy, but it was argued that such a choice was a less likely one for Mr. Saikewicz because of his mentally retarded state.

> Evidence that most people choose to accept the rigors of chemotherapy has no direct bearing on the likely choice that Joseph Saikewicz would have made. Unlike most people, Saikewicz had no capacity to understand his present situation or his prognosis. The guardian ad litem gave expression to this important distinction in coming to grips with this "most troubling aspect" of withholding treatment from Saikewicz: "If he is treated with toxic drugs he will be involuntarily immersed in a state of painful suffering the reason for which he will never understand. Patients who request treatment know the risks involved and can appreciate the painful side-effects when they arrive. They know the reason for the pain and their hope makes it tolerable."[2]

As a result, the court saw itself as faced with a difficult and delicate problem of determining a close question of "substituted judgment" as to what this particular patient would want for himself, and it was a natural concomitant of that decision to rule that where such questions of "substituted judgment" (rather than just "medical questions") must be decided in the future, they must be decided by a court. However, it might be argued on facts such as those in *Dinnerstein* that *no one* in such a state would want to be kept alive. If that were found as a fact by the court, it would seem to clear the way for the court to authorize doctors to make "no-code" decisions in the future on substantially the same medical facts. The court would have made a substituted judgment determination in advance for a whole class of cases, and the only remaining question in each such case would be the medical one of whether or not the requisite medical facts were present.

I understand Professor George Annas to be making essentially this point in a recent article in which he attempts to reconcile *Saikewicz* and *Quinlan*. He sees *Quinlan* as a case where the substituted judgment question was open and shut—no one would want to continue in Karen Quinlan's type of hopeless vegetative state. Hence the only remaining question was the medical one of whether her diagnosis and prognosis were accurate, and the *Quinlan* referral to a hospital ethics committee was solely for the purpose of having the decision of this medical question second-guessed because of its grave consequences.

He sees *Saikewicz* as consistent with *Quinlan* because it should be interpreted as requiring advance court approval only in cases where the substituted judgment question is not open and shut. Although I think that this is an important and ingenious distinction that may well be adopted by the courts in the future, I am doubtful regarding what I take to be his suggestion that the distinction was before the minds of the two courts when they rendered their decisions. This issue, like the others I have mentioned in this chapter, requires for its resolution future cases that present to the courts facts that make careful resolution of it necessary and proper. And even if it is decided in the way that has been suggested, the courts will still need a rich mix of cases to provide them with a basis for marking out clearly the various sets of medical fact situations that present cases that offer no contest as to the substituted judgment question involved. Only in that way will guidelines be provided to the medical community that they can apply to cases without fear of legal error.

Clearly, the issues that must be faced in developing euthanasia decision guidelines are difficult and disturbing. They seem to raise unanswerable philosophical problems in a context that imposes awesome responsibility and that forces us to face unsettling facts about our own mortality. In light of this, it is not surprising that the medical community has been slow to take on the responsibility for formulating such guidelines for itself. Moreover, individual judges in Massachusetts have made clear that they are less than delighted with the notion that they should be forced to take on this task. However, the job must be done, and it must be done in a fashion that develops and refines principles, based on shared societal values, that are capable of producing answers when applied to the facts of real cases. There is no better way to do this, it seems to me, than through the traditional common-law method of developing and refining the principles as a by-product of the process of actually deciding the individual cases. In default of the availability of such a mechanism within the medical community, the courts have made themselves available for this purpose. Instead of fighting the *Saikewicz* decision, it would seem to me a more sensible course for the medical community to work with the courts in this enterprise by bringing euthanasia cases to the courts in a cooperative effort that can produce principles that are informed by both societal values and medical expertise. Without such cooperation, it is too likely that the near future will provide more opportunities for doctors and nurses to respond to newspaper headlines by saying, "Oh God, somebody got caught."

ACKNOWLEDGMENT

The author takes this opportunity to thank his former student and research assistant, David Zizik, Esq., of the Rhode Island Bar, for valuable assistance rendered in the preparation of this chapter.

REFERENCES AND NOTES

1. Daniels, T., The nurse's tale. *New York Magazine*, p. 37, April 30 (1979).
2. *Superintendent of Belchertown School v. Saikewicz*, 1977 Mass. Adv. Sh. 2461, 370 N.E.2d 417 (1977).

3. Relman, A. S., The Saikewicz decision: Judges as physicians, *N. Engl. J. Med.* **298**:508 (1978).
4. Relman, A. S., The Saikewicz decision: A medical viewpoint, *Am. J. L. Med.* **4**:233 (1978).
5. Buchanan, A., Medical paternalism or legal imperialism: Not the only alternatives for handling Saikewicz-type cases, *Am. J. L. Med.* **5**:97 (1979).
6. *Raleigh Fitkin-Paul Morgan Memorial Hospital v. Anderson*, 42 N.J. 421, 201 A.2d 537, *cert. denied*, 377 U.S. 985 (1964).
7. See *In re Custody of a Minor*, 1978 Mass. Adv. Sh. 2002, 379 N.E.2d 1053 (state may intervene and administer chemotherapy to a minor child suffering from acute lymphocytic leukemia, where parents of said child refuse to administer the treatment, and where the evidence shows that it is the only type of medical treatment that may save the child's life by possibly effecting a cure for his leukemic condition); *In the matter of Shirley Dinnerstein*, 1978 Mass. App. Adv. Sh. 736, 380 N.E.2d 134 (for a discussion of the *Dinnerstein* case see the text accompanying note 8, *infra*); *Lane v. Candura*, 1977 Mass. App. Adv. Sh. 588, 376 N.E.2d 1232 (competent adult patient has the right, subject to the countervailing state interests, recognized by the Supreme Judicial Court in *Saikewicz* to refuse to consent to the amputation of her gangrenous leg, although such procedure is necessary to save her life).
   In the following cases, the probate courts allowed petitions authorizing a guardian to withhold his or her consent to further treatment of the particular patient-ward: *In re Keefe*, Civil No. 513889 (Mass. Middlesex Cty. P. Ct., Oct. 10, 1978); *Damiano v. Haller*, Civil No. 78-F0629-Gl (Mass. Norfolk City. P. Ct., March 11, 1978); *In re Piatrowicz*, Civil No. 1948 (Mass. Essex Cty. P. Ct., Dec. 23, 1977).
   See also *In re Perry*, Civil No. 513889 (Mass. Middlesex Cty. P. Ct., Oct. 10, 1978) (court refused to appoint guardian for the purpose of consenting to blood transfusions of incompetent adult Jehovah's Witness, even though the same was necessary to save her life, where said incompetent had, while competent, expressly stated her disapproval of such treatments and had refused to give consent thereto); *In re Norton*, Civil No. 124,929 (Mass Plymouth Cty. P. Ct., Jan. 6, 1978); *In re Huber*, Civil No. 1630 (Mass Suffolk Cty, P. Ct., Dec. 12, 1977) (both of the latter cases involved the dismissal of petitions for guardianship for want of jurisdiction on the ground that the proposed ward was legally dead within the "brain death" criteria set forth by the Supreme Judicial Court in *Commonwealth v. Golston*, 373 Mass. 249, 366 N.E.2d 744 (1977); *In re Moulton*, Civil No. 177645 (Mass. Hampden Cty. P. Ct., Dec. 7, 1978); *In re Basmajian*, Civil No. CA226 (Mass. Hampshire Cty. P. Ct., Nov. 11, 1978) (in each of the latter two cases, the court granted the guardian *ad litem*'s motion to dismiss on the ground that the facts of the case presented no "life-saving or life-prolonging treatment alternative" and that therefore, the question whether to remove the patient from a respirator "is a decision which should be made by the [patient's] attending physician." These conclusions apparently rested upon the authority of the *Dinnerstein* case, see text accompanying note 8, *infra*. In each case, medical evidence was admitted that supported the conclusion that no possibility existed that the patient would ever recover from his or her comatose condition).
   In *Hall v. Meyers*, Civil No. 32245 (Mass. Suffolk Sup. Ct., Jan. 8, 1979), an inmate (Meyers) of the Massachusetts correctional system who was suffering from severe uremia refused to consent to kidney dialysis, without which his chances for survival were poor. The Department of Corrections brought a declaratory judgment action seeking authorization to administer dialysis without Meyers's consent, including the use of force if necessary. The judge rendered judgment giving the department such authorization. The case is being appealed to the Supreme Judicial Court, SJC No. 1763.
   There are a number of undocumented cases (perhaps 40–50) that have been tried in the Boston juvenile courts involving minors and various "right-to-die" issues, as well as at least 12 probate court cases, also undocumented, that involve petitions for the appointment of a guardian with authority to consent to treatment on behalf of an allegedly incompetent patient.
8. *In re Dinnerstein*, 1978 Mass. App. Adv. Sh. 736, 380 N.E.2d 134 (1978).
9. Baron, C. H., Medical paternalism and the rule of law: A reply to Dr. Relman, *Am. J. L. Med.* **4**:337 (1979).
10. Annas, G. J., Reconciling Quinlan and Saikewicz: Decision making for the terminally ill incompetent, *Am. J. L. Med.* **4**:367 (1979).

14

DOCTORS, DECISIONS, AND THE LAW

ARNOLD S. RELMAN
Harvard Medical School
Boston, Massachusetts 02115

Editor, New England Journal of Medicine
Boston, Massachusetts 02115

So much has been written about the *Saikewicz* decision[1] that any further commentary needs justification. My reasons for continuing the discussion are twofold. First, there is still no consensus on what exactly the Massachusetts Supreme Judicial Court really meant to say. More cases will need to be tried before we have a clear understanding of what the law in this state expects families and physicians to do when life-and-death decisions must be made for incompetent patients with irreversible illness. Meanwhile, it should be useful to delineate the limits of our uncertainty. Second, I believe that in addressing this issue the court took hold of a basic moral question that goes to the heart of one of the most important problems in medical ethics. Even if there had been no *Saikewicz* case, we would still be wrestling with the underlying issue. I therefore want to direct my remarks briefly to the decision itself and then I wish to focus on the ethical questions it raises.

Let us consider first the essential facts of the case. Mr. Saikewicz, a mentally retarded man of 67 with an IQ of about 10, developed acute myelomonoblastic leukemia, a disease that untreated is usually fatal within six months, or a year at most. Chemotherapy offered him about a 50% chance of a temporary remission, which might have added six months or perhaps a year of life, at a cost of modest to moderate physical discomfort. He probably would have had to be restrained to receive the treatment and would have experienced considerable psychological distress as a result. With a similar illness in a competent patient of the same age most hematologists (but by no means all) would have recommended treatment and most (but not all) patients would have accepted that recommendation. However, after weighing the arguments pro and con, the court decided that Mr. Saikewicz should *not* be treated. Applying the doctrine of so-called "substituted judgment," the court took pains to make clear that it was deciding not what a "reasonable person" would have wanted but rather what Mr. Saikewicz himself would have chosen had he been able to make this choice himself.

I have never understood the logic of invoking "substituted judgment" in this case. How anyone could presume to know what a man born with an IQ of 10 would choose under such circumstances, if he had not been mentally defective, is totally beyond my comprehension. Nevertheless, the court opted for withholding treatment not because it was applying any external standard but because it believed it was acting as Mr. Saikewicz's agent—a responsibility that it believes should not be delegated to any other element of society: neither the family nor the physicians in the case nor an ethics committee.

This decision by the Supreme Judicial Court, which upheld the lower court, was handed down over a year after Mr. Saikewicz had died. It was obviously of little importance to him, a point worth recalling when the virtues of the judicial system as a means of making "on-line" decisions for incompetent patients are being extolled. However, my concern here is for the generalizations the court drew from this case. The court said first that, except under special circumstances, any *competent* patient with an irreversible illness has the right to refuse life-prolonging treatment. Second, the court said that *incompetent* patients should have the same right, but it could be exercised only by the judiciary. Third, the court set down a procedure for the future handling of such cases that seemed to require routine prior approval by a probate judge after a full adversarial hearing. The court seemed to be saying that whenever an *incompetent* patient was in a situation in which a *competent* patient might opt for the withholding of a life-prolonging treatment, the decision whether or not to withhold such treatment could only be made by a judge.

The language of the decision was sweeping, at least to physicians, and imprecise. Most legal counselors in this state advised their clients that the court probably meant what it appeared to mean, and so the withholding of most forms of life-prolonging therapy for incompetent patients with irreversible illnesses would henceforth have to be approved in a probate court. This naturally caused great consternation in hospitals throughout the state. There probably were some overreactions, but the fact remains that lawyers were divided in their view of the *Saikewicz* decision. Many people in both the legal and medical profession simply did not know what to think or what to do.

The *Dinnerstein* decision, handed down the following year by the Massachusetts Appeals Court,[2] helped to clarify the situation by stating that physicians do not need prior judicial approval to write "do not resuscitate" orders for incompetent patients if such patients are "terminally ill" *and* have no chance of being restored, even temporarily, to "normal, integrated, functioning, cognitive existence." The *Dinnerstein* court did not define "terminally ill" but implied that a "terminal" illness is one that would inevitably kill the patient within a year.

While physicians in Massachusetts are gratified by this clarification and feel less constrained in carrying out their professional responsibilities than they did before the *Dinnerstein* decision, many questions are still unresolved. Uncertainty remains as to exactly how the "terminal illness" decision is to be made and by whom. Mrs. Dinnerstein herself lived more than a year after legal action was begun, thus refuting the expert testimony at the hearing. As every experienced physician knows, it is difficult to predict the duration of "terminal" illnesses with any precision. Patients sometimes live for months or even years after experts have predicted their demise. How will the courts deal with this problem? And does the *Dinnerstein* decision apply to *any* form of life-prolonging treatment or only to cardiopulmonary resuscitation? Furthermore, what are the limits of the court's definition of "normal, integrated, functioning, cognitive existence"? How much "normal, integrated, functioning, cognitive" potential does the patient need to have before the court would require prior judicial review of any decision to withhold treatment? These are just some of the questions that await resolution in the aftermath of *Dinnerstein*.

I think it obvious that we have not seen the last of the litigation flowing from the *Saikewicz* decision. Other cases will have to be tried, in Massachusetts and elsewhere, before there is any clear judicial consensus on the legal issues. Meanwhile, however, vigorous discussion continues among lawyers, doctors, ethicists, and others interested in the underlying moral dilemma posed by the conditions of the incompetent patient for whom critical life-or-death decisions must be made.[3,4,5]

In my opinion, the courts have no business intruding upon what are essentially private decisions that can best be made by the next of kin or by agents previously appointed by the patient. The goal should be to do what the patient would have wanted, and I do not believe that the courts are in a better position to make that judgment than families or agents in consultation with physicians. Crucial therapeutic decisions for terminally ill incompetent patients are at best difficult to make. Proper management requires the expertise of a physician intimately familiar with the clinical situation, a caring family, or a legally appointed agent who can represent the patient's wishes in discussions with the physician, and often the counsel of others who know the patient very well. The decision-making machinery must remain in close touch with the clinical situation, constantly ready to deal with changing conditions. The judicial process, whatever may be its virtues as a mechanism for the evaluation of evidence in retrospect, is simply not designed to handle decisions on a day-to-day basis.

Such arguments notwithstanding, some lawyers, philosophers, and ethicists believe there is something fundamentally wrong about doctors and families making life-or-death decisions for incompetent patients. They propose instead either a requirement for legal review, in agreement with the *Saikewicz* decision,[4] or some kind of mandatory review by an ethics committee.[6] The basis for this position seems to be the view that doctors should be responsible only for "technical" advice, and that the choice between alternatives is a "moral" question that must be made for the incompetent patient by some properly constituted, impartial body representing "society." Baron thinks only the courts can make such moral judgments,[4] but Buchanan calls this notion "legal imperialism" and suggests that an ethics committee would be a better protector of the patient's rights.[6] Buchanan, however, recognizes the practical limitations of *prior* review and suggests instead that the ethics committee should act on a *post hoc* basis.

I see no justification for this kind of meddling in private affairs. When incompetent patients have no next of kin, courts must appoint a legal agent or guardian to look after the patients' interests. But when there is a family, surely it makes sense to leave the matter in their hands, with the help of a physician. I suppose there are instances in which families and physicians have conspired together against the best interests of the patient. But if so, there are already remedies at hand, namely, the courts and the ethical review bodies of the medical profession. These remedies function only in retrospect, of course, and need to be activated by a complaint. They may not be as vigilant, therefore, as some might wish. However, I believe that deliberation and restraint are only proper in such matters. After all, we are dealing with what must be very rare cases of abuse, in an area of human relations that is terribly private and sensitive.

Some believe that "living wills" offer a solution to this dilemma. Such documents, now legalized in about 10 states,[7] offer the legally competent person an opportunity to state in advance how he would wish to be treated in the event a terminal illness rendered him incompetent. While I subscribe to the basic intent of such legislation, i.e., to increase the autonomy of patients, I have grave doubts about the usefulness of the legislation that has been passed to accomplish this end. Most of these legal statements signed by the patient are so vague that they would be of little real help to physicians in actual clinical situations. No two situations are exactly alike. Each incompetent patient with an apparently terminal illness presents unique problems. The day-to-day decisions that must be made can rarely be anticipated in sufficient detail to write an adequate set of instructions to the physician. No "living will" can substitute for face-to-face discussions between the physician and the patient's agent. What often happens, however, is that the family—which *should* be acting as the patient's agent—is not able to cope with the crisis, is unwilling to take responsibility, or simply cannot agree on what should be done. Anticipating such problems, many people might want to write a "living will" that designates a particular agent who can be trusted to act responsibily and sensitively in accordance with the wishes of the person drawing up the will. Such legislation has been introduced into the Michigan legislature, and I think it has much to recommend it.[7]

In the final analysis, there must be some trust. The patient must be able to trust a friend or a relative to speak for him when he is unable to express his wishes. And the patient's agent must trust the physician in charge of the case to give humane, medically sound advice. Physicians must of course be held accountable, under the law and by their professional peers, for their technical competence and their ethical integrity—just as families and agents must be held legally accountable for what they do in behalf of the incompetent patient. Furthermore, it is important that all medical considerations and all the important decisions made for the patient be fully documented in the medical record. Whenever there is any doubt about medical decisions, medical consultants should be called. All major decisions involving the withholding of life-sustaining treatment should probably have the written concurrence of at least one other physician, simply to document in the record that there is a sound medical basis for the decision, as well as the informed consent of the patient's agent or next of kin. Ethics committees can also be helpful when families or physicians feel the need to consult them. But what we should avoid, I am firmly convinced, is the routine involvement of the judicial machinery of the state, or the routine investigations of bureaucratic committees into these private matters. The care of the dying patient has always been the responsibility of families and physicians, and I see no reason to change that tradition.

REFERENCES

1. *Superintendent of Belchertown State School v. Saikewicz*, 1977 Mass. Adv. Sh. 2461, 370 N.E.2d 417 (1977).
2. *Dinnerstein*, 1978 Mass. App. Adv. Sh. 736.
3. Relman, A. S., The Saikewicz decision: A medical viewpoint, *Am. J. L. Med.* **4**:233 (1978).
4. Baron, C. H., Medical paternalism and the rule of law: A reply to Dr. Relman, *Am. J. L. Med.* **4**:337 (1979).
5. Annas, G. J., Reconciling Quinlan and Saikewicz: Decision making for the terminally ill incompetent, *Am. J. L. Med.* **4**:367 (1979).
6. Buchanan, A., Medical paternalism or legal imperialism: Not the only alternatives for handling Saikewicz-type cases, *Am. J. L. Med.* **5**:97 (1979).
7. Relman, A. S., Michigan's sensible "living will," *N. Engl. J. Med.* **300**:1270 (1979).

15

CONVERSATION WITH SILENT PATIENTS

ROBERT A. BURT
Yale Law School
New Haven, Connecticut 06520

In March 1976 the New Jersey Supreme Court ruled that Karen Quinlan's father could be appointed her guardian in order to direct termination of her treatment because, although Karen was an adult, she was comatose and her physicians believed there was "no reasonable possibility" she would ever regain consciousness.[1] Mr. Quinlan had sought advance judicial approval for this course; the court left unclear whether any judicial participation in such decision would be necessary for future cases or whether a person's comatose state and the physicians' gloomy prognosis would themselves justify such decision by immediate family without fear of subsequent civil or criminal liability.* In July 1976 the Massachusetts Supreme Judicial Court ruled that such a decision to withhold treatment should be made directly by a judge in a declaratory proceeding rather than by some judicially appointed guardian.[2] In this case, Joseph Saikewicz was a profoundly retarded, virtually uncommunicative 66-year-old man diagnosed as suffering from an incurable leukemia. His physicians sought and obtained judicial authorization to withhold an apparently painful course of chemotherapy that might have extended his life for six months to a year.

These two cases do not raise new issues for the medical profession; the question of others' decision making for wholly uncommunicative patients has always attended medical practice. But *Quinlan* and *Saikewicz* do provide strikingly new answers to this question. In the past, it seems that most physicians assumed authority themselves to withhold

---

*The court was eager to remove any threat of subsequent liability from the immediate decision-makers in order (as they said regarding physicians) "to free [them] ... from possible contamination by self-interest or self-protection concerns which would inhibit their independent medical judgments for the well-being of their dying patients." The court was equally eager to keep the judges away from issuing declaratory judgments in specific cases on the grounds that "a practice of applying to a court to confirm such [treatment termination] decisions would generally be inappropriate ... [and] impossibly cumbersome."

treatment from such persons, although no systematic data are available. In some instances, family members would be directly consulted; in others, perhaps most instances, physicians would act alone. There are, however, virtually no recorded criminal prosecutions anywhere in this country against physicians for such actions or against family members acting in collaboration with physicians.[3] Thus physicians have had practical but not formal immunity from legal sanction in these matters.

This implicit resolution no longer seems adequate to many physicians and laymen. Some directly challenge it on grounds that physicians are abusing their authority and must be brought under formal legal control. Others, particularly physicians, respond to this challenge by claiming that their traditional implicit immunity from legal sanction in these matters should now be made explicit.[4] For a complex of reasons, physicians' traditional modes of responding to uncommunicative patients no longer seem adequately justified; the courts' willingness in *Quinlan* and *Saikewicz* to spell out justifications is a response to this perception.

The belief in the existence of some unambivalently correct, some indisputably justified response from physicians to these uncommunicative patients is a fallacy. Belief in this fallacy, however, has led physicians and family members into court, at least in New Jersey and Massachusetts, and has equally led courts in those states to offer absolution to these supplicants. This course will have destructive consequences. The protection that the courts appear to offer to the uncommunicative patients, and implicitly the comfort offered to their physicians and family, will be a sham. In order to protect both these patients and the psychological integrity of everyone else affected by them courts should refuse to offer clear, advance immunity from any prospect of subsequent liability for physicians and family who contemplate withholding treatment from silent patients.

Detailed consideration of the facts in both the *Quinlan* and *Saikewicz* cases and the tangled efforts of the participants to come to terms with those facts will illuminate this prescription. In Karen Quinlan's situation, confusion came from her anomalous status as neither recognizably alive nor dead. At a party with friends in April 1975, she suddenly fell unconscious and stopped breathing for at least two 15-minute periods. The causes for this were never identified. Karen was taken to a hospital, where a series of emergency measures were employed, including a mechanical respirator to assist her breathing. It soon appeared to the physicians that her initial breathing cessation had led to destruction of those portions of her brain cortex that regulate consciousness but that the subcortical or "vegetative" brain segments remained substantially intact. This lower brain controls autonomic nervous system functions such as heartbeat, breathing, and reflexes. Thus, by application of both the traditional cardiac/respiratory and the newer brain-death standards,[5] Karen was clearly alive. She was not alive, however, by the conventional standards of communicative sociability or conscious self-reflection and it seemed highly probable to the physicians that she would never regain these capacities, that she was, at best, permanently comatose (in "a persistent vegetative state," as one expert testified) because the presumably destroyed cerebral cells could never regenerate.

But all physicians acknowledged that, to some degree at least, her prognosis was uncertain both because the origin and extent of brain damage could not be adequately identified and because data on long-term comatose persons have not yet been collected in a manner that would satisfy rigorous scientific standards of objective observation.[6] Almost all comatose persons who appear like Karen, as described in medical literature or within the experience available to physicians testifying in her case, die within six months or so after falling into unconsciousness. But medical professionals commonly understand, although they rarely advertise this publicly, that many of these persons die because their treating physicians abandon hope for regained consciousness and withhold antibiotic treatment from the recurring bronchial infections that regularly afflict comatose persons. Anecdotal reports exist, however, of persons comatose even for years who suddenly regain consciousness. Thus the attending physicians were guarded in making clear-cut predictions for Karen.

In addition to her prognosis, there was one other critical uncertainty that affected everyone's perception of Karen: that is, whether she "felt pain" in her coma. Karen's family clearly believed she experienced considerable recurrent pain. They derived this belief quite reasonably from her periodic groans, grimaces, and accompanying increased muscular tension and perspiration. These episodes could be stimulated by someone's pinching her—one of the tests invoked to assay brain death. But, notwithstanding the commonsense conclusion that she felt pain because she groaned when pinched, it is reasonable to believe that her responses were simply reflexive actions, empty shells without even rudimentary awareness from Karen's perspective. This was the opinion of the testifying physicians with extensive experience in treating comatose persons.[7]

The critical importance of these differing opinions regarding Karen's pain perception is underscored for our analytic purposes by comparing Joseph Saikewicz's situation. Unlike Karen, he had never participated in the conventional life of social communication. Joseph was 66 years old when his leukemia was discovered in a routinely administered blood test. He had been resident in a state institution for the retarded since age 14; before that, he had lived with his parents, who, according to institutional records, had obtained some formal diagnosis of Joseph as retarded when he was only eight months old and "had numerous contacts with various authorities during his childhood because of supposed neglect, abuse and ill health." A social worker's report gives the flavor of social interactive perceptions of Saikewicz a few months before his leukemia was discovered:

> [He] was in his recreational room when I met him. He was engaged in observing others watching t.v. while he uttered loud declaration [sic] sentences, which while emphatic were completely unintelligible to me. He shook my hand when I presented it, but otherwise did not relate to me or to what I said.[8]

In court testimony he was described as "profoundly mentally retarded" with a measured IQ of 10. The court found that "in the course of treatment for various medical conditions during [his] residency at the school, he had been unable to respond intelligibly to inquiries such as whether he was experiencing pain."

Even more than for Karen Quinlan, the question of Saikewicz's pain perception was the central explicit issue in the deliberations on his treatment. Physicians testified that, although his leukemia had no known cure, they knew of no mentally normal person with this illness who had declined the chemotherapy and its life-prolonging possibility. Nonetheless, they recommended withholding this treatment from Joseph both because it brought painful nausea and other discomfort that could only partly be relieved by additional medications and because the treatment required intravenous administration that appeared possible for Joseph only through physical bed restraints. The physicians thus concluded that the pain induced by the treatment and the enforced immobility would terrify Saikewicz.* This pain would apparently only directly accompany the administration of the chemotherapy during approximately five to seven days of each month that Saikewicz lived.

The ascription of experienced pain to Joseph Saikewicz and to Karen Quinlan was critical in the courts' resolutions because both courts insisted that the treatment decisions could only be reached through an imagined construction of Joseph's or Karen's individual perceptions and wishes. In its opinion, the Massachusetts Supreme Judicial Court ruled that the trial judge must choose either for or against the chemotherapy only by considering "the singular situation viewed from the unique perspective of the [incompetent] person. . . ." The New Jersey Supreme Court held that Karen's father had "no parental constitutional right that would entitle him to a grant of relief"; he had standing only to claim Karen's right to control her treatment, to "render [his] best judgment . . . as to whether she would exercise it in these circumstances." But no one could read the intentions of these utterly silent persons without losing the conventional grasp on distinctions between self and other, without utterly confounding his pain with theirs. The courts insisted that others attempt to fix this delineation between themselves and Karen or Joseph, but they failed to appreciate both the magnitude of this difficulty and the consequences of pretending otherwise.

The New Jersey Supreme Court itself attempted to imagine what Karen might decide "if [she] were herself miraculously lucid for an interval (not altering the existing prognosis of the condition to which she would soon return) and perceptive of her irreversible condition. . . ." It concluded that in this lucid moment she "could effectively decide upon discontinuation of the life-support apparatus. . . ." But many people who recognize that they suffer from an imminently terminal illness currently seek techniques such as body freezing to preserve them in suspended animation with the hope that some future cure might be found for them. Why would Karen reject this course when her coma had apparently achieved what these people so fervently pursue?

The court was wrong that Karen would be "perceptive of her irreversible condition"; she would perceive this only if she misread the physicians' testimony as the court did. The physicians testified that her prospects for regaining consciousness were remote but that they could not conclusively rule out the possibilities of spontaneous cure or discovery of some new treatment technique to this end. These possibilities did not depend on some "miracle" from physicians' perspective but rather could not be excluded on the basis of their science's understanding of her situation.

---

*Beyond this, the physicians suggested that the chemotherapy itself might even shorten rather than prolong his life because his enforced total immobility would increase the likelihood and complicate the treatment of respiratory infection, which frequently accompanies this chemotherapy for all patients.

But if Karen's ultimate recovery was improbable, would she take account of the pain she suffered while waiting for this unlikely event? The court also addressed this question and again misread the testimony regarding her condition. It suggested that maintaining her life "only to vegetate a few measurable months with no realistic possibility of returning to any semblance of cognitive or sapient life" was to "compel Karen to endure the unendurable," and continued:

> We perceive no thread of logic distinguishing between such a choice on Karen's part and a similar choice which, under the evidence in this case, could be made by a competent patient terminally ill, riddled by cancer and suffering great pain. . . .

But the physicians had testified that Karen most likely perceived no pain while in her coma, even though it might seem otherwise to those who witnessed her reflexive grimaces and groans.

I am not suggesting that it would be unreasonable for Karen to decide, if momentarily lucid, that her treatment and life should be ended. But if she neither experienced pain in her coma nor would have any memory of suffering if she regained consciousness, avoiding pain would not be a reasonable ground for such decision. Her continued coma did create pain that she, if lucid, could feel. But this pain, the only indisputable pain, was suffered by those around her—by family and friends, physicians and nurses—and not by her. If Karen were determined to be utterly self-centered in her deliberations, disdainful of the consequences to others in pursuit only of her most selfishly conceived interests, continuation of the life-support apparatus would be the most rational course for her. In other words, if she determined to think like the economists' stereotyped rational calculator, the embodiment of self-centered individualism,[9] she would choose continued treatment. But if Karen vowed to act unselfishly—to conceive herself foremost as a loving daughter, sister, and friend—she could reasonably conclude that the remote possibility of her recovery was not worth the suffering that safeguarding this eventuality imposed on others.

Karen's father could not view the treatment decision through her eyes, as the court directed, without deciding whether Karen would want to view her decision through his eyes. He could not avoid asking whether she was narrowly self-centered and wished to be seen as such or whether she would want to put (and be seen to put) alleviation of other's suffering ahead of her more narrowly conceived interests. He could not, that is, sustain the clear-cut delineation between self and other that the court enjoined when it rejected his claims as parent to decide for Karen and instead described his role as guardian to think as if he were she. Neither he nor anyone else as guardian could sustain this delineation without, paradoxically enough, losing the capacity to imagine how Karen might define her individual identity.

The difficulties for anyone attempting to comply with this same directive are even more patent in Joseph Saikewicz's case. Karen at least had once been part of ordinary communal life; Joseph never had a conventionally recognizable identity. The Massachusetts court's effort to identify a treatment decision "in . . . step with the values and desires of the affected individual" must founder against this reality. If Joseph did feel considerable pain and even terror during the chemotherapy administration, it is not at all clear that he would remember these feelings or dread their recurrence. If he had no capacity to recall or foretell but conceived himself as living only in each moment for itself, how then might he evaluate one week of terror during each month of life? To conclude that Joseph would surrender three weeks of his customary painless life in order to avoid one week of extraordinary pain, a judge must assume that Joseph shares some conventional attitudes toward past and future. Why should this judge then not assume that Joseph shares the conventional attitude that leads all "mentally normal" leukemia sufferers to opt for the chemotherapy in question? Although the trial judge decided otherwise for Joseph, for himself he stated, "I'd rather take the treatment than just take the chance of dying tomorrow or next week."

The difficulties that everyone inevitably confronted in sustaining conventional delineations of self and other in addressing both Joseph and Karen thus could not be wished away by an adamantly asserted categorical imperative. These difficulties instead had the following consequences: (1) others' ascription of vast, even omnipotent, power to the apparently impotent person; (2) others' consequent unwillingness to engage directly in a test of strength; (3) others' retreat toward greater impotence and denial of any transactional participation, which they mask by assertions of exclusive choice-making capacity. The participants lost their intrapsychic balance, their grasp on conventionally recognizable reality, because of their inability to find contrapuntal pressure from a conventionally presented "other" acknowledged as such in Karen Quinlan or Joseph Saikewicz.

This psychology explains one of the most extraordinary aspects of Karen Quinlan's case: that she remains alive though still comatose as I am writing almost four years after she was stricken and three years after the court in effect authorized withholding the treatment that was allegedly critical to sustain her. The Quinlan family and both the trial and appellate courts assumed that Karen would die without the assistance of a mechanical respirator.[10] They were mistaken, but this mistake was not inevitable. The trial witness generally acknowledged as the most experienced and expert physician available in the treatment of comatose persons like Karen had explicitly testified that she did not need the respirator to sustain her life. He stated that while examining her to prepare his testimony he had removed her from the respirator for several minutes—long enough to confirm his belief in her potential capacity for independently sustained respiration.[11] The other testifying physicians disagreed with this prognosis, however, and both the trial judge and the supreme court chose to believe these physicians.[12]

After the supreme court ruling, the treating physicians removed Karen from the respirator by gradual steps and thus confirmed that their testimony had been erroneous. I do not believe that their testimony had been consciously deceptive or that the courts chose consciously to deceive themselves in accepting that disputed testimony. It is, however, striking that this one question was critical to everyone's conception of the central issue in the case, that it was capable of more extensive and rigorous confirmation than had been obtained before the trial, and that this confirmation was not obtained, notwithstanding the clear conflict in the evidence offered. The result at least suggests that the family, treating physicians, attorneys, and judges all tacitly agreed to debate the question of Karen's continued life on grounds that would not address her actual power to sustain life so that, if they decided on termination, she would remain more powerful than they. Only one testifying physician broke this tacit agreement, by actually taking the measure of Karen's independent power, and his testimony was disregarded.

I would speculate further that the underlying reason for this tacit agreement was the reluctance of the family, treating physicians, attorneys, and judges to be drawn into direct conflict with Karen, to test whose power was in fact greater. This motive appears to underlie the later report published of the reasons offered by Karen's principal treating physician for his initial unwillingness to accede to the family's request to remove the respirator and his subsequent gradual weaning her from it after the supreme court decision:

> In general, what seemed to happen to Dr. Morse was a concern that turning off the respirator for Karen Ann might result in a rather gruesome death scene. Turning if off, he felt, was not going to result in a three- or four-minute ordeal—it might be prolonged. And he said he was worried about the feelings and reactions of some of the family, especially the younger ones. He thought it would be unpleasant, and that they might not soon forget it. [13]

It was, however, not clear why Dr. Morse assumed that Karen's family, "especially the younger ones," would be gathered around her hospital bed when he chose to turn off the respirator, particularly if the event might be as "gruesome" as he anticipated. The possibility that he might move to protect the family from this experience and bear the burden of this witness alone did not seem an attractive alternative to Dr. Morse, at least in this account by another physician of a conversation with him.

Unwillingness to enter into a direct struggle with an apparently dying person was an explicit and, ultimately, the most clearly dispositive reason offered by Joseph Saikewicz's physicians for their decision to withhold treatment from him. The transcript of the trial proceeding makes this clear. [14] At the end of this proceeding, the trial judge determined that chemotherapy should be withheld. But virtually moments before he reached this determination—in the typed version, on page 43 of the total transcript length of 45 pages—the judge stated his contrary inclination. After hearing the formal testimony, the judge worked toward his conclusion in this way:

THE COURT: There is evidence that chemotherapy treatment is apparently the only treatment, but by giving it to him he may have some discomfiture at the time of the treatment that may prolong his life.

DR. DAVIS: That is one thing and if they don't give it to him at all, then he may die in a matter of days or weeks.

THE COURT: That is the choice I have to make.

DR. DAVIS: That is it. I don't know. I don't have that deep knowledge.

THE COURT: I am inclined to give treatment.

DR. JONES: One thing that concerns me is the question about his ability to cooperate. I think it's been made clear that he doesn't have the capability to understand the treatment and he may or may not be cooperative, therefore greatly complicating the treatment process. . . . That was to be weighed, whether [the treatment] could be administered.

THE COURT: Dr. Davis, do you agree?

DR. DAVIS: I think it's going to be virtually impossible to carry out the treatment in the proper way without having problems. You have to see him. When you approach him in the hospital, he flails at you and there is no way of communicating with him and he is quite strong; so he will have to be restrained and that increases the chances of pneumonia, to restrain him if he can't be up and around.

MR. MELNICK [Court-appointed guardian for Saikewicz, who concurred in the physicians' determination to withhold treatment]: With no treatment he may live longer; with treatment, the treatment itself may terminate his life sooner. There is some risk because of the toxic nature of the treatment, so in effect, by ordering the treatment there is a possibility that you may shorten his life and there is a chance that you may be prolonging it.

THE COURT: Maybe I should change my judgment.

DR. DAVIS: One other factor. Though we will get a remission, we are not through at that point. He'd have to be under medical care weekly and continue treatment and he may be in the hospital for four to five weeks initially and will have to be coming back on a regular basis. That enters the picture.

MR. ROGERS [Attorney for the state institution]: The issue boils down to, as Dr. Davis said, the quality of life now and when he goes through it and he certainly will suffer. The low probility [sic] that he will go into remission has to be measured against any person's life and to gamble for success which he can't do for himself.

THE COURT: Do I have to form a written judgment?

MR. ROGERS: Yes, I will draft it.

THE COURT: After a full hearing with medical specialists and doctors being present and their testimony being taken, the Court determines and adjudges that chemotherapy treatment should not be given at this time.

This proceeding occurred on May 13; the Massachusetts Supreme Judicial Court affirmed the trial court's judgment on July 9; and Saikewicz died on September 4, 1976. The decision to withhold chemotherapy during this time may have saved Saikewicz from needless pain and perhaps even quicker death. But this decision was reached with no effort to test a central factual premise on which it rested—whether Saikewicz's acquiescence in the treatment could somehow be obtained, perhaps through heavy sedation or through continuous intensive efforts by institutional staff familiar to him to keep him calm. All such efforts might have failed. But, at least according to the available record of the proceedings, none was tried.

Most fundamentally, this omission reflected everyone's unwillingness to enter into sustained interaction with Joseph Saikewicz, everyone's wish to absent himself from any transaction with him. The trial transcript shows this if we attempt to identify from it precisely who decided to withhold treatment from Saikewicz—the doctor or the judge. The judge claimed power to decide, to which the doctor deferred on the ground that he lacked "that deep knowledge"—until the judge suggested that his decision would require the doctor to treat. The doctor then objected, "You have to see him. . . . [H]e flails at you and there is no way of communicating with him and he is quite strong." The judge had thus succeeded in obtaining a highly explicit recommendation from the doctor and then encircled his decision to withhold treatment with the rhetorical flourishes "after a full hearing with medical specialists and doctors being present and their testimony being taken, the Court determines and adjudges. . . ." Who then was responsible for this decision? No one was prepared to speak the final word with Joseph Saikewicz, to cut off any imagined conversation with him. Death was inflicted as if it took force on its own.

But if this effort by others to enter the mind of Joseph Saikewicz or Karen Quinlan leads only in such circles, can a sensible path be found by abandoning the fiction of substituted judgment and by frankly acknowledging that, because their purposes and interests are inscrutable, others' interests regarding them should explicitly prevail?[15] The trial judge in Saikewicz's case appeared to adopt this view in one portion of his ultimate written order by stating that "the quality of life possible for him even if the treatment does bring about remission" counted against the treatment. The appellate court struggled, somewhat unconvincingly, to interpret this observation as referring only to Saikewicz's continued suffering from the direct effect of the chemotherapy, notwithstanding remission, and thus to avoid the implication that the trial judge viewed the life of this profoundly retarded man as lacking sufficient "quality" to warrant treatment efforts. But if Joseph's perspective, or Karen's, is inherently unknowable, why shouldn't courts acknowledge that they face a conundrum and turn to a calculus that would account for the identifiable emotional and financial costs to others of the continuation or termination of treatment?

This move is not advisable because it does not avoid the conundrum of inscrutability; it merely restates it. This so for two reasons: first, that the psychological forces potentiated in the confrontation with Joseph or Karen will inevitably press others to perceive the costs to themselves of either course as inscrutable, as fundamentally incommensurate; and, second, that this perception will be most intense and unsettling when those others conceive themselves as the exclusive and invulnerable choice-makers and when the decision is concomitantly conceived as occurring at a fixed, final, and irreversible moment.

Anyone's sense of psychological integrity, of sanity, is kept intact essentially by holding firm to a belief in well-defined "others" seen as such and seen as invulnerable to destruction solely by force of the observer's thoughts as such.[16] This is the belief that utterly uncommunicative persons such as Joseph Saikewicz and Karen Quinlan urgently call into question for all who confront them.

The critical role of the belief in the other seen as such and as indestructible is visible in the religious context that played a significant role in the Quinlan case. The Quinlan family were devoutly practicing Catholics; in coming to their initial decision to withhold treatment, they relied on the advice of their priest and church doctrine.[17] This doctrine, in common with other religious and secular traditions, offers distinctions between "ordinary" and "extraordinary" means of sustaining life or "active" and "passive" withholding of treatment.[18] Refusing treatment is authorized only in the latter categories. If the diseased person himself refuses "ordinary" treatment or "actively" causes his death, he is committing suicide; if others make such decision, even for a comatose person, they are committing murder.

These categorizations are at best opaque. If a mechanical respirator is extraordinary, for example, what are massive dosages of antibiotics? If antibiotics are extraordinary, what is chemical nutriment provided only through intravenous means? And so on. The coherence of these categorizations clearly rests on a belief in some force that both transcends and works actively within the dying person. Some have conceived this force to be the process of dying itself, invoking the idea of "natural death" as actively claiming a person when death's time has come. But the notion of biological death itself as an active force cannot explain when individuals should use or withhold power within their biological capacities to arrest death.[19] None of these ideas cohere without some underlying belief that the dying individual has an indissoluble relationship with some force or being beyond himself and his biological death—an afterlife, perhaps, or a Deity.

If those who apply these categorical distinctions to a comatose person believe in man's immortal soul and indissoluble relation to a Deity, they are able to hold fast to a belief that power and impotence alternate between them and the dying person. If the diseased person believes in these premises, he can convincingly ascribe power to some force or being other than himself even at the moment he chooses (or, in his terms, acquiesces in) his death. Through these ascriptions, no one believes himself a wholly omnipotent choice-maker for a wholly choiceless other. Each sees himself as alternatingly choice-maker and choiceless in withholding treatment and inviting death. Each thus can hold to beliefs equivalent to the dynamic conception of human equality, resting on alternating ascriptions of omnipotence and impotence to all actors, that I described earlier.

These distinctions, such as between ordinary and extraordinary treatment, cannot, however, be convincingly applied by a person who believes in the individualistic secular premises common to our time. In the context of these premises, such distinctions become as empty and unconvincing as the courts' attempts to discern Karen Quinlan's or Joseph Saikewicz's intentions. These distinctions contain no force in this person's mind that interrupts his self-referent omnipotent beliefs, no convincing reassurance that his wish to destroy the apparently impotent other is futile. His premises of secular individualism instead draw this person more and more into battle to ensure the utter impotence of this more and more threatening other—the battle that more and more insistently threatens destruction of people's sense of integrity, their belief that they cohere as integers, their sanity.

This is not an argument in favor of religious faith but rather a description of the psychological effect and importance of religious faith.[20] There are secular means to approach this same goal. This is the purpose of my attention to the format for legal regulation of these decisional processes. The law can embody the secular community of others to whom each individual is held and can hold himself accountable. But this embodiment is lost when legal institutions purport to answer questions that cannot in their nature yield univalent "correct" answers, when those institutions issue authoritative-seeming pronouncements to silence disputatious conversation whose indefinite prolongation alone can preserve the psychological balance of self-confronting-other on which individual and social repose ultimately depends.

The law must not vest Karen Quinlan's father or physicians with invulnerable future immunity to decide whether she lives or vest such power in a judge, as in the Saikewicz case, so that the "correct" treatment decision can be made.[21] The psychological integrity of all of the participants in those cases, including society at large, will be best safeguarded by assuring that everyone, on confronting Karen or Joseph, hears in them some other's voice, recognizable as such and indestructible as such. Neither can speak, of course. The idea that a family member, physician, or judge can speak both to them and for them at the same moment likewise cannot give them a voice. But if everyone knows that when he participates in the treatment decisions, the law retains authority to review his participation after the treatment decisions have been made and acted upon, that recognizably and indestructibly "other" voice for Karen and Joseph can be assured.

The most likely consequence of this prescription is that all of the participants in the Quinlan and Saikewicz cases would remain extremely uncomfortable in reaching any treatment decision and that this discomfort would most likely prompt them toward prolonging treatment efforts. One powerful motive, at least, toward this end is the prospect of later criminal prosecution, which, by my prescription, cannot be avoided by the advance immunization wrongly provided by the New Jersey and Massachusetts courts. Absolute immunity from prosecution could only be obtained by assuring that Karen or Joseph remained alive no matter what other harmful consequences appeared to follow from this treatment course. Family and physicians would thus find themselves trapped by a rigid legal rule that seems unsatisfying, inattentive to the individual complexities of their situation.

Others might withhold treatment in apparent violation of the legal mandate. These others would know only that some adverse legal consequence could follow from their actions, but they could not know the precise measure of that adversity before they acted. Because declaratory relief would be withheld, family and physicians could not know whether a prosecutor might seek a criminal indictment, whether a jury might convict, whether a judge might sentence harshly or leniently for their actions withholding treatment.[22] These actors could withhold treatment and minimize these risks only by the most intense collaboration with one another, and with intense individualized attention to the uncommunicative patient, in order to build a defensible record that every treatment effort had been made and that the futility and painfulness of these efforts appeared so palpable to so many different people involved in the decision that all were prepared to take the risk of prosecution for criminal conspiracy rather than continue heartless compliance with the apparent letter of the law.

A realistically sensible attorney, consulted in a case like Quinlan's or Saikewicz's, could guide physicians or family to assess with reasonable assurance the likelihood of criminal liability if they decided to withhold treatment. This attorney would pose a series of questions for both physicians and family. "Do you all agree that withholding treatment is right? Whom have you consulted in reaching this decision—other physicians who have corroborated your views of prognosis and treatment alternatives; friends and mentors to corroborate your views of the moral propriety of this course? What other, wider collaborations have you sought in the professional technical literature and the writings of religious or secular ethicists that have guided and confirmed your decision?"

The attorney would ask the participants to imagine that their decision would be challenged later. "How comfortably, how fully, could you defend yourselves and, most importantly, how reliably could you see yourselves defending one another—family and physicians in mutual defense, each member of the family and of the medical team (including treating physician and hospital chief, nurse, and intern) testifying for one another?" In order most fully to guard against any liability, the attorney would advise that treatment should be prolonged unless the face-to-face participants were unanimously agreed otherwise and could support this agreement on a wide base of technical and ethical opinion in published literature and elsewhere. Accordingly, the objections or even hesitant misgivings of any one participant would be greatly magnified for the others, and each would have increasingly intense motivation to talk, to probe, to plead, to argue with the others.

In these uncomfortable, even painful discussions forced by the prospect of liability, no one could see himself as the exclusive choice-maker for an utterly choiceless other, whether treatment was continued or withheld. Each could be led to see the futility of attempting to pretend that anyone was powerless in this decision—even including the apparently silent partner, the comatose or retarded person whose true psychological power over each of the others would be given voice by the cacophony of voices consulted for mutual defense against some later possible retribution. In this intense collaboration with one another, and implicitly with the silent patient, each of the participants could work to reassert his own sense of identity, his own grasp on self-coherence, that the confrontation with pain and death inevitably calls into question.

Legal institutions cannot guarantee that this collaboration will occur or that it will succeed in obtaining ultimate reassurance. The law can only provide a format for deliberative processes that makes this result more likely than not. This collaboration is painful; many people quickly move to avoid this pain by conceiving themselves through rigid sterotypes and this avoidance brings isolating and abusive consequences for everyone. To counteract these destructive forces, social arrangements, including legal institutions, must provide strong incentives to prompt this collaborative process, for mutual acknowledgment of alternating power and powerlessness and thereby of fundamental equality among people.

The New Jersey Supreme Court, in its *Quinlan* opinion, reached toward this position by mandating collaboration between family and physicians and by requiring both to present their decisions to newly created hospital "ethics committees" before a comatose person's treatment might be discontinued. But the court appeared to promise immunity from civil or criminal liability to all these actors when they unanimously agreed on treatment termination. By my prescription, however, if every participant had followed the process advised by my realistically sensible attorney, and all had agreed to withhold treatment, he could assure them almost to a certainty that no prosecutor would seek to indict, or if indicted no jury would convict, or if convicted no judge would imprison or fine any of them for criminal conspiracy. But he could assure only *almost* to a certainty. The attorney could not, and the law must not, promise any of these people that there is nothing to fear, that they are invulnerable against later retribution, when they have all agreed to dispense death or to acquiesce in its imminence.

The *Quinlan* court's solution and other current proposals to legalize and regularize death's dispensation, to identify some among us—diseased person, doctor, judge—as ultimate and exclusive decision-maker, aborts the very process of communal collaboration by which each of us sustains individual identity.* The law should not offer to relieve everyone's mutual distress if some among us will assume an identity, as exclusive choice-maker for all others, that none can coherently exercise. This fictitiously bestowed identity cannot effectively allay anyone's distress. It can provide a seductive route for temporarily holding that distress away from conscious awareness. But this temporary holding action only increases the apparent costs, the perceived potential devastation, for all actors of ever consciously acknowledging their stressful confusion. Thus, rather than look toward one another for mutual support and recognition, everyone is pressed to look away.

This aversion can take many forms. Imprisonment in mammoth, socially isolated institutions in the name of "treatment" emerged as the characteristic modality from the last century. More direct physical obliteration of unconventional people in the name of "individual liberty" may be emerging as the contemporary modality. Joseph Saikewicz offers an ironic instruction here. His leukemia was discovered during the administration of extensive health examinations to all residents at the Massachusetts state retardate institution; these examinations were ordered by a court in response to litigation invoking the new constitutional doctrine of "rights to treatment" for institutionalized persons. That litigation had demonstrated that these residents suffered the common institutional history of brutal abuse and neglect from state custodians, that their physical health and psychological capacities for conventional social interactions were left undeveloped or directly assaulted from the moment they had been consigned to these warehouses.[24] Saikewicz's history seems unusual only in that his family kept custody of him until he was 14 and, at least according to state records, directly subjected him to abuse and neglect rather than rely on state surrogates for this purpose from the first moment he had been identified as "mentally abnormal." The final episode of state response to Saikewicz can thus be seen as an explicit avowal of everyone's previously unspoken assumption—that he was too difficult to engage and must instead be obliterated from sight.

Joseph Saikewicz did create terrible difficulties throughout his life for anyone who wished to engage and to cherish him. But that difficulty, that terror, cannot reliably be abated by attempting to obliterate him; it can be suppressed but only with recurrent and increasingly taxing psychic effort by all involved. The escalating difficulty and accompanying destructiveness of this effort to preserve rational order explains why the 19th-century custodial institutions grew to hold more and more people for longer and longer terms. This same psychological dynamic can feed the impulse that leads from authorizing the death of comatose people to retarded people already suffering from terminal physical

---

*Recently enacted and proposed statutes that give binding legal force to a person's previously executed "living will" are, in effect, efforts to keep the now-silent patient as the exclusive decision-maker in the transaction.[23] The statutes appear to make the treatment termination decision wholly mechanical for the comatose person's physicians and family; if the person had previously "signed on the dotted line," that choice is construed as binding and immunizing regarding others' later actions leading to his death. I doubt the good sense of enshrining in statutory format an individual's attempt to act as if the prior directive left them choiceless. No statute should require that family or physicians automatically implement a person's earlier choice for death without regard to their current personal misgivings and interpersonal distress.

illness to retarded people suffering only from retardation. It can lead from old people who have become too burdensome to sustain to anomalous newborns whose sustenance is too burdensome to anticipate.

The connection is not fanciful. Physicians involved in the care of such newborns have recently called for laws that give explicit immunity to them and to parents who decide for termination.[25] Tragic burdens are clearly borne by everyone in these situations, and it does seem harsh to add the prospect of some criminal liability as yet another burden. But legal immunity can too readily feed the destructive dynamic already visible in some physicians' and families' reaction even to minimally deformed children.[26] Law reformers are now campaigning vigorously to empty custodial institutions that have done terrible damage to people suffering from apparent mental abnormalities, whether retardation or psychotic illness. These institutions were created with initial proclamations of benevolent intentions. The current reformist intentions to take care of helpless, abnormal people by removing them from brutal institutions can readily become translated into an intention to fill up grave sites with them.

Each step toward this end has its logic. Each step seems to promise that an ordered, socially conventional life will be made easier for the survivors. But this ordered normality will be felt as increasingly constricted for the survivors. They will perceive increasingly magnified threats to its preservation, thus perpetually feeding the sense of disease, disorder, and injustice that the initial steps were meant to appease.

The family members and physicians who petitioned the courts in the *Quinlan* and *Saikewicz* cases sought assurance that their wishes to end disease and its attendant suffering were wholly righteous, correct, justified. The judges might have understood these wishes as an expression of the terrifying disorder that the petitioners felt in confronting alien experience. Legal institutions play a critical role, with other social mechanisms devoted to the furtherance of rational ordering, in allaying this disorder for everyone. But to accomplish this, the question of technique is critical. The judges in those cases embraced a technique that could ultimately only reinforce rather than allay everyone's sense of disorder, of injustice.

The hallmark of the judges' error was in their willingness to assume the role of exclusive choice-maker for everyone in the transaction and thus to permit all others to conceive themselves as powerless and choiceless. They thus invited all others to suppress conscious awareness of their own power and the power of the other apparently impotent participants in the transaction—the comatose, the retarded, the silently dying persons. The Massachusetts Supreme Court was most explicit in assuming this role for itself:

> We do not view the judicial resolution of this most difficult and awesome question—whether potentially life-prolonging treatment should be withheld from a person incapable of making his own decision—as constituting a "gratuitous encroachment" on the domain of medical expertise. Rather, such questions of life and death seem to us to require the process of detached but passionate investigation and decision that forms the ideal on which the judicial branch of government was created. Achieving this ideal is our responsibility and that of the lower court, and is not to be entrusted to any other group purporting to represent the "morality and conscience of our society," no matter how highly motivated or impressively constituted.

The court here posed a false alternative for itself, as if the role of omnipotent choice-maker on "questions of life and death" must either fall to physicians or to some other identifiable persons. This is the fallacy, the ultimate pretension, of the conception of the individual as socially isolated and rationally self-controlling. The court is correct that this individualistic ideal finds expression in the constitutional role of the life-tenured judge responsible only to his vision of the "morality and conscience of our society." But the court assumes that this ideal is attainable in a statically fixed embodiment. It is not. The ideal is only amenable to ceaseless pursuit. The judges' failure to see themselves as participants with only limited, though inescapable, power in controlling themselves and others was mirrored in their failure to see comatose or retarded persons as equally powerful, though limited, participants in addressing ultimate questions of life and death.

REFERENCES AND NOTES

1. *In re Quinlan*, 355 A.2d 647 (1976), from which all unreferenced quotations regarding this case are drawn.
2. *Superintendent of Belchertown State School v. Saikewicz*, 370 N.E.2d 417 (1977), from which all unreferenced quotations regarding this case are drawn.
3. See Veatch, R., *Death, Dying, and the Biological Revolution: Our Last Quest for Responsibility*, p. 79, Yale University Press, New Haven (1976).
4. See Annas, G., After *Saikewicz*: No-fault death, *Hastings Cent. Rep.* 8(3): 16 (1978).
5. See Black, P. McL., Brain death, *N. Engl. J. Med.* 299(7-8): 338, 393 (1978).
6. See the trial judge's summary of the physician's testimony, 348 A.2d 801, 810-12, 818-19 (1975).
7. Ibid., p. 819: "[T]here is no evidence [Karen] is now in pain. Dr. Morse describes her reacting [t]o the noxious stimuli—pain—as reflex but not indicative that she is sensing the pain as a functioning human being does. The reaction is described as stereotyped, and her reflexes show no adjustment that would indicate she mentally experiences pain."
8. Report of Marcy T. Pitkin, M.S.S., February 5, 1976, in record of proceedings in the Hampshire County Probate Court.
9. See Arrow, K. J., *Social Choice and Individual Values*, 2nd ed., p. 114 ("Extended Sympathy"), Yale University Press, New Haven (1963).
10. See, generally, Quinlan, J., and Quinlan, J., with Battelle, P., *Karen Ann: The Quinlans Tell Their Story*, Doubleday, Garden City (1977).
11. A. . . . I took her off the respirator, and . . . she breathed without the respirator quite regularly, although she increased her rate. . . .

    Q. Doctor, is there any significance to the fact that the removal of the respirator was for a period of approximately four minutes? . . .

    A. Well, certainly, if she was able to maintain a perfectly normal level for four minutes, one has to be able to say that the function is potentially there.

    Q. Function for what?

    A. For spontaneous breathing, is potentially there. And it would take a much longer period of observation, which wasn't appropriate under the circumstances, to find out whether she needs the respirator at all. I wouldn't be competent from my knowledge to comment on that point.

    Testimony of Dr. Fred Plum, *In the Matter of Karen Quinlan: The Complete Legal Briefs, Court Proceedings and Decision in the Superior Court of New Jersey*, pp. 481, 483. University Publications of America, Arlington, Va. (1975).
12. See the trial judge's conclusions at 348 A.2d 809, 817.
13. Note 10, Quinlan and Quinlan, op. cit. *supra*, p. 311.
14. *In re Joseph Saikewicz*, No. 45596, Hampshire County, Mass., Probate Court, May 13, 1976.
15. See Ramsey, P. *The Patient as Person*, p. 161, Yale University Press, New Haven (1970); note 3, Veatch, op. cit. *supra*, p. 94.

16. This proposition is developed, and applied in the context of doctor–patient relations generally, in my book, *Taking Care of Strangers: The Role of Law in Doctor–Patient Relations*, Free Press, New York (1979).

17. See, generally, note 10, Quinlan and Quinlan, op. cit. *supra*; see the trial judge's account at 348 A.2d 813.

18. See, generally, note 3, Veatch, op. cit. *supra*, p. 77.

19. Ibid., p. 278.

20. See Ramsey, P. *Ethics at the Edges of Life: Medical and Legal Intersections*, p. 9, Yale University Press, New Haven (1978); Loewald, H. W., *Psychoanalysis and the History of the Individual*, p. 53, Yale University Press, New Haven (1978).

21. Compare the invulnerability to sanction that the Supreme Court assumes must attach to a judge's action no matter how outrageous or patently contrary to law. *Stump v. Sparkman*, 98 S. Ct. 1099 (1978).

22. This position would not violate the current constitutional norm proscribing vagueness in criminal laws since all would have ample notice of the possibility that their conduct would involve liability. Some have argued in various contexts that predictive certainty in both applying criminal law norms and dispensing criminal sanctions should be more stringently valued. See Goldstein, J., Police discretion not to invoke the criminal process, in: *Crime, Law and Society* (A. Goldstein and J. Goldstein, eds.), Free Press, New York (1971); Morris, N., *The Future of Imprisonment*, University of Chicago Press, Chicago (1974); Gaylin, W., *Partial Justice: A Study of Bias in Sentencing*, Knopf, New York (1974). Courts have approached these claims cautiously, with explicit concern that the values in predictive certainty not obscure countervailing values in idiosyncratically individualized application of law. See *Gregg v. Georgia*, 428 U.S. 153 (1976); *North Carolina v. Alford*, 400 U.S. 25 (1970). Whatever its merits in the contexts of these cases, the imperative of predictive certainty in doctor–patient relations can be overvalued with seductive ease and destructive consequences.

23. See California Health and Safety Code, div. 7, p. I, chap. 319, secs. 7185–95; note 20, Ramsey, op. cit. *supra*, p. 318.

24. See, for example, *Wyatt v. Aderholt*, 503 F.2d 1305 (5th Cir., 1974); *New York State Association for Retarded Children v. Rockefeller*, 357 F. Supp. 752 (E.D.N.Y., 1973).

25. See Duff, R. S., Campbell, A. G. M., Moral and ethical dilemmas in the special-care nursery, *N. Engl. J. Med.* **289**:890 (1973).

26. See my article, Authorizing death for anomalous newborns, in: *Genetics and the Law* (A. Milunsky and G. Annas, eds.), p. 435, Plenum, New York (1975).

DISCUSSION

Dr. DAVID I. TODRES (Massachusetts General Hospital): It is a pleasure to hear a dialogue taking place between lawyers and physicians. I think this is a step forward to clarifying many of the dilemmas that confront us. Following the confusion that came out of the *Saikewicz* decision, I think we are starting to see some benefits beginning to surface. As a result of the *Saikewicz* decision we had the *Dinnerstein* case, which helped to clarify some of the issues relating to resuscitation orders. Although the Supreme Judicial Court of Massachusetts may see fit to overrule the *Dinnerstein* decision, we should continue to work toward clarification of these issues, rather than see these decisions as confusing the issues. Decision making today is complex. Most lifesaving or life-prolonging situations take place in the hospital environment, often in intensive care settings where we have large numbers of physicians, nurses, and other health care personnel involved in the patient's care. The close, trusting doctor–patient relationship is often not achieved.

Decision making may at times become theoretical, with discussions of the legal, moral, and philosophical aspects. But let us not forget the role of the physician, who is more than a technologist, who cares with compassion for his patient, who does not only see the diseased organ but the whole patient—an individual who is a member of society.

Our society today has created a health care system that has become very specialized and in doing so has isolated itself. We have achieved a sophisticated degree of expertise in diagnosis and therapy but our ability to resolve the difficult decision-making dilemmas has lagged far behind. Today the patient says, "I need a doctor who looks at my heart because this is what is wrong with me—I need a cardiologist." The patient is then disturbed to find that he does not have a doctor who looks at the "whole me."

Today the doctor–patient relationship is often not the trusting relationship of the past. Medicine is not alone in being a profession that is criticized by society. We should recognize that this is happening in our society with most professions. The lawyers have been subject to it and today the nuclear engineers are getting it! We should listen to the criticism and do everything we can to improve a trust relationship that has eroded between the patient and physician. This trust relationship takes effort and time in today's complex medical care system.

I think physicians tend to overlook the ingredient of time in the healing process. We should remember to introduce the element of time in our dialogues between doctors and lawyers in order to resolve these issues on a practical level.

Physicians should be cognizant of the legal issues in the care of their patients so that they may recognize where the law may need to change, and, in concert with their legal friends, work toward this change. Medicine cannot be practiced in isolation. I am delighted to hear this morning's dialogue, which I trust will set the stage for further positive interaction between lawyers and physicians in the future.

Dr. LAWRENCE P. ULRICH (University of Dayton): In spite of Professor Burt's rather persuasive and impassioned plea, I would like to support Dr. Relman on one of his contentions. I sit on an ethics committee at a hospital in Dayton. I am a philosopher and a theologian myself and the rest of this committee are doctors. For the past two years we have been wrestling with policy and procedures for terminating life-support systems. We have succeeded in coming out with a policy which I think is a very good one. I have found that despite some popular opinions, doctors really are very concerned about moral issues. They are, however, equally concerned about malpractice suits, and I think therein lies the tale.

The law may be useful in many issues but it is important for us to realize that the primary relationship between human beings is a moral and not a legal relationship and therefore I would support Dr. Relman in saying that these questions ought to be decided on the moral level, in an informal relationship. That entails a number of problems and considerable time would be needed to examine what moral relationships are. Part of it has to do with mutual responsibility and taking risks. We sometimes think that if the law decides something, then no injustice will be done. It may be the case that no legal injustice will be done but a moral injustice may be perpetrated. I don't know whether pushing some of these cases into the law courts is going to help all that much to solve the problem of avoiding mistakes.

Moral decisions are not easily made. Dr. Relman said that physicians have to make them continually. I am concerned that many physicians are not equipped to make moral decisions. It is true that many of our young physicians coming out of medical school now have had training in making moral decisions. The training is minimal. Some of the older physicians have never had any training in making moral decisions. I mean genuine philosophical training.

Do you see this as a real problem or do you think that physicians have some sort of an innate ability to make sound, careful, and considered moral judgments?

DR. ARNOLD S. RELMAN: The question implies that you have to be a professional moral philosopher, or have the competence of a moral philosopher, to make moral decisions. And furthermore, it suggests that we'd better include philosophy, particularly ethics, in the medical or the premedical curriculum.

That's an interesting proposition, but I am not sure it is valid or practical. In their everyday life, in their most intimate and personal relations, people are making moral decisions all the time. And it seems to me that these moral decisions are not to be reserved only for those who are interested in the philosophy of morality.

I don't think that doctors have to have any special training in moral philosophy. They have to know their business very well, they have to have experience, and they have to be sensitive to the human condition of their patients.

Moral decisions must be made as part of the practice of medicine. The moral aspects of the decisions cannot be separated from the medical parts of the decisions. There is no way of dividing these decisions up into technical ones and moral ones. They are all tied together. Since a human being is not a machine, whatever you do as a physician will have moral and ethical implications.

I think doctors have to do the best they can with the moral problems they must face. They have to be responsible for what they do. They have to be willing to defend what they do in court. They have to confer with their colleagues when there is any legitimate difference of opinion. And above all, they have to be sure that what they are doing is what their patients would want them to do.

MR. MITCHELL KLING (Student, Harvard Medical School): I agree with Dr. Relman in theory about medical technical decisions being inseparable from moral decisions. It would be nice to be able to entrust decision making about an individual patient to the physician. But from my limited experience on the wards, difficult moral decisions arise involving certain unresponsive patients, who, in my view, may not be treated as persons. I think there should be a protocol at each hospital for what to do in such circumstances and there should be a means of holding people responsible for their decisions.

MR. DANIEL E. GABRIELS (Student, Albany Medical College): I have regularly encountered physicians who could not consistently make explicit the reasons for their decisions to withhold or extend therapy and who could even upon occasion be dislodged from their carefully thought-out positions by an aggressive neophyte.

As an illustration of the capricious nature of determinations made, I vividly recollect one morning on surgery rounds in the intensive care unit. A surgeon covering for his partner, who had departed that morning on vacation, approached the bedside of a somnolent patient and ordered that all medications and hyperalimentation be discontinued and that the patient be placed on room air.

In the continuing absence of a medical consensus which affords the patient a good idea of what to expect *in extremis*, it seems entirely appropriate that the law implicate itself in the clarification of the decision-making process.

PROF. RICHARD BERQUIST (College of St. Thomas, St. Paul, Minnesota): I am another moral philosopher, basically in sympathy, though, with Dr. Relman's position, especially in terms of the immediate practical utility of much moral theorizing. It tends to be unfairly abstract frequently. However, there is one distinction which is fairly significant in moral theory and since philosophers have made confusions, I suppose, in this area, only philosophers can resolve them. And that is the distinction between active and passive euthanasia, which I think is very, very central.

Obviously, the law is going to be concerned with any intentional desire on the part of any physician to terminate a life. So, as we all know, active euthanasia or actually killing a patient is first-degree murder and is bound to be a serious concern in any legal system.

One of the difficulties is that so-called passive euthanasia, that is, the withholding of medical treatment in some cases, is taken to be like a kind of watered-down case of active euthanasia. I suspect this climate of opinion contributes to the fact that when people think that doctors make decisions to withhold treatment, they are to some extent making decisions on whether the patient would be better off dead or not.

It strikes me a more traditional understanding of the problem of withholding treatment could do a great deal to eliminate concern of the legal system for such matters. The older idea was that withholding treatment was *not* done precisely for the purpose of bringing about the death of the patient, and was not based on some decision that the patient would be better off dead, but rather on the idea that the treatment itself imposed a rather extraordinary or heavy burden on the patient or on others, or that it was in some sense useless under the circumstances. There wasn't any kind of indirect intention to hasten anyone's death.

I do think that distinction of moral theory elaborated somewhat can do a great deal to help in this area.

MR. G. MICHAEL SMITH (Attorney, Atlanta, Georgia): I would like to preface my remarks by saying that I am in active practice and I realize we are all burdened down by an overabundance of regulations. But I would like in a very simple fashion to suggest possibly not a solution but a direction toward helping the doctors, the patients, and the law in this particular dilemma.

I understand that there are various boards in various hospitals around the country that review the question of whether or not a patient should be taken off a life-support system. However, the dilemma the physician faces is the threat of a lawsuit if he makes a mistake or an action by the district attorney for murder. This is a reality of life that we all have to deal with daily.

I suggest having the boards make a determination for the doctor where there is a very serious question and the doctor finds himself in a dilemma. This apparently is being done. But I suggest that we go a step further and, like common law, we use precedent and allow the boards to be made up of not just physicians but maybe of theologians or philosophers and people from various disciplines. These decisions of the board could be recorded and there could be meetings around the state or country, where established precedent could be reviewed. In this way a doctor in making a determination could either rest on precedent or approach the board.

DR. RICHARD RATZAN (University of Connecticut Medical School): About three years ago there was an article in the *New England Journal of Medicine* outlining their policy at Massachusetts General concerning decisions about management of the critically ill or dying patient. It considered the quality of life and allowed one person to take the responsibility, namely, the private physician. It allowed for appeal if house staff, family members, or others on the team felt that there was an injustice. And it would be subject to Professor Burt's idea of a decision having been made, then standing up on its own merits should someone want to adjudicate it.

Has this policy been successful at Massachusetts General? Have there been many legal problems with appeal to the courts? Has it worked and what is your opinion of it?

DR. RELMAN: I can't answer that question in any detail.

With all due respect, I don't think that the policy you have described is quite as comprehensive as you suggest. But I recognize that there is a lot that can be done within a hospital to make sure that decisions of this sort are orderly and thoughtful. In the last analysis, these decisions are so personal and so individualized that neither precedent nor law nor regulations will really help when you come down to it.

I am afraid I must resort to an unsatisfactory statement, namely, that you have to be there. You have to be involved in making these decisions to see how personal and how individual they are. I don't really believe precedent helps very much.

PROF. CHARLES H. BARON: My involvement in this area began with an article in which I criticized the legal profession and criticized the courts in the *Saikewicz* decision for not going far enough in making sure that there decisions were not sham decisions, as I think they have been accurately described as potentially being by Professor Burt.

I am very much concerned that these decisions not be sham. That is the reason why it is important for them to be truly adversary, why it is important that there be advocates in each case who will aggressively press the arguments on each side of each issue.

With respect to interpreting legal decisions, let me say that there are definite problems here just as there are problems with interpreting medical conclusions. This is a brand new area. The things that the courts have said in this area are naturally inchoate. There is just the beginning of elaboration of principle here.

But what is needed in order to make these principles clearer is precisely what is needed to be able to make medical diagnoses and medical principles clearer, and that is more clinical material. We need more cases. If Dr. Relman wants to get a better idea of what the Supreme Judicial Court thinks with respect to any given issue, his solution is to bring more cases before the Supreme Judicial Court and give it an opportunity to elaborate what it believes to be the appropriate societal response to each in terms of what it believes is the public morality of this country.

We may all have our own ideas about personal morality but there is a public morality which is supposed to find its embodiment in law, however inadequately the courts may do this at times. The way to elaborate upon this public morality is by providing more and more cases which give the courts an opportunity to elaborate a common law response.

Although there will be errors and there will be embarrassments in this process as it develops, the genius of the court system, of the law in this country, is that although judges make mistakes, they are forced to make them in public. When doctors make mistakes, they are able to make them in the privacy of intensive care units in a way which may never come to public light.

PROF. AVERY: Professor Baron, before you leave that plea for more decisions to be made, I have a very practical question as one on the front lines. How much would such a proceeding cost and who pays the bill?

PROF. BARON: It is impossible to say exactly how much it would cost. I think that the bill should be paid by the public. I don't think that it should be paid by the relatives of the individual who is involved. I have been involved in two of the few cases which have been brought in the wake of *Saikewicz* and I can tell you that there were hearings in those cases which went on, in one case, for perhaps six hours. In the other case the hearing was very brief because the hospital really had no evidence to present for the position that it was taking, and, as a result, the matter was quickly dismissed. But in the other case, where the hospital did have evidence for the position that was being pressed, the hearing went on for six hours.

All I can tell you is that after a while, these cases, many of them, will not need to be brought. But the reason why they will not need to be brought will be because the court will begin to mark out areas in which it will have made advance decisions by developing principles which can be applied by doctors thereafter without court approval. But until these cases are brought before the court to mark out these areas, it seems to me all of them must be brought to court.

PROF. ROBERT A. BURT: Let me briefly comment on Professor Baron's last remarks.

It is true that judges act publicly and doctors act privately. But this fact does not necessarily prove that judges are the preferred decision-makers in the matters we've been discussing. The process by which judges come to public decisions on these matters has adverse consequences that themselves have to be attended to.

Let me give one illustration. In 1923 only approximately two states had compulsory sterilization laws with regard to the mentally retarded. Proposals were abounding throughout the country, as a reflection of the growing success of the eugenics movement. But many people felt terribly uncomfortable about these proposals. Then, in 1923, a case reached the Supreme Court of the United States. In *Buck v. Bell* Justice Holmes said, in effect "It is not my decision as a Supreme Court justice; this sterilization law is the decision of the state legislature of Virginia. While I would not necessarily agree with them, that is within their discretion to act and therefore I approve the constitutionality of their action."

Accordingly, a public norm was set. Within the next ten years or so, some twenty-five states adopted compulsory sterilization laws on the strength of this principled ruling by the Supreme Court of the United States that appeared only to say, "While you are not obliged to pass this statute, there is nothing wrong with it."

The concern that I have in these cases is that a momentum gathers that makes people feel more comfortable than they should feel about a whole range of difficult decisions. If these decisions are left in relatively low visibility, with occasional public judicial review on an *ad hoc* basis, in an after-the-fact process, we won't get tidy sets of principles that will satisfy the moral philosophers among us in terms of rigorous clarity or publicity or complete consistency. But it seems to me that we will have a decisional process that is truer to the complexity of these matters and truer to the fact that in these very difficult decisions, almost every way that one turns is wrong. To go back to the *Saikewicz* case, whichever decision was made, whether for treatment or against treatment, was a wrong decision. A decision had to be made, to be sure. But, still, each course of action had enormous costs to it.

I am concerned about setting up a social mechanism in judicial review that obscures the fact that we are dealing with a choice among wrongs, not a choice of a clearly right decision. And that is why, in this area, I simply can't agree with Professor Baron's impulse, as attractive and lawyerlike as it is generally.

16

PROLONGING LIVING AND PROLONGING DYING
A Distinction That Is Not Decisive

ROBERT M. VEATCH
The Hastings Center
Institute of Society, Ethics and the Life Sciences
Hastings-on-Hudson, New York 10706

Some time ago I was involved in a particularly agonizing decision about the care of a baby born with a serious genetic problem. There were conspicuous physical abnormalities: deformed hands and misshapen ears. His arms were rigid, fingers tightly flexed. The baby was also in respiratory failure and appeared to have a heart defect. He was placed immediately on a respirator. The physicians consulted with the family. After a few days, it became clear that death was inevitable. Serious heart and respiratory problems meant that at the most the baby had another week or two to live. The parents, knowing that the baby's life-stream was flowing inevitably and rapidly toward death, decided that the respirator was useless. It was turned off. The baby died rapidly. The diagnosis of trisomy-18 was confirmed.

At first this case appears similar to many others involving chromosomal abnormalities, where parents in consultation with medical personnel have to make decisions about whether to continue medical treatment. There seems to be a resemblance to the now famous series of cases involving Down syndrome or trisomy 21 in conjunction with duodenal atresia.[1] Both cases involve an extra chromosome. Both involve children who will die if not treated. Both are what all admit are tragically serious genetic afflictions. Yet, there is an important difference. The Down syndrome children with the intestinal atresias are not inevitably dying, as was that trisomy 18 baby. The opening of the intestinal blockage in the Down syndrome babies could be said to prolong living. The use of the respirator in the trisomy 18 baby merely prolonged an inevitable dying course.[2] We are left with a crucial moral question. Does it really make any difference whether the dying is inevitable? How much should we make of this distinction between treatments that merely prolong an inevitably dying process and those that would prolong living, even if the life that is prolonged is at a quality that is compromised to the point that some would consider it a life not worth living?

The distinction appears to be emerging as a crucial one in many recent discussions.

A now famous opinion authorized nontreatment of the leukemia of Joseph Saikewicz, the 67-year-old severely retarded patient at Belchertown State School not far from Boston. The court made a great deal of the inevitabilit of death and argued that "the most significant of the asserted state interests is that of the preservation of human life. Recognition of such an interest, however, does not necessarily resolve the problem where the affliction or disease clearly indicates that life will soon, and inevitably, be extinguished."[3] The court goes on to argue that "there is a substantial distinction in the state's insistence that human life be saved where the affliction is curable, as opposed to the state interests where, as here, the issue is not whether, but when, for how long, and of what cost to the individual that life may be briefly extended."[4]

In Joseph Saikewicz's case, death was anticipated without the treatment within a few weeks. With the treatment, there was somewhere between a 30 and 50% chance of a remission for 2 to 13 months. The *Saikewicz* court finesses the question of how soon one would have to die in order to be considered inevitably dying. The implication is clear, however, that they considered a person who could live on a few months to be one who was inevitably dying, like that trisomy 18 baby, rather than one whose life could be prolonged indefinitely, like the Down syndrome cases.

The matter becomes much more complicated in the light of the *Dinnerstein* case, which was decided in 1978 by the Massachusetts Appeals Court. This case involved a 67-year-old woman with Alzheimer's disease, another incurable condition. After suffering a massive stroke, she was immobile, speechless, unable to swallow without choking, and barely able to cough. A decision had to be made about continuing the treatment. The *Dinnerstein* court makes explicit the use of the distinction between prolonging living and prolonging dying, but in a way that seems only to confuse matters more. The *Dinnerstein* court authorizes a decision not to resuscitate. In fact, it authorizes the decison in such cases without advance approval by the courts. In doing so, it says that "this case does not offer a life-saving or life-prolonging treatment alternative within the meaning of the Saikewicz case."[5] Thus the *Dinnerstein* court, like the *Saikewicz* court, seems to assume that the distinction between patients who are inevitably dying and those who could live if treated is important, but implies that the *Dinnerstein* case somehow differs significantly from *Saikewicz* on just this distinction. For the *Dinnerstein* court, it is within the competence of the medical profession to determine what measures are appropriate to ease the imminent passing of an irremediably terminally ill patient.

We are left in utter chaos. Both courts seem to imply that the distinction between prolonging living and prolonging dying is crucial. Yet, Dinnerstein's court sees their patient as differing from Saikewicz in this regard, in spite of the fact that both had about a year to live. On that basis, the *Dinnerstein* court makes the strange assumption that in cases when patients are irremediably terminally ill, the medical profession has some special competence in deciding what counts as appropriate treatment.

The California Natural Death Act makes use of exactly the same distinction.[6] That act, which became law in 1977, gives patients the legal authority to execute a directive requiring that life-sustaining procedures be withheld when one is in a terminal condition. Terminal condition is defined so rigorously that only those on death's door will qualify. It means for California's law an incurable condition that, regardless of the application of life-sustaining procedures, would within reasonable medical judgment produce death and where the application of life-sustaining measures serve only to postpone the moment of death of the patient. Furthermore, the treatment may be stopped only at the point where death is imminent. To make matters more complicated, the physician is required to follow the instructions only in cases where a patient has been diagnosed as terminally ill for at least 14 days.

Why this incredible fascination with the sharp distinction between those who are inevitably dying and those who are not? Why is it that we seem so much more comfortable with the decision of the parents of the trisomy 18 baby than that of the parents of the Down syndrome children? Will the weight given this sharp distinction between those who are inevitably dying and those who could live if treated stand up to moral scrutiny? I think not.

The Moral Basis of Treatment Refusal

Traditionally, two criteria have been put forward to justify treatment refusals: the uselessness of the treatment and the burden of the treatment.[7] The uselessness of the treatment is the easiest criterion to apply. If the treatment cannot help the trisomy 18 baby in any significant way, it is foolish to insist on offering it. Those who make the distinction between prolonging living and prolonging dying decisive, as has happened in Califonria and in the *Saikewicz* and *Dinnerstein* opinions, come down exclusively on the criterion of uselessness. Sometimes they go on to make the serious mistake of assuming that uselessness is a purely technical, medical question. Thus the *Dinnerstein* court makes the mistake of saying that this presents a question peculiarly within the competence of the medical profession.

Is, however, the question of the uselessness of a medical treatment a purely technical, medical decison? Before I began working directly on questions of the ethics of death and dying, I studied pharmacology. I learned what those of you who are medically trained know: the usefulness of a proposed intervention is often very debatable. It will depend to a great extent on the values held by those considering the use of the intervention. The respirator in the case of that trisomy 18 baby was useful for extending its life a few more days, but not for anything more. The chemotherapy for Mr. Saikewicz's leukemia would have been useful for getting a few more weeks of life, but not for prolonging his life indefinitely. There is no way that a judgment can be made, even about the usefulness of a treatment, without incorporating some evaluative framework about what would count as a useful treatment. That is why it is so wrong to defer exclusively to medical professionals for making such judgments, even in the case of the patient who is inevitably dying.

Making the sharp distinction between the inevitably dying and those who could live if treated requires something more than that particular set of value judgments, however. It requires the exclusion of the second traditional criterion that has authorized treatment stoppage. It excludes the consideration of the burden of the treatment. If uselessness turns out upon examination to be an evaluative judgment, then certainly burdensomeness will as well. The real question is not how common the treatment is or how often it would be used with patients have a particular diagnosis, but whether the treatment is fitting given the patient's condition. That is why Pope Pius XII said so wisely that morally one is held to use only ordinary means and then went on to define what he meant by saying that ordinary means were to be determined "according to the circumstances, persons, places, times, and cultures, that is to say means that do not involve any great burden on oneself or another."[8]

The only way that the distinction between being inevitably dying and being able to live if treated can be decisive is if this second criterion, the burdensomeness of the treatment, is totally ignored. Admittedly, there are complications when the second, more slippery criterion is introduced. For instance, do we really want to open the door that Pope Pius XII did when he permitted burden to others to be included in the calculation, or do we want to hold to the more traditional medical ethic that includes only burden to the patient? Certainly, the safer and easier course is to limit the relevant consideration to burden to the patient. That would mean that in the Down syndrome cases where patients would suffer very little, if at all, treatment could not be refused by guardians on this basis. If, however, a particular baby with a genetic affliction, even Down syndrome, were to suffer greatly as a result of the treatment, it would be morally a very different matter. For instance, in one baby with which I am familiar, the Down syndrome existed in the presence of a severe structural defect in the heart. This defect would have required perhaps as many as 100 painful operations during the early years of life, and even then successful life prolongation was in doubt. In this case, in contrast to the atresia cases, burden to the patient may become so excessive that a reasonable parent might decide that the treatment ought not be delivered.

Some recent commentators have tried to place limits on the decision to provide or refuse medical treatment by trying to determine what is "medically indicated." Paul Ramsey has recently proposed that we can determine what treatments are required by asking what is medically indicated.[9] Perhaps this is a wishful thinking on the part of one who does not know the enormous ambiguity of the term. I frankly do not know anymore what it means for something to be medically indicated. Does it mean that the intervention would prlong life indefinitely? If so, then treatment is required for somebody like Karen Quinlan. It would also be required for the hemodialysis patient and others for whom continuing treatment is an excruciating burden. Does it mean that medicines will produce a desired effect if one happens to be desired? If so, abortifacient medicines are "indicated" for women with unwanted pregnancies. The mere fact that a drug will produce a biological effect cannot make that drug indicated, unless one makes the moral judgment that the effect will be morally right and appropriate as well as desired. Does "medically indicated" mean that the intervention would tend to be approved by medical professionls? That seems to be implied in the *Dinnerstein* decision. If so, then any intervention that physicians have by custom approved and used, whether it is helpful or moral or tolerable by the patient, is "indicated." The mere fact that professionals tend to approve an intervention or that the intervention would indefinitely prolong life cannot make the intervention justified or right.

Calling something medically indicated seems to give it an objective, almost scientific, endorsement. I am afraid, however, that once labels of being medically indicated or useful are scratched to see what lies beneath the surface, we find ethical and other value judgments, pure and simple judgments that ought to be rooted in the familial systems of religious or philosophical belief. If that is the case, then such judgments ought to come within the framework of familial integrity, in cases when patients cannot themselves make such judgments. They ought to come within that framework until such time the family is seen as being irresponsible or malicious or simply foolish.

Certainly we cannot distinguish between the inevitably dying and those for whom life could be prolonged indefinitely on the basis of any such pseudoscientific term as *medical indication*. We are thus frustrated in our quest for some objective basis for limiting decisions about medical treatment of the seriously ill and genetically afflicted. We cannot seek out objectivity by the presumption that technical, medical facts will provide a basis for sorting the cases. We cannot find it in the attempt to draw a sharp line between those who are inevitably dying and those who will live if treated. That leaves cases where living will be prolonged and cases where only the dying will be extended.

Perhaps there is an objective and moral basis for making such judgments. This is particularly important for most critical cases in genetics because the more subjective standard is so dangerous. Except for those rare genetic diseases that express themselves in adulthood, such as Huntington's chorea, the genetically afflicted patient will never have been competent and will never have had a chance to express his own wishes about medical treatment. Some more objective standard is essential for evaluating the judgment of guardians who must make these choices for the genetically afflicted and other patients who have never been competent. I am convinced that we at least have an objective basis for setting limits on parental or guardian decisions in such cases. The objectivity, however, cannot come from the brute facts of medicine. It cannot come from the distinction between the inevitably dying and those who could live if treated. It will have to come from reasonable people gathering together to decide what the rational limits are in deciding what is useful treatment and what is one that is so gravely burdensome that it is expendable. These are inevitably judgments of value, but that does not exclude the possibility that there may be a rational basis for a consensus on setting the limits for such judgments.

It is at this point that we might salvage some significance for the distinction between prolonging living and prolonging dying. It may well be that if the patient is inevitably dying, then we as a society can accept the judgment of the responsible guardian that any treatment that merely prolongs the dying is expendable. Being in a condition where one is inevitably dying may be a sufficient condition for justifying treatment stoppage, but it cannot be a necessary condition. To make it a necessary condition would be to turn our backs on the tradition that has considered gravely burdensome treatments extraordinary or expendable, even if they do prolong life indefinitely. It would be turning our backs on the agony of that most vulnerable group of citizens, those who have never been competent and are unable to speak for themselves. Sometimes guardian treatment refusals will have to be acceptable even when the genetically afflicted patient is not dying inevitably.

References and Notes

1. Veatch, R. M., *Case Studies in Medical Ethics*, p. 337, Harvard University Press, Cambridge (1977).
2. Veatch, R. M., *Death, Dying, and the Biological Revolution*, p. 114, Yale University Press, New Haven (1976).
3. *Superintendent of Belchertown State School v. Saikewicz*, 1977 Mass. Adv. Sh. 2461, 370 N.E.2d 417 (1977), p. 425.
4. Ibid., p. 425.
5. *In the Matter of Shirley Dinnerstein*, 1978 Mass. App. Adv. Sh. 736, 380 N.E.2d 134, p. 139.
6. California Health and Safety Code, Section 7185–7195.
7. Veatch, *Death, Dying, and the Biological Revolution*, p. 105.
8. Pope Pius XII, The prolongation of life, in: *The Pope Speaks* 4:393 (1958).
9. Ramsey, P., *Ethics at the Edges of Life*, p. 177, Yale University Press, New Haven (1978).

17

ETHICS AND THE COSTS OF DYING

JOHN C. FLETCHER
Clinical Center
National Institutes of Health
Bethseda, Maryland 20205

INTRODUCTION

The development of ethical issues in the biomedical fields since 1966 is marked by the transformation of issues from moral problems that confront the individual physician or patient to large social-ethical issues that face the whole society. For example, earlier discussions of moral problems in terminal illness centered upon when and how to tell the patient that the prognosis is terminal,[1] when to suspend treatment in the dying patient,[2] when the patient is really "dead,"[3] the ethics of research in a dying patient,[4] or who should be involved in making decisions in such situations.[5] These moral problems contain an element of simplicity as "one-on-one" problems, i.e., dilemmas confronting individual physicians and patients. Yet each of these problems take place within the context of a larger reality, such as the social-economic setting of dying, that has a great influence on the whole experience of dying in our culture. There is a tendency in writings on ethics to study the particular moral dilemma and neglect a larger framework that determines many more features of decisions than we prefer to admit. Where people die and how much it costs to die in the United States are issues that shape many elements of decision making in the particular case.

Thesis

Little is known in a definitive sense about the costs, direct and indirect, of dying in the United States. What is known to date will be reported here and leads to a thesis that the costs of dying constitute a significant ethical problem in that (1) the largest expenditures are associated with institutions oriented toward cure rather than care of individuals who are terminally ill, (2) the needs of terminally ill patients and their families are not directly met through the predominant purposes for which expenditures are made, and (3) ethical principles (justice and love) require that there be more options for the care of the terminally ill that will be expressive of these values.

Definition of Terminal Illness

Following Mushkin's earlier work,[6] *terminal illness* will be defined here as an illness that results in death, in which the patient is given a short time to live and has restrictions placed on normal living because of the illness. The cancer patient who will die three months from now and has been forced to leave a job fits the definition; the person who will collapse and die of a heart attack at work does not. Several stages in the development of a terminal illness are discussed by physicians: an initial stage where treatment is given, a second stage where there is no reasonable hope of a cure but the patient is still able to function, and a final stage of care in which the patient becomes debilitated. The ensuing discussion assumes that terminal illness is experienced in the second and third stages. A working group of the National Institutes of Health[7] developed an operational definition of terminal illness: "A state of disease characterized by a progressive deterioration with impairment of function and survival limited in time, usually from several days to a few months."

BACKGROUND FACTS ON HEALTH COSTS

To appreciate the significance of the costs of dying, certain facts should be established as background about the rapid escalation of health costs as a source of concern. The costs of dying occur with expenditures for health escalating by degrees so large as to cause deep concern to political and medical leadership and usher in a new era of "cost-containment" proposals and experiments. The following facts are dependent upon the work of Russell's[8] economic study of medical advances and their diffusion.

The first three tables illustrate the problem in the largest numbers. Table I illustrates the total national expenditures for health care. Table II shows the percentage of gross national product occupied by health costs. A statistic that has considerable meaning to the individual, the cost of one day in the hospital, is contained in Table III.

The most important contribution to rising costs, according to Russell, is the growth of the "third-party payers," e.g., private health insurance and public programs such as

TABLE I. Total National Expenditures for Medical Care

Year	Expenditure
1950	$12 billion
1965	$39 billion
1976	$139 billion
1978	$192 billion

TABLE II.   Health Costs and
Gross National Product

Year	Percentage of gross national product
1950	4.5%
1965	5.9%
1976	8.6%

Medicare and Medicaid. Table IV shows the increase in third-party payments for hospital costs 1950–1976. The power of ideas is never so clearly demonstrated as in the case of the philosophy underlying third-party payments. The idea that the cost of medical care should not be a barrier to the delivery of life-saving or life-preserving care lies behind third-party payments. As a consequence, however, the sense of economic responsibility is not shared by the average individual in the population. Persons, including physicians, expend resources as if they cost nothing or were being paid for by others. The belief that cost should be no barrier to medical care, particularly in saving human life, is the greatest cause of rising health costs. The high costs of specific medical technologies and the readiness of citizens to consume health care are also contributions to rising costs.

BACKGROUND FACTS ON COSTS OF ILLNESS, CAUSES, AND LOCATION OF DEATH

A second level of background facts is needed prior to more specific discussion of the cost of the care of the terminally ill. Evaluation of the cost of illness, according to methods of Rice[9] and the Georgetown University Public Services Laboratory,[10] includes not only direct costs for the care and treatment of a disease but the costs of lost earnings in the context of losses to the economy. Table V illustrates the hierarchy of the most expensive diseases in 1975.[11] The leading category is disease of the circulatory system, including stroke. Table VI shows that the leading cause of death in 1973 was also diseases of the circulatory system, and Table VII confirms that such will be increasingly the case by the year 2000. Causes of death are predominantly functions of the aging process in our

TABLE III.   Cost of One Day in Hospital[a]

Year	Cost
1950	$14
1976	$151
1978	$203

[a] "This is the cost per patient day after subtracting the costs of outpatient visits. Without this adjustment, the numbers are $16 and $173, respectively. The data necessary to make the adjustment were first published for the year 1965, so the $14 estimate for 1950 has been derived by assuming that the ratio of inpatient to outpatient care was the same in 1950 as in 1965" (Russell, p. 1).

TABLE IV. Third-Party Payments for
Hospital Costs

Year	Percentage of total costs
1950	40–50%
1976	90%

current and future society. Since 80% of those who die in the United States die after they reach the age of 65, new cures for cancer would mean only a slight reduction in the incidence of death. In reality, the causes of death would be rearranged and older persons would die of some other cause, probably of circulatory disease. Population studies show no abatement of the progressive aging of the U.S. population. Table VIII depicts the increase in the numbers of individuals in the United States who are 65 years of age and over from 1900 to 2030. One lesson from these figures is that the problem of caring for terminally ill individuals will occur increasingly in older persons.

TABLE V. Economic Cost of Illness According to Type of
Costs and Disease Category: United States Fiscal Year 1975[a]

Disease category	All costs (Amount in $ millions)
All diseases	238.875
Circulatory	45.687
Accidents, violence	27.482
Digestive system	21.660
Unallocated	19.126
Cancers	18.933
Mental	18.890
Respiratory	18.714
Nervous system	14.049
Connective tissues	12.651
Oral	8.123
Genitourinary	7.985
Other	7.262
Endocrine, metabolic	6.307
Cerebrovascular	6.088
Ill-defined conditions	5.956
Eye	5.022
Infective, parasitic	4.238
Pregnancy	3.631
Skin	2.574
Congenital anomalies	1.524
Blood	1.153
Perinatal morbidity	1.053

[a] SOURCE: Pariner, L., Berk, A., and Mushkin, S., *The Economic Cost of Illness, Fiscal Year 1975*, Report BIA, p. 7, Georgetown University Public Services Laboratory, Washington, D.C. (May 12, 1977).

TABLE VI.   Causes of Death (1973)[a,b]

Cause of death	Total reported
Major cardiovascular	841,000
Malignant neoplasms	356,000
Cerebrovascular diseases	214,000
Accidents	116,000
Influenza/pneumonia	63,000
Other respiratory diseases	55,000
Diabetes	38,000
Genitourinary diseases	27,000
Suicide	25,000
Homicide	20,000
Infective and parasitic diseases	16,000
All others	202,000
Total	1,973,000

[a] Of those who die in the United States, 80% die after they reach age 65.
[b] SOURCE: The Kennedy Institute, Georgetown University, Q. Rep. 2:7 (1976).

Table IX shows the significant increase in the hospital as the location for death from 1946 to 1976. Public health facilities are currently the location for dying for 70% of those who die. Of the some 2 million Americans who will die in the present year, 980,000 will die in hospitals, 420,000 in nursing or convalescent homes, and 600,000 at home or at the scene of accidents.

TABLE VII.   Number and Percentage Distribution of Deaths and Causes Projected for Year 2000[a]

Disease category	Total deaths	Percent of overall total
Total all diseases	2,607,090	100
Circulatory	1,483,349	55.1
Cancers	467,040	17.9
Accidents, violence	189,345	7.2
Respiratory	166,062	6.3
Digestive	95,054	3.6
Endocrine, metabolic	57,672	2.2
Ill-defined	41,805	1.4
Genitourinary	38,558	1.4
Congenital anomalies	38,519	1.4
Nervous system	22,736	0.8
Infective, parasitic	20,856	0.5
Blood	7,271	0.2
Connective tissues	6,975	0.2
Skin	3,014	0.1
Pregnancy	566	—

[a] SOURCE: Public Services Laboratory, Georgetown University, Cost of disease and illness in the United States in the year 2000, Public Health Rep. 93:548 (1978).

TABLE VIII.   U.S. Population 65 Years
and Over[a]

Year	Number of persons 65 years old and over (in millions)
1900	3.1
1940	9.0
1965	18.5
1975	22.4
2000	30.6 (projected)
2030	51.6 (projected)

[a] SOURCE: Special Committee on Aging, U.S. Senate, Part 1,
Developments in Aging: 1977, Report No. 95-771, p. 6, U.S.
Government Printing Office, Washington, D.C. (1978).

DIRECT COSTS OF TERMINAL ILLNESS

There are three direct costs of terminal illness that must be paid by family, third-party payers, or the individual's estate: hospital costs, physician costs, and the cost of burial. No definitive recent studies exist for the first two types of costs in large numbers of patients and their families, yet there are a sufficient number of studies with smaller numbers that can be superimposed upon older material to yield some useful information. The most extensive and reliable study of the funeral industry, completed in 1978 by the Federal Trade Commission's Bureau of Consumer Protection,[12] showed that average funeral and burial costs were over $2000. Bereaved family members are vulnerable because they are forced to make a major purchase decision while traumatized by the death, under pressure of time, and usually without itemized price and other relevant information.

Older National Studies of Costs of Terminal Illness

In 1973 Mushkin[6] estimated that 22% of all hospital expenditures (except psychiatric hospitals) were for terminal illness as defined here. She based this estimate in part upon

TABLE IX.   Where Do We Die?[a]

Place of death	Percentage of total deaths 1976	Percentage of total deaths 1946
Public health facilities	70%	49%
General hospitals	70%	
Nursing, convalescent homes	30%	
Home, scene of accidents[b]	30%	51%

[a] SOURCE: Ryder, C. F., and Ross, D. M., Terminal health care—Issues and
alternatives, *Public Health Rep.* 92:22 (1977).
[b] Of total patients polled, 80% prefer home care to hospitalization.

TABLE X.   Estimated Hospital Costs for Terminal Illness:
1973, 1977[a]

Year	Total hospital expenditures	Cost of terminal illness
1973	$19 billion	$ 4 billion
1977	$73 billion	$16 billion

[a] Estimate: 22% of all hospital (except psychiatric) and nursing home expenditures.

two government-sponsored studies of hospital expenses in the final year of life.[13,14] A study by the National Center for Health Statistics of the annual average 1,649,000 adults aged 25 years or older who died in 1964 or 1965 found that 73% had some hospital or institutional care in the 12 months before death. When cause of death was considered, there were marked differences in the amount and expense of care. Almost 92% of the decedents whose deaths were attributed to cancer received hospital or institutional care during the last year of life. Of those who received such care, 53% had bills of $1000 or more. The second study was of costs of those enrolled in Medicare who died in 1967 and 1968. Of the 21 million persons who were enrolled in 1967, 5% died, yet 22% of all reimbursements under the program were made on behalf of these decedents. The $1157 average reimbursement for decedents was over twice that paid out for survivors. The Third National Cancer Survey,[15] published later than the Mushkin review, found that of the 2863 patients dying within 24 months of diagnosis, there were 2088 with admissions to a hospital resulting in death. Average lengths of stay and payments for admissions that resulted in death were higher than the averages for all patients combined. On the average, first admissions that resulted in death were longer by as much as one week and costlier by as much as $735 than first admissions for all patients combined.

Using Mushkin's estimate as a bench mark, Table X illustrates the national significance of hospital expenditures for terminal illness in 1973 and 1977.

Studies of Costs of Terminal Illness in Patients and Families

The pioneer study of costs of terminal illness in families was published by Cancer Care, Inc., in 1973.[16] A social service agency in New York that modestly supplements the income of some of its client families who have a member with cancer taught families to keep records of costs. The study included 115 families with a member diagnosed with cancer. The median age of the patient was 60 and involved 84 women and 31 men, 80% of whom were married at the time of the study. The median income of the families in the sample was just over $8000, while the median income for families in NewYork State in 1970 was $10,870. Low incomes reflect, among other things, the loss of income due to the illness. The median duration of illness was 24 months. Table XI shows the distribution of the total cost of illness among the families, with the median cost being $19,054. Table XII shows that the median cost of the illness was 2⅓ times the median income of the families. Table XIII shows that while 84.4% of the families incurred total costs in excess of $10,000, only 39.1% received medical insurance payments of that amount or

TABLE XI.   Total Cost of Illness

Cost of illness	Number	Percent
Under $5,000	3	2.6
$5,000–9,999	15	13.0
$10,000–14,999	28	24.4
$15,000–19,999	16	13.9
$20,000–24,999	18	15.7
$25,000–29,999	12	10.4
$30,000–39,999	16	13.9
$40,000–49,999	4	3.5
$50,000 and over	3	2.6
Total	115	100.0

more. The families reported that expenses in excess of costs were met by cashing in life insurance, borrowing from relatives, gifts, deferral of college education, and loans from banks and other lending institutions. More than one-third (43) of the families were in debt 6 months after the death of the patient.

In the context of research conducted with a public health nurse-directed program of home care for children dying of cancer, Martinson and colleagues first estimated costs of home care as opposed to hospital care for a group of unmatched controls.[17] No randomization was used for selection of participants, but careful records were kept of costs. A home care group of 36 children was compared with a hospital care group of 22. Table XIV illustrates Martinson's findings. The mean costs for children dying at home ($810) were estimated on the basis of the costs for the research program to keep a public health nurse on call 24 hours per day to advise and work with the parents in caring for the child. Physicians served as consultants to nurses. The difference between the original estimate of the mean of hospital care, $5880, and the actual mean cost of $13,016 was due to the researchers' underestimate of the use of laboratory work, sometimes on the day of death. Striking differences were reported between the morale of survivors in home care and in hospital care, although these were gathered impressionistically. In addition to the findings on costs, the study does show that parents can successfully care for a dying child at home with the proper guidance.

A third study of patients and families ($N = 71$) was recently done by Kassakian and colleagues in a rural setting in Vermont.[18] Two groups of oncology patients were studied, those receiving nurse practitioner visits within the home and those not receiving such visits. By review of charts and recrods, the costs of health care in the last month of life were reconstructed. Classification was made according to place of death. "Home deaths" were patients who died at home. "Home–hospital deaths" were those hospitalized for not greater than three days before death. "Nursing home deaths" were those who died in nursing homes. All patients were expected to die and were no longer being treated with chemotherapy or radiation. Figure 1 depicts a comparison of costs for the last month of life, including a projection of costs associated with "hospice" care. "HHA" and "VNA" represent Home Health Association and Vermont Nursing Association, respectively.

TABLE XII. Total Cost of Illness in Relation to Annual Family Income

Cost of illness	Annual family income				Total families	
	Under $5000	$5000–9999	$10,000–14,999	Over $15,000	Number	Percent
Under $10,000	5	7	4	2	18	15.6
$10,000–19,999	15	11	11	7	44	38.3
$20,000–29,999	7	9	9	5	30	26.1
$30,000–39,999	4	6	4	2	16	13.9
$40,000–49,999	—	1	3	—	4	3.5
$50,000 and over	1	—	—	2	3	2.6
Total	32	34	31	18	115	100.0
Percentage	27.8	29.6	27.0	15.6	100.0	

TABLE XIII.   Amount Received from Medical Insurance

Amount received	Number	Percent
No medical insurance	3	2.6
Under $2,500	7	6.1
$2,500–$4,999	19	16.5
$5,000–$9,999	41	35.7
$10,000–$14,999	20	17.4
$15,000–$19,999	10	8.7
$20,000–$24,999	8	7.0
$25,000–$29,999	2	1.7
$30,000–$34,999	2	1.7
$35,000–$39,999	2	1.7
$40,000 and over	1	0.9
Total	115	100.0

Other measures were made of social and interpersonal aspects of the patient-family experience. The authors concluded that the home care group was benefited not only economically but psychologically as well, because more independence can be nurtured at home by relatives as opposed to the relative isolation encouraged by hospital rules and schedules.

The small empirical evidence available suggests that the greatest costs by far are hospital costs, and that in some cases the cost of burial will exceed the physician costs. Least is known empirically about physician costs in the care of the terminally ill. The Cancer Care, Inc., study found that costs for physicians and surgeons accounted for 12.3% of the total accumulated costs. Rice and Hodgson[19] found that the cost of cancer care in 1975 was divisible into the following categories:

Hospital care	78%	$4.100 billion
Physician services	13%	0.671 million
Other professional services, drugs, nursing home care	9%	0.477 million
Total	100%	$5.248 billion

Although these figures do not apply directly to the cost of terminal illness, it is plausible to believe that the cost of physician services in caring for cancer patients is in the same range as physician costs in caring for the terminally ill. The actual mean cost of physician services in caring for the terminally ill is probably lower than 13% of the total cost, but the issue needs much more careful study.

Indirect Costs of Terminal Illness

The social and economic costs of terminal illness include both direct and indirect costs. The latter category can be subdivided as follows: (1) lost wages and earnings of

TABLE XIV. Estimated and Actual Costs for Home Care and Hospital Care

Group	N	Duration of final care (days)			Cost estimate			Actual total hospitalization cost		
		Mean	Median	Range	Mean	Median	Range	Mean	Median	Range
Home care	36	32.4	21	1–104	$810[a]	$633	$65–$2,620			
Hospital care	22	29.4	21.5	1–89	$5,880[b]	$4,300	$200–$17,800	$13,016	$8,326	$68[c]–$58,833

[a] Based on the cost of nursing services at the rate of $10/day to be on call 24 hours a day and for telephone consultation, $45/home visit, and $10/clinic visit.
[b] Based on the cost of nursing services, room, and board at the rate of $200/day.
[c] Patient died in emergency room.

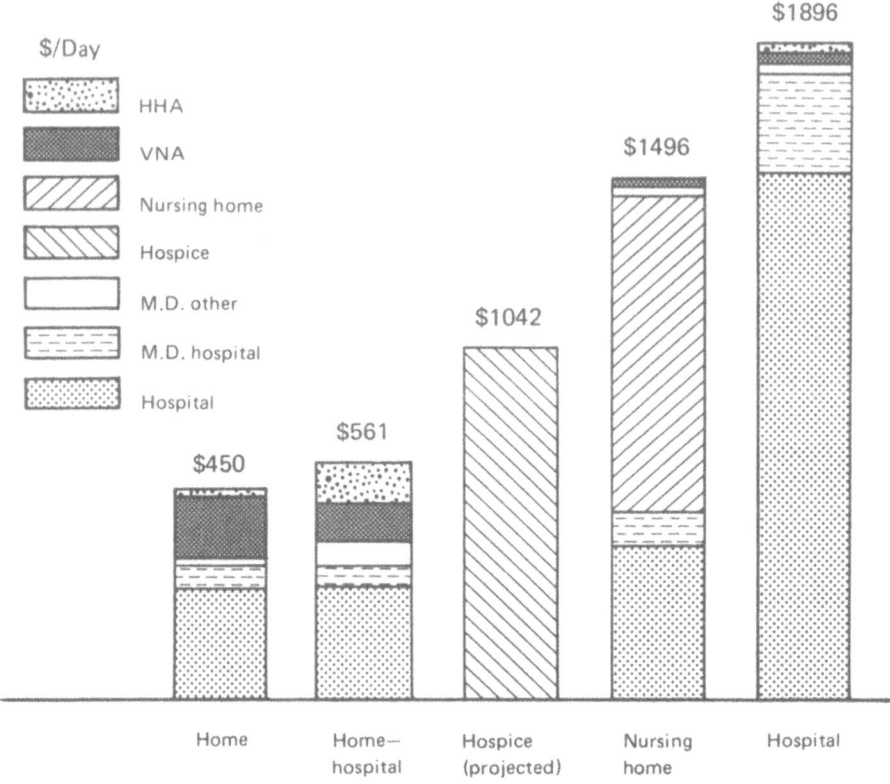

FIGURE 1. Comparison of home and hospital—relative cost of place of death. [SOURCE: Kassakian, M. G., Bailey, L. R., and Rinker, M., *et al.*, The cost and quality of dying: A comparison of home and hospital, *Nurse Pract.* 4:20 (1979).]

family members; (2) transportation, equipment for home; (3) pain, loneliness, fatigue; (4) behavior problems in surviving children; (5) morbidity and mortality of bereavement.

The Cancer Care, Inc., study found that each case of cancer (115) directly affected 333 other persons in the family or a proportion of 3 for 1. The "involvement factor" in the society must be at least 2 persons directly affected by the terminal illness of each 1 person. Family members' wages lost from work, investments deferred, and the costs of servicing loans are three indirect economic costs that have never been studied. Families in the Cancer Care, Inc., study reported that more than half (61) lost income as a direct result of the terminal illness, either through the patient's inability to work or because another family member had to stop or cut back on work to care for the patient. The same agency studied 88 children in families with a member dying of cancer.[20] While less than half the children had demonstrated deviant behavior before the parental illness, more than two-thirds were experiencing behavior problems at the time of the study. Parents reported school problems, attention-getting physical symptoms, sleep disturbances, eating disturbances, acting-out, aggressive behavior, and tearfulness.

Lynch[21] extensively reviewed the relationship of heart disease and other diseases with forms of loneliness such as experienced by the widowed, the divorced, those maritally separated, the old and single young living alone, and children from broken homes. A higher rate of morbidity was shown for each category. Jacobs and Ostfeld recently reviewed the literature on the effects of illness and death of one family member on the spouse or conjugal partner for the National Institutes of Health.[22] They concluded that "the findings on the mortality of bereavement together with studies of the morbidity of bereavement establish that there exists a significant threat to the health of one marital partner when death occurs to the other. . . ." Rees and Lutkins[23] surveyed a semirural area of Wales to determine whether bereavement produced an increased mortality among close relatives, defined as a spouse, a child, a parent, or a sibling. It was found that 4.6% of bereaved close relatives died within one year of bereavement compared with 0.68% in the control group. This is a sevenfold increased risk, and the risk was especially great for widowed persons. Risk of bereavement was doubled when the site of death was a hospital compared with home.

Summary

The previous section has surveyed relevant sources on the costs of terminal illness. More careful research must be supported in this area that accounts for costs in the last months of a terminal illness. The most outstanding and problematic cost of terminal illness is associated with hospital care. When this cost is matched with the increasing trend toward hospitals as the location of death, the question must be raised about whether the society is being well served by these institutions in the medical care of the dying. Dr. Robert Butler, director of the National Institute of Aging, has stated that "currently, less than one percent of the hospitals in the United States have any type of program related to the care of the dying."[24] This fact constitutes a final doorway to the many ethical issues associated with the cost of terminal illness. The following section will analyze and discuss some of these issues.

ETHICAL ISSUES AND THE COSTS OF DYING

There are three levels of ethical issues with respect to consideration of the costs of terminal illness. The first level stems from the question: What are the basic ethical principles that should inform any consideration of costs in the medical care of the dying? The second level relates to the problem of inequity between large expenditures on hospital costs and small benefits to dying patients and their families. The third level stems from policy choices about alternatives to hospitals in their current form as the predominant location for medical care of the dying. A discussion of these issues will follow an introduction to aspects of the question of costs of terminal illness.

What Kind of Question Is the Question of the Costs of Dying?

The inquiry has raised a question about the appropriateness of the costs of dying in the United States. This question can be taken strictly in an *economic* sense. Is the cost so high as to be prohibitive in the light of other needs? What ways can be found to economize? Do alternative forms of medical care of the dying really *save* money or will their start-up costs and hidden costs exceed even the size of current expenditures? The question about appropriateness of costs can also be approached in the strict *institutional-administrative* sense. Will it really *work* to make changes in the current forms of payment for medical services to the dying? Will such individuals and their families really receive better medical care? To give either of these questions top priority at this stage of moral conflict about the costs of dying does not do justice to the significance of the question. The question about appropriateness of the costs of dying is also a question about the *rightness*, in an ethical sense, of introducing the notion of costs at all into considerations of the care of the dying. In short, there are ethical considerations involved in raising the question of costs.

There has been an obvious reluctance to discuss publicly the issue of the cost of terminal illness. The paucity of the literature on costs in contrast to the growing literature on the experience and understanding of death and dying is an example. Part of the reluctance can be explained by the philosophy underlying third-party payments that was mentioned earlier, i.e., money should be no barrier to medical care. Yet there is more than displacement of fiscal responsibility behind this reluctance. When questions are raised about the costs of dying, there are responses of fear. The issue stirs up images of comparison of economic and social value of different persons, selection of persons for "dying wards," and also programs of euthanasia. There are serious moral problems of long standing associated with death and dying about which there is continuing debate, such as voluntary euthanasia[25] and selection of patients for lifesaving therapy.[26] Perhaps the lack of closure and social consensus on these issues allows them to "leak" into the issue of costs and cloud the picture. The ethics of the costs of terminal illness can be addressed independently of other moral problems in death and dying, but the question of motivation for addressing costs is related to the whole "field" of moral problems in death and dying. The actual conditions that cause excessive costs cannot be reformed apart from attention to ethical beliefs that relate to decision making in the care of the dying.

The costs of terminal illness are, in addition to being economic and political problems, a *moral* problem. Very great suffering is added to other forms of suffering by the size of costs involved and the stress of debts that extend beyond the death of a family member. The suffering can be understood economically, but it also must be seen in terms of increased morbidity and mortality of close relatives of the terminally ill. There is a moral duty to relieve suffering that is at the center of the morality that informs medicine. Therefore, moral reasons justify taking steps to reexamine the social practices that underlie the forms of medical care of the terminally ill that have evolved. An ethical evaluation should precede and accompany the economic and political considerations involved in a reexamination.

Functions of Basic Ethical Principles

There are many ethical principles within the moralities of the social system. Among them, the principles that should inform any discussion of the costs of terminal illness are justice and love. Before further discussion of their relevance, some functions of ethical principles as such will be briefly described. The most general function of ethical principles is to raise the deliberations about any specific conflict of obligations or duties to a general level of consideration as to what constitutes the best interests of the individual and the community. Ethics, pursued as principled reflection about concrete cases of moral conflict, is one of the highest human attempts for conflict resolution, although many moral conflicts are so tenacious as not to yield immediately to good arguments. Some compromises of ideal principles must usually be made to bring some "closure" to moral conflicts.

A second function of ethical principles is to furnish members of society with ideals for critical appraisal of moral rules and the moralities of specific groups. Morality, like other human institutions, requires reform and reconstruction, especially in times of rapid social change. The moral rules that apply to the medical care of the terminally ill have been under serious review for a major part of the century.

Third, ethical principles furnish grounds for validating moral rules and for following the obligations expected in application of the rules. Moral rules tell us what to do in specific situations, e.g., "killing is wrong," "harming others is wrong," "misusing social institutions is wrong," or "help others in need," "be fair." But these injunctions do not tell us *why* the rules are valid. Nor do they tell us what to do in cases of conflict of rules that produce great moral suffering, such as the conflict between the rule against killing and the rule to help the one in need that often conflict in the care of the dying. Ethical principles furnish a source of higher social authority to relate questions about why one ought to act in a certain way or what to do in cases of conflict. To answer such questions with "because the rules say so," or in a conflict of rules to refer only to one's own situational self-interest for a resolution, is considered wrong in this society. It is also wrong to refer to some supramoral authority ("God told me to do it"). Ethical principles provide higher standards available to a commonly shared need for reasons to back up and explain the moral rules.

Fourth, ethical principles have a symbolic function because, as ideal concepts, they point beyond themselves to the cooperation and reciprocity that are required for the social purpose of morality to be achieved, i.e., the resolution of conflicts of interests and desires. To be moral at all—that is, to overrule self-interest in cases where following it would be harmful to others—requires that one be able to "reverse" roles in behavior and imagine what it is like to be at the "giving and receiving end of particular actions."[27] Reciprocity, the ability to give and take mutually, presumes that one can imagine oneself in the other's place. Moral rules do not of themselves furnish the motivation to reciprocate. As ideal constructions, ethical principles do not furnish the self-respect or respect for others that is the presupposition of morality.[28] Ethical principles point beyond themselves to an "ethical spirit" that shapes attitudes of self-respect and respect for others that must be internalized in the person. The sources of inspiration for "ethical spirit" are many, referred to as "postethical" frameworks of belief about the world, human destiny, and human existence. Because there are many and varied belief systems in a pluralistic society that influence the shaping and selection of basic ethical principles, there is a need for common ground on which persons of differing beliefs can stand to cooperate in the practical tasks of reasoning about the great and small issues of moral conflict. The search for such common ground is beyond the scope of this chapter, but the need for sources of ethical unity in the midst of diversity of beliefs is especially poignant for those who work daily in the physical, moral, and spiritual tasks of companionship with those making final passage to death.

Justice and the Costs of Dying

There are at least two points at which the principle of justice is relevant to the costs of terminal illness. The first concerns what is due to individuals or their "desert." As explained by Beauchamp and Childress, "one has acted justly towards a person when that person has been given what he is due or owed, and therefore has been given what he deserves or can legitimately claim."[29] To be treated justly means to be treated according to rules of fairness—for example, not to burden an individual unfairly because that person happens to be vulnerable or have exploitable needs. The medical care due a person in the second or third stages of a terminal illness is not the same as that due a person in the first stage where curative therapy may be reasonable. The predominance of hospitals' acute care facilities as the site of death in the United States, and the lack of organized approaches to the dying in these units, makes it likely that the dying patient will be burdened unfairly at a basic level; i.e., an unfitting mode of care will be offered as the only choice. The Subcommittee on Terminal Illness of the National Institutes of Health[7] cited earlier introduced its final report as follows:

> The 1960's and 1970's saw the rise of highly developed medical technology and significant medical treatment advances for many diseases. There has been a trend towards dying in hospitals and similar institutions. . . . the health care delivery system of today is geared to episodic, acute, and curative interventions, usually of a sophisticated nature. Some believe that the modern hospital is highly depersonalized. Many authors have pointed out that the terminally ill patient in the modern acute hospital may encounter uncontrolled physical symptoms, isolation, and depersonalization instead of receiving sympathetic understanding and expertise in meeting his medical and emotional needs. The family may have limited access to the dying patient in the acute care hospital. Anecdotal reports suggest that the proper use of analgesics by the medical and nursing profession may need updating. Finally, traditional nursing home care has been described as spotty and episodic. (p. 1)

The burden of costs is a burden added to the underlying burden of an inappropriate mode of care. The American people probably would not mind paying for expensive care as long as it reached its intended goal and relieved the suffering for which the expenditure was intended. An analogy can be made to a community with a very expensive fire department. The community is willing to pay a great deal for protection from fires. However, the fire department does not extinguish fires as effectively in one section of the community as in others. The fire fighters are very effective in many areas, but in others the fires rage on. Yet the costs of the fire department continue to rise. When criticized by offended citizens, the fire department responds by saying that they have the best methods or fire fighting and are making every effort, etc. What must be changed first is the inappropriate mode of care predominantly offered to the dying in the acute care setting of the modern hospital. Reducing costs of dying may follow such efforts, but cost reduction should not be made the primary motivation for change in medical care of terminal illness.

PRINCIPLES FOR TERMINAL CARE.   What is due to the dying patient and family? Focusing on care of the terminally ill, the NIH Subcommittee[7] enumerated these principles for terminal care:

1. The patient and the family, not the disease, is the unit of care.
2. Treatment of symptoms (including psychological and social) of terminal illness is a prime importance in patient comfort.
3. Follow-up of survivors (bereavement care) has benefit and should be included in all programs of terminal care.
4. Home care as well as institutional care must be part of the program for the terminally ill and must be available 24 hours a day.
5. Terminal care requires a multidisciplinary approach involving physicians, nurses, social workers, dieticians, clergy, volunteers, etc.
6. Regular medical evaluation is required to determine if disease-oriented treatment would be of value to the patient and to provide necessary management of symptoms. (p. 3)

Although very little evaluation research has accompanied the first efforts to deliver terminal care in accord with these principles, what has been published is encouraging. Martinson reported significant differences in attitude and family solidarity among home care families as contrasted with families whose child died in the hospital.[30] Kassakian and her colleagues in a study cited earlier used the Karnofsky Performance Status to measure the quality of life among patient groups who died at home, in a hospital, or in a nursing home. Findings were not dramatically different in these groups, perhaps reflecting more than anything else the rural setting in which the study was conducted. Patients at home were found to be more independent and mobile, and those who were hospitalized less troubled by pain. The study does not report on the approach to pain taken in the home care cases. Hospice, Inc., of New Haven, Connecticut operated, until recently, only a home care and outpatient program on the principles cited above, which was supported by the National Cancer Institute. The organization is now housed in a physical facility for the care of the dying. In its final report to the Cancer Institute,[31] a study was reported comparing the psychological effects of hospice care with hospital care of the dying. Thirty-nine patients with only hospice care were compared to 35 patients woh had spent at least 50% of their terminal care in an acute care unit in a hospital. The study showed significant differences in that hospice patients had less anxiety, depression, and hostility and more feeling of solidarity with the family. A series of reports from Great Britain[32-34] and Canada[35,36] present encouraging results for physician, staff, and family in a reformed setting for terminal care ordered around the principles cited above.

COMPARATIVE COSTS OF TERMINAL CARE.   There are no studies with large numbers comparing the costs of unreformed terminal care in a hospital with other settings capable of applying the principles of terminal care. Beside the Martinson and Kassakian studies cited earlier,[17,18] what is known is sketchy but worth reporting. St. Christopher's Hospice, London, reported in 1978[37] the inpatient costs for three hospices as compared to those for a general teaching hospital (see Table XV).

The Palliative Care Unit of the Royal Victoria Hospital of Montreal reported that their approach had saved the Canadian Health Service almost $400,000 in 16 months, largely due to reduction in laboratory work.[38] A study of 59 hospice organizations by the

TABLE XV.   Inpatient Costs

Institution	Weekly rate (in pounds)	Percent of general hospital cost	Daily rate (in pounds)
St. Christopher's Hospice	177	51	25
St. Barnabas Hospice	170	48	24
St. Joseph's Hospice	120	34	17
General teaching hospital	350	100	50

Government Accounting Office[39] could find none with detailed reports of the number of services provided and their costs. Presumably, the lack of attention to costs reflects the stresses of the early start-up period in a new organization. Other sources of reluctance to study costs in the early stages of reform of terminal care may relate to ethical concern about the suggestion that cost reduction is the main purpose of reformed terminal care. In seeking to reduce physician and hospital opposition to reform efforts, those who try to make reforms would logically restrain opportunities to study costs. However, a competent and scientific study of the comparative costs of care is an outstanding requirement for policy-makers in this field.

JUSTICE AND INSTITUTIONAL APPROACHES TO TERMINAL CARE.   The second point at which the principle of justice is relevant to the costs of terminal illness is in the distribution of proper terminal care. The distributive aspect of justice relates to conditions of scarcity and the proper distribution of burdens and benefits. There is a scarcity of the proper mode of medical care for the terminally ill and their families. The ethical problem now facing policy-makers is how the most fitting approach can be developed and most equitably distributed among those who most need it. Present efforts could be described as "experimental" in terms of testing the delivery of terminal care within institutional approaches that vary widely. Seven organizational approaches to reformed terminal care are being used: (1) a hospital with a home care program; (2) a palliative care unit within a hospital; (3) a home care and outpatient program; (4) a "freestanding" hospice with physical facilities and a home care program; (5) a hospital with a team of experts who work with dying patients and families; dying patients are not separated from others; (6) a hospital whose leadership attempts to reeducate staff and establish the "hospice approach" within the whole hospital; no special team, and patients are not separated; (7) a nurse-directed home care program; physicians serve as consultants to the nurse.

There are many unanswered questions about whether reformed care in terminal illness can be offered side by side with acute care in the typical modern hospital setting. Will not the curative organizational approach of the current hospital overwhelm the effort to reform? Is it necessary to make a new institutional beginning on separate territory? These questions are directly related to costs, since the GAO study[39] did find that the amount of funds needed to establish a hospice is related to the types of services provided: ". . . the highest need is associated with those providing inpatient services in new facilities; the second highest, with those providing inpatient services in converted facilities; and the lowest, with those providing only home-based services" (p. 20).

A study of care of elderly persons that should make policy-makers cautious about the human and economic consequences of the separation of home care and institutional care of terminally ill persons was based on data from 1609 persons 65 and over from the Cleveland area.[40] As illustrated by Figure 2, the study found that as one becomes more impaired, the costs or values of home services increase and the proportion of care provided by families and friends also increases. At the "greatly impaired" level where the break-even point in cost is reached, families and friends are providing over 70% of the value of services received by older people. Families and friends spend $287 per month in services for every $120 being spent by agencies. The study clearly supports the idea that home care will never be enough for most elderly persons and that families cannot be expected to carry the burden completely at the greatly impaired level. This study should serve to "deromanticize" home care, especially since many elderly persons do not have intact families.

For the purpose of the ethics of policy making in this field, it is too early to make any more than a commitment to assist the evaluation, research, and planning that will be needed to make preferential decisions later. Among the institutional approaches, all of these institutional options should now be understood as being capable of embodiment of the principles of terminal care. None should now be especially preferred on the policy level. Each community and region should be encouraged to examine its own needs and the most fitting local approach should be tested in that community. Federal and state funds could be made available to assist in the planning process, as well as to encourage the

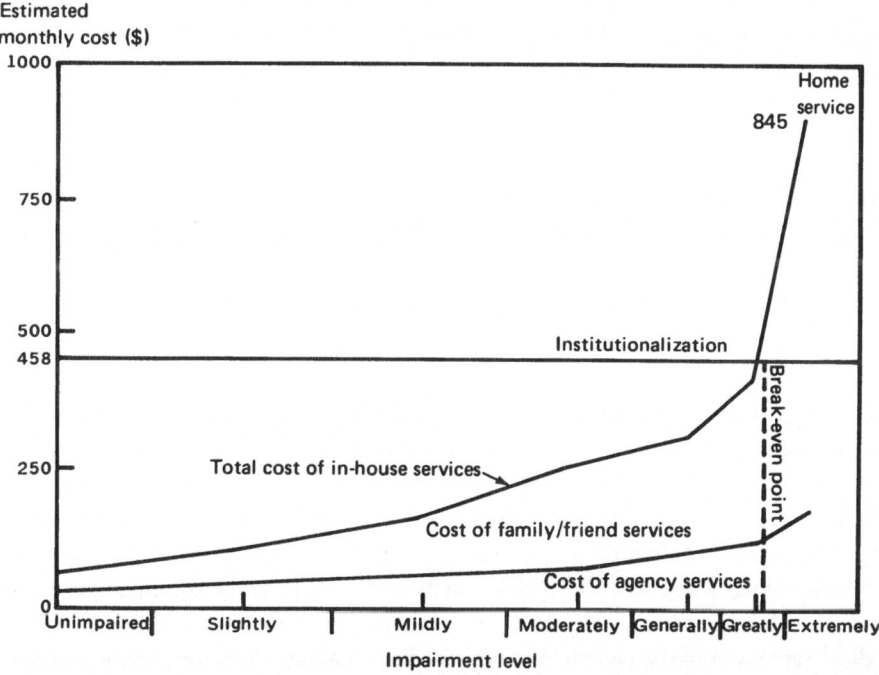

FIGURE 2. Comparison of estimated monthly cost of home services and institutionaliztion at each impairment level per individual. [SOURCE: Comptroller General of the United States: *Home Health—The Need for a National Policy to Better Provide for the Elderly,* HRD 78–19 (December 30, 1977).]

vitally needed social and economic evaluation of the comparative costs and benefits of the different approaches.

JUSTICE AND THE DISTRIBUTION OF THE COSTS OF TERMINAL ILLNESS.   There is also a scarcity of funds to pay for approaches that differ from the inpatient hospital care of the terminally ill. Until recently, large third-party public and private insurers did not reimburse for services except on an inpatient basis or for home care and skilled nursing services that were tied to a previous hospitalization for the particular illness. The Health Care Finance Administration of the Department of Health, Education and Welfare has inaugurated the Medicare and Medicaid Hospice Demonstration Program to study costs and utilization of services on an outpatient and home care basis. Prohibitions against paying for certain costs will be waived for participating hospice organizations in the project. Several private health insurance programs are also experimenting with such reimbursements. These are welcome first steps in remedies needed to redistribute the costs of terminal illness.

Also, protection from the costs of terminal illness is unfairly distributed. Not all Americans are equally insured against catastrophic and terminal illness. Although more than 90% of all Americans are either privately insured or enrolled in a public program of protection from losses in medical care, there are many Americans with "shallow" insurance coverage that does not adequately cover terminal illness.[41] The precise number of these thinly protected individuals is unknown, but it is known that in 1978 roughly 9% of all families had out-of-pocket medical expenses that exceeded 15% of their income. The uninsured are from lower-income families (income under $10,000) and are young. Unemployed individuals who are young adults account for one-third of the uninsured. Protection from the costs of catastrophic illness, which certainly include many terminal illnesses, needs to be more fairly distributed through the population as a whole.

Love and the Medical Care of the Dying

Within the framework of an ethical analysis of the moral problems of death and dying, the cost of terminal illness is a second-level concern vitally related to other moral problems of first-level import. To see the cost question in ethical perspective, one should define an ethics of medical care of the dying that is inclusive of the costs question. The kind and quality of care needed by the dying is the priority question that has been outlined in an earlier section. Within the spectrum of other moral problems in death and dying, there must be an adequate account of how decisions are made that move patients from the "sick role" to the "dying role."[42] The decision to suspend curative attempts for only palliative care is one that requires the most careful attention to informed consent of patient and family. When there is also an issue of physically or psychologically moving from one setting to another, the decision making can be exacerbated. Once the decision has been made to suspend treatment, a new question arises about the *consistency* of medical care owed to the patient. Periodic medical evaluations of the patient's symptoms and condition need to be made to ensure that the original decision to suspend treatment was valid. Decisions around the medication of the patient and the patient's activities are also value-laden.

In the view of the author, an ethics of medical care of the dying can best be understood by guidance from the principle of love. Love means *actively* to seek the good of the individual and the community. There is a note of urgency and immediacy in the love principle that is lacking in discussions of beneficence or nonmaleficence, other ethical principles that direct the seeking of good the or avoidance of evil. The author attempts to keep faith with a theological tradition of interpretation of love and justice as the most universal ethical principles, but he also employs other ethical principles and traditions in the practical resolution of moral conflicts. The author does not believe, as do some religious thinkers, that all other ethical principles can be derived from the love principle.[43]

The principle of justice does not inform us completely as to *why* the society should be concerned to remedy the underlying burden placed upon the dying and their families. Our beliefs about justice are offended by the uneven punishment inflicted on individuals by excessive expenses for a natural and unavoidable event. But beyond the injustice is the restriction on the need for love between the dying, their families, and others that is typical of the predominant mode of care. In the present mode, the patient is isolated from family for long amounts of time, immobilized, and often with restrictive rules about communication with children. There are many tasks to be accomplished when the patient is dying, not the least being the opening of communication between members of the family, which often requires expert help. There is much to be done in creative living, if there is time. Old hatreds and grudges can be faced and forgiven. Gratitude and thanks for a life well lived need to be expressed. Financial and legal affairs need clarifying and settling. The family needs to know itself in its strength and limitations. The patient's and family's need for religious help is especially great as they move together to the ultimate limit of life. To work together, cry together, ask for help, and give it are acts of love that increase the well-being of the individual and the community.

The expenditures in the medical care of the terminally ill and their families ought to be used to support a mode of care that increases the opportunities to express love between family members and others in the society involved with the family. The standards of justice are needed to remedy an unfair situation so that the needs of love may be better served.

The love that needs to be expressed between patients and families is a good in itself. This good is worth seeking entirely on its own. There may be, however, other benefits that extend from this good. Persons who return to their daily tasks from bereavement following the death of a family member that was characterized by acts of love are probably more likely to benefit the community and be less dependent on it for care. Persons who withdraw from family life and society even further as a result of the depression that can accompany loss that is characterized by unrelieved guilt and helplessness are less likely to benefit the community.

The quality of medical care of the dying and the quality of decision making in the moral problems of death and dying should be placed in a framework of the ethics of love, i.e., actively seeking the good ("best interests") of the individual and the community. The love principle also points to a source of self-respect and respect for others, even in the face of death, that nourishes the spirit of those whose task it is to care for the dying. Depression and resignation often characterize those who are in charge of medical care of the dying. It is difficult to keep one's heart in the battle when all of the patients die. Beyond ethical principles is the realm of commitment, the realm of the heart. Commitment to the belief that death is not defeat but an end that is an opportunity for growth for all concerned can strengthen the human capacity to love that is a primary source of meaning in human relationships. A far more expensive price can be paid by continued withdrawal from the ethics of love in the care of the dying than by continuing escalation of the costs of dying, although the latter is influenced in many ways by the former.

ACKNOWLEDGMENT

The author acknowledges gratefully the contribution of Nelson O. Chipchin, a volunteer worker at the Clinical Center, who assisted with the research for this article.

REFERENCES

1. Fletcher, J., *Morals and Medicine*, p. 34, Princeton University Press, Princeton (1954).
2. Sowle, C. R., When should patients be allowed to die? in: *Ethics in Medical Progress* (G. Wolstenholme and M. O'Connor, eds.), p. 199, J. and A. Churchill, London (1966).
3. Ayd, F. J., When is a person dead? *Med. Sci.* 18:33 (1967).
4. Beecher, H. K., Ethical problems created by the hopelessly unconscious patient, *N. Engl. J. Med.* 278:1425 (1968).
5. Williamson, W. P., Life or death—Whose decision, *J. Am. Med. Assoc.* 197:793 (1966).
6. Mushkin, S. J., Terminal illness and incentives for health care use, in: *Consumer Incentives for Health Care* (S. J. Mushkin, eds), p. 183, published for the Milbank Memorial Fund by Prodist, New York (1974).
7. Interagency Committee on New Therapies for Pain and Discomfort, Report of the Subcommittee on Terminal Illness, p. 2, National Institutes of Health, Bethesda, Md. (mimeo) (February 23, 1979).
8. Russell, L. B., *Technology in Hospitals*, Brookings Institution, Economic Studies Program, Washington, D.C. (May 1978).
9. Cooper, B. S., and Rice, D. P., The economic cost of illness revisited, *Soc. Sec. Bull.* 39:21 (1976).
10. Parniger, L., Berk, A., and Mushkin, S. J., *The Economic Cost of Illness, Fiscal Year, 1975*, Report B1A, Georgetown University, Public Services Laboratory, Washington, D.C. (1977).
11. U.S. Department of Health, Education and Welfare, *Health, United States*, p. 427, DHEW Publication No. (PHS) 78–1232 (December 1978).
12. Bureau of Consumer Protection, Federal Trade Commission, *Funeral Industry Practices*, Final Staff Report to the Federal Trade Commission and Proposed Trade Regulation Rule (16 CFR Part 453) (June 1978).
13. National Center for Health Statistics, DHEW, *Expenses for Hospital and Institutional Care During the Last Year of Life for Adults Who Died in 1964 or 1965*, United States (Series 22, No. 11) (March, 1971).
14. Piro, P. A., and Lutins, T., *Utilization and Reimbursements Under Medicare for Persons Who Died in 1967 and 1968*, DHEW, Social Security Administration, Office of Research and Statistics, DHEW Publication No. (SSA) 74–11702 (1973).
15. Scotto, J., and Chiazze, L., *Third National Cancer Survey: Hospitalizations and Payments to Hospitals*, DHEW Publication No. (NIH) 76–1094 (March 1976).
16. Cancer Care, Inc., *The Impact, Costs, and Consequences of Catastrophic Illness on Patients and Families*, One Park Ave., New York (1973).
17. Martinson, I. M., Armstrong, G. D., Geis, D. P., *et al.*, Home care for children dying of cancer, *Pediatrics* 62:106 (1978).
18. Kassakian, M. G., Bailey, L. R., Rinker, M., *et al.*, The cost and quality of dying: A comparison of home and hospital, *Nurse Pract.* 4:18 (1979).

19. Rice, D. P., and Hodgson, T. A., *Social and Economic Implications of Cancer in the United States*, National Center for Health Statistics, DHEW (mimeo) (June 1978).
20. Cancer Care, Inc, *Listen to the Children* (A study of the impact on the mental health of children of a parent's catastrophic illness), One Park Ave., New York (1973).
21. Lynch, J. J., *The Broken Heart*, Basic Books, New York (1977).
22. Jacobs, S. C., and Ostfeld, A. M., *The Effects of Illness and Death of One Family Member on Other Family Members: A Review with Some Recommendations for Research and Medical Practice*, p. 45, Departments of Psychiatry, Epidemiology, and Public Health, Yale University School of Medicine, (September 1978).
23. Rees, W. D., and Lutkins, S. G., Mortality of bereavement, *Br. Med. J.* 4:13 (1967).
24. Butler, R. N., A humanistic approach to our last days, *Dallas Med. J.* 64:509 (1978).
25. Kohl, M. (ed.), *Beneficent Euthanasia*, Prometheus Books, Buffalo (1975).
26. Swinyard, C. A. (ed.), *Decisionmaking and the Defective Newborn*, Charles C Thomas, Springfield, Ill. (1978).
27. Baier, K., *The Moral Point of View*, p. 108, Random House, New York (1966).
28. Aiken, H. D., *Reason and Conduct*, p. 82, Knopf, New York (1962).
29. Beauchamp, T. L., and Childress, J. F. *Principles of Biomedical Ethics*, p. 169, Oxford University Press, New York (1979).
30. Martinson, I. M., Facilitating home care for children dying of cancer, *Cancer Nurs.* 1:41 (1978).
31. Hospice, Inc., New Haven, Ct., *A Continuum of Care—At Home Program for Patients with Terminal Cancer and Their Families*, Final Report NCI-CN-55053 (1977).
32. Saunders, C., Terminal patient care, *Geriatrics* 21:70 (1966).
33. Saunders, C., The challenge of terminal care, in: *Scientific Foundations of Oncology* (T. Symington and R. L. Carter, eds.), Heineman, London (1975).
34. Twycross, R. G., Choice of strong analgesic in terminal cancer: Diamorphine or morphine, *Pain* 3:93 (1977).
35. Mount, B., The problems of caring for the dying in a general hospital, *Can. Med. Assoc. J.* 115:119 (1976).
36. Mount, B., Use of the brompton mixture in treating the chronic pain of malignant disease, *Can. Med. Assoc. J.* 115:122 (1976).
37. St. Christopher's Hospice, *Analysis of Nursing Costs*, Forty-First Newsletter, 51–53 Lawrie Park Rd., London (1978).
38. Palliative Care Service, Royal Victoria Hospital, Report, McGill University, Montreal (October 1976).
39. Comptroller General of the United States, *Hospice Care—A Growing Concept in the United States*, Report to Congress, HRD-79-50 (March 6, 1979).
40. Comptroller General of the United States, *Home Health—The Need for a National Policy to Better Provide for the Elderly*, HRD 78-19 (December 30, 1977).
41. Congressional Budget Office, Congress of the United States, *Profile of Health Care Coverage: The Haves and Have-Nots*, Background Paper, U.S. Government Printing Office, Washington, D.C. (March 1979).
42. Osmond, H., and Siegler, M., The doctor and the dying role, *Practitioner* 216:690 (1976).
43. Frankena, W., *Ethics*, p. 56, Prentice-Hall, Englewood Cliffs, N.J. (1973).

Discussion

PROF. RALPH B. POTTER (Harvard Divinity School): My remarks are really an appendage to the entire morning's discussion. I would like to offer reflections on the recurrent issue of "who shall decide" brought to the fore by the jurisdictional dispute of this morning.

To me, it is crucial to ask why the question "who shall decide" seems to be an important question. I gather the answer is that because we believe that what is at stake is *how* issues shall be decided. At the root of the jurisdictional controversies is the plausible assumption that many very important issues will be decided differently according to disciplinary training and professional vocation and formation of those who decide.

It is taken for granted in many cases that in deciding what is to be done, a scientifically trained geneticist would be likely to see things somewhat differently from a philosopher; that a lawyer will fail to appreciate considerations important to a physician; that a government official will have a perspective different from an immediately affected patient or his or her family.

Although interested parties may hold some values in common, they are seldom willing to trust fully to the common sense of others, which seems in our society to be fragmented. So we have this issue about who shall decide, which rests on the conviction that cases will be decided differently depending on which group is accorded jurisdiction, whose particular expertise and human sensibility is allowed to prevail.

Some systematic differences and patterns of formulating and acting upon matters of policy I think should in themselves be studied systematically with the aid of such disciplines as descriptive ethics, sociology of knowledge, and sociology of the professions. And I wish to summon such resources to our aid for the sake of turning attention to the question of how decisions are indeed made by particular groups and how peculiar biases might be compensated for in the effort to do greater justice to the concerns of all those affected by decisions in cases relating to genetics.

The type of study that I would like to see done could be performed by heavily funded research teams but more happily can be advanced simply by enhancing our own everyday powers of observation and recollection. I would have people proceed inductively, listening carefully to debates on occasions such as this but also every day, and seeing what factors each party actually considers to be relevant. Having thus gathered the elements of deliberation, one could ask a number of simple but edifying questions. For example, what type of question do people actually treat in trying to decide what is to be done?

Having worked on that kind of question in relation to other topical areas, I believe there are at least four recurring areas or types of issues. One always is an empirical factual set of questions pertaining to the particular case at hand, but alongside that there is a second area of judgments concerning whose good is first to be served in this case; and the mode of arguing that kind of question differs from the confirmation of factual elements.

But there is a third complicating element: there are beliefs regarding what makes actions right or wrong or ways of life good or bad. And finally, all these are seen in a broader context of fundamental perceptions concerning the meaning of life, which may vary from person to person and group to group.

I think this complicated structure of any debate about policy can shed light on why we have such sharp jurisdictional disputes. The different evidence and different logic, different rules of inference must be pursued in each of these several components. One implication is that there is no nonmoral view—vindicating, in a sense, Dr. Relman's insistence that medical decisions are indeed moral decisions. At the same time, equally, there is no approach which does not make some assumptions regarding facts which may fall most clearly within the expertise of a given field; so there can be no obvious monopoly of wisdom, which, after all, is that which is to be desired as we go about making our decisions in matters that deeply affect ourselves and others.

The second kind of question that follows on this isolation of the elements of decision would be to ask how the understanding of these several elements may be systematically weighted, biased, warped, given different interpretations by persons who have received particular forms of professional training and experience. Do philosophers habitually leave out of account considerations which physicians know to be important? Or do political scientists give undue weight to factors of special interest to their discipline? What forms of skewing of our perception can be attributed to patterns of professional training, to the isolation of some forms of professional practice, or to other identifiable sociological factors?

One could elaborate at great length—and I will desist from the temptation—about all of the kinds of questions one could pursue in this area to make ourselves more self-conscious about how our judgment is affected if not determined by the circumstances of our existence. The exercise, I believe, would be important in expanding and correcting our own judgments.

I want to end on a happier note but a more difficult note still: No one of us need be limited by our professional role. We bring to the deliberation whatever peculiar personal resources we have developed, and so we can ask beyond these questions and the sociology of knowledge and the sociology of profession what type of personal experience within and outside of professional life would it be useful to cultivate in order to increase the likelihood of wisdom and thoughtful decision, so that we can counterbalance our professional formation or deformation by nonprofessional involvements, mindful that it is the whole person who finally must decide.

I want to try to illustrate with one simple element. There is great debate about what constitutes a moral problem and there are, again, very regular controversies concerning even the definition of what the domain of the moral may be. Sociologists and those attuned to their way of thought tend to take "moral" to mean that which people actually resort to in trying to decide a given issue, and they reconstruct out of the observation of particular decisions a whole pattern of moral choice which they then describe to be acting. Whereas in another setting, psychological thought may tend to make "moral" mean that which has come to have a heavy psychological impact. So one can say, for example, that I formerly treated this as a matter of indifference, now it has become a moral problem; now I feel that the very core of my being is at stake and actively interested in this matter.

Philosophers habitually take "moral" to mean that which may be universally prescribed for all persons and must conform to some rule of generalized ability and so forth.

Over and above those three views there is a fourth everyday view which has to do with seeing a moral problem as one that treats the distribution of benefits and harms, that makes a difference, that changes the payoffs for various people involved.

I think I adhere to this last view. But it involves one in a very complicated pattern of arguing. Because in order to determine what is a benefit or what is a harm, one must clearly have a view of what life is about. And that is a rather grand order and it is rather more than we have been willing to debate in public for several generations.

I don't see how that debate can be kept out of the public forum again. We are going to have to open up to discuss the issues that were central to my field of ethics for centuries and have now been pushed to the sideline as being unmanageable and awkward, but I would say inescapable. So I hope that we can uncover resources to help clarify the debates that will go on endlessly, I fear, about who shall decide, treating them rather as questions of how ought questions to be decided.

18

Recent Developments in Abortion Law

Leonard H. Glantz
Department of Socio-Medical Sciences and Community Medicine
Boston University School of Medicine
Boston, Massachusetts 02118

At the First Symposium on Genetics and the Law held in May 1975 there was no paper presented whose sole purpose was to discuss abortion laws. *Roe v. Wade*[1] and *Doe v. Bolton*[2] were sufficiently new so that little legislative and judicial action had been taken. Since then there has been an outpouring of both legislative and judicial activity. Legislation has, for the most part, focused on ways to limit the impact of *Roe* and *Doe*, while the judiciary has been concerned with assuring that legislation not invade the zone of privacy created by *Roe*, as well as by previous and subsequent cases. Legislative activity can be described as ingenious at times and lawless at other times. The tension that exists between prochoice and antichoice (or prolife/antilife or proabortion/antiabortion, depending on which side of the issue you stand) factions is amply reflected in the tension between legislatures and the courts. Individual legislators feeling the pressure from well-organized antiabortion groups may vote for legislation they do not personally believe in, or that they know will be struck down in the courts.[3] The situation we find is that the prochoice viewpoint is litigated with scant private funds, while the antiabortion point of view is litigated with state funds through attorneys-general offices. Given this, the American Civil Liberties Union, Planned Parenthood, and other groups have amassed a remarkable record of successful litigation in defense of the rights of individuals to make procreative decisions without undue state interference.

In this chapter we will review how some states have tried to limit the scope of *Roe* and *Doe*, and how the courts have responded.

Viability

Since *Roe* held that a state can prohibit abortions after the fetus becomes viable unless the life or health of the woman is endangered by continuation of the pregnancy,

213

states have tried to define that point as early as possible. In *Roe* the court held that a fetus becomes viable when it is "potentially able to live outside the mother's womb, albeit without artificial aid."[4] Missouri, in its post-*Roe* statute, defined viability as "that stage of fetal development when the life of the child may be continued indefinitely outside the womb by natural or artificial life-supportive systems."[5]

Planned Parenthood of Missouri brought suit to challenge this and other sections of the statute.[6] The concern was that the term *indefinitely* could mean a very short period of time, and therefore this definition would markedly move up the point of viability. The Supreme Court found that the definition of viability did not violate the terms of *Roe*, since the point at which life could be sustained indefinitely may well occur later in pregnancy than the point where the fetus is potentially able to live outside the mother's womb.

The Court went on to say that it is not the "proper function" of a court or legislature to place viability at a specific point in the gestation period. It was recognized that the point of viability varied with each pregnancy, and the determination or viability in a particular case must be a matter of judgment of the attending physician.

In *Colauti v. Franklin*[7] the Supreme Court, in striking section 5(a) of the Pennsylvania Abortion Control Statute,[8] further clarified a vague aspect of the definition of viability in *Roe*. In *Roe*, at one point, viability is defined as the point at which the fetus has potential for "meaningful life."[9] Although this could be construed as requiring a "quality of life" judgment, the Supreme Court in *Colauti* stated that "meaningful life" means something other than "momentary survival."[10] Thus the assessment of "meaningful life" requires a temporal rather than a qualitative judgment.

Colauti also indicates that a physician may not be held criminally liable for making a good faith error in determining viability. In this case the Supreme Court tries once and for all to establish the primary role of the physician in making the viability determination, and excluding as much as possible the legislature and judiciary from establishing the point of viability. The Court states:

> Viability is reached when, in the judgment of the attending physician on the particular facts of the case before him, there is a reasonable likelihood of the fetus' sustained survival outside the womb, with or without artificial support. Because this point may differ with each pregnancy, neither the legislature nor the courts may proclaim one of the elements entering into the ascertainment of viability—be it weeks of gestation or fetal weight or any other single factor—as the determinant of when the State has a compelling interest in the life or health of the fetus. Viability is the critical point. And we have recognized no attempt to stretch the point of viability one way or the other.[11]

This powerful language should put an end to any legislative or judicial tampering with the term *viability*.

CONSENT AND INFORMED CONSENT

Some legislatures have tried to adapt (or corrupt) the concepts of consent and informed consent, concepts with the purpose of ensuring autonmy and self-determination, to deny autonomy and self-determination to certain classes of women, especially minors and married women.

In *Danforth* the Supreme Court ruled that a state may require a woman to sign a form stating that her consent is informed, free, and uncoerced. By requiring a woman to give an informed consent to an abortion, the court reasoned, the state assures that a woman's decision is made knowledgeably. This requirement also doesn't single out or unduly burden the decision to abort, since informed consent must also be given prior to undergoing other types of medical procedures.

In a footnote the Court accepted as a definition of informed consent:

> The giving of information to the patient as to just what would be done and as to the consequences. To ascribe more meaning than this might well confine the attending physician in an undesired and uncomfortable straightjacket in the practice of his profession. [12]

Although this standard was set forth by the Court in 1976, in 1978 Louisiana passed a statute requiring a physician to give every patient desiring an abortion eight specific "facts." [13] These include the following:

3. That the unborn child is a human life from the moment of conception and that there has been described in detail the anatomical and physiological characteristics of the particular unborn child at the gestational point of development at which time the abortion is to be performed, including, but not limited to, appearance, mobility, tactile sensitivity, including pain, perception or response, brain and heart function, the presence of internal organs and the presence of external members.

5. That abortion is a major surgical procedure which can result in serious complications, including hemorrhage, perforated uterus, infection, menstrual disburbances, sterility and miscarriages and prematurity in subsequent pregnancies, and that abortion may leave essentially unaffected or may worsen any existing psychological problems she may have, and can result in severe emotional disturbances.

This is not designed to inform the woman as to what will be done and the consequences, but to require a physician to describe to the woman the Louisiana legislature's view of fetal development and abortion. The statute is currently under attack in the courts[14] and should be struck down. A less restrictuve informed consent provision was found to be unconstitutional by an Illinois Federal District Court. [15]

The issue of spousal consent has been authoritatively decided in the *Danforth* case. There the Court held that a married woman's right to receive an abortion may not be conditioned upon her husband's consenting to the abortion. Or, to put it another way, a husband may not be given veto power over his wife's decision to abort.

Although the *Danforth* case also holds that parents may not be given veto power over their minor daughter's consent to an abortion if she is capable of giving an informed consent to the abortion, there may still be some role for parents in a minor's decision to abort.

In *Belotti v. Baird*[16] the Supreme Court hinted that a statute that prefers parental consultation and consent, but that permits a mature minor capable of given informed consent to obtain, without undue burden, an order permitting the abortion without parental consultation, and further permits even a minor incapable of giving informed consent to obtain an order without parental consultation where there is a showing that the abortion would be in her best interests.[17] might be constitutional.

On the basis of this and other opinions the Seventh Circuit Court of Appeals[18] held several parts of the Illinois Abortion Law[19] to be unconstitutional, and further held that:

1. A mature or emancipated minor must be allowed to receive an abortion without parental consent;
2. That there can be no rule that parents be notified in every case, but that parents can be so notified when such notification would further the minor's best interests;
3. If a minor must go to court to receive permission to obtain an abortion because she is not mature or emancipated the court must make its decision based on the minor's best interests;
4. That the state may not require refusal of parental consent as a precondition to such a judicial proceeding;
5. That any court procedure must be "speedy," "non-burdensome," and "anonymous."

Although the Supreme Court will be deciding a case involving a minor's right to consent to abortion, and the role of parents in this process, we can predict that the controversial nature of this area will lead to years of legislation and litigation.

PAYMENT

In regard to the rights of poor women, no issue is more important than determining the extent of the obligation of the state to finance their abortions through the Medicaid system.

The Supreme Court has decided three cases concerning this subject. In *Beal v. Roe*[20] the issue was whether or not the federal Medicaid law requires states to pay for "unnecessary" abortions. Under the challenged Pennsylvania Medicaid regulations, the state would pay for an abortion if

1. There is documented medical evidence that continuance of the pregnancy may threaten the health of the mother;
2. There is documented evidence that an infant may be born with incapacitating physical deformity or mental deficiency; or
3. There is medical evidence that a continuance of a pregnancy resulting from legally established statutory or forceable rape or incest, may constitute a threat to the mental or physical health of a patient, and
4. Two other physicians concur with these findings.

In a footnote[21] the Court pointed out that in *Doe v. Bolton* it was stated that "whether an abortion is necessary is a professional judgment that may be exercised in the light of all factors—physical, emotional, psychological, familial, and the woman's age— relevant to the well-being of the patient. All these factors may relate to health. This allows the attending physician the room he needs to make his best judgment."[22] All the parties to the litigation in *Beal* agreed that the regulations in Pennsylvania permitted the physician to take into account the factors cited in *Doe v. Bolton*. Given this definition of "necessary," the Court held that the Medicaid statute did not require states to pay for "unnecessary" abortions. The Court was not asked to decide whether it was permissible to require two other physicians to concur with the attending physician's judgment, but its permissibility is very questionable. The holding in this case is very narrow—as long as a state pays for abortions that physicians' exercising their professional judgment deem to be necessary for a very wide range of reasons, the state is acting in accordance with the federal Medicaid statute.

The second case, *Maher v. Roe*,[23] involved the constitutionality of a Connecticut Medicaid regulation that provided that the state would pay for any abortion that the attending physician judged to be medically necessary, including psychiatric necessity. As in *Beal*, this leaves the attending physician with wide discreation to decide what is necessary treatment for his patient. The plaintiffs argued, and the district court agreed, that exclusion of nontherapeutic abortion from a state program that subsidizes the medical expenses incident to pregnancy and childbirth violates the Equal Protection Clause of the Constitution.

Unfortunately, the Court, in reversing the district court's decision, did not simply say that since the Medicaid program does not fund other "unnecessary" procedures it need not fund elective abortions. Instead, the Court found that not funding such abortions does not impinge on a fundamental right, even though it makes the exercise of that right impossible. It also stated that *Roe* "implies no limitations on the authority of a state to make a value judgment favoring childbirth over abortion. . . ."[24] This is clearly not the case. If there were "no limit" on such authority, a state should be able to outlaw abortion, something *Roe* specifically prohibits. It is not only the holding of the case but its rather broad language that makes readers wonder about the Court's statement that "our conclusion signals no retreat from *Roe* or the cases applying it."[25]

Once again it must be stated that this case is rather limited. It was decided on a set of facts that left the determination of medical necessity up to the judgment of physicians. Further, the Court held that a state could favor childbirth over abortion where the woman would suffer no physical or mental harm from the continuation of the pregnancy.

In its third opinion, the Court in a *per curiam* opinion held that a city hospital need not perform nontherapeutic abortions.[26] This opinion was handed down with, and based upon, *Beal* and *Maher*.

It appears that the future of such litigation will involve determining the extent of the state's obligation to fund medically necessary abortions, and defining what constitutes medical necessity. Although the Supreme Court has not decided a case directly on point, lower federal courts have confronted the issue.

In Massachusetts, the Medicaid program provided that the state would pay for an abortion only if it was necessary to prevent the death of the mother, or to treat a victim of forceable rape or incest.[27] This was challenged in *Preterm v. Dukakis*.[28] The petitioners argued that the federal Medicaid law and regulations require states to fund all medically necessary abortions. Federal Medicaid regulations stipulate that a state Medicaid agency may not deny or reduce the amount, duration, or scope of a required service to an otherwise eligible recipient solely because of diagnosis, type of illness, or condition.[19] The court of appeals found:

> When a state singles out one particular medical condition—here, a medically compli-
> cated pregnancy—and restricts treatment for that condition to life and death situations
> it has, we believe, crossed the line between permissible discrimination based on degree
> of need and entered into forbidden discrimination based on medical condition. . . . We
> know of no other instance where a legislative decision to pay for medical care is based
> on the distinction between life and death. If only those suffering at death's door from
> sickle cell anemia or syphilis could receive publicly provided medical care, but not
> those condemned to a lifetime of dependency, one would be hard put to discern any
> rational social objective being thereby served. In addition, [the Massachusetts law] is
> inconsistent with the Act, which provides for a central role for the physician in
> determining proper treatment, by circumscribing his professional judgment so drasti-
> cally.[30]

The court therefore found that the Massachusetts law contravened the federal Medicaid law.

However, this was not the end of the case. Congress had passed a rider to the Health, Education and Welfare and Labor Appropriations bill that has come to be known as the Hyde Amendment. The Hyde Amendment states:

> None of the *funds in the Act* shall be used to perform abortions, except when the life of
> the mother would be endangered if the fetus were carried to term; or except for such
> medical procedures necessary for the victims of rape or incest, when such rape or
> incest have been reported promptly to a law enforcement agency or public health
> service or except in those instances where severe and long-lasting physical health
> damage to the mother would result if the pregnancy were carried to term when so
> determined by two physicians.[31] (emphasis added)

The petitioners argued that this amendment did not change the Medicaid law insofar as it requires states to pay for medically necessary abortions. Instead, they argued, this only prohibits the use of *federal* funds for such abortions that fall within the scope of the Hyde Amendment. Not only does this argument appear to be correct because of the wording of the statute, but it is bolstered by the fact that congressional rules provide that no appropria-tion bill shall change existing law.[32]

The appeals court found, however, that the legislative intent was to change the Medicaid law, and therefore the Hyde Amendment limited the scope of the Medicaid act. But the court did find that state law could be no stricter than the Hyde Amendment, and therefore Massachusetts would be required to pay for abortions where long-lasting and severe physical health damage to the mother would result from continuation of the pregnancy.

Although another court has found that the Hyde Amendment does not abrogate the state's obligations to fund medically necessary abortions,[33] a definitive resolution of the issue awaits decision by the Supreme Court.

The constitutional question remains to be answered.[34] It would appear that there is no rational distinction between a state paying for other necessary medical care, especially that care related to childbirth, and medically necessary abortions. One federal appeals court has already pointed out that funding doctors' services for women who will suffer physical injury if the pregnancy is continued but refusing to pay for abortions for women who will suffer mental injury under the same circumstances is a distinction that is "nothing less than absurd."[35]

ADVERTISING

Some states have passed laws restricting or prohibiting the advertising of abortion services. Louisiana, for example, absolutely prohibits the placing or carrying of advertisements publicizing the availability of abortion services. Violation of the law carries penalties of up to one year imprisonment with or without hard labor and/or a fine of up to $5000.[36]

In 1973 the Massachusetts Bay Transportation Authority refused to accept an advertisement from a clinic offering abortions services. The U.S. District Court enjoined the MBTA from refusing such advertising.[37]

It seems beyond dispute that any state action prohibiting abortion advertising violates the Constitution. In *Virginia State Board of Pharmacy v. Virginia Citizens Consumer Council*,[38] the Supreme Court held that a state may not completely suppress truthful information about a lawful activity even when that information would be considered "commercial speech." A later case, *Carey v. Population Services*,[39] dealt with a New York statute that prohibited the advertisement or display of "any instrument or article, or any recipe, drug or medicine for the prevention of conception. . . ."[40] Citing the Virginia State Board of Pharmacy, the Court struck down the New York statute as violating the First Amendment. The Court, in responding to one of the state's arguments, found that even if some individuals would be embarrassed or offended by contraceptive advertisements, this fact does not justify the suppression of truthful speech.

ZONING

In 1975 an abortion clinic entered into a lease in Southborough, Massachusetts, in order to acquire premises in which to operate their abortion clinic. After taking the appropriate steps to obtain a Certificate of Need, the town selectmen voted to amend their zoning bylaws to prohibit all abortion clinics. Every other type of medical facility was permitted to locate in the town, and the only other universally prohibited uses for all districts were "trailer camps," "commercial racetracks," "junk yards," and "piggeries or fur farms." The Massachusetts Supreme Judicial Court held that such action constituted an abridgement of a fundamental right and was therefore invalid.[41] The court also held that even if abortion services were available in adjacent areas, the town could not act as it did. It found that one town could not deprive a person of a fundamental right because it was available in another town.

In a similar type of case, Cleveland, Ohio, prohibited the further issuance of licenses permitting abortion services. However, in this case there existed other abortion clinics in the same city. As a result, the court found that this ban did not "unduly burden" either the abortion decision or the doctor–patient relationship, and therefore the ordinance was valid.[42]

It seems clear that a locality may not ban the establishment or operation of all abortion clinics through restrictive zoning and licensure ordinances. Unlike the situations in the Medicaid cases discussed earlier, by banning abortion clinics in a particular locale, there exists state action that totally prohibits the obtaining of an abortion. Indeed, if local governments were permitted to ban abortion facilities, and all local governments in a state exercised this authority, then the local governments would have effectively outlawed abortions in the entire state. This, of course, was explicitly prohibited in *Roe v. Wade*. It can be safely stated that as a result of this, zoning laws will not be an effective means of depriving women of their right to receive abortions.

Conclusion

This chapter has discussed a number of mechanisms that states have tried to use to restrict women's access to abortions, and the response of the courts. This discussion has included what I think are the most important issues decided to date. There are others— states have passed extensive record-keeping and reporting provisions (which are constitutional when the privacy of the woman is adequately protected and the record-keeping and reporting requirements are not unduly burdensome),[43] provisions depriving a woman of custody of her child if the abortion produces a live birth,[44] and high annual licensing fees for abortion clinics.[45] There are other means of state harassment that are used, and undoubtedly legislatures and antichoice groups will think of new ones. As in the case of desegregation, another extremely emotional issue, we can expect years, if not decades, of legislative and judicial action.

References and Notes

1. 410 U.S. 113 (1973).
2. 410 U.S. 179 (1973).
3. See the remarks of Massachusetts state senator Louis Bertonazzi in *Dilemmas of Dying: Policies and Procedures for Decisons Not to Treat*, G. K. Hall, Boston (1980).
4. 410 U.S. at 160.
5. H.C.S. House Bill No. 1211 §2(2).
6. *Planned Parenthood of Missouri v. Danforth*, 428 U.S. 52 (1976).
7. 439 U.S. 379, 47 U.S. Law Week 4094 (1979).
8. Pa. Stat. Ann., Tit. 35 §6605(a).
9. 410 U.S. at 163.
10. 47 U.S. Law Week at 4096.
11. 47 U.S. Law Week at 4097.
12. *Supra* note 6 at fn. 8.
13. La. R.S. 40:1299.35.6.

14. Fight brewing over louisiana abortion law, *Medical World News*, p. 42, Sept. 4 (1978).
15. *Wynn v. Scott*, 449 F. Supp. 1302, 1316 (N.D. Ill. 1978).
16. 428 U.S. 132 (1976).
17. Ibid. at 145.
18. Ill. Rev. St. Ch. 38, § 81–51 *et seq.*
19. *Wynn v. Carey*, 582 F. 2d 1375 (7 Cir. 1978).
20. 432 U.S. 438, 45 U.S. Law Week 4781 (1977).
21. 45 U.S. Law Week 4781 n. 3.
22. Citing, *Doe v. Bolton*, 410 U.S. 179, 192.
23. 432 U.S. 464, 45 U.S. Law Week 4787 (1977).
24. 45 U.S. Law Week at 4790.
25. Ibid.
26. *Poelker v. Doe*, 45 U.S. Law Week 4794 (1977).
27. Chapter 367, §2, Item 4402–5000 of the Massachusetts Acts of 1978.
28. No. 78–1324 (U.S. Court of Appeals, 1st Cir. 1978).
29. 42 C.F.R. §440.230(c) 1978.
30. *Supra* note 28 at 10.
31. Section 209 of Pub. L. 95–205; 91 Stat. 1460 (Dec. 9, 1977).
32. House Rule XXI (2).
33. *Zbaraz v. Quern*, Cir. No. 77C 4522 (N.D. Ill. May 15, 1978).
34. But see *Woe v. Califano*, 460 F. Supp. 234 (S.D. Ohio 1978), holding an earlier and more restrictive version of the Hyde Amendment to the constitutional. The short opinion appears to base its holding on lack of violation of due process, and does not discuss the equal protection issue.
35. *Supra* note 28 at 21.
36. La. R.S. 14:88.
37. *Preterm, Inc. v. MBTA*, Civ. No. 74–159-M (D.C. Mass. 1974), discussed in *Fam. Plann./Popul. Rep.* 3:76 (August, 1974).
38. 425 U.S. 748 (1976).
39. 431 U.S. 678 (1977).
40. New York Education Law §6811(8).
41. *Framingham Clinic v. Bd. of Selectmen of Southborough*, 367 N.E. 2d 606 (Mass., 1977); and see *Planned Parenthood of Minnesota v. Citizens for Community Action*, 538 F.2d 861 (8th Cir. 1977).
42. *West Side Women's Serv. v. City of Cleveland*, 450 F. Supp. 796.
43. *Planned Parenthood v. Danforth, supra* note 6.
44. See *Freiman v. Ashcroft*, 584 F.2d 247 (8th Cir. 1978).
45. La. R.S. 40:1299.35.17 ($1000 per year for each facility plus an additional fee of $500 per year for each physician who performs abortions).

LAW AND THE CONTROL OF GENETIC DISEASE

MODERATOR: AUBREY MILUNSKY

19

THE POTENTIAL PLAINTIFF
Preconception and Prenatal Torts

MARGERY W. SHAW
Medical Genetics Center
The University of Texas Health Science Center at Houston
Houston, Texas 77025

INTRODUCTION

It is well established in American law that a fetus is not a "person" in the constitutional sense.[1] Personhood occurs at the moment of live birth.[2] Fetuses are not counted as U.S. citizens in the decennial census; there is no income tax exemption for the unborn or the stillborn; states are under no legal obligation to provide welfare funds or food stamps for fetuses (although they may, if they wish, increase the benefits of a pregnant woman).[3]

Nevertheless, there are new developments in common law, statutory law, and constitutional law providing legal protection of the fetus as a "potential person." In addition, a body of law is growing on the yet-to-be-conceived person. A potential right to inheritance by unconceived or unborn heirs has long been recognized.[4] A legal cause of action for injuries to the fetus ripens at live birth.[5] The U.S. Supreme Court, in the 1973 abortion decision, recognized the state's interest in protecting the fetus during later pregnancy—the period of extrauterine viability.[6] But if the mother's life or health is endangered after viability, the fetus, as a not-yet-person, may lose its potential right to life.[7]

This dichotomy between maternal rights and fetal rights has created new tensions in our society. The courts are struggling with novel issues and applying tort law principles in their attempts to stabilize the disequilibrium created by new genetic and reproductive technologies.[8]

I shall briefly review a few cases of preconception and prenatal torts in order to highlight some of the legal problems encountered and I shall try to discern trends evolving in dealing with the roles of the parents, the child, the medical profession, and the state in our pluralistic society.

PRECONCEPTION TORTS

A long line of cases extending over at least 45 years has firmly established the principle that negligence in performing a sterilization operation followed by the birth of an unplanned child is recognized as a cause of action in common law.[9] Without arguing about semantic inconsistencies, the courts have variously referred to these torts as "wrongful birth,"[10] "wrongful pregnancy,"[11] and "wrongful conception."[12]

The judicial reasoning in the unplanned-baby cases has evolved from blessings to benefits to burdens. The Blessings Theory holds that any child is "a joy and a blessing" to its parents. As one judge phrased it: "Who can place a price tag on a child's smile or the parental pride in the child's achievement?"[13] Another judge said, "The plaintiff has been blessed with the fatherhood of another child."[14] Still another court echoed this sentiment, concluding that the "cost incidental to such birth was far outweighed by the blessing of a cherished child. . . ."[15] In these cases, there were no damages allowed to the parents for the malpractice of failed sterilization.[16]

The Benefits Rule gradually supplanted the Blessings Theory. Here the trier of fact may consider the expenses incurred and possibly nonmonetary injuries such as emotional pain and suffering, but these damages are partially or wholly offset by the tangible and intangible benefits of parenthood.[17]

In addition to sterilization, contraception is an alternative method for preventing birth. In the case of *Troppi v. Scarf*,[18] a pharmacist allegedly dispensed tranquilizers instead of oral contraceptive pills. The court held that, on remand, the jury could balance the positive and negative effects of having a healthy, unwanted child in fixing damages, using the Benefits Rule.[19]

The Burdens Rationale is based on the simple premise that if the parents took active steps (e.g., sterilization)[20] to prevent the birth of a child, and if, through the physician's negligence, they are burdened with the cost of rearing and educating that child, then the physician may be liable in damages.[21]

These preconception cases have involved the parents' desire to prevent the conception and subsequent birth of a healthy child. The trend here suggests the strengthening of parental rights to decide whether or not to conceive a child. The courts have become increasingly willing to protect the parents' right to choose not to conceive.[22]

More important to geneticists, however, is the prevention of conception of defective children. Genetic screening and counseling is useful in warning high-risk couples of possible genetic defects.[23] Physician awareness is critical to informing persons of the risk of gonadal exposure of mutagens (such as radiation and chemotherapy).[24]

Since the thalidomide tragedy, doctors and mothers have become increasingly aware of the teratogenic effects of drugs taken during pregnancy. In a recent case, the New York Supreme Court allowed the parents to claim emotional harm after the mother took a drug (Delalutin) to prevent a threatened miscarriage and later gave birth to a child without limbs.[25] But the Court of Appeals of New York has consistently denied damages for the parents' emotional injury when a defective child is born.[26] Whether this stance will be maintained, however, is doubtful.

In the case of *Renslow v. Mennonite Hospital*,[27] a 13-year-old girl received the wrong Rh-type blood transfusion. Eight years later she gave birth to an infant with erythroblastosis fetalis. The infant plaintiff sustained permanent brain damage. The Illinois Supreme Court, in 1977, upheld the child's cause of action for preconception negligence by the physician and the hospital.

In *Park v. Chessin*,[28] the parents were allowed a cause of action against a physician for allegedly failing to provide them with proper genetic counseling after their first child died of infantile polycystic kidney disease and before the second child was conceived. The risk of recurrence was one in four. Unfortunately, the second child was also affected and died 2½ years after birth. Although the physician was exonerated by the jury on remand to the trial court,[29] the principle is now well established that health providers have an affirmative duty to offer genetic counseling to couples who are at high risk of having a child with genetic disease.

In *Jorgenson v. Meade–Johnson Laboratories*,[30] the father sued a pharmaceutical company, claiming that the oral contraceptives his wife had taken before giving birth to twin daughters with mongolism caused damage to the chromosome structure within the mother's body, thereby inducing the disease.[31] The Tenth Circuit Court of Appeals held that the complaint stated a cause of action.[32]

Justice Oliver Wendell Holmes once stated that a person could incur "a conditional prospective liability in tort to one not yet in being."[33] The cases cited above have upheld that notion as it applies to physicians, hospitals, and drug companies. The ultimate question, then, is whether the courts will find a similar duty that parents refrain from conceiving a defective child when they are informed of their genetic risks. This would require no new elements of tort law. The corollary of the right to procreate is the duty not to harm another person by the exercise of that right. John Stuart Mill's espousal of individual liberty has been interpreted by some to mean that a person may choose in any matter that has a direct adverse effect on no one but himself.[34] But a federal appeals court judge disagrees with the distinction of harm to others and harm solely to oneself because in actuality others are affected by virtually any action that an individual takes or fails to take.[35]

There are two obstacles to a recognition of negligence actions against the parents of wrongful conception by genetically defective gametes, resulting in the birth of an abnormal child. The first is the concept of "wrongful life," which Professor Capron has discussed in Chapter 9.[36] This concept embodies the argument that an infant in wrongful life cases claims he should never have been born, which to some is a logical and legal absurdity[37] but to me is perfectly valid reasoning. The second stumbling block is the public policy of favoring intrafamilial immunity.[38] In order to preserve and promote family harmony, courts have been reluctant to allow spouses and children to sue each other.[39] But intrafamilial immunity has been eroded in other settings,[40] and some conditions may be burdensome enough to the defective infant and to society that the parental right to reproduce becomes instead a privilege to produce, insofar as possible, healthy infants.[41]

Prenatal Torts

After conception, the mother is given an unfettered right, during the first trimester, to choose whether or not to carry her fetus to term.[42] With few exceptions the decision is hers alone, unhampered by intervention by the state, the spouse, or the parents of pregnant minors.[43]

Failure of the physician to invoke a reasonable and timely diagnosis of pregnancy may result in a lawsuit. For example, a New York court in 1974 upheld the parents' complaint because their physician failed to diagnose pregnancy in time for an elective abortion of a presumably healthy but unwanted fetus.[44] But in the same year a Wisconsin court denied the parents' claim for the costs of rearing a healthy child after the physician failed to diagnose the pregnancy in time for an abortion, relying on the Blessings Theory that the doctor would pay the bills and the parents would have the pleasure of the child.[45]

Failure to make a proper intrauterine diagnosis in instances of defective fetuses denies the mother the right to selective abortion. In *Jacobs v. Theimer*, a physician failed to diagnose maternal rubella during early pregnancy and therefore failed to warn the mother of the risks of severe congenital defects in her child.[46] The parents were allowed to recover reasonable and necessary expenses for the care and treatment of their child's severe physical impairment.[47]

In another case of missed diagnosis, the physician not only failed to test the parents for the carrier state of Tay–Sachs disease but failed to offer amniocentesis to determine that the fetus was afflicted.[48] In still another example, the physician failed to offer amniocentesis to a pregnant woman with increased risk because of advanced maternal age, and an infant with Down syndrome was born.[49]

The tort of wrongful life has been asserted by the child in prenatal as well as in preconception cases.[50] Under what conditions would a child be justified in asserting such a claim? Some courts have allowed the claim against physicians, only to be overturned on appeal. No court has yet recognized a claim for damages by the child against the parents, for the reasons mentioned above.

I would proffer, however, that once a pregnant woman has abandoned her right to abort and has decided to carry her fetus to term, she incurs a "conditional prospective liability"[51] for negligent acts toward her fetus if it should be born alive. These acts could be considered to be negligent fetal abuse resulting in an injured child. A decision to carry a genetically defective fetus to term would be an example.[52] Abuse of alcohol or drugs during pregnancy could lead to fetal alcohol syndrome or drug addiction in the infant, resulting in an assertion that he had been harmed by his mother's acts.[53] Withholding of necessary prenatal care, improper nutrition, exposure to mutagens and teratogens, or even exposure to the mother's defective intrauterine environment caused by her genotype, as in maternal PKU,[54] could all result in an injured infant who might claim that his right to be born physically and mentally sound had been invaded.[55]

The fetus that is obliged by his parents and obstetrician to become a live-born infant should be protected by the state under the *parens patriae* doctrine. Courts have ordered blood transfusions to pregnant women who refused them, acting on behalf of both the mother and her fetus.[56] States have required Rh and VD testing of pregnant women in the interests of the fetus.[57] But courts and legislatures should not stop there. They should, through tort law and statutory law, take all reasonable steps to ensure that fetuses destined to be born alive are not handicapped mentally and physically by the negligent acts or omissions of others. As this review of recent cases illustrates, both the legislative and judicial branches of government have taken cautious steps in this direction. The pace is quickening, and perhaps, when we meet for the Third Symposium on Genetics and the Law, we will see the gap closing between science and technology on the one hand and legal responses on the other.

ACKNOWLEDGMENTS

The work presented in this chapter was supported in part by USPHS Medical Genetics Center Grant GM 19513.

REFERENCES AND NOTES

1. "All this, . . . , persuades us that the word 'person,' as used in the Fourteenth Amendment, does not include the unborn." *Roe v. Wade*, 410 U.S. 113, 158 (1973).
2. The "moment of live birth" is a convenient point in time that can be ascertained easily, witnessed, and verified, whereas the moment of conception, the moment of quickening, and the "period of viability" are impossible to determine with certainty.
3. The U.S. census does not enumerate fetuses. *Abele v. Markele*, 351 F. Supp. 224, 229, n.9 (Conn. 1972). The U.S. Supreme Court has held that states are not required to provide AFDC benefits for unborn children. *Burns v. Alcala*, 420 U.S. 575 (1975). But the New York Court of Appeals upheld the state's regulation allowing eligible women to claim AFDC benefits from the fourth month of a medically certified pregnancy. *Bates v. Toia*, 5 F.L.R. 2004 (N.Y. Ct. App., Oct. 16, 1978).
4. *Carroll v. Skloff*, 202 A.2d 9 (Pa., 1964). "If the fetus is never born, it does not and cannot have an estate from which others may take." Ibid. at 11.
5. *Bonbrest v. Kotz*, 65 F. Supp. 138 (D.C. 1946). This was the first case allowing recovery for prenatal injuries not resulting in death. See Annot., 40 A.L.R.3d 1220 (1971) for a review of liability for prenatal injuries. Wrongful death actions involving a fetus after viability resulting in stillbirth have met with some success, but since wrongful death actions are purely statutory it depends on the wording or the interpretation of the statute. See *Stern v. Miller*, 348 So.2d 303 (Fla. 1977), *rehearing denied*, for a case where the court held that a viable fetus is not a "person" within the intent of the Florida statute on wrongful death.
6. "For the stage subsequent to viability, the State in promoting its interest in the potentiality of human life may, if it chooses, proscribe abortion except where it is necessary, in appropriate medical judgment, for the preservation of the life or health of the mother." *Roe v. Wade*, 410 U.S. 113, 164–165 (1973).
7. Ibid.
8. See Shaw, M. W., Perspectives on today's genetics and tomorrow's progeny, *J. Hered.* 68:274 (1977).
9. An excellent review of these cases may be found in Robertson, G. B., Civil liability arising from "wrongful birth" following an unsuccessful sterilization operation, *Am. J. L. Med.* 4:131 (1978).
10. Of 51 "wrongful birth" cases involving sterilization failure, reported before mid-1978, 31 arose since 1973. Ibid. at 136.
11. In *Bushman v. Burns*, 47 U.S.L.W. 2155 (9-12-78), after a failed vasectomy, the parents narrowed their claim to the medical expenses and pain and suffering of pregnancy and delivery and did not claim damages for raising their child. The court designated this case "wrongful pregnancy," and stated that any benefits conferred by the child would not be used to offset the damages. See also *Coleman v. Garrison*, 327 A.2d

757 (Del. Super. Ct. 1974), ". . . view the action as one for 'wrongful pregnancy' rather than one for 'wrongful life.'" Ibid. at 761.

12. See *Sherlock v. Stillwater Clinic*, 260 N.W.2d 169 (Minn. 1977). "We hold that in cases such as this an action for 'wrongful conception' may be maintained. . . ." Ibid. at 170.

13. *Terrell v. Garcia*, 496 S.W.2d 124, 128 (Tex. Civ. App. 1973, *writ ref'd n.r.e.*).

14. *Christensen v. Thornby*, 192 Minn. 123, 126, 255 N.W. 620, 622 (1934).

15. *Ball v. Mudge*, 64 Wash.2d 247, 250, 391 P.2d 201, 204 (1964).

16. See Robertson, *supra* note 9 at 148 for a discussion of the controversial nature of damages in wrongful birth cases. Some courts claim that the intangible benefits of parenthood are difficult or impossible to assess, while other courts conclude that while it is possible to compare the economic burden of rearing a child with the benefits of parenthood, the outcome, *in every case, as a matter of law*, must be that the benefits outweigh the burdens. Ibid at 148.

17. The Benefits Rule, as it appears in the Restatement of Torts § 920 (1930), states: "Where the defendant's tortious conduct has caused harm to the plaintiff or to his property and in so doing has conferred upon the plaintiff a special benefit to the interest which was harmed, the value of the benefit conferred is considered in mitigation of damages, where this is equitable."

18. 31 Mich. App. 240, 187 N.W.2d 511 (1971).

19. Ibid. at 521.

20. Sterilization is the second most popular contraceptive choice in the United States, just behind oral contraceptives. Westhoff, C. F., and Jones, E. F., Contraception and sterilization in the United States, 1965–1975, *Fam. Plann. Perspect.* 9:153 (1977). Over 30% of all married couples in the U.S. had chosen sterilization in 1975. Ibid. at 154.

21. Large awards have been made in some cases. In *Bowman v. Davis*, 48 Ohio St.2d 41, 356 N.E.2d 496 (1976), a woman who gave birth to twins after a tubal ligation was awarded $450,000. See Robertson, *supra* note 9 at 137, note 20.

22. The landmark U.S. Supreme Court decision, *Griswold v. Connecticut*, 381 U.S. 479 (1965), protects a married couple's right to use contraceptives. This right was extended to unmarried couples in *Eisenstadt v. Baird*, 405 U.S. 438 (1972).

23. Reilly, P., Genetics, Law and Social Policy (1977). For an in-depth discussion of genetic counseling with liberal references, see Note, Father and mother know best: Defining the liability of physicians for inadequate genetic counseling, *Yale L.J.* 87:1488 (1978).

24. See Estop, S. D., and Forgotson, E. H., Legal liability for genetic injuries from radiation, *La.L. Rev.* 24:1 (1963), and Comment, Radiation and preconception injuries: Some interesting problems in tort law, *Sw. L.J.* 28:414 (1974). No long-term studies have yet been reported on chemotherapeutic agents that are known to be mutagenic, so that the risk to the germ cells in the gonads has not been assessed.

25. *Vaccaro v. Squibb Corp.*, 412 N.Y.S.2d 722 (1978), relying on *Karlsons v. Guerinot*, 57 A.D.2d 73 (1977).

26. See *Howard v. Lecher*, *infra* note 48; *Becker v. Schwartz*, *infra* note 49; *Park v. Chessin*, *infra* note 28.

27. *Renslow v. Mennonite Hospital*, 67 Ill.2d 348, 367 N.E.2d 1250 (1977), *aff'g* 351 N.E.2d 870 (1976). See Wachsman, H. F., Doctor and hospital sued successfully by child born eight years after injury to mother, *Legal Aspects of Medical Practice* 6:41 (July 1978), and Grumet, B. R., The ovarian plaintiff, *Legal Aspects of Medical Practice* 6:44 (July 1978).

28. *Park v. Chessin*, 387 N.Y.S.2d 204 (Sup. Ct. 1976), *aff'd*, 400 N.Y.S.2d 110 (App. Div. 1977), *modified*, 386 N.E.2d 807 (N.Y. 1978). Cohen, M. E., Park v. Chessin: The continuing judicial development of the theory of wrongful life, *Am. J. L. Med.* 4:211 (1978), provides a survey and analysis of the tort of wrongful life. See also, Note, Torts prior to conception: A new theory of liability, *Neb. L. Rev.* 56:706 (1977), for an excellent discussion of *Renslow*, *supra* note 27, and *Park*.

29. Oelsner, Jury clears two doctors of birth-defect liability, *New York Times*, B, p. 2, col. 1, April (1978).

30. 483 F.2d 237 (10th Cir. 1973).

31. Ibid. at 239. See also Carr, D. H., *Chromosome studies in selected spontaneous abortions: Conception after oral contraceptives, Can. Med. Assoc. J.* 103:343 (1970). Carr reports that 48% of 54 abortions conceived within 6 months after discontinuing oral contraceptives were chromosomally abnormal compared to 22% of 227 controls. Ibid. at 344. Three out of five abortuses conceived 8 to 12 months later carried one extra chromosome (mongolism is due to an extra chromosome). Ibid. at 345.

32. 483 F.2d at 241.

33. *Dietrich v. Inhabitants of Northampton*, 138 Mass. 14 (1884), cited and quoted in *Zepeda v. Zepeda*, 190 N.E.2d 849, 853 (Ill., 1963).
34. Mill, J., *On Liberty*, p. 22, Henry Holt, London (1859).
35. *Winters v. Miller*, 446 F.2d 65, 73–74 (2d Cir. 1971), Moore, J., concurring and dissenting; *cert. denied*, 404 U.S. 985 (1971).
36. Capron, A. M. The continuing wrong of "wrongful life", Chapter 9, this volume.
37. Tedeschi, G., On tort liability for "wrongful life," *Israel L. Rev.* 1:513 (1966).
38. Cohen, *supra* note 28, at 231 has argued that an intrafamilial wrongful life suit should be beyond the scope of judicial review since it turns on the moral question of whether the parents' rights to have a child should be invaded, and this question is protected by the constitutional right to privacy in *Griswold v. Connecticut*, 381 U.S. 479 (1965) and *Roe v. Wade*, 410 U.S. 113 (1973).
39. In two cases, illegitimate children brought suit against their fathers, claiming that they had been injured by the stigma of bastardy. *Zepeda v. Zepeda*, 41 Ill. App.2d 240, 190 N.E.2d 849 (1963), *cert. denied*, 379 U.S. 945 (1964) (court refused to allow damages); *Pinkney v. Pinkney*, 198 So.2d 52 (Fla. Dist. Ct. App. 1967) (cause of action disallowed). In *Burnette v. Wahl*, 47 U.S.L.W. 2483 (2-6-79), the Oregon Supreme Court refused to recognize a cause of action on behalf of child criminally neglected by his parents, causing emotional and psychological injury. A recent newspaper article reports that a 25-year-old man brought suit against his parents, complaining of "malpractice in parenting" causing him to need psychiatric care. The district court judge dismissed the suit. The *Houston Post*, § A, p. 28, col. 1, March 29 (1979).
40. The doctrine of intrafamilial immunity for intentional torts (such as beatings and rapes) and for negligence (such as automobile accidents) has been abandoned in most states.
41. See Shaw, M. W., Genetically defective children: Emerging legal considerations, *Am. J. L. Med.* 3:333, 340 (1977).
42. *Roe v. Wade*, 410 U.S. 113, 183 (1973). This choice, however, is undermined by the unwillingness of some states to provide Medicaid funds for elective abortions. The U.S. Supreme Court has ruled, in *Beal v. Doe*, 432 U.S. 438 (1977) and *Maher v. Roe*, 432 U.S. 464 (1977), that Title XIX of the Social Security Act does not mandate funding of nontherapeutic abortions, but the court does not speak to the issue of whether states can deny abortions if the fetus is defective but the mother's life or health is not endangered.

  In *Poelker v. Doe*, 432 U.S. 519 (1977), the U.S. Supreme Court held that the Constitution does not prohibit a state from expressing a preference for normal childbirth but the court did not differentiate between normal mother and normal fetus. It was ruling on a policy directive of the mayor of St. Louis and the staff practice of two city-owned hospitals that prohibits abortions except for "grave physiological injury or death to the mother." A Utah statute, which goes even farther because it limits funding to Medicaid recipients for abortion only when the mother's life is endangered, was upheld by a federal district court in *D.R. v. Mitchell*, 456 F. Supp. 609 (Utah, 1978). If a defect in the fetus could have been detected, but abortion funds were denied to the indigent mother, would she or her infant have a cause of action against the state rather than the physician? See Shaw, *supra* note 41 at 339.

  The Fourth Circuit Court of Appeals clarified the regulations in Virginia for funding Medicaid abortions by stating that the requirement that a physician certify that the recipient mother's life was endangered was contrary to the state's expressed objective of eliminating funding for nontherapeutic abortions. It ordered that the provisions include other factors such as physical, emotional, and psychological health of the mother as well as other familial factors and maternal age. *Doe v. Kenley*, 47 U.S.L.W. 2275 (10-31-78). Presumably this directive might encompass fetal defects. Medicaid policies in California specificaly allow reimbursement if amniocentesis has shown that the fetus will be born with a severe mental or physical abnormality. *Fam. Plann./Popul. Rep.* 7:61 (1978).
43. See *Planned Parenthood of Central Missouri v. Danforth*, 428 U.S. 52 (1976); *Singleton v. Wulff*, 428 U.S. 106 (1976); and *Bellotti v. Baird*, 428 U.S. 132 (1976).
44. *Ziemba v. Sternberg*, 45 A.D.2d 230, 357 N.W.S.2d 265 (1974).
45. *Rieck v. Medical Protective Co.*, 64 Wis.2d 514, 219 N.W.2d 242 at 244 (1974).
46. 519 S.W.2d 846 (Tex. 1975).
47. *Ibid.* at 850. See also, Kass, M., and Shaw, M. W., The risk of birth defects: *Jacobs v. Theimer* and the parents' right to know, *Am. J. L. Med.* 2:213 (1976), for a discussion of the duty of physicians and genetic counselors to keep their patients informed and recent progress in medical genetics and prenatal diagnosis likely to affect future court decisions.
48. *Howard v. Lecher*, 53 App. Div.2d 420, 386 N.Y.S.2d 460 (1976), *aff'd*, 42 N.Y.2d 109, 366 N.E.2d 64,

397 N.Y.S.2d 363 (1977). In another Tay–Sachs case, an amniocentesis was performed but the test results were mistakenly reported as negative. A cause of action was upheld for the parents' emotional pain and suffering but not for the child's claim of wrongful life. *Gildiner v. Thomas Jefferson University Hospital*, 451 F. Supp. 692 (E.D. Pa., 1978).

49. *Becker v. Schwartz*, 400 N.Y.S.2d 119 (App. Div. 1977), *modified* 386 N.E.2d 807 (N.Y. 1978). This is the only case in which an appellate court has upheld the parents' cause of action "for the expenses of raising and institutionalizing the afflicted child." 400 N.Y.S.2d 119 at 120.

50. For a listing of 19 cases where wrongful life claims have been made by the child, see Robertson, *supra* note 9, at 133, n. 3.

51. See text accompanying note 33, *supra*.

52. The elements of duty to the fetus, breach of that duty, an injured infant after live birth, and causation of that injury by failure to abort are all present. But see the obstacles discussed in text accompanying notes 36–41, *supra*.

53. A criminal charge against a heroin addict was brought in California for endangering her fetus by continuing to use heroin during the last two months of pregnancy, which resulted in the birth of twin boys addicted to heroin. The court found that the California Penal Code, § 273a(1), prohibiting the endangering of a child does not include an unborn child. *Reyes v. State*, 141 Cal. Rptr. 912, 915, 75 Cal. App.3d 214, 219 (1977).

54. Of 97 children born to 32 mothers with phenylketonuria (PKU), only 3 had IQ scores above 90, while 56 scored below 70. Perry, T. L., Hansen, S., Tischler, B., *et al.*, Unrecognized adult phenylketonuria: Implications for obstetrics and gynecology, *N. Engl. J. Med.* 287:395 (1973).

55. In a prenatal tort case, an infant was entitled to recover for injuries sustained in an auto accident caused by defendant's negligence. Judge Proctor stated that "justice requires that the principle be recognized that a child has a legal right to begin life with a sound mind and body." *Smith v. Brennan*, 157 A.2d 497, 503 (N.J. 1960).

56. In *Raleigh Fitkin Memorial Hospital v. Anderson*, 42 N.J. 421, 423, 201 A.2d 537, 538 (1964), *cert. denied*, 337 U.S. 985 (1964), the court ordered a blood transfusion for a pregnant woman who was a Jehovah's Witness, to save the lives of the mother and the unborn child.

57. Other examples of state intervention are cited in Shaw, M. W., Procreation and the population problem, *N.C. L. Rev.* 55:1165 (1977), suggesting that there is a compelling state interest in protecting the welfare of future generations and preventing the birth of children who are destined to be burdensome to the state. Ibid. at 1180.

DISCUSSION

MR. RICHARD E. VODRA (Cystic Fibrosis Foundation): I am extremely troubled with the prospect of allowing a child to sue her or his parents for wrongful life and with the whole concept that we discussed yesterday with Dr. Fletcher of the idea that the state has an interest in somehow compelling the nonbirth of genetically handicapped individuals.

I have three major concerns. The first is how would those disorders selected for abortion be defined? Prognosis in some diseases may be so variable as to invalidate such selections.

The second concern is in terms of negligence. Right now in cystic fibrosis there is no prenatal diagnostic test for the disease. But we do know that the sibling of someone with cystic fibrosis has two chances in three of being a carrier. If two siblings of different people with CF marry, odds are high that one of their children would be affected by cystic fibrosis.

Now, with high risks do you say the parents are negligent if they allow that child to be born?

Finally, who is the child going to recover from, since the child's support is based on the parents being able to care for the child. If the child is recovering from the mother, where is the monetary source? What is the benefit to the child taking the mother's resources and putting it in a trust fund for the child when that would take the resources away from the mother that are needed to care for the child?

It seems to me the whole concept of allowing the child to sue the parents for wrongful life fails on a number of tests and is really not a terribly fruitful line of further discussion.

PROF. MARGERY W. SHAW: I think if we were to take a vote in this room, there would be an overwhelming majority in favor of your position on wrongful life. Most people that I talk to bring up the same issues and arguments that you have so eloquently stated.

In terms of the negligence theory and the foreseeability, I think the courts have handled these kinds of things in many other situations. Whether one considers an injury an injury or not, we have all kinds of injuries that are compensable by law, from minor injuries to major injuries to death; and the courts deal day in and day out with these shades and degrees of injury.

In terms of proximate cause, if one is speeding at ninety miles per hour and no accident occurs, one is not held liable in negligence. But if one could calculate that there was a one in thirty-six risk that a fatal accident would occur at ninety miles per hour on a certain stretch of highway and then that risk materialized (in other words, in your case the cystic fibrotic child is born), then proximate cause can be claimed in court.

So I think these are not unique problems that the courts deal with. They don't deal with them in the genetic context.

MR. IAN KENNEDY (Lawyer, British Broadcasting Corporation, London): As an Englishman commenting on things going on in the United States, it can easily be said that I just don't understand them. I bring to them different cultural attitudes and so on; but it seems to me the conversation concerning the appropriate response of the legal system to problems usually creates problems itself.

It seems that perhaps in this respect, your views could well be bringing lawyers into the kind of disfavor one heard yesterday and this morning, because isn't it a case of following logic somewhere where one ought not to follow it? The proposition is that a child is born with some handicap. Surely the question is: What does one do

233

about that child's handicap? And if it is that one's major aim is to provide some sort of resources so that that child will be able to get through life, then the next question is: Where do the resources come from?

As I take it, madam, your response is to say one goes to the tort system. My view would be that that would be an inappropriate response and that there are much more creative devices known to the legal system and capable of being created by the legal system. One looks around the world and one can see schemes in England, in New Zealand, and in Australia for doing it. And the tort system is perhaps the least effective device. It falls down on so many difficulties. The correlation of proof, of costs. The only people who benefit are the lawyers. And I speak as a lawyer.

If one is to use the tort system, it must be madness to contemplate a child suing its mother. What is the possible benefit for that child, given the premise that one wants to build an institutional framework whereby the child is to be benefited? Where is the money coming from?

In the case of rape or assault, one is criminally liable with regard to the crime; and the other case of motor accidents—of course, there is insurance. In the context we are talking about, there isn't insurance for the most part. Where is the child to get his money from? Merely cutting the cake a different way.

It seems the whole fabric of your argument is built upon premises that we should reexamine if, on your ending note, we are to consider appropriate legal responses to genetic developments and technological developments in general.

PROF. SHAW: I would welcome alternative appropriate legal responses. I have considered criminal negligence, statutory law, and so forth as alternative responses. Fines and imprisonment are something that are anathema to our ideas of the freedom to reproduce.

But I think that this is what the symposium is all about, that we do talk together as Marjorie Guthrie has so eloquently pleaded for; that we explore the possible ranges of solutions and that we tackle them one by one.

DR. DANIEL STOWENS (Utica, New York): As a representative from a rural area, your concept of prenatal tort negligence threatens our whole way of life.

Now, a farm woman living in the typical American scene is exposed daily to herbicides, fertilizers, insecticides, diesel oil fumes. Furthermore, she is often so far away from any available medical care that she is willy-nilly negligent in not getting proper care in the first place.

Are we really supposed to have women who get pregnant move out of cities so as to avoid the pollutants there, and do we really depopulate our farms? If this concept holds, what are you going to do about it?

PROF. SHAW: I think I will limit any comments on environmental pollution until tomorrow's discussion.

A PARTICIPANT: Concerning immunity of parents to suits from children, the Supreme Court of Oregon recently decided that abused children have no cause of action against parents in civil suits based on parents' violation of statutes that are intended to insure minimal care and support of children. We should just think about the difference between court-imposed liability on parents and other parties, and that parents can with immunity do to their children what would be crimes or torts if they were committed on other people.

PROF. SHAW: Of course, we have child abuse statutes in many states where that immunity does not hold, and I might cite a criminal charge against a heroin addict that was brought in California for endangering her fetus by continuing to use heroin during the last two months of pregnancy, which resulted in the birth of twin boys addicted to heroin. The court held that the California penal code, which prohibited the endangering of a child, did not include the endangering of an unborn child.

Ms. BONNIE DANK (Washington, D.C.): I would like to rephrase the question concerning a child bringing a wrongful life suit against a parent. What should the state responsibility be in assuming the care of an individual who was born under such circumstances or perhaps after prenatal diagnosis had shown that it was going to have Down syndrome? What would be the state's position in suing the parent for wrongful life?

PROF. SHAW: I have no idea.

MR. DANIEL GABRIELS (Albany Medical College): It seems to me that those couples at risk who choose to reproduce without providing for more safeguards or utilizing those tests available have affirmed, and as a group are responsible for, the statistical reality of suffering occasioned by the decision of their class. I would regard a person firing into a crowd as equally morally aberrant whether or not the shot connected with its vulnerable target.

Perhaps afflicted children should be afforded a cause of action and recovery against that group of persons who bear corporate responsibility for the existence of their number, whether or not the irresponsible running of risks produced a victim in the particular instance.

DR. AUBREY MILUNSKY: Does the possibility of fetal abuse, as you described it, contradict the rights to procreate and found a family?

PROF. SHAW: Law is made difficult because many cases revolve around a right against a right. And here we are talking about the right to procreate versus the right to be born mentally and physically healthy.

An example, of course, is the right of freedom of the press versus the right to a fair trial. We have seen a lot of that in the newspapers lately.

Easy law is made when you are dealing with a right against a wrong. But when you deal with a right against another right and decide which right is more fundamental or which right is paramount, then you have problems; and the courts struggle with these all the time.

So that the right to procreate, yes, has been reiterated by the Supreme Court in many different settings and it will continue to be. I am suggesting that the right to be born mentally and physically healthy insofar as possible by the parents' and obstetricians' acts should be recognized also and considered and weighed and balanced along with the parents' rights to reproduce.

20

GOVERNMENTAL RESPONSIBILITIES IN GENETIC DISEASES

ARNO G. MOTULSKY
Center for Inherited Diseases
University of Washington
Seattle, Washington 98195

All modern governments include among their responsibilites the protection of public health of their citizens. The term *public health* is often broadly drawn and includes the promotion of good medical care with access of the entire population to medical practitioners and hospitals. The specific measures that have given citizens of developed societies their current long life-span consisted largely of good hygiene, which cleaned up environmental contamination of water and milk and made available preventive immunization and antibiotic agents. These procedures, together with the provision of sufficient food to prevent malnutrition in the majority of the population, have led to better health of developed societies than was ever known previously in human history. With current attainments, however, we have not yet reached the optimum in good health. With the conquest of most infections and nutritional diseases, developed societies continue to face birth defects and genetic diseases as significant health problems.

What do we expect from the government vis-à-vis genetic diseases? The issues are more complex than those faced in the prevention and management of infectious diseases. The spread of genetic diseases is quite different from that of infectious diseases. All infectious diseases spread *horizontally*. We would like to protect the present-day population against their ravages and we accept governmental intervention as a requirement for maintenance of public health. Genetic diseases spread *vertically*, i.e., from one generation to the next. The benefits of prevention are therefore not immediately apparent. Yet the responsibility to protect the good health of future descendents is generally acknowledged.

Concern with genetic diseases soon leads to problems of value judgments. What is normal? What is abnormal? Where does a genetic "disease" start as contrasted with a deviation from the normal? While everyone agrees that smallpox and infantile diarrhea are bad diseases, to be avoided, there would be less agreement about Turner's syndrome, about mild mental retardation, and many other conditions often classified as genetic diseases. Different ethical standards about certain reproductive practices such as abortion exist in our society. Since a positive intrauterine diagnosis of a genetic disease is sometimes the reason for abortion, government involvement in genetic disease tends to get embroiled in the abortion controversy with all its bitter implications. Not least, there are horrid historical precedents in the misapplication of genetic knowledge and pseudoknowledge in the hands of a criminal government, such as that of Nazi Germany.

In view of these considerations, any prescription for government involvement in genetic disease therefore must consider the various problems unique to this group of diesases. While our inclination is to state that (1) genetic diseases are common, (2) not enough money is spent on their prevention and management, and (3) the government therefore should provide these funds, we must realize that increasing governmental involvement in genetic diseases as a special and unique group of illnesses may pose possible dangers in *any* society. Whoever pays the piper calls the tune! The most well-intentioned government in the face of limited resources and exponential escalation of medical costs may curb civil liberties concerning reproduction and reward or punish its citizens financially or otherwise, depending upon whether certain reproductive decisions are favored or not. Forced sterilization may be one such consequence.

DEFINITION OF GENETIC DISEASE

A classification of genetic disease may be useful for further consideration of governmental responsibilities in medical genetics. Such diseases can be classified in four categories: (1) cytogenetic diseases, (2) monogenic or Mendelian diseases, (3) multifactorial disorders or interaction of multiple genes with environmental factors, and (4) maternal-fetal incompatibility such as Rh hemolytic disease of the newborn.

Not all cytogenetic abnormalities or Mendelian traits produce disease. We pragmatically include as genetic diseases those condtions that clearly cause illness or produce impairment of function recognized by both physicians and laymen alike. Thus most everyone will consider trisomy 21 or Down syndrome as a disease. Similarly, cystic fibrosis, sickle cell anemia, and muscular dystrophy are Mendelian conditions clearly accepted as disease by all observers.

Genetic factors play a role in many common birth defects such as cleft lip and palate and congenital heart disease; in diseases of middle life such as coronary heart disease, diabetes, cancer, and others; and in the common psychoses: schizophrenia and manic depressive illness. Such diseases are not usually classified as "genetic diseases." The cutoff point between genetic disease and nongenetic disease is blurred. Indeed, genetic factors play a role in most diseases. When considering the role of the government vis-à-vis genetic diseases, however, only unambiguous genetic conditions, i.e., those caused by chromosomal aberrations or Mendelian mutations, will be considered. The legislation of the recent Genetics Disease Act emphasizes Mendelian diseases over chromosomal aberrations.

Are there any other characteristics that distinguish genetic diseases from other sickness? Genetic diseases often cause disease from birth on. Many are incapacitating; others lead to mental retardation. Their impact in terms of social, emotional, and financial costs is usually greater than that of other illness. However, some genetic conditions, such as Huntington's chorea, may start later in life. Medical and rehabilitative treatment of genetic diseases as a group is not particularly different from management of other illnesses; i.e., genetic diseases at the present time do not offer a unique form of treatment not available for other illnesses. If gene therapy ever were to become available, special genetic techniques would be necessary to isolate the required genes, but the general approach of treatment of such patients is not likely to differ principally from current approaches to therapy.

It is generally agreed that the government has the responsibility to provide its incapacitated and sick citizens with good care regardless of financial status. How to provide such care is a matter of controversy. Third-party insurers as a group are not too happy to provide prolonged expensive care, and governmental support in one form or another is usually necessary. Such aid often takes the form of crippled children's services, mental retardation programs, and various maternal and child health plans.

There are certain unique features in the management of genetic diseases apart from medical and surgical treatment that make them a more complex problem. The new and special modalities of management of genetic diseases include the following: (1) genetic counseling, i.e., the provision of advice regarding recurrence risks and related care including intrauterine diagnosis for some conditions; (2) screening of various populations and subpopulations to detect, treat, or prevent genetic diseases.

GENETIC COUNSELING[1]

The crux in the context of this chapter is the ultimate outcome of genetic counseling. Most counselors feel that parents should have the right to select that course of reproduction which they prefer rather than following what is considered best for them by the counselor, society, or the government. Such parental rights include (1) taking of high risks that may lead to a child affected with genetic disease, (2) not choosing intrauterine diagnosis, or (3) not selecting abortion when the intrauterine diagnosis indicates a genetic disease. Yet, when the government provides funds for genetic services, such support is obtained under the premise that genetic disease will be prevented. Can we reconcile freedom of reproductive choice with such a goal? Fortunately, most parents will in fact select a course of reproductive action that coincides with that thought appropriate by society. The higher the risk, the fewer will be the number of children born to the involved couples.[2] Better education of the population regarding genetic disease and human biology undoubtedly will increase the number who make the "right decision" from the point of view of society. Yet the decision to bring a sick or a potentially sick child into the world must be respected. We should go out of our way to make sure that this civil liberty is not taken away or curtailed directly or indirectly.

The incorporation of all genetic services into a general medical health plan, such as national health insurance, that includes preventive medicine services of all kinds should make it easier to curb possible governmental abuses aimed at genetic diseases selectively. Appropriate education of the medical and public health professions and of legislators and the institution of national health insurance should ultimately make it unnecessary to set aside special funds for genetic diseases. With current reimbursement plans, however, governmental financial aid is required.

GENETIC SCREENING

Genetic screening currently involves (1) screening of those who will develop a disease and can be treated—phenylketonuria is the cardinal example, and (2) screening of pregnant women of advanced maternal age for chromosomal abnormalities. These types of screening programs ultimately need no special genetically oriented governmental schemes. Screening of newborn infants for phenylketonuria and selected diseases is already done by medical facilities and requires no special genetic techniques. It is incidental that PKU and other inborn errors are *genetic* diseases!

Screening of pregnant women of advanced maternal age for chromosomal abnormalities should be initiated by obstetricians as part of modern obstetrical care. Decisions following positive tests for chromosomal abnormalities that raise the question of abortion are individual decisions of the family. Certainly, every patient should be made aware of all reproductive options including intrauterine diagnosis and abortion. The ultimate parental choice must remain a private matter!

Another type of screening, such as used for sickle cell and Tay–Sachs carriers, requires prospective studies. Screening here involves widespread testing of the prereproductive or reproductive population. However, if *all* medical practitioners were to test Jewish pregnant women for Tay–Sachs heterozygotes* followed by screening of their mates if the test is positive, population programs would no longer be required. Strong efforts to include the Tay–Sachs testing program in medical practice are required. There are many difficulties with such an approach but ultimately such a scheme will raise fewer problems than current screening of the entire population of Ashkenazi Jews, who carry this gene in high frequency.

The sickle cell screening program in the black population has suffered from many problems.[3] Again, the development of intrauterine diagnosis by fetal blood sampling[4] and the newer schemes of a linked nucleotide sequence polymorphism[5] should allow sickle cell screening to be incorporated into the medical management of pregnancies.

The limitation of Tay–Sachs and sickle cell carrier testing as preliminary steps to intrauterine diagnosis can be contested since some carriers might want the option of selecting a spouse who does not carry the identical genetic trait. In other families, couples who are both carriers might prefer not to have children at all. Under such circumstances, more widespread screening would be required to identify those who are carriers.

---

*I realize the technical problems with testing of pregnant as opposed to nonpregnant women.

Society will need to decide whether large-scale screening programs should be instituted under governmental sponsorship. Alternately, education regarding common genetic traits causing disease in the affected population should lead to frequent requests of pre-pregnancy testing. However, such selective voluntary schemes are not likely to encompass as many individuals as would a broad population-based program. However, serious misconceptions about the meaning of the carrier state with possible psychologic harm and about the costs of population screening programs cannot be entirely dismissed.

OTHER PROBLEMS

It certainly is a governmental responsibility to ensure that all genetic services, particularly intrauterine diagnosis, are available equally for all members of the population, particularly poor and minority people. At the present time, such services are largely used by middle- and upper-class people.

Currently, third-party insurers usually refuse to pay for genetic counseling because it is a preventive and not a curative care. Such a stance is nearsighted. Strong efforts must be made to include preventive measures of all sorts in any new medical practice schemes.

There are no private practitioners in medical genetics. All genetic services are given by institutionally based professional genetic personnel. The development of medical genetics, which started as a pure research specialty, to one of practice ultimately needs to be followed by the creation of more service-oriented practitioners in the field, as is true for all other medical specialties.[6] The full recognition of medical genetics as a specialty will be helped by accreditation, as is currently under development. The existence of more practitioners of medical genetics alongside a new profession—that of paraprofessional counselors-genetic associates—should allow rational and cost-effective use of professional talent.

PROBLEMS WITH GOVERNMENTAL CONTROL IN MEDICAL GENETICS

To make sure that governmental funds are properly expended, all government programs need to set up regulatory rulings. Unfortunately, rules often lead to overregulation with its attendant bureaucratization, excessive personnel, much red tape, and increased costs. With the best intentions, much good work may get strangled in red tape! Working in a large university for many years, I have seen the development of more and more administrative layers, often ascribed to needs to conform to federal rules. The tremendous increase in law school enrollment in recent years also does not augur well for the future in this regard. More lawyers need more jobs and the government's new programs provide outlets that will lead to more regulation.

At the same time, the involvement of the government allows quality control and the establishment of standards. The new programs should ultimately bring genetic services to all geographic regions and to all social classes. Here, new ways of attracting genetic professionals to underserved areas of the country need to be devised. Distribution of government monies to every state regardless of the existence of talent and laboratory skills is not appropriate and squanders resources. Financial or other inducements need to be considered for optimal distribution of genetic services. Genetics is not unique—the problem of maldistribution exists with other specialized medical services as well and, in fact, with all of medical care.

CURRENT GOVERNMENTAL TRENDS

A popular method of examining medical expenditures is the use of cost–benefit analyses. If more money can be saved with a certain health scheme, such a plan, it is argued, should be instituted. Often, such analyses are carried out much too simplistically and there is no general agreement regarding the appropriate quantitative parameters.[7,8] Cost-benefit analyses in medical genetics are particularly difficult because benefits of the next or even of future generations often need to be considered. In general, one should be quite skeptical of such analyses and a full consideration of other than economic benefits in planning new programs should always be carried out.

Computerization of data is becoming a useful procedure. Governmental support is often tied to exhaustive data keeping that creates extensive computerized lists of patients and families with genetic traits or diseases. The possibility of abuse with computerization of personal data is well recognized but is even more acute with genetic diseases.

The fact that some genes are more frequent in certain ethnic groups often raises delicate issues and for some aspects of sickle cell screening has even caused the fear of genocidal intentions. Considerably more public education regarding human genetics will be required to obviate the many sensitive problems that arise from these facts.

FUTURE PROBLEMS RAISED BY GENETIC VARIABILITY

It is likely that more and more genes will be identified that make individuals more susceptible to environmental insults of all sorts.[9] The problem of environmental cancer may largely resolve as one of differential metabolisms of carcinogenic substances or of differences in DNA repair enzymes in various members of the population. Otherwise stated, the same environmental agents will have different effects depending upon an individual's genetic constitution. The field of ecogenetics deals with these matters. We need to know our genotype to practice optimal preventive medicine. The ideal society must provide that particular educational and occupational niche which best suits one's genotype! Legislation and regulations that recognize such genetic variability, however, can raise many problems.[10]

Some examples:

1. Should bread be supplemented with iron to help with widespread iron deficiency in the population if some individuals (i.e., those with hemochromatosis and thalassemia) might be injured by absorbing too much iron and will develop iron storage disease?

2. Should we encourage milk drinking in blacks if 80% of blacks are lactose-intolerant for genetic reasons and develop abdominal discomfort or diarrhea upon drinking too much milk?

3. What if some traits (i.e., alpha antitrypsin inhibition deficiency) cause occupational disease by genetic susceptibility to some irritant substances in industry? Who decides whether such a person should work? Could industry misuse such findings by blaming the worker for disease and not cleaning up the environment?

4. What if some chemicals that women are exposed to in their occupations are teratogenic? Can potentially pregnant women be excluded from the work force?

The scientific basis to answer most of these questions in imperfect and is particularly prone to abuse. However, human biochemical genetics uncovers more and more variation and some of this variation is likely to have health consequences.

WHAT ARE CURRENT GOVERNMENTAL RESPONSIBILITIES?

Research certainly is required to resolve many issues and needs governmental support. Such research involves basic and applied research in human and medical genetics as well as in fundamental genetics. The federal government particularly needs to foster good research in human genetic variation and its relationship to health and disease.

The federal government, in distributing funds for genetic training and particularly with the recently released funds for the Genetics Disease Act, has chosen to deal with state governments only. State governments often lack the necessary expertise in genetic services that are best represented currently in the large academic medical centers. Some tension between state health departments and the medical centers occasionally has occurred. Health departments have not yet fully realized that an increasing percentage of medical care, and not only research and teaching, is being provided by the academic medical centers. Almost 10% of all practicing physicians in this country are paid members of clinical departments of our medical schools![11] This service function of medical schools, whether one likes it or not, needs to be recognized by the federal and state governments in distributing funds from the Genetics Disease Act.

At the same time, much attention needs to be devoted to providing optimal and quality service at the least expenditure. Use of nonmedical genetic associates and experimentation with new and less expensive ways of providing genetic services therefore need to be encouraged.

CONCLUSION

1. Care of many cases of genetic diseases is not different from management of other diseases. The management of genetic diseases ideally should be incorporated in broad general health plans. Newer health schemes such as national health insurance should place a strong stress on prevention, including prevention of genetic disease.

2. As more genetic disease prevention and treatment is incorporated into the general health schemes, there will be less opportunity for possible abuses involving abrogation of reproductive freedoms.

3. Education of the public and the health professions in the principles of human and medical genetics will aid in individual decisions that are likely to be similar to those thought best by most informed members of society.

4. Human genetic diversity leads to differential responses to environmental agents. The societal and legal implications of these findings are considerable and deserve close attention by geneticists, lawyers, and the public at large.

ACKNOWLEDGMENT

The work reported herein was supported by USPHS Grant GM 15253.

REFERENCES

1. Motulsky, A. G., Genetic counseling, in: *Cecil Textbook of Medicine* (15th ed.) (P. B. Beeson, W. McDermott, and J. B. Wyngaarden, eds.), p. 56, Saunders, Philadelphia (1979).
2. Carter, C. O., Roberts, J. A. F., Evans, K. A., *et al.*, Genetic clinic. A followup, *Lancet* 1:281 (1971).
3. Murray, R. G., Jr., Public health perspective on screening and problems in counseling in sickle cell anemia, in: *Genetics Issues in Public Health and Medicine* (B. H. Cohen, A. M. Lilienfeld, and P. C. Huang, eds.), p. 264, Charles C Thomas, Springfield, Ill. (1978).
4. Kan, Y. W., Prenatal diagnosis of hemoglobin disorders, *Prog. Hematol.* 10:91 (1977).
5. Kan, Y. W., and Dozy, A. M., Antenatal diagnosis of sickle-cell anemia by DNA analysis of amniotic fluid cells, *Lancet* 2:910 (1978).
6. Motulsky, A. G., Postdoctoral education and training in medical genetics, in: *Service and Education in Medical Genetics* (I. H. Porter and E. B. Hook, eds.), p. 409, Academic Press, New York (1979).
7. Inman, R. P., Editorial, On the benefits and costs of genetic screening, *Am. J. Hum. Genet.* 30:219 (1978).
8. Nelson, W. B., Swint, J. M., and Caskey, C. T., A comment on the benefits and costs of genetic screening, *Am. J. Hum. Genet.* 30:663 (1978).
9. Omenn, G. S., and Motulsky, A. G., "Ego-genetics": Genetic variation in susceptibility to environmental agents, in: *Genetic Issues in Public Health and Medicine* (B. H. Cohen, A. M. Lilienfeld, and P. C. Huang, eds.), p. 83, Charles C Thomas, Springfield, Ill. (1978).
10. Motulsky, A. G., Bioethical problems in pharmacogenetics and ecogenetics, *Hum. Genet. (Suppl.)* 1:185 (1978).
11. Committee on a Study of National Needs for Biomedical and Behavioral Research Personnel (National Research Council), Personnel Needs and Training for Biomedical and Behavioral Research, 1978 Report, National Academy of Sciences, Washington, D.C. (1978).

DISCUSSION

Ms. MARJORIE GUTHRIE: When we talk about government, I like to think that government is "we." In order to make people feel that there is an opportunity to get the government, meaning us, to do what is right for families with these kind of problems, I just want to bring in the word *hope*.

A number of people have come up to me and said "Why is it that families with a disease like Huntington's disease make an appointment and don't show up, or if they do come, never return again for the further counseling that would make them feel that they can manage the problem of Huntington's?" and my answer is that, somewhere along the line, the word *hope* has been missing. If people would take the attitude that was expressed by Dr. Motulsky, and others, then there would be the opportunity to do the education that might lead to a better result at the end of a long period of discussions.

But when people feel hopeless, they are not going to come, because they feel that there is no role to play. They are sure you are going to say, "Of course, you mustn't have children."

This is the attitude that has been perpetuated. I think that your sensitivity, which I certainly appreciate, must be shared with the public. Somewhere there might be a role that we should play through the government, through the National Genetic Diseases Act, to educate people to come. Because there is hope for discussion and for the future of their families if the professional people will share those kinds of feelings.

DR. AUBREY MILUNSKY: Examination of the abysmal appropriations for the National Genetic Diseases Act leads one to the view that much energy expenditure might be useful if the medical genetics community and the public at large began to educate their congressmen and senators about the importance of genetic disease as a public health problem.

Dr. Motulsky, what kind of master plan could be devised? All of us have had experience with legislators, both in state and federal office, who become catalyzed to unbelievable action as soon as their families are touched.

PROF. ARNO MOTULSKY: Since you are dealing with legislators, this becomes a political process. Patients with certain diseases who can mobilize a large group will get more of a hearing than a single individual. Unfortunately, in medical genetics we are dealing with many different diseases so there is no clearly defined genetic disease lobby that legislators look to. If in fact such a lobby could be organized, I think we would have better results.

In the long term, I feel that *not* singling out genetic diseases as something special would be better. I plead for de-emphasis of the unique nature of genetic disease and prefer that genetic diseases by considered within the broad framework of care and prevention of *all* diseases.

MR. MITCHEL KLING (Student, Harvard Medical School): There is just one thing that bothers me about government programs to prevent genetic disease. What might it do to free choice? For example, a person with $alpha_1$-antitrypsin deficiency, more susceptible to various acquired lung diseases, might be kept from doing certain jobs simply because the company doing the work would consider that person a bad risk. I wonder what suggestions you might have to offer for that problem.

PROF. MOTULSKY: I believe much of this will be discussed tomorrow.

It's a very thorny issue. Ultimately we will have to face the fact that we all possess different genotypes, and certain "environments" are better and others are worse for each individual. How to deal with this problem by legal regulations and to prevent abuse is going to be very difficult. There is no simple answer.

But I think it is wrong to say that there is no problem. There isn't a problem yet. To wish it away and say that government can abuse such concepts is not justified. We must get together and come forward with scenarios that might happen. I think this conference is a good place to do that and tomorrow afternoon when these matters will be discussed in detail some answers hopefully will be forthcoming.

PROF. Y. EWARD HSIA: I want to ask Dr. Motulsky to clarify one problem and comment on another. The one problem I would ask him to clarify is if we want to de-emphasize the difference in genetic diseases, should we then no longer have any emphasis on genetics? Should we go back to the status quo, where most physicians are unaware of any need for consideration of genetic problems as being at all distinct from nongenetic problems? In other words, should there be a subspecialty of medical genetics?

The second problem is that in the subspecialty of medical genetics counseling, I admit that I am a parasite and that I have been heavily subsidized. But it probably costs me about five hundred dollars per family to give good counseling. Now, if we are to maintain this kind of quality within the private medical community, costs would be prohibitive.

PROF. MOTULSKY: You put your finger on some important practical problems. I certainly would not de-emphasize genetics teaching or attempts of understanding of genetics among the medical profession or other health professions. I am very strongly interested in such teaching. We have to do a lot in order to make physicians and others understand more about genetics and how genetic factors produce diseases.

My message is that for administration of genetic disease programs we should try to incorporate prevention and treatment of genetic disease into broader health schemes of various sorts and work toward the idea that genetic counseling is an important component of disease prevention.

I would hope that management and prevention of genetic disease can be placed within that framework and can be paid for appropriately. I realize very well that the provision of genetic services is very expensive. We must find ways of decreasing the cost of genetic counseling such as by use of more nonmedically trained genetic counselor-associates. Various new schemes need to be instituted which will decrease such costs.

The way we provide genetic services now with its high costs will be difficult to justify for the entire population. Furthermore, there just would not be enough personnel available.

A PARTICIPANT: Given the expense of genetic counseling and given what I know is your long-standing interest in the whole question of genetic susceptibility, there is a whole range of diseases that we don't commonly think of as being genetic at all. Do you look forward at some foreseeable time to the disappearance of genetic counseling as a medical specialty?

PROF. MOTULSKY: No. In fact, genetic counseling will become more accepted. At the present time this field is considered esoteric, special, unique, and out of the mainstream of medicine. What I hope to see is that treatment of genetic disease becomes an integral part of medical practice and genetic counseling an integral part of prevention of disease.

DR. MILUNSKY: By placing medical genetics where it really should be, as part of the fabric of medicine, does one not run the risk of removing its fundamental uniqueness, in terms of prevention?

PROF. MOTULSKY: At the present time, unfortunately, insurance companies, the public, and the medical profession are largely care-oriented and prevention is not given as much emphasis as it should have. But there are enough forces at work to stress that much of future improvement of public health (i.e., cancer and heart disease) will lie in better health habits and prevention of these diseases. As soon as these facts are recognized, and they are increasingly recognized, genetics will become a harminous part of the various facets involved in disease prevention.

21

PUBLIC PARTICIPATION IN GENETIC POLICYMAKING

The Maryland Commission on Hereditary Disorders

NEIL A. HOLTZMAN
The Johns Hopkins University School of Medicine
Baltimore, Maryland 21205

Division of Hereditary Disorders
Preventive Medicine Administration
State of Maryland Department of Health and Mental Hygiene
Baltimore, Maryland 21201

The Second Symposium on Genetics and the Law would not have taken place if the control of genetic disease depended only on counseling about the recurrence risks of a few rare diseases, and on abstinence from procreation of those with affected offspring. To these quaint options, however, others have been added. As a result of populationwide screening, some diseases need no longer devastate a family even once; presymptomatic affected individuals can be identified in families without a positive history and treatment started in time to prevent disability (for example, phenylketonuria). Through screening, carriers can be identified before they have any affected offspring (for example, sickle cell or Tay–Sachs diseases). Prenatal screening (for example, maternal serum alpha-fetoprotein for neural tube defects) and diagnosis, and abortion of affected fetuses, as well as artificial insemination and *in vitro* fertilization can also reduce the burden of genetic diseases.

Mechanisms are poorly developed for deciding when, how, and which new technologies should be widely disseminated, or for maintaining their effectiveness and safety under routine use. Nor are there adequate procedures for educating providers and consumers in order to assure appropriate and voluntary use of these technologies, or for protecting confidentiality of information and for determining who has legitimate access to

The author is a former member of the Maryland Commission on Hereditary Disorders.

it. Because everyone is a potential screenee, because these technologies are applicable to a large number of conditions whose combined incidence is not rare, because they often depart from traditional medical practice, and because they may create conflicts between the individual and society, the establishment of policies dealing with their use should involve the public.

In this chapter, I will first analyze several mechanisms for the dissemination of procedures for the detection and management of *specific* genetic disorders. Next, I will describe the efforts of one state, Maryland, to establish policies dealing with a *wide range* of hereditary disorders. Finally, and again paying particular attention to Maryland, I will consider the possibilities of obtaining meaningful public participation in genetic policymaking.

Market Incentives

In the marketplace, commercial pressures could result in excessively rapid diffusion of new technologies. Drug companies and commercial laboratories have recently discovered the profitable market for prenatal and newborn testing: approximately 3 million pregnant women and infants each year! Several companies are eager to market reagents for serum alpha-fetoprotein screening of pregnant women, even though experience with this test in the United States has been restricted to a few pilot programs.[1]

Physician reticence might constitute a counterforce to commercial eagerness and reduce the rate of diffusion below acceptable levels.[2] Practicing physicians have little training in genetics[3] and a substantial minority do not believe that neonatal PKU screening is worthwhile.[3,4] Recently, one commercial laboratory advertised its newborn hypothyroid screening test directly to the public, presumably so that mothers would exert pressure on their physicians to order it.[5]

Although the countervailing forces of commercial pressure and physician reticence might ultimately result in optimal use, three factors argue against the laissez-faire solution. First, tests provided on a fee-for-service basis (the usual market mechanism) might be too costly for some; inequities in the utilization of prenatal diagnosis already exist.[6] Second, individuals who would benefit from genetic services might not seek physicians' services. Third, other forces are unlikely to wait until the market reaches equilibrium.

Categorical Compulsory Legislation

When physicians were slow to adopt neonatal PKU screening in the early 1960s, the National Association for Retarded Children and its local chapters, aided by a few zealous professionals, lobbied successfully for mandatory PKU screening laws in many states.[7] At the time, the validity of the screening test and the effectiveness of treatment had not been established. These laws have many defects. Few of them address the problem of assuring reliable laboratory procedures. In at least one state with a mandatory screening law for seven diseases, health department resources are inadequate to follow up all of the infants with abnormal test results.[8] In some states many commercial laboratories perform the test, raising costs to parents and making quality of screening and follow-up difficult to monitor. Only a handful of states with compulsory laws reimburse parents for the costs of treating affected infants.

Nor are health care providers always capable of interpreting the results of mandatory tests. A decade after the PKU laws were passed, almost half of practicing pediatricians and family practitioners surveyed could not interpret an abnormal test result correctly.[9]

The trend of compulsory categorical legislation continued into the early 1970s, when sickle cell screening was made a prerequisite for admission to public schools.[7] The benefits of this requirement escape definition, and doubts remain about what sickle cell screening accomplishes.[10]

COMMUNITY-BASED PROGRAMS

Communities have initiated mass screening programs for genetic conditions like sickle cell, Tay–Sachs, and thalassemia, which have high frequencies in well-defined ethnic groups. Although the organizers frequently include those who have had direct experience with the specific disease and understand the reasons for screening, this knowledge may not be adequately conveyed to others in the community. Participants have limited understanding of risk.[11] Those with positive tests may suffer diminished self-esteem and feel stigmatized.[12-14] Peer-group pressures may subtly coerce some to participate in community or school mass programs.[15]

MALPRACTICE SUITS

The courts may influence the rate at which genetic screening, diagnosis, or counseling is adopted by the medical profession by ruling that failure to provide these services constitutes malpractice. An increased number of older pregnant women may be offered prenatal diagnosis as a result of the ruling of the New York Court of Appeals that doctors might be negligent by not informing patients of the availability of the technique.[16] This is a slow and uncertain mechanism, however, for establishing public policy.

COMPREHENSIVE APPROACHES

As is characteristic of the American way, each of the approaches discussed thus far deals with only a segment of genetic services, usually just a single disease or a few diseases. None of them establishes a set of principles applicable to a wide range of genetic conditions or to a variety of services, such as screening, diagnosis, counseling, and treatment.

None of these approaches allows for comprehensive planning, for dispassionately setting priorities among programs. Instead, those that are most profitable, or that represent the interests of the most politically effective lobby or the best organized community, are the ones to which resources are applied.

In the early 1970s, concern that problems already encountered in genetic screening would recur as the field expanded resulted in the publication of guidelines for genetic screening.[7,17] The establishment and subsequent operation of the Maryland Commission on Hereditary Disorders demonstrated that these principles could be translated into effective public policy.

ORIGINS OF THE MARYLAND COMMISSION ON HEREDITARY DISORDERS

In 1972 the Maryland General Assembly passed the Sickle Cell Anemia Information and Prevention Act requiring that couples coming for marriage licenses be informed about sickle cell anemia and that pregnant women be tested for it. The Maryland Health Department convened a committee of medical specialists, community physicians, and consumers to implement the act. Many participants doubted that "crash" legislation dealing with specific disorders was the ideal way to proceed. Rather than recommend implementation of the Sickle Cell Act, the committee recommended its repeal and, in its place, new legislation that would establish general principles for programs dealing with a wide range of hereditary disorders.[18] Under the leadership of the senator who introduced the sickle cell legislation, the General Assembly in 1973 repealed the Sickle Cell Act and established the Commission on Hereditary Disorders. In so doing, the assembly recognized that "specific legislation designed to alleviate the problems associated with specific hereditary disorders may tend to be inflexible in the face of rapidly expanding medical knowledge. . . ."[19]

The Commission is empowered to "promulgate rules, regulations and standards for the detection and management of hereditary disorders for the State of Maryland." These must be in accord with a number of principles including (1) consultation with experts and laymen in public hearings; (2) consideration of the "medical psychological, ethical, social, and economic effects of each program"; (3) informed consent; (4) availability of test results to screenees or their parents; (5) availability of counseling, which is nondirective and does "not require restriction of childbearing"[20]; (6) confidentiality of information pertaining to individual subjects; (7) regular review of procedures and programs. The Commission also has the authority to investigate unjustified discrimination resulting from identification of a person as a carrier of a hereditary disorder.[19]

To ensure further that regulations promulgated by the Commission were in the best interest of the public, the legislation provided for a consumer majority on the Commission. Five of the 11 voting members represent the public. Appointed by the governor, they cannot be health care providers. Two Commission members represent the legislature, and the remaining 4 represent the medical societies and medical schools in the state. These voting members have staggered four-year terms, renewable once. The Maryland Department of Health and Mental Hygiene has *ex officio* representation on the Commission.

WORK OF THE COMMISSION

Turning to the work of the commission, let us consider whether it is consistent with the principles just enumerated, and to what extent it has been influenced by consumer participation.

The Commission has overcome bureaucratic inertia, accelerating the institution of new health department programs and improving coordination between units within the department. It initiated pilot screening programs for neonatal hypothyroidism and galactosemia, and when the Commission's Committee on Hypothyroid Screening, which included several endocrinologists, was satisfied that problems concerning laboratory reliability and follow-up had been eliminated, the Commission voted to make hypothyroid screening routinely available.

Toward the end of 1977 the Commission began to examine the benefits and risks of maternal serum alpha-fetoprotein (AFP) screening for fetal neural tube defects. After information and opinions from laymen and professionals were collected, and after assuring adequate availability of follow-up procedures, the Commission endorsed a pilot program of maternal serum AFP screening. Obstetricians and parent groups have already been contacted, the health department is currently setting up the laboratory procedures, and screening begins in 1980. As part of the program, the health department will make personnel available to counsel women with confirmed abnormal test results in their obstetrician's offices.

As the principles embodied in its founding legislation called for informed consent for all testing, the Commission requested the General Assembly to repeal an earlier statute that made PKU screening compulsory except when parents objected on religious grounds. In 1975 the assembly complied and the Commission promulgated its own regulations, which required informed parental consent.

The Commission's activities and regulations do not deal only with programs for which the state health department has primary responsibility. The Commission promulgated regulations requiring that any facility in the state that offers sickle cell screening, whether public or private, must inform the Commission of its programs in detail. In addition, each program must involve the community being screened in planning and operation; provide educational information describing the disease, its genetics, and the purpose of the program to individuals before they are screened; obtain informed consent; treat all results as confidential medical records; and provide follow-up when necessary. The Department of Health is responsible for enforcing the regulations.

The Commission has also been concerned about inadequacies in third-party coverage for genetic disorders. Working with the Hemophilia Foundation, it obtained from Maryland Blue Cross/Blue Shield an agreement to reimburse for the home use of concentrated clotting factors on a pilot basis. The Commission recently obtained permission from PKU families whose medical insurance does not cover low-phenylalanine formulas to negotiate with their insurance companies to modify this anachronism. As a result, several companies have agreed to cover these formulas. The Commission is currently surveying the availability of prenatal diagnostic services and of reimbursement for them.

The Commission holds public hearings before deciding how to act. Both lay and professional Commission members have appeared on radio talk shows to discuss genetic issues and the work of the Commission.

Recognizing that public policies cannot be appropriately maintained without assessing public response, the Commission has actively encouraged studies to evaluate the impact of new regulations and programs. The effects of the Commission's requirement for parental informed consent before newborns can be screened for PKU and other disorders is currently under investigation by a group at Johns Hopkins. This group will also assess the education, consent, and counseling interventions of the AFP screening pilot program. The reliability and validity of the screening test itself will be determined separately. Only if the pilot phrase demonstrates that the program is of net benefit will the Commission support AFP screening throughout the state.

Thus the Commission has applied the general principles set forth in its founding legislation to a host of problems. Moreover, since the Commission began, the legislature has refrained from passing legislation dealing with genetic diseases. Bills dealing with hemophilia and mandating neonatal hypothyroid screening were withdrawn when the Commission indicated that it was dealing with these issues.[21]

On several occasions the Commission has asked the legislature to exclude the abortion of seriously defective fetuses, discovered by prenatal diagnosis, from any restrictive legislation. Thus far, the legislature has complied.

PUBLIC PARTICIPATION

Although the Maryland legislation was adopted as a model by the Council of State Governments,[22] and some of its provisions have been applied elsewhere, no state that I know of has given a body on which consumer representatives predominate the regulatory authority vested in Maryland's Commission. The reason, I believe, stems from the reluctance of professional groups to share power. In Maryland, community involvement in initiating and supporting the legislation offset the opposition of some professionals. When the General Assembly was considering the establishment of the Commission, one health department official expressed concern that it "would have the power to dictate program operation," and interfere with the health department's traditional role. A practicing physician objected to the consumer majority on the Commission because it would "clearly dilute the ability of knowledgeable and concerned physicians from guiding medical policies for our citizenry." A full-time academic physician asked: "Could not this weight of consumers jeopardize the professional and medical input into decisions, and the political nature of the program supercede the humanitarian and practical approaches of the bill?"

Not all of us would agree that professionals always know best. Modern history is replete with examples of misuse of scientific advances for which professionals must accept some responsibility—from nuclear energy to coronary bypass surgery. With increasing specialization, professionals may see the trees but not the forest; when benefits are tangible but risks unknown, when alternate ways of spending public funds for health are seldom considered, professional wisdom has its limits.

Rather than diluting professional opinion, the question is whether consumer participation in public policy making, as on the Maryland Commission, can counterbalance it. The few studies of boards on which consumers participate suggest inherent weakness in the power of consumer representatives.[23,24] This stems from the selection process, from a lack of awareness on the part of the consumers of whom or what they represent,[25] and from the stature of professionals as experts that enhances their power of persuasion.[24] Nevertheless, the presence of consumers may have a salutory effect on professional board members. Exposed to consumers' scrutiny, professionals become directly accountable and may examine the opinions of their colleagues more closely. This may have played a role in the one issue on which professional members of the Commission differed with organized professional groups. The Maryland Chapter of the American Academy of Pediatrics asked the Commission to modify its requirements for written parental informed consent for neonatal screening because it feared that informed consent would be extended to other routine procedures and also that many parents would refuse to allow their infants to be screened. Both professional and consumer members of the Commission responded that until it was demonstrated that parents could not be informed about the benefits and risks of screening or that a substantial number of parents who had been adequately informed refused screening, informed consent should be obtained. As I have already indicated, these questions are currently under study.

Consumer participation is critically influenced by professional attitudes toward it.[24] By and large, the professionals on the Maryland Commission have been sympathetic to consumer participation. In contrast to other consumer boards where, despite majorities on paper, actual consumer participation was weak,[26] the consumers as a group attended 78% of Commission meetings; the professionals attended only 67%. Considerable time has been spent informing consumers about the issues before any action is taken. Additional consumer input is facilitated by holding meetings at the same time and place each month.

I do not believe that this high level of consumer involvement, which has contributed to the success of the Commission, will inevitably continue. Mechanisms to provide greater assurance that consumers appointed to the Commission can speak effectively on behalf of the public must be found. The Consumer Council of Maryland, established by the legislature to advise the attorney general, has recently undertaken to orient new members of boards and commissions on their responsibilities as public representatives. The council would also like to improve the process of selection of consumer representatives.[25]

Public interest will also be stimulated by encouraging consumers to bring problems to the Commission and responding constructively to them. In response to consumer concerns, the Commission recently established a committee to examine teratogenic and mutagenic effects of radiation and the adequacy of current regulations and procedures for dealing with the problem.

As the alternatives that I described earlier should make clear, consumer participation in public policymaking in genetics is important. Moreover, if we believe that individuals can and should be educated about genetics and have a right to decide whether or not to participate in screening, then surely citizens acting collectively have the right to determine the benefits and risks of genetic screening and other services and whether public funds should be allocated. As the private sector will be unable to provide high-quality screening and adequate follow-up to all who could benefit, the necessity for allocating public funds will increase.

Professionals cannot be the sole makers of policies that involve the deployment of public resources and that may infringe on the liberties of the individual. This principle was recognized by Congress in the Genetic Diseases Act, when it called for "community representation . . . in the development and operation of voluntary genetic testing or counseling programs."[27] As with the other provisions of the act, this one will have to be implemented at the state or regional level. The accomplishments of the Maryland Commission on Hereditary Disorders demonstrates that it can be done.

References and Notes

1. The Food and Drug Administration (FDA) has the legal authority to restrict the sale, use, and distribution of new devices, including reagents for laboratory tests, in order to assure safe and effective use. Thus far the FDA has been concerned only with the reliability of reagents *per se* and not with the likelihood that they might be inappropriately used due to deficient provider understanding or inadequate follow-up. If the FDA should interpret its restrictive power more broadly, market incentive will be curtailed and another mechanism for appropriate dissemination of new technologies will have been introduced.
2. Obstetricians have been slow to recommend mid-trimester amniocentesis to women at risk of delivering offspring with genetic or chromosome diseases, and in the 1960s physicians were reluctant to agree to programs to assure that their newborn patients would be screened for phenylketonuria. Both innovations were reported in widely read professional journals. Whether physicians would have responded more rapidly to advertisements and other promotional devices of pharmaceutical companies remains speculative.
3. Rosenstock, I. M., Childs, B., and Simopoulos, A. P., Genetic screening, a study of the knowledge and attitudes of physicians, National Academy of Sciences, Washington D.C. (1975).
4. Chwalow, A. J., Faden, R., and Holtzman, N. A., Informed consent for newborn genetic-metabolic screening, *Am. J. Hum. Genet.* 30:108A (1978).
5. Metpath Advertisement, This year the neonatal T_4 test can save 600 babies from mental retardation, *New York Times*, Week in Review, Nov. 19 (1978).
6. Bannerman, R. M., Gillick, D., VanCoevering, R., *et al.*, Amniocentesis and educational attainment, *N. Engl. J. Med.* 297:449 (1977).
7. Committee for the study of inborn errors of metabolism: Genetic screening: Programs, principles and research, National Academy of Sciences, Washington, D.C. (1975).
8. Grover, R., Wethers, D., Shahidi, S, *et al.*, Evaluation of the expanded newborn screening program in New York City, *Pediatrics* 61:740 (1978).
9. Holtzman, N. A., Rare diseases, common problems: Recognition and management, *Pediatrics* 62:1056 (1978).
10. Stomatoyannopoulos, G., Problems of screening and counseling in the hemoglobinopathies, in: *Birth Defects: Proceedings of the Fourth International Congress* (A. G. Motulsky, W. Lenz, and J. Ebling, eds.), p. 268, Excerpta Medica, Amsterdam (1974).
11. Childs, B., Gordis, L., Kaback, M., *et al.*, Tay–Sachs screening: Motives for participating and knowledge of genetics and probability, *Am. J. Hum. Genet.* 18:537 (1976).
12. Childs, B., Gordis, L., Kaback, M., *et al.*, Tay–Sachs screening: Social and psychological impact, *Am. J. Hum. Genet.* 28:550 (1976).
13. Clow, C. L., and Scriver, C. R., Knowledge about and attitudes toward genetic screening among high-school students: The Tay–Sachs experience, *Pediatrics* 59:86 (1977).

14. Kenen, R. H., and Schmidt, R. M., Stigmatization of carrier status: Social implications of heterozygote genetic screening program, *Am. J. Public Health.* **68**:1116 (1978).

15. Holtzman, N. A., Genetic screening: For better or for worse? *Pediatrics* **59**:131 (1977).

16. *Becker v. Schwartz*, 46 N.Y.2d 401, 413, N.Y.S.2d 895 (1978).

17. Lappe, M., Gustafson, J. M., and Roblin, R., Ethical and social issues in screening for genetic disease, *N. Engl. J. Med.* **286**:1129 (1972).

18. Holtzman, N. A., Lapides, J. L., and Clarke, G. J., Commission on Hereditary Disorders, *Am. J. Hum. Genet.* **26**:523 (1974).

19. Annotated Code of Maryland 1957 (1978 Suppl.), Article 43, Sections 814–821.

20. This phrase was inserted to placate antiabortion legislators. For the same reason, a statement in the original bill, "that the extremely personal decision to bear children, should remain the free choice and responsibility of the individual, and ... should not be restricted by the State," was eliminated.

21. In 1979, without objection from the commission, the legislature modified the insurance code, so that insurers could not refuse to insure or permit any differential in life insurance policies "solely because the applicant or policyholder has the sickle-cell trait, ... or any genetic trait which is harmless within itself, unless there is actuarial justification for it."

22. Council of State Governments, Hereditary disorders act, *Suggested State Legislation* **33**:227 (1974).

23. Veatch, L., Community boards in search of authority, *Hastings Cent. Rep.* **5**:23 (1975).

24. Metsch, J. M., and Veney, J. E., Consumer participation and social accountability, *Med. Care* **14**:283 (1976).

25. Stellmann, C. B. Consumer representatives, Department of Licensing and Regulation, Offices of the Attorney General, Consumer Council of Maryland (1977).

26. Bellin, L. E., Kavaler, F., and Schwarz, A., Phase one of consumer participation in policies of 22 voluntary hospitals in New York City, *Am. J. Public Health* **62**:1370 (1972).

27. PL 94–278, 94th Congress, HR 7988 (April 22, 1976).

Discussion

Ms. Tabitha M. Powledge (Hastings Center, New York): You may be right that the reason why the other states haven't adopted the Commission model is essentially a turf dispute. How was that kind of professional suspicion overcome in Maryland and the legislation put into effect? Given the fact that the Commission has now been in operation for several years, and there is a history and the professional fears have not been realized, what sort of prospects do you see for establishing this model in other places in this country?

Dr. Neil A. Holtzman: I would think that the kind of documentation we can now provide, showing that consumers can be coped with, might provide some impetus for other states to go along the same lines.

I am obviously being facetious there because I think, as I hope my remarks demonstrated, consumers play a a very critical role. They must be there and must eventually develop a point of view which they will protect from the excesses of professionals, which I think are much greater in many instances than the excesses of consumers.

Insofar as how it came about in Maryland, I think part of it again was community pressure. It turns out that we had this sickle cell law and it was really community physicians more than any other group who expressed concern about singling out sickle cell disease. I think they had the great wisdom to be able to generalize the point that sickle cell disease was like other genetic diseases. They successfully led the fight for the repeal of that law, and passage of this much broader legislation.

There were a couple of political tricks. That is, the group that was involved went to the legislators who were involved in supporting the sickle cell legislation and persuaded them of the error of their ways and got their support for this new legislation.

Mr. Herbert Friedman (Attorney, Boston): I would be interested in how you defined consumers. Did you define them as simply non-M.D.'s and nonproviders? Did you define them as at-risk parents, as affected persons?

Dr. Holtzman: The law simply states that they cannot be employed as health providers.

I think the question of defining consumers is a very difficult one. I indicated they are appointed by the governor. The governor of Maryland at that time had many other faults which have been discovered. But the basis for his appointments was quite reasonable. I think that is, however, a tenuous way of establishing consumer representatives.

It is interesting that the legislature of Maryland also formed something called a Consumer Council, whose members are also appointed by the governor. But the Consumer Council has recently recommended that it become a repository for consumers' interest in serving as representatives on boards and commissions, and that it develop, with the assistance of the attorney general's office, out of which it operates, mechanisms for screening and promoting consumers, so that at least those who serve have an interest. And, presumably, one can see it operating that way: the ability to represent a defined constituency.

The council is also interested in developing training programs for consumer representatives so that they understand the purposes of the commissions and boards, since it is not only the Hereditary Disorders Commission on which they serve.

Mr. H. Friedman: Do you happen to know whether in fact you have as consumers persons interested and, if so, did their attitudes on issues differ from the attitudes of those not identified with those issues?

Dr. Holtzman: We had consumer representatives—not all of them—who in their families, as many of us have, had individuals with severe handicaps, not necessarily genetic. We also had, interestingly, one consumer member of the Commission, who was not among the first consumer members, who was a regular attender of Commission meetings and was there early to lobby for a particular screening program. Interestingly, as this individual sat and participated at hearings, his views underwent considerable change and I think he became tremendously concerned about the problems of screening and somewhat modified his views about the ratio of benefits to risk in screening programs.

Ms. Marjorie Guthrie: I just wanted to say that Dr. Holtzman is being very modest. The fact remains, as I know it, that people like him and Dr. McKusick and some other professional people were very wonderful in involving the consumers by their own encouragement. They made people feel that they had a role to play. Members that I knew and went to see and hear felt that the professional people opened the door in a wonderful and positive way and made the consumer feel welcome, willing to learn, willing to listen, and to participate on an individual level so that they could communicate together. I think that was a very important part of the success of the consumer being involved.

Dr. Holtzman: That's very kind of you, Mrs. Guthrie. I think that it is precarious, though, in that one must keep open the question of assuring that there is consumer participation and that when it is there, it is meaningful. That is why I was pleased to have that opportunity to raise the question of developing panels of consumers, specifically having programs for consumer orientation.

Dr. Aubrey Milunsky: You don't really believe that you can obtain informed consent for genetic screening on a mass basis, do you?

Dr. Holtzman: The question of informed consent for screening, mass screening or otherwise, still consists of each individual coming to be screened. And until it can be shown that it is not possible to obtain informed consent, I think it should continue to be administered.

I think that is a researchable question. We are in the process of evaluating the quality of informed consent in every hospital in Maryland on newborn screening and the results are just being evaluated now. It is very interesting that there are marked differences from one hospital to the next in how informed consent is administered.

The other aspect of this study for which we haven't looked at the data yet was whether the informed consent procedure itself did result in the communication of any information to the mothers. If it should be shown that that is not possible a significant proportion of the time, then I think the question is a legitimate one, that perhaps it is not possible to obtain informed consent. But I think for us to state that it is not possible without making an effort to inquire is not the appropriate way to go.

Dr. Milunsky: Perhaps the time has come to develop an atmosphere in which high school children grow up with the view that it is important to know, that it is important to recognize one's genetic accountability and, endowed with that information, be equipped to make the decisions we all have to make, without government intervention.

Mr. J. McInerney (Biological Sciences Curriculum Study, Boulder, Colorado): Dr. Holtzman's and Dr. Milunsky's statements really crystallize the problems that we are facing in public education in terms of dealing with a number of these issues for our junior high school and high school students.

Dr. Holtzman's comments raise generally the problem of enlightenment and informed citizenry capable of participating effectively in matters of public policy. I think this group is a good place to begin to address some of the problems that high school teachers and other public educators are facing to get high school and junior high school students to deal with some of these issues. The constraints are many and they range from political pressure on the part of special interest groups to the failure of administrators actively to support high school teachers when they begin to deal with issues similar to those we have been discussing here for two days. We receive more and more requests for information dealing with the biosocial aspects of advances in genetics and other areas of reproductive biology. But until there is some mechanism for active support of educators, I don't think we are going to get very far. And I would urge this gathering actively to pursue those kinds of support mechanisms when you leave here.

MR. D. GABRIELS: I would like to elaborate on a question I raised earlier and ask if any member of this body could suggest precedents or an approach utilizing the extant legal hardware. Such risks run counter to all decisions to embrace the percentage of tragedy in genetic disease by reproducing. Diabetic mothers subject their progeny to an increased risk of congenital malformations. Yet it seems a disproportionate burden to require the unfortunate mother who does bear a victim to shoulder the entire cost of the infant's malady, as though she had constructed the child from scratch with certainty of its affliction. She has done neither more nor less than the mother who uneventfully disregarded the risk.

My intuition is that they should be equally accountable for the harm produced, though I realize our legal system in general requires a harm caused through the behavior of the defendant in order to be set into action.

22

SEX AND THE SINGLE CHROMOSOME

Rights and Obligations in the Uses of Genetic Technology

JAMES R. SORENSON
Department of Socio-Medical Sciences and Community Medicine
Boston University School of Medicine
Boston, Massachusetts 02118

JUDITH P. SWAZEY
Department of Socio-Medical Sciences and Community Medicine
Boston University School of Medicine
Boston, Massachusetts 02118

Medicine in the Public Interest, Inc.
Boston, Massachusetts 02110

INTRODUCTION

Gender identification and sex selection, we would be the first to agree, are not the most salient topics in the panoply of medical, legal, social, and ethical issues surrounding the uses of genetic technology. But, to paraphrase our clinician colleagues, they are "interesting" issues, ones that we have discussed often during the several years that we have examined the field of applied human genetics as social scientist researchers and as teachers of medical sociology and medical ethics. For, both in and of itself, and as a case study that raises more generic issues, the capability to detect prenatally the presence or absence of a single Y chromosome raises important ethical and social policy questions about the rights and obligations of those providing and utilizing the knowledge and techniques of applied human genetics.

In this chapter we will consider, first, the types of issues that are raised by the question of whether or not to disclose gender information that is acquired incidental to prenatal testing for a chromosomal or genetic disorder. Second, we will explore some of the questions surrounding the use of prenatal diagnosis specifically to determine gender in a presumptively normal fetus, a usage that seems to shift a medically embedded technology, developed for a "patient"-oriented diagnostic model, into a nonmedical consumer-oriented framework.

GENDER DISCLOSURE IN DISEASE-FOCUSED PRENATAL DIAGNOSIS

Prenatal diagnosis is a diagnostic procedure that very infrequently finds what it is looking for. Stated differently, in the vast majority of cases prenatal diagnosis will remove or greatly reduce concern about the presence of some specific defect in a fetus. In about 95% of all prenatal diagnostic cases, negative results will be achieved, and the prospective parents can generally expect the birth of a normal child with the same level of risk that any couple faces.

Rather than removing or lessening uncertainty about the normalcy of the fetus, however, prenatal diagnosis sometimes may restructure uncertainty. For example, pre-natal diagnosis can be used to diagnose the sex of a fetus in cases where the mother is a designated carrier of an X-linked disorder, such as hemophilia. In these situations prenatal diagnosis may not remove uncertainty about the health status of the fetus. Instead, if the fetus is diagnosed as male, the odds of having a child with hemophilia have been doubled, from 25 to 50%. And there is still uncertainty about the actual health of the fetus.

Prenatal diagnosis also may sometimes create uncertainty. For example, a major reason for undertaking prenatal diagnosis is to look for chromosomal anomalies associated with advanced maternal age. Usually, of course, a normal karyotype is found. However, sometimes a karyotype may reveal not the usual chromosomal anomalies associated with advanced maternal age but some other anomaly, such as an XYY chromosomal comple-ment. In such situations, given present-day uncertainty about the clinical significance of such a chromosomal complement, difficult questions are raised about whether to tell prospective parents, and if so what to tell them.

In virtually all cases of prenatal diagnosis in which the procedure is undertaken because of some variously defined elevated risk for an anomaly, the gender of the fetus will be identified in the process. This raises questions about the policy of gender disclosure in prenatal diagnostic situations in which knowledge of the sex of the fetus is incidental to the main purpose for undertaking the procedure. Do patients have a right to this information, given that they are paying for it and it is a normal yield of prenatal disgnosis? If one perceives such a right, is there a corollary obligation on the part of the provider to disclose fetal gender? Or, to what extent does a user of medical technology have an obligation to see that that technology is limited to its intended medical purposes—in this case the diagnosis of disease and abnormality?

One can also approach the issue of a patient's right to gender information in this situation by raising questions about the extent to which a karyotype constitutes part of the medical record of the patient. Ethically and legally there seems to be growing consensus that patients have a right to information in their medical record. To the degree that a karyotype, or an interpretation of a karyotype, is or is not viewed as part of the medical record, one can raise questions about the right of patients undergoing prenatal diagnosis for genetically indicated reasons to information concerning the gender of their fetus. One also needs to ask, in this context, to whose medical record the karyotype belongs: Is the patient the mother or the fetus/child-to-be?

More generally, should the provider of prenatal diagnosis disclose gender informa-tion (1) routinely, (2) only upon parental request, or (3) only at his or her discretion? And, if (1) or (2), to whom should the provider disclose or not disclose? Also, should the gender disclosure policy be the same or different when the fetus is diagnosed as normal or abnormal?

Our work with over 80 genetic clinics throughout the United States[1] suggests to us that counselors differ in their approach to gender disclosure in situations where knowledge of the sex of the fetus is incidental to the purpose of undergoing amniocentesis. For example, we know that some clinics readily make gender information available to parents. Other places, however, are less willing to report gender. Anecdotally, for example, we have been told that some providers will not disclose gender in a "normal" fetus at least until the pregnancy is too far advanced for an abortion.

The strategy of disclosure of fetal gender along with disclosure of health status in many ways seems to be the least problematic posture for a provider of prenatal diagnostic services. Yet, upon closer examination, one can begin to raise questions about such a position. For example, while we know of no statistics on the issue, some parents-to-be report not wanting to know the sex of their child before birth. After undergoing amniocentesis, one woman in our genetic counseling study told us, "I didn't want to know the baby's sex . . . for a while I was tempted to call [the physician] and find out, but I'm glad I didn't do it." The anticipation and surprise of learning the sex of the child at birth thus may be something that a sizable number of parents want to retain as part of the birth process, even in this age of the new genetics. This is one uncertainty some parents may not want reduced by prenatal diagnosis. On these grounds, then, one can argue that parents-to-be have a right *not* to be told fetal gender as a routine consequence of prenatal diagnosis.

To the extent that one endorses such a position, a strategy of negotiating gender disclosure with parents prior to knowing diagnostic results would seem a requisite part of preamniocentesis counseling. This strategy of establishing a "contract" about the nature and objectives of prenatal diagnosis is the second one we would like to consider. In the situations we are considering here, parents-to-be ostensibly come for prenatal diagnosis because they or their primary prenatal care provider is concerned about possible fetal disease or disorder. In this context fetal gender, except in the case of sex-linked disorders, is a medically irrelevant piece of information acquired as part of a medically oriented prenatal diagnosis. Granted this, and because some parents may consider gender disclosure at birth an integral part of birthing, one might argue that gender disclosure should be at the discretion of the parents.

This position, in turn, has some interesting implications. For example, at present the primary provider of prenatal care usually is not the person performing the actual laboratory work associated with prenatal diagnosis. Parents are referred to a secondary or tertiary level specialist for the diagnostic work, and it is this specialist who is in charge of disseminating information about the fetus, including its gender. If one endorses the idea that disclosure of gender ought to be negotiated with the parents, one can then raise the issue of to whom, if anyone, the prenatal specialist ought to disclose gender without parental consent. If in fact fetal gender is medically irrelevant, then there would appear to be no persuasive argument for disclosure of fetal gender even to the primary provider of prenatal care, if parents prefer nondisclosure.

The notion of establishing a "contract," we feel, is one that ought to be considered more generally by providers of diagnostic procedures in medical genetics, within the broader context defining the rights and obligations of both providers and patients. In prenatal, newborn, and adult testing, unanticipated findings, or findings that are of questionable or unknown clinical significance, may be found. Thus, for example, fetal karyotyping may be negative for trisomy 21 but may reveal an XYY chromosomal complement or a possibly anomalous chromosomal banding pattern, and newborn metabolic screening, while negative for an established problem such as PKU, may show biochemical variations of unknown clinical significance. Providers often make difficult and unilateral decisions about whether—and if so, what—to disclose about "unexpected findings."[2] Many of the problems generated by this mode of decision making, we feel, could be obviated if a "contract" was negotiated between provider and patient prior to testing, in which the types of findings that the test could produce are set forth, and both parties agree about what types of information will be disclosed to the patient or to parents.[3]

The third strategy listed above consisted of purposively not disclosing gender to parents, or of delaying disclosure to such a degree that it is virtually impossible for parents to act on it. Of the three disclosure strategies we are outlining, this is the one that we feel raises some of the most interesting social, ethical, and legal issues.

In adopting a strategy of nondisclosure, a prenatal care specialist could make one of two arguments. First, he or she might argue that the parents-to-be sought the service because of concern about a medical issue. Because gender identification is irrelevant to the reason for the seeking of prenatal diagnosis, the provider of the diagnostic services has no obligation to disclose fetal gender to the parents, nor presumably to the primary provider of prenatal care. A second argument, not totally divorced from the first, is that the provider is concerned about the possibility of parents using gender information to terminate a fetus of the undesired sex. If this is the case, then such a provider might argue, as before, that prenatal diagnosis is a medical procedure, and that to use it for "nonmedical purposes" such as sex selection is to misuse it. This, of course, raises questions about the proper bounds of the physician's role as a gatekeeper of the information yielded by contemporary genetic technology, a topic we explore in the second section of this chapter. It also poses questions about the extent to which personal values, such as those regarding abortion, ought to enter into a physician's rights and responsibilities in his role as a professional. Another question is whether delaying disclosure of gender information constitutes lying, and, if so, whether or not it is a morally permissible lie.[4] These and issues raised by a unilateral nondisclosure policy, like those linked with a unilateral disclosure policy, could be eased by a contract-setting process in which both parties articulate their beliefs and wishes.

Access to Prenatal Diagnosis for Gender Identification

Prenatal diagnosis, as we have noted, was developed and is being used as a medically embedded technology, provided by medical professionals and utilized by patients. The extent to which pregnant women might want to use this technology for the essentially nonmedical purpose of gender identification in a presumptively normal fetus is empirically unknown. But, because this use of prenatal diagnosis exists in principle and, we have been told anecdotally, has been employed, it poses an important series of ethical and social policy issues. These issues include the rights and responsibilities of individual providers and utilizers, and institutional and societal decisions about the proper or just allocation of medically developed technologies.

Sociologically, the use of prenatal diagnosis solely for gender identification seems to shift the employment of this technology from a physician–patient medical model to more of a service provider–consumer model. If, within the latter model, a physician provides the diagnostic service, his function would seem akin to the "engineering" or "plumber" role described by Veatch, in which the physician sees his task as simply to dispense the knowledge or techniques at his disposal.[5] If this is the role, one can then raise the question of whether a physician is needed to dispense the technology. While any appropriately trained technician might suffice, however, one needs to consider the type of consumer contract that could or should be established, and whether that contract involves an obligation to disclose any abnormalities detected in the process of gender identification. If so, one can then argue that, at some point, the physician or a comparably trained professional is a necessary participant in an essentially nonmedical use of prenatal diagnosis.

Moving back a step, one needs to ask whether the physician's present role as provider of prenatal diagnosis establishes him as the technology's gatekeeper, determining who shall have access and by what criteria. Legally and, many would argue, morally, the physician does not have an obligation to do or provide *anything* a patient wishes. Such a role obligation, ethicist Paul Ramsey argues, would turn "physicians into 'animated tools' (to use Aristotle's definition of a slave) that simply assist a patient to attain anything he wishes."[6] If one accepts the legal and moral arguments for a physician exercising a gate-keeping role, the question then becomes one of what constitutes appropriate criteria for allowing access to the service in question. All such criteria, we would argue, have value components that need to be recognized and articulated, whether they are framed in terms of risk–benefits assessments, allocation of scarce resources, or, more obviously value-laden, judgments about the appropriateness of an abortion decision based upon gender preference.

Comparable questions need to be structured and examined for the utilizer of prenatal diagnosis. To begin with, is a woman seeking a test solely for gender information a "patient" in any conventional sense, as opposed to a "consumer" who wants to purchase a bit of technologically generated information? The answer, we think, is that such a utilizer is not a "patient," which in turn raises questions about consumer rights and obligations and the nature of this type of contractual relationship. To say that a given provider does not have an obligation to perform whatever service a patient or consumer desires does not, in turn, mean that a patient/consumer has no right to seek that service. In the case of testing for gender identification, however, arguments about consumer rights are complicated by the fact that access to the service may not affect the consumer alone: it may, directly or indirectly, affect the fetus or potential person.

In this context, arguments about the "right" to obtain gender identification testing may vary according to why that service is sought, and the values one employs in deciding whether it is an appropriate reason. The principal reasons for seeking prenatal gender identification would seem to be (1) satisfying parental curiosity about the sex of their developing offspring, and (2) making an abortion decision based on gender preference. For both cases, one can spin out complex arguments for and against a "right" to obtain prenatal gender identification.

One can argue against a right to satisfy parental curiosity, for example, in terms of benefits and risks: amniocentesis subjects the fetus to a statistically small but nonetheless existent risk of injury, which some may view as outweighing the benefit of satisfying parental curiosity. On the other hand, one can argue that, given the centrality of becoming a parent in many people's lives, they ought to be able to do that which, for them, will enhance the experience of pregnancy. Here, one again needs to factor in potential risk to the fetus and, if data are available, the positive or negative effects on maternal bonding of knowing gender in advance.

The values that dictate one's position on abortion in general or for particular reasons, ranging from detection of a major defect to gender preference, are both complex and familiar ones that we will not pursue in this chapter. We would, however, raise one question that to us merits serious thought. Abortion based on gender preference may be a decision that few persons currently would opt for. It also is a decision that many, even those not opposed to abortion on principle, may argue against as an improper reason for terminating a fetus. Abortion based on gender preference, one may argue, is an "immoral" act in terms of personal values, or it should be opposed because (1) it conceivably could upset the male–female sex ratio in society, or (2) it might alter the birth-order pattern, creating a predominantly male-first birth order. Conversely, one could argue for the availability of gender selection based on factors such as pressures to limit family size. Does a couple's meeting of this societally determined obligation entail a reciprocal right to exercise gender preference in having children? If so, should society, in turn, make prenatal gender identification tests available on demand?

Questions of value can enter into assessment of even the most common reason for amniocentesis today—advanced maternal age. It is fairly well accepted medical practice now to tell pregnant women 35 and older of their increased risk for having a child with a chromosomal aberration, and to offer amniocentesis. The reason for this position has been that historically women 35 and older, while having only a small percent of all pregnancies, gave birth to a majority of children with Down syndrome. Hence, by screening pregnant women 35 and older one could identify many Down syndrome fetuses.

Recent evidence, accumulated in a variety of places, suggests that there has been a substantial decrease in the age of mothers of infants with Down syndrome. For example, whereas in the recent past mothers 35 and older in one state, Massachusetts, had only 13% of all pregnancies but 50% of infants with Down syndrome, more recently such mothers in one major hospital in that state, while accounting for 7% of all pregnancies, have had only 35% of all infants with Down syndrome.[7] This means that 65% of the infants with Down syndrome were born to mothers less than 35. In light of such evidence, should amniocentesis be offered to women younger than 35, and if so, how much younger? And, what considerations go into making such an assessment? Should one compare the risk of an infant with Down syndrome to the risk of amniocentesis, or what?

These questions move us into ethical and social policy questions about the allocation of resources, questions that turn upon how we define the common good and the most just means of distributing resources to achieve that good. At present there is discussion and debate about the medical "underutilization" of prenatal diagnosis, and an inability to meet the demand for this service. Given the medical needs for prenatal diagnosis, and the financial and personnel costs involved in the provision of this technology, should it be allocated for essentially nonmedical uses? This is but one of innumerable decisions about the allocation of resources that institutions, communities, and society as a whole must begin to grapple with, however uneasily.

As we confront these allocation questions, whether they involve particular technologies or medical/health care costs and services more generally, we need to surface the values that underlie the types of allocation decisions we are considering. Which concepts of social justice underlie present criteria for allocating prenatal diagnosis, and which ought we employ in the future?[8] At present, access to prenatal diagnosis is based primarily upon criteria of need, with the determination of need defined principally by the gatekeepers rather than by the potential utilizers of the service, as is true of medical care more generally. However, a patient- or consumer-defined "needs" principle, akin to that advocated by Pellegrino and others for first contact primary care,[9] could substantially change access criteria to prenatal diagnosis. Other canons of social justice according to which goods and services can be allocated include meritarian conceptions, principles of equity, and conceptions of societal contributions. How we choose to allocate our resources, as Veatch has shown in his analysis of the values implicit in various national health insurance proposals, will differ in fundamental ways according to which of these different notions of social justice is given moral primacy.[10]

The current ability to identify fetal gender by amniocentesis also suggests the wisdom of engaging in anticipatory discussion about the social and ethical pros and cons of other less invasive and more readily accessible modes of gender identification and sex selection. For example, technological developments might permit us to identify fetal gender by means of a simple test (e.g., blood or urinalysis), largely removing resource allocation as an issue. Or, as forecast in a Delphi study conducted for the National Commission for the Protection of Human Subjects of Biomedical and Behavorial Research, we might develop a safe and effective commercially marketable method for preselecting the sex of one's children. In the National Commission Delphi Study, the scenario described the development of a "Select-A-Boy, Select-A-Girl" vaginal douche and foam kit. The Delphi panelists were asked to assess the social, ethical, legal, and policy implications of such a development, and to decide whether the "XYZ Company" should be allowed to market the kits for over-the-counter sale. The range of issues and conflicting opinions that such a potential development engenders is reflected by the panelists' response to this scenario.

The 96 "experts" (ethicists, lawyers, medical scientists, social scientists, and public interest representatives) were split as a whole, and by discipline: 47 held that the company should be allowed to market the kits, and 49 that they should not be so allowed. [11]

To end our provisional sketching of the issues generated by the use of prenatal diagnosis for gender identification, it also occurs to us that such a use of this technology raises interesting questions about how we ought to interpret and act upon the now prevalent belief that health care is a "right." We argued at the beginning of this section that the use of prenatal diagnosis for gender identification in a presumptively normal fetus moves this technology out of the medical arena. If, however, one accepts the 1946 World Health Organization definition of health (a definition about which we personally have grave reservations), there would seem to be reasonable arguments for considering gender identification as an aspect of health care (if not medical care) services. "Health," according to the WHO, "is a state of complete physical, mental, and social well-being and not merely the absence of disease or infirmity." The WHO definition, Callahan has observed, was an ambitious one, "giving medicine an enormous mandate and theoretically excluding no human problem from treatment by medical means or from interpretation in medical or biological categories. . . . Moreover, it is patently the case that the combination of an all-encompassing concept of health together with biomedical advances has created a situation in which nearly anything that anyone wants from medicine can find legitimation as a health need."[12] Prenatal diagnosis for gender identification, we think, provides a provocative example of the issues that must be addressed and resolved in defining the meaning of health, and the role of medicine in relation to that definition.

REFERENCES AND NOTES

1. With support from the National Foundation—March of Dimes, we are completing a national study of foundation-supported genetic counseling clinics in the United States. In addition to looking at clinic policies and staff, we are prospectively assessing the impact of genetic counseling on patients receiving counseling in these clinics. Reports from this study are forthcoming.
2. See, for example, The unexpected chromosome, in: *Case Studies in Medical Ethics* (R. M. Veatch, ed.), p. 137, Harvard University Press, Cambridge (1977).
3. The notion of contract setting has been discussed most fully in terms of establishing the mutual expectations between a primary care provider and a patient for longitudinal and integrative care. Many elements in primary care notions of a contract, it seems to us, can be transferred to other areas of medical services, such as prenatal diagnosis.
4. See Bok, S., *Lying: Moral Choice in Public and Private Life*, Pantheon Books, New York (1978); Veatch, R., *Death, Dying, and the Biological Revolution*, Ch. 6, Yale University Press, New Haven (1976).
5. Veatch, R., Models for ethical medicine in a revolutionary age, *Hastings Cent. Rep.* 2:5 (1972).
6. Ramsey, P., *Ethics at the Edges of Life*, p. 158, Yale University Press, New Haven (1978).
7. Holmes, L., Genetic counseling for the older pregnant woman: New data and questions, *N. Engl. J. Med.* 298(25):1419 (1978).
8. See the papers on Justice and Health Care Delivery in: *Ethics and Health Policy* (R. Veatch and R. Brown, eds.), Ballinger, Cambridge (1976).
9. Pellegrino, E., The social ethics of primary care: The relationship between a human need and an obligation of society, *Mt. Sinai J. Med.* 45:593 (1978).
10. Veatch, R., What is a "just" health care delivery? in: *Ethics and Health Policy*, op. cit.
11. Policy Evaluation Instrument-2, Synthesis Paper on Select-A-Boy, Select-A-Girl Marketable Kits, A Com-

prehensive Study of the Ethical, Legal, and Social Implications of Advances in Biomedical and Behavorial Research and Technology, Center for Technology Assessment, New Jersey Institute of Technology and Policy Research, Baltimore. The synthesis paper Amniocentesis Becomes Routine in the same Policy Evaluation Instrument also touches on gender identification and selection.
12. Callahan, D., Biomedical progress and the limits of human health, in: *Ethics and Health Policy*, op. cit., p. 160.

Discussion

Ms. Tabitha M. Powledge (Hastings Center, New York): I am very uncomfortable with couching this argument in terms of allocation of scarce resources; Dr. Milunsky and I got into this a little bit yesterday. For one thing, what happens when those resources are no longer scarce? You then don't have a leg to stand on.

I am uncomfortable with your use of the word *interesting*. It seems to me this is more than an interesting issue. It is a tough issue. And let me outline the reasons why I think so.

In the March PNAS, researchers at Stanford published their work on fetal cell sorting. It is now possible to sort, with the help of a laser, fetal from maternal cells in maternal blood. What the article doesn't emphasize and what the reports on it have not emphasized is that it also gives you an instant readout on whether or not the fetus is male; no waiting for cells to grow. And the earliest date on this is at fifteen weeks.

It is quite conceivable that for some women at least you might be able to tell fetal sex in the first trimester. It strikes me that once you have got a situation where you are not talking about an abortion at twenty weeks but an abortion considerably earlier than that, that the current picture where fetal diagnosticians are getting only occasional requests for sexing of fetuses would change very rapidly. And I am really rather struck by the way that, when I talk to people in this profession about this problem, we get words like, "Oh, isn't that interesting?"

It strikes me that *interesting* is maybe not the word you apply here. I don't want to contend that it is a crisis situation, but it really is going to come upon you people before too long and it is not clear to me that there is any kind of thought-out response.

Prof. Judith P. Swazey: We were using "interesting" in quotes very much tongue in cheek because it is a word in clinical medicine that portends a case with a lot of dimensions, and we clearly don't see it as a frivolous set of issues. We think they are both in and of themselves very important and tough issues.

I think I would argue in terms of framing it in allocation of resources in which it belongs, at least in part, but not exclusively there, because I think all health care or medical care at bottom is a scarce resource and decisions have to be made about where we are going to put our priorities, our personnel, and our finances.

Ms. Powledge: But the ultimate outcome of the argument you are making in this paper is that we are going to take it out of the medical realm and maybe have somebody set up a street corner shop where he does prenatal diagnosis but he doesn't call himself a doctor, just for anybody who wants to know fetal sex. It strikes me that would be a move I would not like to see.

Prof. Swazey: I think that is one feasible scenario and, again, I think that imposes interesting legal regulatory licensure issues. I think the National Commission a few years ago did a Delphi study and one of the scenarios they had was one's reactions to marketing Test-a-Boy, Test-a-Girl kits, which is exactly this type of thing of moving it into the commercial realm. So I think it is a possibility, without endorsing it.

Prof. James R. Sorenson: I concur with Prof. Swazey's position. We don't view it just as interesting. We do think that it surfaces a variety of generic problems in this area that raise many very serious questions, and I agree that fundamentally at some point you have to couch some of the issues as issues of the allocation of scarce resources. But what we tried to do at the same time was surface a variety of other types of ways in which one could couch the problems and intentions of using the technology for this purpose.

23

GENETICS, ADOPTION, AND THE LAW

JOHN R. BALL AND GILBERT S. OMENN
Office of Science and Technology Policy
Executive Office of the President
Washington, D.C. 20500

INTRODUCTION

The law is both fluid and flexible: fluid in that it changes over time, and flexible in that it bends to meet the circumstances of individual cases. Law is especially fluid when it is a reflection of social policy, and it is reasonable that it should change as social policy changes. The views of society toward adoption are changing as a result of people's interest in the roots of their individuality. In many cases, adoptees are asserting a right to information about their birth parents, and in several states legislation has been considered that would open to their inspection parts of the adoption record that traditionally have been closed. This interest of the individual in his or her personal and family history potentially conflicts with other societal interests. Society, through an emerging body of law, seeks to protect the interests of individual privacy and of the confidentiality of certain sensitive information. Thus the asserted "right" of an individual to know his genetic heritage conflicts with the right of the birth parents to individual privacy.

Concurrently with these social changes, scientific advances in medical genetics have occurred. New knowledge and new techniques have permitted more choices by prospective parents in deciding whether to attempt to conceive children and whether to carry pregnancies to term. In making possible these new choices, medical genetics in part exacerbates certain dilemmas that are addressed by the law, including individual privacy and the confidentiality of medical information. Medical geneticists have an opportunity to inform social policy and through it, the law. Because of the advances in medical genetics, it is necessary to reconsider laws affecting the flow of medical information and to put the law on a firmer scientific footing, consistent with developing social policy.

The views expressed are those of the authors and do not necessarily represent views or policy of the United States government.

Adoption

Adoption, since Babylonian times, has been the method of specifying by law the relationship of parent and child.[1] Referred to in the Code of Hammurabi, adoption has a long history among primitive tribes throughout the world. Early Greek and Roman culture had extensive legal provisions for adoption, the principal objective being a mechanism by which to pass both the religious leadership of the family and the estate.

Across the world, there is great variation among countries in reasons for adoption, frequency of adoption, and the legal provisions regulating adoption.[2] A basic division in the legal status of adoptee and adoptive parents exists among countries. For example, in Greece adoption is considered primarily a means for satisfying the desire of childless couples for children. In contrast, Polish law stresses the interest of the adoptee as primary. In those nations whose legal system derives from Roman law or the Code Napoléon, the legal links of the adopted child and his birth parents are maintained. The child not only may inherit from them but also may call upon them for support. In other countries, including England and the United States, the adoptee's legal relationship with the birth parents is severed. The single factor commonly shared by all countries that have adoption legislation is the requirement of participation by some judicial or administrative process.

Most modern law dealing with adoption is a product of legislative enactments of the 20th century. England had no provision for legal adoption until 1926; before that time there could be only informal adoption, and the birth parents could recover the adopted child at any time. The first English adoption statute did not provide for inheritance by the adoptee of the adoptive parents' estate. Nevertheless, English society and law had long made provision for disadvantaged children, and the English law of adoption rapidly began to reflect the primacy of the adoptee's interest.

The law of the United States in general is derived from English law. The United States Constitution reflects English common law, which was the law in effect in England in the late 18th century. Where the Constitution is silent, and where states have passed no specific legislation, the construction under which courts determine individual rights is that of common law. However, since adoption was unknown in common law, all adoptions are controlled under the authority of a specific statute in the state having jurisdiction.[1] In addition, since there was no common law history to inform and expand on the statutes, early cases held that for an adoption to be valid, there must be strict compliance with the literal provisions of the law.[3]

In the years since Massachusetts passed the first adoption statute in 1851, the courts have become more "liberal," and the present view is that while the law must be complied with substantially, an adoption will not be defeated when the intent is clear and there has been compliance with material provisions.

Adoption as social policy and through law developed in a unique way in the United States. Adoption grew to be primarily for the benefit of the adoptee and the adoptive parents. Thus even earliest adoption laws mandated a severing of all legal ties between adoptee and birth parents.[2] This ethic both facilitates and blocks the flow of information concerning adoptee and birth parents.

The statutes of American jurisdictions all require a judicial procedure to effect an adoption.[4] Although there is considerable variation in specific provisions, most statutes provide for: (1) a petition to a designated court, (2) notice to specified parties, (3) consent of the natural parents and/or other specified persons or institutions, (4) an investigation and report to the court as to the suitability of the prospective adopters, (5) a hearing by the court, after which the court either decrees or refuses to decree the adoption, and (6) provision for appeal by the parties.

All U.S. jurisdictions provide for the sealing of records relating to adoption proceedings after entry of the final decree. Although such secrecy provisions represent a departure from the general policy of full disclosure of public documents, they are not unique to adoption. The purpose of the secrecy provisions is to protect the people at each point in the "adoption triangle": the adoptee, the birth parents, and the adoptive parents. The adoptive parents and the adoptee are to be protected from interference and disturbance of the developing family relationship, and the birth parents' privacy is to be protected so that they may overcome the factors that forced them to give up the child for adoption.[5,6] Thus most states require that an amended birth certificate be issued, substituting the names of the adoptive parents for these of the birth parents.

Adoption law in the United States reflects a social policy that attaches primary worth to the individual rather than to inherited relationship. In provisions for the judicial adoption process, the law recognizes the rights of each individual in the adoption triangle but the interests of the adoptee child are always considered foremost.[6,7] Early in the adoption process, the rights and interests of the birth parents, particularly the mother, are considered next in importance and are commonly held superior to those of the prospective adoptive parents[8]; hence the strict requirements for informed consent and the safeguards against undue pressure on the birth parent.[9] Finally, however, at a crucial point, the interests of the adoptive parents assume primary importance over the interests of the birth parents; with the adoption decree, all rights of the birth parents to the adoptee are severed.[10] In return, the law assures the birth parents, through the sealing of the adoption record, that their right to privacy shall be upheld.[4] The adoptive parents, the adoptee, and the birth parents then are freed as much as the law can accomplish to establish and to reconstitute their life patterns.

PRIVACY

There is no mention in the United States Constitution of the word *privacy*, but there is substantial enumeration of rights fundamental to the individual. The right of privacy is an expression of the ethic of individual rights. The right of privacy is particularly related to adoption, since the landmark cases establishing that right in law concern procreation and the marriage relationship.

The legal concept of a right of privacy was enunciated in 1888 ("the right to be let alone"),[11] but it was not until 1965 that the right was specifically labeled as one supported by the Constitution. Justice Douglas, in his opinion in *Griswold v. Connecticut*,[12] explained the Court's holding, striking down a state statute that prohibited the use of contraceptives by married couples, by stating that there was a zone of privacy within the "penumbra" of the Constitution, protected by the First, Third, Fourth, Fifth, Ninth, and Fourteenth amendments. Justice Harlan, in his concurring opinion, wrote, "I believe that the right of privacy in the marital relation is fundamental and basic—a personal right retained by the people." In 1972, in *Eisenstadt v. Baird*,[13] the Supreme Court expanded the right of privacy to include contraceptive use by unmarried persons. Justice Brennan stated, "If the right of privacy means anything, it is the right of the individual . . . to be free from unwarranted governmental intrusion into matters . . . fundamentally affecting a person. . . ." In the last six years, decisions by the Court regarding abortion have extended the right of privacy further, to include the right of pregnant women to control their own bodies, at least until the stage of extrauterine viability of the fetus.[14]

More recent cases pointed out by Margery Shaw[15] suggest that the constitutional zone of privacy continues to expand, and now encompasses more than procreation and the marriage relationship. Some of these cases concern unreasonable interference with the parent–child relationship, the right of privacy of minors, and the right to privacy or nondisclosure of medical information. It should be noted in each case, however, that the right of privacy is not absolute. The protected zone of privacy may expand or contract depending upon the nature of the private act or information and upon the counterbalancing state interest. For example, although a woman may decide to have an abortion, the state may limit that right to apply only before the stage of fetal extrauterine viability.[16] Assertions of a right of privacy are likely to be upheld at this point only if they involve "fundamental" individual rights and if there is no compelling counterbalancing state interest.

CONFIDENTIALITY

The right to expect that private information be held confidential is rooted in the right of individual privacy but is also based on practical concerns. The expectation of confidentiality with regard to medical information is ancient, dating from the Hippocratic oath as the duty of the physician to maintain strict confidentiality.[17] Certain classes of individuals, because of their businesses or professions, have the need and opportunity to obtain personal information from other individuals. Personal problems may be recounted to priests, rabbis, or counselors; difficulties with legal authorities are discussed with lawyers; medical and psychiatric histories are given to physicians. In each of these cases, the professional receives information that the individual giving it expects to be used to alleviate his or her problem, and for no unrelated purpose. The ethics of the profession, and in many cases the law, support this expectation; the free exchange of information between the individual and the professional is given greater value than open disclosure of information. Thus law as a reflection of social policy fosters the private nature of the relationship and the confidentiality of the information transmitted.

Few rights are absolute, however. The state may override the individual's right that information be kept confidential if it can show a substantial counterbalancing interest, and in many cases the state has been able to show such an interest. In general, if there is a substantial threat to third parties that may be alleviated by release of information, the state may compel release. With regard to medical information, the state has compelled reporting of infectious diseases, child abuse, and gunshot wounds in order to protect the public health, protect individual health and welfare, and enhance the administration of justice.[18]

In addition, the duty to disclose personal information has been held to extend to the duty to disclose threats of personal harm to the one threatened. In *Tarasoff v. Regents of the University of California,*[19] the court held that when disclosure is necessary to prevent harm to others, the confidential character of personal information must yield. The case involved a psychologist–patient relationship in which the patient stated his intent (later carried out) to kill his former girl friend. As the court stated, "the protective privilege ends where the public peril begins."

In addition to personal information obtained by professional people for the purpose of aiding the individual giving it, much personal information is collected for purposes other than direct help to the individual. For example, banks and credit agencies require personal financial information on opening accounts or on taking out loans. Government organizations collect information for tax, intelligence, and statistical purposes. Researchers carry out surveys on attitudes, beliefs, and life-styles. All these data are to some extent personal, given by the individual with the expectation of a certain degree of confidentiality, and private to the extent that the individual might be embarrassed if they were to become widely known.

Because of the expanding uses of personal information, and because of the mushrooming capabilities of data collection and storage both inside and outside government, there is a growing public and private concern about the dangers to individual liberty and privacy.[20] At the national level, numerous legislative enactments have attempted to strike a reasonable balance between the federal government's legitimate needs for personal information and the individual's legitimate concern that personal information be kept confidential. The two most important acts have been the Freedom of Information Act (FOIA) of 1966 (amended in 1974)[21] and the Privacy Act of 1974.[22] The FOIA requires disclosure, with certain exceptions, to any person of substantive and policy information maintained by federal agencies. As a result of this right of access, individuals are also able to obtain access to records about themselves. The Privacy Act is Congress's first attempt to incorporate comprehensive privacy protections into the records management practices of the federal government. It requires public notice of agency records systems, provides for individual access to personal records, sets up procedures for an individual to correct records about himself, limits disclosures of records, and establishes certain policies of fair information practice.

In addition to these statutes, Congress has further regulated information practice through the Fair Credit Reporting Act (1970), the Family Educational Rights and Privacy Act (1974), the Equal Credit Opportunity Act (1974), the Fair Credit Billing Act (1974), the Fair Debt Collection Practices Act (1977), and the Right to Financial Privacy Act

(1979). Each of these statutes represents an attempt to balance legitimate but conflicting interests; each, however, indicates a preference for individual privacy in the absence of substantial or compelling countervailing interests.

GENETIC COUNSELING

Advances in medical genetics, especially prenatal diagnosis of genetic disorders[23] and genetic screening,[24] have given the genetic counselor greater knowledge and have facilitated a much greater role in the management of genetic diseases. With that role there is a greater responsibility to provide information and counseling to those who might be affected with a disease. A problem arises when the genetic counselor discovers that the affected or carrier individual is an adoptee or has given up a child for adoption. That problem involves locating and counseling the birth parent or the adoptee.

Omenn, Hansen, and Hall have reported[4,25] five cases in which the family history revealed that a child given up for adoption several years earlier was at 50% risk for inheritance of Huntington disease, myotonic dystrophy, Fabry disease, von Willebrand disease, or Waardenburg syndrome, respectively. In each case, the clinic professional staff were able to work through the courts and the adoption agencies to make contact with the adoptee, the adoptive parents, or the birth parents. In some cases, direct contact was by the adoption agency staff. In other cases, contact was through a medical professional familiar with the family. Each case, however, required considerable effort and negotiation with the specific court and adoption agency. Not all the outcomes in each case were considered optimal by the counselors. The legal obstacles encountered in the search for the adopted child and the birth parents are thus of practical importance. The complementary situation has also been encountered (Omenn, unpublished). Adopted persons have presented with a clinical history and biochemical findings of coproporphyria in one case and with possible hereditary nephritis in a second case. Information about biologic parents was sought and obtained for proper evaluation of the patients.

INFORMATION NEEDS: THE SEARCH FOR GENETIC HERITAGE

The development of adoption law and the evolution of the concept of a right of privacy, while not necessarily parallel, share similar bases. Both share an ethic of the value of individual rights, both are grounded in the fundamental relations of the family, and both seek to protect the confidentiality of certain personal information. Indeed, both share roots as an expression of social policy. Social policy by its nature changes, however. There are changes taking place in society's attitudes toward adoption and toward the confidentiality of information, and the law is in the process of change as well. Recent advances in medical genetics illustrate that science can inform and influence the further development of the law of adoption.

The statutes of all American jurisdictions provide in some way that records of adoption proceedings be confidential. This general provision is reflected in the Uniform Adoption Act, drafted by the National Conference of Commissioners on Uniform State Laws and approved by the American Bar Association in 1953 (revised in 1969, amended in 1971).[26] The Uniform Adoption Act has been adopted by only four states (Montana, New Mexico, North Dakota, and Oklahoma), but its approach has been followed in other recent state laws. With regard to the confidentiality of proceedings, the act states:

> All papers and records pertaining to the adoption shall be kept as a permanent record of the court and withheld from inspection. No person shall have access to such records except on order of the judge of the court in which the decree of adoption was entered for good cause shown.[27]

The definition of the requirement of "good cause" is found in no adoption statute, and the meaning of the phrase has proven elusive to the courts as well. Hansen and Omenn,[4] reviewing state adoption statutes in 1976, found that approximately one-half the states allow opening of the adoption record by order of the court upon a showing of good cause. Four states require a showing of benefit to the adoptee, while the remainder do not specify criteria to allow opening by court order. In fact, there is little difference between the requirement of a court order without criteria and the requirement of a court order upon a showing of good cause, since the discretion of the court is by implication usually invoked only for good cause. Hansen and Omenn found six cases in the literature in which individuals other than the adoptee sought to open the adoption records. In five of the six, the court denied the application; in the sixth, the only case with an appellate decision, there was (apparently) fraud of sufficient magnitude to vacate the adoption decree.[4] Of those cases in which opening of the record was denied, two involved the attempt to prove adultery of the spouse, one was an attempt to defend a paternity action, one involved a grand jury's blanket request for all adoption records, and the fifth concerned a charity seeking access to determine beneficiaries of a trust fund. None of these cases involve what the courts would term such a compelling reason as to overcome the right of privacy the adoption statutes seek to uphold.

It has become especially popular recently for individuals to delve into their genealogies, to search for their roots. The search is especially poignant and important for many adoptees wishing to learn more of their personal history and genetic heritage. Often, that search takes them to the courts to attempt to gain access to the adoptive record and to the legislatures to obtain laws enabling such access. In 1976, only eight states allowed access to the adoptive record by the adoptee after obtaining majority.[4] In a suit brought in New Jersey because the state adoption statute did not allow automatic access by the adoptee, the provision of the statute requiring good cause in order to unseal the record was held constitutional. The court in that case, *Mills v. Atlantic City Department of Vital Statistics*,[28] discussed the interests to be protected by the secrecy provision. After considering the right of privacy of the birth parents and the right of the adoptive parents to freedom from interference, the court concluded that intangible benefits to the adoptee from knowing the information in the birth record did not outweigh injury to the interests of the birth parents and adoptive parents. Considering specifically the assertion that the constitutional right of privacy extended to the right to obtain information regarding heritage, background, and physical and psychological heridity, the court held that the information sought was not so intimately personal (fundamental) as to be constitutionally protected. Even though the court's holding meant that the state's denial of *automatic* access to the adoptive records did not contravene the constitutional right of privacy, it did not mean that the adoptee could not obtain the information sought if good cause were shown. In fact, the court indicated that access to family medical history may be considered good cause if evidence were presented of justifiable need.

Two other cases illustrate the status of the developing law of access to medical information in the adoptive record. The first, *In re Adoption of Female Infant,*[29] involved a District of Columbia statute that allowed inspection of the record only when the court was satisfied that the welfare of the child would thereby be promoted or protected. In that case, an adoptee wanted the adoption record unsealed in order to obtain, among other things, medical information to guard her children against any hereditary diseases. In spite of the recommendations of experts to open the records if the information obtained were accompanied by appropriate counseling, the court refused the request. Its decision was based upon a balancing of the interests of the individuals in the adoption triangle, and upon the view that the applicable statute required a stricter standard for disclosure than those statutes allowing disclosure for good cause. On balance, it determined that it was not in the best interest of all the individuals involved to unseal the record.

An opposite result was arrived at by the court in *Chattman v. Bennett.*[30] In that case, the applicable New York statute allowed disclosure for good cause, and the court granted access to medical information in the record so long as individual identifying information was deleted. Importantly, part of the rationale for the holding was that the adoption statute allowed the medical history of the adoptee (including family medical history) to be given to the adoptive parents during the adoption process. To have denied the adoptee access to the records would have been inconsistent with the apparent intent of the statute.

It would appear that the best view is that adoptees' requests for medical information should be granted so long as the identity of the birth parents is withheld.[3,4] The granting of the request would serve both the public policy interests of the availability of the most complete medical information for the individual and the social policy favoring the right of privacy of the birth parents. A number of states are moving in that direction through recently enacted statutes, or in considering legislation. Minnesota in 1977 passed a statute enabling an adoptee, through a careful administrative process, to obtain the information on the original birth certificate, including even the identities of the birth parents.[31] The District of Columbia in 1979 considered, but did not pass, a bill that would allow adult adoptees, after appropriate professional counseling, to learn the identities of their birth parents.[32] Perhaps the most specific enactment relating to medical information is a recent Iowa statute, which reads:

> Notwithstanding any other provision in this section, the juvenile court or court may, upon competent medical evidence, open termination or adoption records if opening is shown to be necessary to save the life of or prevent irreparable physical harm to an

adopted person or the person's offspring. The juvenile court or court shall make every reasonable effort to prevent the identity of the natural parents from becoming revealed under this subsection to the adopted person. The juvenile court or court may, however, permit revelation of the identity of the natural parents to medical personnel attending the adopted person or the person's offspring. These medical personnel shall make every reasonable effort to prevent the identity of the natural parents from becoming revealed to the adopted person.[33]

An important companion to this provision is another section of the Iowa adoption statute that relates to the preadoption investigation and report. That section requires that the report include:

(1) the complete family medical history of the person to be adopted, including any known genetic, metabolic, or familial disorders and
(2) the complete medical and developmental history of the person to be adopted.[34]

Few states are so specific as to what medical information is to be included in the preplacement report. No other state is so specific with regard to the medical indications for reopening the record. Both sections of the statute, however, indicate a recognition of the changes in social policy on the one hand and of the medical advances in genetic diseases on the other. Those sets of changes and the trends in the law suggest a recommendation for approaching the issue of the need for medical information in the adoption record.

A RECOMMENDATION

Recent advances in medical diagnosis and treatment in general and in medical genetics in particular have made it ever more useful for the physician to know fully the medical history of the patient and the patient's genetic relatives. In an era when nothing could be done for nearly all genetic diseases and when none could be diagnosed *in utero*, the need for such information was much less. Now, however, many genetic diseases have available methods of treatment and many diseases can be diagnosed *in utero*. Even in those cases for which there is not yet appropriate treatment, awareness of the genetic history may well prevent needless, expensive, and possibly harmful diagnostic evaluation. In many cases, knowledge of the genetic background of the patient is sufficient to raise the physician's level of suspicion so that appropriate diagnostic tests may be carried out.

The medical professional should be guided by three principles with regard to his or her responsibility in the adoption triangle: (1) Disclosure of any information should be dependent upon the proband/patient's consent; and (2) the need for the disclosure should meet a substantial private need; or (3) the need for disclosure should provide an overriding public or private benefit, sufficient to offset any possible detriment. It is possible for the law to balance the rights and interests of all parties to the adoption triangle while allowing the free flow of useful medical information. We propose a statutory approach consisting of three appropriate steps:

First, information with regard to the adoptee's medical and genetic background should be carefully, routinely, and completely gathered during preplacement proceedings. That information should be placed in a report, separate from the adoption report, given to the adoptive parents at the time of adoption and available to the adoptee later upon request. Any information that would allow identification of the birth parents should be deleted.

Second, provision should be made for addition to the adoption record and to the medical record of any significant relevant medical or genetic information obtained subsequent to the adoption.

Third, provision should be made to allow for contact with the appropriate persons (adoptee, adoptive parents, birth parents) by medical professionals in case significant relevant information were uncovered.

These suggestions reflect appropriate social policy. They would not require any person to divulge information to any other, nor would they require medical professionals actively to intervene in any specific case. What they would do is remove certain existing prohibitions from access to adoption records, and allow appropriate flow of relevant medical information. At the least, states should recognize the legitimate need for medical information as "good cause" to open the adoption record under present statutes.

References and Notes

1. Holder, A. R., *Legal Issues in Pediatrics and Adolescent Medicine*, p. 211, Wiley, New York (1977).
2. Clark, H. I., *Social Legislation* (2nd ed.), p. 303, Appleton-Century-Crofts, New York (1957).
3. Carter, J. R., Confidentiality of adoption records: An examination, *Tulane L. Rev.* 52:817 (1978).
4. Hansen, K. D., and Omenn, G. S., Genetic counseling: The search for the adopted child, *J. Legal Med.* 8AA-8FF (November/December 1976).
5. *In re Reinius*, 55 Wn.2d 117, 345 P.2d 672 (1958).
6. Colo. Rev. Stat. C.4,§4-1-1 (1964).
7. *In re Berth*, 79 C.A.2d 221, 179 P.2d 572 (1947).
8. *In re Christie*, 98 C.A. 158, 276 P. 1045 (1929).
9. N.Y. Dom. Rel. L. §111 (McKinney 1954).
10. Del. Stat. Ann. §919 (1953).
11. Prosser, G. W., *Law of Torts* (4th ed.), p. 802, West Publishing, St. Paul (1971).
12. 381 U.S. 479 (1965).
13. 405 U.S. 438 (1972).
14. Reilly, P., *Genetics, Law and Social Policy*, p. 132, Harvard University Press, Cambridge (1977).
15. Shaw, M. W., The legal aspects of genetic screening and counseling, *Excerpta Med. Int. Congr. Ser.* No. 411 (1976).
16. *Roe v. Wade*, 410 U.S. 113 (1973).
17. Shaw, M. W., Confidentiality and privacy: Implications for genetic screening, in: *Tay–Sachs Disease: Screening and Prevention* (M. M. Kaback, D. L. Rimsig, and J. S. O'Brien, eds.), p. 305, A. R. Liss, New York (1977).
18. Annas, G. J., Problems of informed consent and confidentiality in genetic counseling, in: *Genetics and the Law* (A. Milunsky and G. J. Annas, eds.), p. 111, Plenum, New York (1976).
19. 131 Cal. Rptr. 14, 551 P.2d 334 (1976).
20. Powledge, T. M., and Fletcher, J., Guidelines for the ethical, social and legal issues in prenatal diagnosis, *N. Engl. J. Med.* 300:168 (1979).
21. 5 U.S.C. §552 (1976).
22. 5 U.S.C. §552a (1976).
23. Omenn, G. S., Prenatal diagnosis of genetic disorders, *Science* 200:952 (1978).

24. National Academy of Sciences, *Genetic Screening—Programs, Principles, and Research*, p. 9, National Academy of Science, Washington, D.C. (1975).

25. Omenn, G. S., Hall, J. G., and Hansen, K. D., Genetic counseling for adoptees at risk for specific inherited disorders, *Am. J. Med. Genet.* **5:**157 (1980).

26. Uniform Adoption Act, *Uniform Laws Annotated* (master ed. vol. 9) (1971).

27. Ibid. §13(2).

28. 148 N.J. Super. 302, 372 A.2d 646 (1977).

29. 105 Daily Wash. L. Rep 245 (D.C. Super. Ct. 1977).

30. 393 N.Y.S.2d 768 (Super. Ct. 1977).

31. Minn. Stat. Ann. 144.1761 (1977 Supp.)

32. The *Washington Post*, p. A19, col. 5, Apr. 24 (1979).

33. Iowa Code §600.16(3) (1978 Supp.).

34. Iowa Code §600.8(1)(c) (1978 Supp.).

Discussion

Prof. Y. Edward Hsia: The difference between anonymity and confidentiality is very useful in most issues relating to adoption, with one possible exception, which fortunately is more common in romantic novels than in reality but nonetheless has been brought forward in consideration of use related to artificial insemination. That is if the adoptee should as an adult become romantically related to a close relative, because then the genetic risks cannot be covered purely by anonymous medical information.

Dr. John R. Ball: I am sure that Professor Annas is going to speak to that tomorrow. There is a recent article in the *New England Journal of Medicine* on the problems of record keeping and the possibility for, essentially, inbreeding because of the difficulties with record keeping in artificial insemination.

The only legal case that I have found that relates artificial insemination to adoption is a New York case several years ago. The statute required the consent of the father to adoption and the court held that the genetic father—that is, the artificial insemination donor—was not the father under the definition of the statute, but the father was the person who was married to the mother at the time of conception.

Dr. Theodore F. Thurmon (New Orleans): As a matter of practicality, I think family history is usually not available, at least in many cases, until the diagnosis is suspected in the child. So it would be of very little use to have a history of nothing.

What is needed is access to the parents, however anonymous they may be, for examination to determine if they have signs of the illness or if they are the carriers.

So I think that something is missing there in terms of assuming that just the record at the time of the adoption will necessarily tell you anything at all. I think that often it wouldn't. In the few cases I have been associated with, it has been quite useless, but it has been highly fruitful to at least get a photograph of the parents if not to be able to directly contact them.

Dr. Aubrey Milunsky: Of course, that point is made more cogent because some twenty or thirty years may have elapsed since the event and of course genetic eventualities that may have been delayed in their onset might then have occurred.

Dr. Thurmon: Yes. And new technology too.

Ms. Deborah Eunpu (Children's Hospital, University of Pennsylvania): I just have one comment to make based on a private adoption agency that we had occasion to deal with at both ends. In the questionnaires that are provided to parents during the time when they are considering putting up a child for adoption, the questions regarding genetic history are significantly different from the questions that they ask prospective adoptive parents. I think this illustrates that there needs to be some sort of unified program.

Dr. Ball: True. A suggestion that you might take—and this group might be an appropriate one to do so. The Department of Health, Education and Welfare now has a commission looking at adoption and with partial attempts to draft a model adoption statute. One worthwhile thing to do is to express those sorts of concerns to the commission in hopes that that sort of thing may be drafted into their act.

MR. IAN KENNEDY: The English laws were changed on adoption so that the adopted child may have access to public records so as to discover the natural parents. It may only be the mother, if the natural parents were not married. This was not, in my view, for genetic reasons but rather for social reasons. It wouldn't necessarily follow from that that the child would gain access to medical records. The child might gain access to information which would allow him or her to build up a medical history, but of course the records are vested in the National Health Service, are confidential, and wouldn't necessarily be made available.

24

GENETIC TECHNOLOGY AND THE SOLUTION OF CRIME

Forensic Genetics

PETER V. TISHLER
Channing Laboratory
Department of Medicine
Harvard Medical School
and the Peter Bent Brigham Hospital Division
  of the Affiliated Hospitals Center, Inc.
Boston, Massachusetts 02115

The field that one might call "forensic genetics" involves the application of genetic knowledge and methodology to the solution of crime. Clearly, forensic genetics plays a role in the solution of certain crimes, although this role is underexploited at the present time. The purpose of this chapter is to acquaint you in overview form with selected aspects of this field. I hope also to point out some problems in the field, and to suggest remedies to make it more productive.

The three areas that I shall discuss all focus on the identification of the individual from one or another body fluid. In order to accomplish this, the forensic scientist must exploit factors that have the potential for uniqueness and that thus may identify an individual in a manner similar to fingerprints. The forensic scientist has appreciated the significance of the findings of human geneticists in especially two areas. These are the determination of nuclear sex and the identification of heritable polymorphisms.

Nuclear sex determination involves the ascertainment of the unique male or female chromosome in randomly obtained, nucleated cells, without recourse to a chromosome analysis. In the female, one of her two sex chromosomes, the X chromosomes, may be found as an oblong, darkly staining body (the X chromatin body) adjacent to the nuclear membrane.[1] In the male, the sex-determining Y chromosome possesses the remarkable property of intense fluorescence when exposed to certain acridine dyes. This fluorescence may also be found in nucleated cells, in which it exists as a small, round fluorescent body (Y body).[2]

The many thousands of proteins in the body are the products of the action of genes. Two genetic mechanisms exist for diversification of proteins within both populations and individuals. In populations, a number of different forms of the same protein have appeared, largely as the result of mutational events during evolution. These variant forms differ structurally but maintain a common function. For reasons that are largely unknown, the frequency of many of these related proteins, and the genes that control them, has increased substantially above the rate of mutation. These related, relatively common proteins constitute a protein polymorphism.[3] The well-known hemoglobins A, S (sickle), and C constitute a polymorphism for three different forms of the beta chain of human hemoglobin.

In the individual, genes exist in pairs, and the synthesis of each protein is controlled by the action of the two homologous genes. Although there may be many genes coding for a particular set of polymorphic proteins in a population, the individual can have only two. Individual variation is achieved by the fact that these two genes can code for the same protein (homozygosity), which can be any one of the members of a protein polymorphism, or for two different proteins (heterozygosity).

The study of these very common genetic polymorphisms[4] has led to a quantitative appreciation of the scope of genetic diversity in the individual.[3] This is clearly important for forensic genetics, which, I must point out, has also contributed its share of basic knowledge to this field. In addition, the techniques that have been used to detect genetically variant substances—those that point out immunologic differences (as in blood group antigens) or those that separate related proteins that differ in their structure (e.g., electrophoresis)—have also been appropriated by the forensic geneticist. These techniques are rapid and easy and are readily adaptable to screening samples in the crime laboratory.

DISPUTED PARENTHOOD

Perhaps the best known area in which genetic technology is exploited is in paternity (or maternity) exclusion. This topic was reviewed extensively in the First National Symposium on Genetics and the Law.[5] Until recently, *exclusion* of parenthood was the only result of such testing that was recognized in a court of law.[6,7] Exclusion of parenthood can be achieved with absolute certainty. However, determining that a putative parent was a true parent can only be approximated statistically, and sometimes with only a modest probability.

The classical method of testing for parenthood involves the examination of the red blood cells of the putative parent, the true parent, and the child for a series of polymorphic blood group antigens, most of which are common.[8] The list in Table I of these antigenic systems excludes those that are too rare to be of routine value and those whose use is in dispute,[9,10] but more complete tabulations may be found elsewhere.[6,9] The results of analysis of these antigens can exclude parenthood in five ways[9]:

1. The putative parent lacks a specificity that is present in the child and absent in the other parent (e.g., child is blood group AB, mother is group A, putative father is group O).

TABLE I.   Probability of Exclusion of
Parenthood for Useful Red Cell Antigen
Systems[a]

Antigen	Mean probability of exclusion	
	Black	White
ABO	0.1774	0.1342
MNSs	0.3206	0.3095
Rh	0.1859	0.2746
Duffy	0.0420	0.1844
Kell	0.0049	0.0354
Kidd	0.1545	0.1869
Lutheran	0.0368	0.0311

[a] Combined exclusion rate for all antigens, calculated by method
of Wiener[11]: 0.6468 for blacks, 0.7312 for whites. Table
modified from reference 9.

2. The child lacks a specificity found in the putative parent (e.g., child is group O,
   putative father is group AB).
3. The child is homozygous for a specificity not found in both parents (e.g., child is
   homozygous for the Kell blood group, KK; mother is Kk or KK; putative father is
   kk).
4. The child lacks a specificity for which the putative parent is homozygous (e.g.,
   child is kk, putative father is KK).
5. An indirect exclusion may obtain when the study of relatives of the true and/or
   putative parent more clearly defines their genotypes.

Each of the red cell antigen systems has a certain statistical probability, based on the
frequencies of its component genes and the possible genotypes, of excluding parenthood
by one of these five means. This probability, which is easily calculated,[12] is included in
Table I. In addition, when tests are carried out on a number of independent red cell
antigens or other heritable polymorphisms, one may calculate a combined probability of
exclusion of parenthood. This calculation gives the cumulative probability that at least
one of these tests will exclude parenthood for a falsely accused individual.[9] By exploiting
all seven red cell antigenic systems (Table I), one can achieve a combined probability of
exclusion of parenthood of 0.65 and 0.73 for black and white individuals, respectively.

To these red cell antigens may be added a number of polymorphic proteins that are
found in either the red cell or the blood plasma. Although the number of such polymorphic systems is large,[4] some experts believe that only a few have been studied sufficiently
systematically to be considered legally acceptable testing systems.[10,13] Four such plasma
proteins that appear to have gained legal acceptance are listed in Table II. The combined
probability of exclusion of parenthood with these four proteins is 0.42 and 0.52 for blacks
and whites, respectively. Combining these proteins with the aforementioned seven red
cell antigens yields a total combined probability of exclusion of 0.80 and 0.87 for blacks
and whites, respectively.

TABLE II.   Probability of Exclusion of Parenthood for
Certain Plasma Proteins[a]

| | Mean probability of exclusion | |
Plasma protein	Black	White
Gm serum groups	0.2071	0.2275
Group-specific component	0.0731	0.1661
Ag lipoprotein	—	0.0813
Haptoglobin	0.1873	0.1834

[a] Combined exclusion rate[11] for all plasma proteins: 0.4207 for blacks, 0.5167 for
whites. Table modified from reference 9.

In practice in the Commonwealth of Massachusetts, one major laboratory that offers
these tests examines 26 blood groups and 6 plasma protein polymorphisms, with a combined probability of exclusion of about 0.95.[14] Despite this impressive statistic, a surprising number of individuals who have been unjustly accused fail to take advantage of this
procedure. Sussman has pointed out that only about 10% of defendants in paternity
proceedings requested blood grouping tests to substantiate their denial, even though about
40% were in fact not the fathers of the children in question.[15] He states: "To fail to
demand a blood grouping test is indeed to disregard a most important defense, and thus
fail to provide the most substantial evidence of nonpaternity."[15]

The advent of testing for tissue compatibility (histocompatibility) antigens (the HLA
system in man) has provided a highly discriminating method for both the exclusion and
the determination of parenthood. The HLA antigens comprise a series of glycoproteins
coded for by allelic genes at four tightly linked loci on human chromosome 6.[9,16] Extreme diversity of HLA types exists in humans, and many different alleles may be found at
each of the four gene loci in the population. The frequency of the gene for any specific
HLA type is thus quite low.[9] The result of this genetic diversity is that the HLA types of
virtually all individuals are rare, and a high degree of discordance for HLA types prevails
in the general population. The finding of concordance for HLA types between child and
putative parent constitutes strong evidence for parenthood.[17] Conversely, a wrongly accused putative parent can usually be excluded because the child has inherited a different
rare HLA type from the actual parent.

The HLA antigens at two of the four loci (A and B) have been used for several years
for paternity testing. These two groups of antigens achieve a mean probability of exclusion
of paternity of about 0.80.[9] The most powerful demonstration of the discriminatory power
of this antigenic system has been provided by Terasaki,[18] who applied HLA testing to 1000
individuals who were not excluded from paternity by tests of the red cell ABO antigens.
The HLA A and B antigens permitted the definitive exclusion of 25% of these individuals.
Moreover, 64% of the remaining individuals had a high probability of actual paternity
(>0.90). In about one-third of these, the probability of paternity was ≥ 0.98. In the
Scandinavian countries, probabilities of paternity based on HLA typing are now accepted
by the courts if they are very high (>0.95) or low (<0.05).[19] This has been in effect for

about six years.[18] Acceptance in this country appears to be more limited,[6,7] but court admissibility of HLA data as evidence *for* paternity has recently been granted by a California appeals court.[20]

In summary, three different types of polymorphisms can be exploited in dealing with questions of parenthood. These include the red cell antigens, plasma and red cell proteins, and the histocompatibility antigens. Although there is currently some debate concerning the order in which these should be performed, there is little doubt that all three, and particularly the HLA system, provide efficient means for answering the questions.

ANALYSIS OF BODY FLUIDS IN VIOLENT CRIME

Bloodstains

A similar exploitation of genetic polymorphisms may be of great value in identifying the origin of bloodstains associated with violent crime.[21] Red and white blood cells and solutes from the plasma theoretically persist in bloodstains, so the study of stains is not limited by the number of polymorphic markers. However, technical problems assume great importance. These include the quantity of blood in the stain, the age of the stain, the conditions under which it was made and preserved, and the effect of the material on which it was made. They impose severe restrictions on the use of methods that are otherwise routine for fresh whole blood.[4]

Methods for the study of genetic polymorphisms in bloodstains have been developed primarily by chemists at London's Metropolitan Police Forensic Science Laboratory and other labortories in the United Kingdom.[21] The genetic markers (antigens, serum proteins, enzymes) that are reported to be detectable in bloodstains under most conditions are listed in Table III. The frequencies of the genes for many of these polymorphisms are sufficiently favorable in general populations, or in the case of certain markers such as glucose-6-phosphate dehydrogenase or sickle hemoglobin in an especial population,[4,28] that one may achieve a very high combined probability of discriminating between two blood samples.[29]

The successful determination of the sex of the donor of a bloodstain has been reported by a number of investigators.[30-32] All methods depend upon detecting the fluorescent Y body in white blood cells. A major problem in the use of bloodstains for this purpose is the limited architectural integrity of leukocytes in the absence of appropriate preservation. Nonetheless, sufficient numbers of intact leukocytes can apparently be obtained from bloodstains, and the fluorescent Y body seems to maintain its integrity sufficiently that stains can be sexed.

Currently, in the United States, the laboratory of the Federal Bureau of Investigation and those of several large cities are equipped for the genetic analysis of bloodstains. At the FBI laboratory, this analysis routinely includes the red cell antigens, hemoglobin, acid phosphatase, and phosphoglucomutase-1, and plasma haptoglobin.[33] The probability of discriminating between two blood samples with these tests is greater than 0.95.[29] Further studies of esterase D, adenylate kinase, and peptidase A are carried out upon indication.

TABLE III.   Genetic Polymorphisms That Can Be Recognized in Blood Stains

---

Red blood cell antigens: ABO, MNSs, Rh, Kell, Kidd[22,23]

Hemoglobins: A, F, S, C, etc.[22]

Enzymes: Phosphoglucomutase-1, adenosine deaminase, pseudocholinesterase-1 and -2, acid phosphatase, esterase D, peptidase A, glucose-6-phosphate dehydrogenase[22,24−26]

Plasma proteins: haptoglobin, Gm, Inv[22−27]

The FBI has had no problem with the admissibility of these data as evidence in the courts.[33] Y body data are apparently used as evidence in the United Kingdom, but only to ascertain maleness.[32] In the United States, this type of evidence has been introduced by the prosecution in at least one trial.[34] However, the defense argued that the test had been conducted without satisfactory prior experience, methodology, or control observations. The judge thus threw out the evidence, concluding that it represented "questionably accepted scientific testimony."[34]

Semen

Studies by several protein chemists, particularly Sensabaugh and co-workers,[35] have established the presence of many polymorphisms in either seminal plasma or sperm (Table IV). Thus, matching semen with an individual is theoretically as possible as is matching in the previous two examples. However, the problems interposed between this information and its actual application to the study of semen stains or vaginal contents in cases of rape are formidable. Many of the enzymes listed in Table IV are present in a concentration that is insufficient for routine detection.[36] Semen contains potent enzymes that degrade proteins rapidly at body temperature, making the early collection of specimens imperative.[35] Finally, vaginal secretions may also contain the same factor that is under study in semen, and means must be developed to differentiate one from the other, at least for certain markers. None of these problems seems to be insurmountable. Currently, in the United Kingdom, only secretor status, the ABO antigens, and the isoenzymes of phosphoglucomutase-1 are determined in forensic science laboratories in investigations of rape.[37] The FBI laboratory in the United States determines only secretor status[33] and has been unsuccessful in clearly determining the phosphoglucomutase types.

The determination of X chromatin bodies has been advocated as a means of detecting female cells in semen stains or on the penis, thus indicating that the individual had recent

TABLE IV.   Genetic Polymorphisms That Can Be Recognized in Sperm

---

Those present in adequate concentration: phosphoglucomutase-1, peptidase A, phosphoglycerate isomerase[36,37]

Those present in suboptimal concentration: adenylate kinase, amylase, esterase D, glucose-6-phosphate dehydrogenase, 6-phosphogluconate dehydrogenase, phosphoglucomutase-3, diaphorase, transferrin, Gm, Inv, protease inhibitor[35,36,39,40]

---

coitus.[41,42] Here again, this technique has not been studied sufficiently to justify its use in the crime laboratory.

Conclusions

There are other aspects of forensic genetics that I shall not consider. These include the use of dermatoglyphic analysis—prints of hands and even feet—and the search for genetic polymorphisms in other body materials, such as saliva and hair roots. The genetics of the former is quite complex, precluding an easy, brief discussion. Work on the latter, while proceeding along the lines already discussed for blood and semen, is in an even more primitive state of data acquisition.

I conclude that forensic geneticists, many of whom are trained in the scientific method, have been quick to realize the significance to their profession of new knowledge in basic human genetics. However, they have been slow to develop methods for the routine and reliable application of this knowledge in the forensic science laboratory. To a large extent, this reflects the number of individuals engaged in research in forensic genetics. Quite simply, the number is very small. A chemist in the laboratory of a local police department expressed great awareness of the potential of these methods for the solution of crime but stated that his department is unable to marshall the manpower and financial support to set up or study these methods.[43] Professor William Curran has noted this also and points out that this is a universal problem, particularly in the United States.[34] He urges that we make a greater effort to exploit the knowledge contributed by the few good laboratories, and believes that we must support effective forensic science research. He states: "We need to improve vastly the quality of legal medicine and forensic science in the United States, for the safety of all of us and for the sake of effective justice for all . . . in our courts."[34] Forensic genetics should be among the first aspects of forensic science to be nurtured, since the chances for payoff in the manner envisioned by Professor Curran seem to me to be most promising.

Acknowledgment

The author appreciates the assistance of Dr. Leonard Atkins in obtaining information for certain aspects of this chapter.

References

1. Hamerton, J. L., *Human Cytogenetics* (Vol. 1), p. 131, Academic Press, New York (1971).
2. Lamborot-Manzur, M., Tishler, P. V., and Atkins, L., Quantitative studies of Y chromosomal fluorescence in human interphase nuclei, *Clin. Genet.* 3:103 (1972).
3. Harris, H., *The Principles of Human Biochemical Genetics* (*Frontiers of Biology*, 2nd ed., Vol. 19), p. 278, North-Holland, Amsterdam (1975).

4. Harris, H., and Hopkinson, D. A., *Handbook of Enzyme Electrophoresis in Human Genetics*, North-Holland, Amsterdam (1976); and Supplements (1977 and 1978).

5. Konugres, A. A., The current status of paternity testing, in: *Genetics and the Law* (A. Milunsky and G. J. Annas, eds.), p. 219, Plenum, New York (1976).

6. Chakraborty, R., Shaw, M., and Schull, W. J., Exclusion of paternity: The current state of the art, *Am. J. Hum. Genet.* **26**:477 (1974).

7. Reilly, P., *Genetics, Law and Social Policy*, p. 222, Harvard University Press, Cambridge (1977).

8. Race, R. R., and Sanger, R., *Blood Groups in Man* (6th ed.), p. 8, F. A. Davis, Philadelphia (1975).

9. Miale, J. B., Jennings, E. R., Rettberg, W. A. H., *et al.*, Joint AMA–ABA guidelines: Present status of serologic testing in problems of disputed parentage, *Fam. L. Q.* **10**:247 (1976).

10. Sussman, L. N., Paternity blood grouping tests using legally unacceptable testing systems, *Am. J. Clin. Pathol.* **69**:649 (1976).

11. Weiner, A. S., and Wexler, I. B., *Heredity of the Blood Groups*, p. 120, Grune and Stratton, New York (1958).

12. Race, R. R., and Sanger, R., *Blood Groups in Man*, op. cit., p. 504.

13. Sussman, L. N., Medicolegal blood grouping tests (parentage exclusion tests), in: *Progress in Clinical Pathology* (Vol. 5) (M. Stefanini, ed.), p. 143, Grune and Stratton, New York (1973).

14. Alper, C. A., Personal communication (1979).

15. Sussman, L. N., Blood grouping tests in non-paternity, *J. Forensic Sci.* **18**:287 (1973).

16. Rosenberg, L. E., and Kidd, K. K., HLA and disease susceptibility: A primer, *N. Engl. J. Med.* **297**:1060 (1977).

17. Terasaki, P. I., Gjertson, D., Bernoco, D., *et al.*, Twins by two different fathers identified by HLA, *N. Engl. J. Med.* **299**:590 (1978).

18. Terasaki, P. I., Resolution of HLA testing of 1000 paternity cases not excluded by ABO testing, *J. Fam. L.* **16**:543 (1978).

19. Forrest, M. F., The legal implications of HLA testing for paternity, *J. Fam. L.* **16**:537 (1978).

20. HLA is latest paternity clue, *Medical World News*, p. 58, Apr. 2 (1979).

21. Editorial, Enzyme electrophoresis in forsenic medicine, *Lancet* **1**:252 (1974).

22. Culliford, B. J., *The Examination and Typing of Bloodstains in the Crime Laboratory*, U.S. Department of Justice, Law Enforcement Assistance Administration, National Institute of Law Enforcement and Criminal Justice, Washington D.C. (1971).

23. Lincoln, P. J., and Dodd, B. E., The application of a micro-elution technique using anti-human globulin for the detection of the S, s, K, Fy^a, Fy^b, and Jk^a antigens in stains, *Med. Sci. L.* **15**:94 (1975).

24. Parkin, B. H., and Adms, E. G., The typing of esterase D in human bloodstains, *Med. Sci. L.* **15**:102 (1975).

25. Neilson, D. M., Shaler, R. C., Stuver, W. C., *et al.*, Simultaneous electrophoresis of peptidase A, phosphoglucomutase, and adenylate kinase, *J. Forensic Sci.* **21**:510 (1976).

26. Grunbaum, B. W., and Zajac, P. L., Differentiation of the genetic variants of glucose-6-phosphate dehydrogenase in blood stains by electrophoresis on cellulose acetate, *J. Forensic Sci. Soc.* **16**:319 (1976).

27. Khalip, S., Pereira, M., and Rand, S., Gm and Inv grouping of bloodstains, *Med. Sci. L.* **16**:40 (1976).

28. Selvin, S., Black, D. M., Grunbaum, B. W., *et al.*, Racial classifications based on blood group protein systems, *J. Forensic Sci.* **24**:376 (1979).

29. Jones, D. A., Blood samples: Probability of discrimination, *J. Forensic Sci. Soc.* **12**:355 (1972).

30. Phillips, A. P., and Webster, D. F., Improved Y chromosome fluorescence in the presence of magnesium ions, *J. Forensic Sci. Soc.* **12**:361 (1972).

31. Aragonés, J., and Egozcue, J., A method of sexing 90-day-old blood stains in absorbent and nonabsorbent materials, *Acta Genet. Med. Gemellol.* **22**:113 (1973).

32. Wigmore, R., Werrett, D. J., King, L. A., *et al.*, The detection of Y chromosomes in bloodstains—A reevaluation, *J. Forensic Sci.* **24**:366 (1979).

33. Beams, R., Federal Bureau of Investigation Laboratory, Washington, D.C., personal communication (1979).

34. Curran, W. J., Forsenic medical science: The continued problems of judges and juries, *N. Engl. J. Med.* **294**:1042 (1976).

35. Blake, E. T., and Sensabaugh, G. F., Genetic markers in human semen: A review, *J. Forensic Sci.* **21**:784 (1976).

36. Blake, E. T., and Sensabaugh, G. F., Genetic markers in human semen. II. Quantitation of polymorphic proteins, *J. Forensic Sci.* **23**:717 (1978).
37. Sutton, J. G., Further alleles of phosphoglucomutase in human semen detected by isoelectric focusing, *J. Forensic Sci.* **24**:189 (1979).
38. Pereira, M., and Martin, P. D., Problems involved in the grouping of saliva, semen and other body fluids, *J. Forensic Sci. Soc.* **16**:151 (1976).
39. Brinkmann, B., and Koops, E., Phosphoglucomutase (PGM) and 6-phosphogluconate dehydrogenase (PGD) isozymes in human sperm cells, *Humangenetik* **14**:78 (1971).
40. Kühnl, P., Langanke, U., Spielmann, W., *et al.*, Investigations on the polymorphism of sperm diaphorase in man, *Hum. Genet.* **40**:79 (1977).
41. Renard, S., Determination of sex of exfoliated epithelial cells and its significance in forensic science, *J. Forensic Sci. Soc.* **11**:15 (1971).
42. Given, B. W., Sex chromatin bodies in penile washings as an indicator of recent coitus, *J. Forensic Sci.* **21**:381 (1976).
43. McHugh, J., Chemistry Laboratory, Boston Police Department, personal communication (1979).

DISCUSSION

DR. AUBREY MILUNSKY: Dr. Tishler, would you comment on hair roots?

DR. PETER V. TISHLER: The use of hair roots in forensic genetics is primarily along the same lines as looking at bloodstains, namely, looking for polymorphic markers. I may have encountered one, possibly two papers which looked for sex chromatin in hair roots. But my impression about use of hair roots is that the technology here is even in a greater state of infancy than it is with bloodstains or seminal stains.

GENETICS AND FAMILY LAW

MODERATOR: Sanford N. Katz

25

PUBLIC ATTITUDES TOWARD THE HANDICAPPED

BARTON CHILDS
Department of Pediatrics
Johns Hopkins University School of Medicine
Baltimore, Maryland 21205

An hour at an airport or in a museum must convince an observer that individuality is the most salient feature of the human species. The people come in so many sizes, shapes, and arrangements of features, express so many moods, behave so diversely, and exhibit clothing, hairstyles, and cosmetic transformations of such variegated texture and taste as to assure us that each person is a product of biological and social forces that have collaborated to make something unique.

But beneath all the variety of color and composition, the observer detects some constraint, some centripetal tendency to conformity. In physical form, while there is much variation, it is always on the human theme; there are no centaurs, satyrs, or mermaids; and even in decoration there is evidence of adherence to rules of fitness of dress having to do with sex, age, and social and occupational stratification. Some of this conformity can be perceived to be expressed in familial resemblances, perhaps most extravagantly revealed in twins, whose physical and behavioral similarity is often topped off by identical clothes. So variation and uniqueness exist in the presence of conformity; all human beings are qualitatively alike in some things, yet each is different, even while resembling some more than others.

An observer with a quantitative turn of mind could not fail to notice that this uniqueness is marked sometimes by great deviation, sometimes by small, and that the former is rare, the latter common. For any feature, the differences are usually rather minor variations on a main theme; major departures are relatively less frequent, and rarely some are so great as to be perceived as qualitatively different.

This variability is an expression of fundamental biological and social forces that exert powerful effects on human behavior. Actually, each of us is making observations of human differences all the time; we are always storing them, rearranging them, and classifying them in various ways. They supply the background experience against which to compare and assort new experiences; they help to fashion our self-assessment and identity. Unfortunately, these selections and comparisons are usually informed by a limited experience as well as by beliefs and prejudices rather than by any knowledge of the biological origins of differences or any understanding of the genetic kinship of one's self with others. A too-frequent result is an aversion to deviance, whether social or physical, and the degree of the reaction is in direct proportion to the perceived extent of the departure. And such responses, which may be violent and wounding to their object, are often directed to single, easily observed characteristics, overlooking all the other qualities that make up a person.

The point is that classifying and affixing labels is as human as apple pie is American, and since society is composed of human beings, classification and labeling will be with us always. The question is how to make use of necessary classifications without harm to the persons they embrace.

CLASSIFICATION

Classifications arrange things in groups according to shared properties. Without them, knowledge would be unorganized and social intercourse would be impossible. The classifications most pertinent to the purposes of this meeting are those that separate human beings into groups called normal and abnormal. Such assortment is essential in a society concerned to provide special help to the abnormal, but in so doing several problems are raised: (1) It is not always possible to distinguish all members of each class; (2) classifications may have effects not envisioned by the classifiers; and (3) the persons classified as abnormal may be harmed more than they are helped.

Normal versus Abnormal

The distinction between normal and abnormal is sometimes very easy to make. An individual may lack something that everyone is expected to have or may express some distortion of conformation so extreme as instantly to be recognized as beyond any definition of normal. It is such cases that inform the general view of abnormality, and when this word is used in contrast with "normal," it gives an impression of categorical distinction, as if all representatives of either of the two classes are easily distinguishable.

But that is not nature's way. Measurable properties are usually continuously distributed, and histograms of such measurements usually give little help in deciding on some point, or even area, distinguishing normal from abnormal. And when the classes are disjoint, as in some birth defects, it is usually the result of some process that has reached a threshold to produce the "either/or" effect, but measurements of the processes themselves reveal the familiar continuous distributions. In addition, since standards are comparative, what is abnormal under some conditions may be quite normal under others. Some examples may be illuminating. A question often asked is: Where does hypertension begin? No one knows. Some people maintain quite high blood pressures without adverse sequelae, while others have renal disease or strokes with pressures that are called abnormal only after the fact. Birth weights are also continuously distributed, so what is an "abnormal" birth weight? The probability of some medical disaster is increased for babies with both very low and very high birth weights, but it is only a probability. And the likelihood of an adverse social and occupational outcome is greater for a person with a low score in a distribution of measurements of reading skills or of IQ than for one with a high score; but it is again only a probability, and many people with scores in the low ranges make a gratifying adjustment to living. The point is that healthy people may be labeled as having this disease or that defect because their measurements fall into some part of a distribution that is called abnormal because of some statistical probability of an adverse consequence.

The continuity of these distributions has a further meaning. It is the biological evidence that opposes the categorical view. It is the evidence that "each is a piece of the continent, a part of the main," the evidence that the handicapped represent only the extremes of a variation that is an indispensable human property, rather than that they are a discretely different species.

The Uses of Classifications

Obviously, classifications are intended to fulfill useful purposes—useful for both individuals and society. Classification of the handicapped is supposed to be useful in circumscribing populations of persons, the better to help them. According to the report of the Project on Classifications of Exceptional Children, classifications are intended to help in evaluating handicapped children so as to (1) open doors to opportunity, (2) provide data to legislators who can guarantee rights and appropriate funds, and (3) provide the means to bring handicapped people and ameliorating programs together.[1]

All these are laudable, necessary ends and unavoidable if society is to discharge its obligations. But classifications can have the undesired effect of imposing upon an individual an official stamp that may relegate him to one of life's backwaters, to the detriment of his unimpaired qualities and his chance to participate. Categorical classes stamp all included as if all were the same, as if, for example, an IQ of 55 were the same as an IQ of 10, since both are called mentally retarded, or as if two people, both with IQs of 55, had, necessarily, the same competences and deficiencies. In this way a mechanism designed to help those in need may have the unintended effect of depriving deserving people of opportunities and of resources needed to foster normal growth and development.

THE IMPACT OF HANDICAP ON THE PUBLIC AND ON THE AFFECTED THEMSELVES

Both the concept of handicap and its reality have effects on public attitudes, effects that may be reflected in the formation of the views of the handicapped toward themselves.

The Public

Classes, labels, and physical appearance have strong effects on popular acceptance. Riches, beauty, or fame, or merely a reputation for these, are likely to elicit an indiscriminate approbation. And there is ample evidence of an equal, but contrary, response to the *idea* of defect as well as to its physical presence. Richardson and his colleagues have studied these attitudes in a series of experimental situations.[2] In one of these, healthy people of various ages were presented pictures of children with and without various handicaps, and the subjects were asked to choose the one they preferred and then to rank the others in the order of decreasing preference. The figures consisted of six children of the same sex as the subject: one normal, one with a leg brace and a crutch, one in a wheelchair, one with an amputated forearm, one with a facial disfigurement, and an obese child. When these figures were shown to children of 8–10 years, the ranking was that given above. Increasing age changed the ranking, but children below 6 years showed no consistent patterns of choice. Other studies show that although the degree of this adverse response is determined by the extent of the handicap, exposure causes the conspicuousness of the abnormality to recede and allows the force of the handicapped person's personality and talents to emerge. Further, children make this adjustment more quickly and completely than adults. Richardson summarizes these observations by suggesting that in the course of development each of us acquires a set of expectations in regard to the appearance and behavior of others that is based on our experiences, and it is the violation of these expectations that causes the immediate revulsion to the handicapped. But exposure brings these expectations into line with the reality of the whole individual who has much more to express than his defect.

Labels, too, can have adverse effects.[3] Abroms and Kodera studied the acceptance by college students of 15 diseases. The results suggested that conditions that best fit the medical model were most likely to be acceptable, while those generally handled outside conventional medical channels by psychiatrists, educators, and social workers, or through other social agencies—for example, mental illness, cerebral palsy, or mental retardation—were least acceptable.

The Handicapped

If attitudes such as these prevail, they must give respectability to an impersonal, dehumanizing concept of handicaps and defects as something outside the normal current of life, and as the responsibility, not of us all, but of tax-supported social agencies. Such attitudes feed upon themselves; the more the handicapped are out of the public sight and mind, and the more they are made to feel incompetent to participate with normal people in the world's business, the more incompetent they become. Talents remain undeveloped, and they become caricatures of the public image of defective people.

REMEDIES

The remedies for these injustices are implicit in their description. If the handicapped are not out of sight, they cannot be out of mind; if they are a part of everyday experience, expectations cannot be violated; and if opportunities are provided, the handicapped will rise to the challenge. That is what recent legislation and the disease-related agencies are trying to accomplish just now, and with a heartening initial success.[4,5] But it is one thing to make the handicapped visible, to lower the curbs, build ramps, and design buses with mobile steps, and another to accomplish the kind of familiar acceptance experienced by normal persons among other normals. That will require abundant exposure to handicapped people in all phases of life and a general recognition that they represent the extremes of human variation rather than a separate category of human being—no doubt an agenda for the next two or three generations.

This acceptance can be hastened, however, if it is given appropriate attention in school.[4] The Rehabilitation Act of 1973, the Education for All Handicapped Act of 1975, and the efforts of the Association for Retarded Citizens and other groups are all intended to cause disabled children to be included as students with special problems rather than to allow them to be isolated physically, socially, and intellectually. This is the current educational philosophy: handicapped students are to be treated in a manner befitting their individuality, not their labels; they are to be accepted as people with the same need to be educated for living as normal children, but with special obstacles to be overcome in getting on with that requirement.

But mainstreaming alone is unlikely to succeed very quickly without some overt efforts to reorient teaching, especially of science and health education, to include a better grasp of such ideas as the origins and extent of human variability and of the origins of disease in the relations between endowment and the conditions of life. At the moment, biology teachers, for example, are struggling to deal with technological changes that have a great impact on living but that fit poorly into their discipline-oriented curricula. Sweeping progress might be made if the curriculum-makers of primary and secondary schools could become a particular target for efforts to promote change. They are the opinion-leaders who, if they can be convinced, can bring about a rapid diffusion of new approaches into the system. If we expect progress to come exclusively through education of adults, we shall wait a long time indeed. Children have been shown to be more adaptable than adults in this acceptance. Some intellectual grasp of the biological and social meaning of handicap could help a great deal to creat a context for a quicker success of mainstreaming and to provide a timely assist in the redress of injustices we have tolerated for too long.

ACKNOWLEDGMENT

This work was supported in part by National Institutes of Health Grant HD 00486.

REFERENCES

1. Hobbs, N., *The Futures of Children: Categories, Labels and their Consequences*, Jossey-Bass, San Francisco (1975).

2. Richardson, S. A., Attitudes and behavior toward the physically handicapped, *Birth Defects: Orig. Art. Ser.* **12**:15 (1976).

3. Abroms, K. I., and Kodera, T. L., Acceptance hierarchy of handicaps, *J. Learn. Disabil.* **12**:24 (1979).

4. Reynolds, M. C., and Birch, J. W., *Teaching Exceptional Children in All American Schools*, Council for Exceptional Children, Reston, Va. (1977).

5. Bergsma, D., and Pulver, A. E., Developmental disabilities: Psychological and social implications, *Birth Defects: Orig. Art. Ser.* **12** (1976).

Discussion

DR. RICHARD RATZAN (University of Connecticut): Professor Childs, there was recently a segment in the Hartford news about mainstreaming. How do you solve the problems of disruptive children, children with extremely poor eyesight, limited learning ability, or decreased attention span? How do you incorporate them into a class?

PROF. BARTON CHILDS: I don't pretend to know the answer to that. I think that some kinds of adjustments will have to be made to keep those children socially included among other children even while taking into account the fact that they do have handicaps which have to be dealt with, at least in part, separately.

I think that no one pushing the mainstreaming idea has it in mind to handicap other children in the interest of the minority of children who have problems.

I think no one yet knows the answers to the questions that you asked. It is going to take a long time, a lot of planning by people of goodwill, and much effort to accommodate everybody in our system. Some places are doing it much better than others. My only experience is in Baltimore, where they have problems but the school system seems to be struggling to solve them.

DR. RATZAN: The news segment just demonstrated the irony of aggressive advocates of mainstreaming that were really counterproductive. It seems that they were mostly just feeding their own cause and really alienating the parents of the normal children by adopting not an "all or none" but an "all" approach, that these children should be in the normal classes, and it seemed very counterproductive.

PROF. CHILDS: I am personally as opposed to categorical treatment on the one side as the other.

MS. MARJORIE GUTHRIE: I share the speaker's feelings, because when I was a little girl I remember reading the newpapers fifty years ago about Pennhurst. I agree that we have a problem.

But I want to go back, if I may, just a little to what Dr. Childs said on the question this young man asked and I want to ask a question. Is it conceivable that we have not understood the word *education*? Is it conceivable that, to many, the education of a so-called normal child to have feelings and attitudes toward a so-called abnormal child might be a more important education than just reading, writing, and arithmetic? Maybe we ought to go back a little and think about that in terms of what is important. The question I am asking of myself is—if people had these attitudes and if they learned them when they were young and understood the value of human relationships, maybe we would have a better world. I am thinking specifically of a young boy that I met at the age of nineteen who already had some signs of Huntington's disease. He spent his days working with children with cerebral palsy.

DR. AUBREY MILUNSKY: The point that Paul Friedman was making about the allocation of resources raises again the difficult question of how are we to educate our legislators and taxpayers about the inevitable fiscal burdens that arise in the case of the disabled and the handicapped. Not that those moneys should not be spent, but that the public at large become more acquainted with the enormous costs of care and the relative benefits of prevention. The lack of success of professional societies at large to educate the public about these relative merits is disappointing.

301

MR. PAUL R. FRIEDMAN: In a court context, for the reasons I explained, it is rather difficult to inject cost–benefit analysis type arguments. That is, when you are litigating on a consitutional theory, basically you are trying to establish violation of a basic constitutional right and then to rely on existing precedent that the lack of funds is not an adequate excuse. When you are trying to change public attitudes or working in a legislative forum, I think these arguments are terribly important.

There is a basic notion that in providing treatment to individuals in the community, some combination of supervised living and service programs will be far less expensive than maintaining someone in an institution for a life-span. There is good information on what it costs to do that.

In my own limited experience, the economic analyses of what it really would cost to provide services in the community are not very good. There are a lot of variables that aren't either accounted for or controlled that make comparisons difficult.

Then there has to be agreement on what level of programs or services we are going to provide. Services are woefully inadequate and reflect public attitudes that retarded people are subhuman and deserve no services at all. Therefore, it is hard to cost out new programs.

PROF. CHILDS: I think Marjorie Guthrie said something very much to the point.

What we are faced with, it seems to me, is a new context. We are not going to be able to solve these problems in the old context. These are new problems that have to be faced with new methods and entirely new ways.

It think the point I was trying to make is that we have got to face the fact that handicapped people exist. We have got to live with them and we have got to figure out the ways to include them in our society. I mean, issues of cost and issues of all sorts of things will inevitably be raised to prevent these things because people will persist in thinking in the old context.

TAMAH SADICK (Providence, Rhode Island): Mrs. Guthrie, we do have a program in Rhode Island that teaches empathy and understanding. They do go into the public schools, usually the second and the fourth grade, and do teach the normal children what it is like to be handicapped, blind, deaf, mentally retarded, or physically handicapped. It is a marvelous program but they have been unable to get funding for it because they are not giving a service to handicapped people; they are giving another kind of service.

MS. FRANCES BIRNBAUM (Johns Hopkins School of Hygiene and Public Health): I also think it is important not only to educate children but also to educate parents and educate providers.

I went to junior high and high school in a public school system in New Jersey and one of my best friends was blind. Her parents had sent her to a school for the blind for her first few years of education so that she could learn Braille typing and other skills. They then took her out of this school against the advice of all the experts and placed her in the public school system.

There was never any question about this girl's intelligence; she was a very bright girl, very eager to learn, and learned very well. But the feelings of parents of other blind children that these people spoke to and of all the experts was that she would not be able to cope on a level with sighted children in a normal public school. This was found to be incorrect and later, after high school, this girl went on to get a bachelor's degree at Sarah Lawrence College.

I think it is very important that the providers also be educated that these things can be done and simply because we don't know how to do them is not sufficient to say, "Well, then we can't try."

MR. FRIEDMAN: These cases revolve around the Education for All Handicapped Children Act, which, for example, embodies constitutional equal protection rulings handed down in the Pennsylvania Association for Retarded Citizens Education case and a case in Washington called *Mills v. Board of Columbia*. These cases simply incorporate the equal protection principle and the due process mechanism into federal law and make it contingent upon the grant of federal funds to schools around the country.

It is easy to understand the somewhat militant reaction by the parents of mentally retarded children, because for years their children were completely excluded from the school system. Any little service or aid was a favor to be begged for and was thought to be at the unfettered discretion of the superintendent.

Now these people are thinking of themselves, quite properly, as citizens with rights. It will be a difficult period but I don't think the inflexibility that you noted and that is occurring in many places around the country is inherent in the law. It is a matter of working things out.

What the law does is establish a principle that wherever possible, to the fullest extent possible, we normalize the lives of these children and we mainstream them. But it provides a due process mechanism to work out the balance and a hierarchy of placements. The notion is that, if possible, place a child in the main classroom even if it means an extra teacher's aide.

But maybe in a particular case it won't work. Then the due process mechanism should explore having the children be in some of the regular school classes, but not in ones where there is too much frustration or disruption, such as a special classroom within the same school, or a special school, but joining together during recreation or gym.

It is meant to be a flexible, balancing mechanism. But we are in the very early stages and, understandably, we do hear about problems from around the country.

26

DEVELOPMENTS IN THE RIGHT TO TREATMENT

PAUL R. FRIEDMAN
Mental Health Law Project
Washington, D.C. 20036

In the space allotted to me, I will discuss briefly the interesting and significant developments in the legal right to treatment or habilitation for mentally disabled persons, beginning with the *Wyatt* decision in 1971 and continuing to the present.

Until this decade, the role of judges and lawyers in the administration of institutions for the mentally disabled had traditionally been limited to decisions about commitment and release. The care and treatment of the people in state mental institutions was regarded as falling within the unfettered discretion of the state's superintendents and commissioners. Only a few pioneers wrote or spoke about what was happening behind the walls of our institutions: Senator Sam Ervin of North Carolina, with public hearings on my model commitment law, or Frederick Wiseman, in his documentary "Titticut Follies," stirred up a brief public concern that quickly abated, leaving little lasting effect upon public institutions. Then, in 1971, a courageous federal district judge in Alabama issued a landmark decision, which clearly recognized that mentally disabled persons have the same constitutional rights as other citizens and which marked the beginning of a radically different status for such persons under our legal system.

After taking evidence during a hearing on the need for emergency relief during litigation of *Wyatt v. Stickney*[1]—a test case seeking to improve the conditions of Alabama's mental institutions—Judge Frank M. Johnson, Jr., found:

> The evidence... has visibly and undisputedly portrayed Partlow State School and Hospital as a warehousing institution which, because of its atmosphere of psychological and physical deprivation, is wholly incapable of furnishing habilitation to the mentally retarded and is conducive only to the deterioration and the debilitation of the residents. The evidence has reflected further that safety and sanitary conditions at Partlow are substandard to the point of endangering the health and lives of those residing there, that the wards are grossly understaffed, rendering even simple custodial care impossible, and that overcrowding remains a dangerous problem often leading to serious accidents, some of which have resulted in deaths of residents.

Although Partlow State School was a very substandard institution, experts in the *Wyatt* case testified that it was no worse than many institutions in some of our largest and richest states.

On the basis of this shocking factual record, the *Wyatt* court held, for the first time, that institutionalized mental patients have a constitutional right to "receive such individual treatment as [would] give each of them a realistic opportunity to be cured or to improve his or her mental condition." The basis for this holding was presumably the Fourteenth Amendment, which states that no person shall be deprived of liberty without due process of law. This provision has been interpreted to require that governmental action affecting individual liberties must be consistent with "fundamental fairness."

In many states a mentally retarded person subject to civil commitment is denied the full range of procedural safeguards made available to criminal defendants and can be confined for an indefinite term rather than for a fixed sentence. But fundamental fairness requires that something more than mere custodial confinement—i.e., treatment or habilitation—is needed to justify such loss of liberty.

Applying the due process clause to the situation of a mentally handicapped person who had been involuntarily confined, the Supreme Court has stated that the nature and duration of the commitment must bear a reasonable relationship to the purpose of that commitment. In affirming the *Wyatt* decision, the Fifth Circuit Court of Appeals accepted this substantive due process reasoning.

However, finding a violation of the constitutional right to treatment was only the first step for the *Wyatt* court. It was easy to see that Partlow State School was inadequate by any standards. But *what* was the constitutionally required minimum?

Was it possible to set objectively measurable and judicially enforceable standards so that such an abstract "right to treatment" could practically be implemented? To answer this question, Judge Johnson called for additional hearings. He invited representatives of the U.S. Department of Justice and of major professional organizations, such as the American Psychological Association, the American Orthopsychiatric Association, and the American Association on Mental Deficiency, to participate as friends of the court and share their expertise. The result of these hearings was a final order setting standards for mental health and mental retardation facilities and presenting a detailed program for implementing these standards.

The minimum standards that the court promulgated after hearing extensive testimony included (1) a provision against institutional peonage; (2) a number of protections to ensure a humane psychological environment; (3) minimum staffing standards; (4) provision for a human rights committee at each institution; (5) detailed physical standards; (6) minimum nutritional requirements; (7) a provision for individualized evaluations of residents, treatment plans, and programs; and (8) a requirement that every mentally impaired person has a right to the least restrictive setting necessary for treatment.

The *Wyatt* decision presented an extraordinary breakthrough for mentally retarded persons confined in institutions. But it did leave some potential loopholes—most importantly, perhaps, that the due process theory depended upon deprivation of liberty, for which treatment was the *quid pro quo*. In many institutions for the mentally retarded, in contradistinction to those for the mentally ill, the residents were "children" who were admitted by their parents and who did not, at least overtly, protest their admission. If the confinement could fairly be characterized as "voluntary," then a due process right to habilitation might not apply. This loophole was convincingly closed by the next important case in this area, commonly known as the "Willowbrook case."[2]

Noted physicians, researchers, and professors, as well as parents, had testified as to the deplorable conditions at the Willowbrook State School for the retarded in New York. They told stories of bruised and beaten children, maggot-infested wounds, assembly-line bathing, inadequate medical care, insufficient clothing, and inappropriate and cruel use of restraints.

The 1975 Willowbrook consent decree and its accompanying district court memorandum were significant for several reasons. First, the federal court ratified a consent decree that provided very extensive relief—in such areas as staffing ratios and individualized habilitation plans—on a right-to-protection-from-harm theory. This theory, which is premised on the Eighth Amendment's prohibition against cruel and unusual punishment, had generally been regarded by advocates for the mentally retarded as less likely to provide major improvements in the conditions affecting the institutionalized retarded than the "right-to-treatment" theories based on the equal protection or due process clauses. The reason was that, historically, courts measuring conditions in institutions against Eighth Amendment prohibitions have acted only to eliminate conditions that are truly barbarous or inhumane or "shocking to the conscience." It was therefore assumed that, under an Eighth Amendment standard, the court might be willing to enjoin the most obviously barbarous conditions but would not order creation of the kinds of affirmative programs that would clearly be part of a constitutional right to treatment.

In this case, however, after extensive expert testimony, the late Judge Orrin Judd accepted the plaintiffs' argument that in an institution for the mentally retarded it is impossible for an individual resident's condition to remain static—that if the functioning of a resident does not improve, it will inevitably deteriorate. In order to keep residents from being harmed, it therefore is necessary to provide the full range of affirmative relief that had been ordered under a right-to-treatment theory in such cases as *Wyatt*.

The *Wyatt* and Willowbrook models have now been replicated in a number of jurisdictions. The essence of these landmark judicial decisions has also been incorporated into both state laws and federal legislation such as the Developmentally Disabled Assistance and Bill of Rights Act of 1975.

A recent extension of such suits has been the demand that states not only maintain desirable conditions and levels of skilled personnel in state institutions but that they create or finance alternatives to those institutions as well.[3] The theory for such suits is found in that part of the constitutional due process guarantee that means (as judicially interpreted) that even if the state's purpose for restricting an individual's liberty is a legitimate one, it must be accomplished in the least restrictive way possible. Therefore, the reasoning goes, people who are deprived of liberty for the sake of treatment must be treated in the least restrictive setting that is suitable to their condition. States, then, cannot offer only the traditional services in massive institutions when community-based treatment would accomplish the same goals as well or better.

Perhaps the most radical case in this continuum is *Halderman v. Pennhurst State School and Hospital*,[4] a case described by some as "anti-institutional" and by others as a "community services" suit. What made the *Pennhurst* case unique was the plaintiffs' uncompromising position that the institution must be entirely closed and replaced by a network of community facilities and services.

Once again, the evidence presented to the court presented a shocking and disturbing picture. As the court noted:

> The physical environment at Pennhurst is hazardous to the residents, both physically and psychologically. There is often excrement or urine on ward floors, and the living areas do not meet minimal professional standards for cleanliness. Outbreaks of pinworms and infectious diseases are common. . . . The environment at Pennhurst is not only not conducive to learning new skills, but it is so poor that it contributes to losing skills already learned. For example, Pennhurst has a toilet training program, but one who has successfully completed the program may not be able to practice the newly learned skill, and is therefore likely to lose it.

On the basis of evidence presented to it, the court noted that all parties in the lawsuit—including the state defendants—"are in agreement that given appropriate community facilities, all the residents at Pennhurst, even the most profoundly retarded with multiple handicaps, should be living in the community." The primary reason Pennhurst residents were not already in the community was determined to be "the failure of the Commonwealth [of Pennsylvania] and its subdivisions to provide sufficient living units, vocational and daycare facilities and other support services at the community level." As the court analyzed the problem, the central legal issue was "whether the Commonwealth system of incarcerating the retarded in an institution known as Pennhurst in any way violates their constitutional or statutory rights." The court took the now well-established constitutional principles of a due process right to adequate habilitation, and then used this right to question the appropriateness of the institution's very existence:

> On the basis of this record we find that minimally adequate habilitation cannot be provided in an institution such as Pennhurst. As the Court has heretofore found, Pennhurst does not provide an atmosphere conducive to normalization which is so vital to the retarded if they are to be given the opportunity to acquire, maintain and improve their life skills. Pennhurst provides confinement and isolation, the antithesis of habilitation. . . . There is no question that Pennhurst, as an institution for the retarded, should be regarded as a monumental example of unconstitutionality with respect to the habilitation of the retarded.

The court found further that "the evidence has been 'fully marshalled' and we find that the confinement and isolation of the retarded in the institution called Pennhurst is segregation . . . [and that] equal protection principles . . . prohibit the segregation of the retarded in an isolated institution such as Pennhurst where habilitation does not measure up to minimally adequate standards."

The court's antisegregation approach, coupled with its order requiring the closing of

Pennhurst and the placement of all residents in community living and services programs under a timetable to be developed by a special master, is regarded by the plaintiffs' attorneys as spelling the end of more than a century of incarceration for mentally retarded persons in the United States. The case has sent shock waves throughout the mental retardation professional community.

Lest this story of apparently steady progress seem naively optimistic and orderly, it should be noted that a backlash traceable to the *Pennhurst* decision was highlighted in recent proceedings of the original landmark *Wyatt* case—still in implementation stages before Judge Johnson in the Middle District of Alabama.

Ironically (but typically), implementation of the *Wyatt* decree and standards has been a tedious and not entirely successful process. Last fall—more than six years after the initial *Wyatt* decree—lawyers for plaintiffs and *amici* found it necessary to return to court, alleging that the defendants still had not complied with those standards most crucial to ensuring adequate care and treatment of class members: the provision of individualized habilitation programming and the requirement that residents receive services in the least restrictive alternative setting consistent with their needs. Plaintiffs and *amici* argued that the defendants would never achieve compliance without the appointment of a special master to assist in and oversee implementation. Although fully expecting vigorous opposition to the motion, plaintiffs and *amici* were surprised by the approach defendants adopted at the recent hearing.

Although Judge Johnson had made it clear that the 1972 decree was the law of the case and of the circuit, and that the limited purpose of the November 1978 hearing was to take testimony regarding the defendants' compliance with that decree, defendants used the proceeding to challenge the philosophical principles underlying the *Wyatt I* decision. The challenge came in the form of a motion filed by defendants requesting modification of several key standards relating to training and habilitation of residents. Accompanied by a lengthy memorandum written by several prominent mental retardation professionals, the motion asserted that as many as 350 to 700 of Partlow's 1100 members of the plaintiff class were essentially untrainable.

Defendants' experts argued that to keep such mentally retarded persons in "ritualistic, meaningless" training was "demoralizing" to staff and "demeaning" to the residents. In fact, defendants' experts suggested that the futility of the daily training regimen might constitute cruel and inhuman treatment to those residents, and that a realistic alternative would be to remove "nontrainable" residents from formal training and place them in an "enriched daily program."

Defendants' experts urged modification of the court's least-restrictive-alternative standard as well, asking the court to adopt the view that deinstitutionalization is an unrealistic and unachievable goal for many and perhaps most of the mentally retarded residents of Alabama's institutions.

As of this writing, Judge Johnson has yet to rule on defendants' motion for modification, and so its impact is uncertain.[5] But if it is granted, defendants will have successfully called into question the most basic assumptions upon which the original *Wyatt* decree and the cases it generated have been based.

It would appear that the dissolution of expert consensus, the increased professional resistance to major court intervention, and the financial pressure of a "Proposition 13" mentality have combined to set up a roadblock across the path of progress.

According to the plaintiffs' lawyer in the *Pennhurst* case, one of the reasons for the success of that litigation was that "nationally as the case reached trial, virtually all retardation experts had reached a consensus that institutional segregation was antithetical to proper habilitation of the retarded." This and other factors led the plaintiffs to throw off the "tactical circumlocutions and circumspections" of right to treatment, "least restrictive alternative," and other deinstitutionalization cases and to proclaim themselves openly as anti-institutional and in favor of the complete closing of Pennhurst.

But the consensus among experts on the harms of institutionalization, which distinguished institutional litigation from *Wyatt I* to *Pennhurst*, was strikingly absent in *Wyatt II*. In their depositions, virtually all of the defendants' experts stated that their willingness to testify was based in part upon their fears that plaintiffs and *amici* wished to duplicate the *Pennhurst* result and close Alabama's institutions.

These fears may have been founded on the belief that the *Pennhurst* precedent was universally applicable. Although Judge Broderick in *Pennhurst* took pains to limit his ruling to the facts developed about that one institution and its residents, the court also recognized that "at its best, Pennhurst is typical of large residential state institutions for the retarded."

While defendants' experts in *Wyatt II* acknowledged that community-based programs and services may be suitable for some mentally retarded persons, they insisted that deinstitutionalization is unrealistic for most profoundly and severely retarded persons and that the majority of Alabama's institutionalized residents will not achieve "significant" gains in adaptive functioning even with intensive training in the most effective habilitation programs available. Their pessimistic assessment would lead ineluctably to the conclusion that the need for institutions will continue.

Defendants' reference to *Pennhurst* may have been a straw man in the *Wyatt II* hearing because neither plaintiffs nor *amici* had urged that Alabama's facilities be completely closed. Nevertheless, when the issues become more subtle and complex, the consensus among advocates for mentally retarded persons and mental retardation professionals regarding basic principles of developmental programming and normalization may break down. And without a consensus of experts upon which to base a ruling, even activist judges may be uncomfortable in ruling for plaintiffs in future right-to-treatment, deinstitutionalization, or anti-institutionalization cases.

The lack-of-consensus issue is related in part to a feared loss of professional discretion. When *Wyatt I* began, mental retardation professionals generally agreed that federal court intervention was essential to improving Alabama's institutions. Support for a constitutional right to treatment, which implied increased institutional resources, came from all of the major mental health professional organizations. Since 1972, conditions in Alabama's institutions have improved to some extent. We have witnessed increases in staff size, overall improvements in cleanliness and maintenance, and the development of behavioral training models. However, plaintiffs and *amici* and their experts argued in *Wyatt II* that these changes are essentially cosmetic and that the state's institutions remain custodial rather than habilitative. Defendants' experts, on the other hand, contended that the improvements are significant and demonstrate the state's compliance with the court's decree. In the defendants' view, the state has earned the right to regain control of its institutions and return their management to the professionals.

At issue is the extent to which the courts, rather than the professionals, should be responsible for defining and measuring the adequacy of services for mentally retarded persons. Because of professional pride, or fear of liability for decisions over which they have no control, mental retardation professionals are alarmed by the trend of lawyers and judges to substitute their judgment for that of the professionals. The literature is becoming replete with complaints of "overdoses of due process" and stories of residents "dying with their rights on." As the legal rights movement continues to address issues that question the judgment of mental retardation professionals, we can anticipate increasingly strong professional, and perhaps public, opposition to both judicial and legislative reform.

Finally, as both defendants and plaintiffs recognized, the critical issues underlying class-action right-to-habilitation cases is one of judicial control of financing. On the appeal of the *Wyatt* decision to the Fifth Circuit, the state of Alabama took the position that the kind of decree rendered by Judge Johnson constituted an illegal infringement upon the functions of the state legislature. The governor and the legislature were upset, first, by the high cost of implementing minimum standards and, second, by the alleged usurpation of a legislative function—the balancing of the needs of mental patients against other equally important demands upon state revenues, such as old-age pensions, welfare payments, and renovated highways. To the plaintiffs, the question was whether constitutional rights were to be vindicated.

Many constitutionally based decisions require reallocation of financial resources, but lack of adequate financial resources has never been accepted as a valid justification for the denial of a constitutional right. The cases so far have made clear that whether a state decides to run a mental health or a mental retardation services system is entirely within its own discretion but, once it decides to undertake this function, it must do so in a manner that does not violate constitutional rights.

As the recent *Wyatt* hearing suggests, we have now entered an era of heightened concern about diminishing resources—an era in which both judges and legislators, reflecting public opinion, may move more cautiously toward issuing decrees or enacting laws that heavily burden the public purse.

A decision to allocate resources in accordance with cost–benefit analyses would have very serious consequences for the more severely retarded residents of institutions. Even in the present economic climate, a court is unlikely to permit an institution to reduce or terminate programming for profoundly retarded residents who are trainable in order to increase efforts on behalf of moderately or mildly retarded persons. But suppose that the experts argued that severely and profoundly retarded persons are not trainable or can benefit only minimally from habilitation efforts that would lead to dramatic improvements in less severely retarded persons. Might a concern—even if unarticulated—for maximum returns on investments alter expert opinions about an individual's potential benefit from habilitation or other services?

The defendants' argument that the severely and profoundly mentally retarded persons at Partlow should receive an "enriched" environment without active habilitation and training is essentially the same argument made by defendants in the early right-to-education cases to justify exclusion of mentally retarded children from the public schools. It also echoes the traditional justifications for withholding medical treatment from infants with serious birth defects.

The extent to which both economic factors and intraprofessional rivalries and tensions will limit the articulation and implementation of the right to treatment in institutions and in the community will be a matter of great interest to mentally disabled persons and their families, friends, and advocates in the coming years.

REFERENCES

1. 325 F. Supp. 781 (M.D. Ala. 1971), 334 F. Supp. 1341 (M.D. Ala. 1971), 344 F. Supp. 373 and 387 (M.D. Ala. 1972), affirmed *sub nom Wyatt v. Aderholt*, 503 F. 2d 1305 (5th Cir. 1974).
2. *New York State Association for Retarded Children v. Rockefeller*, 357 F. Supp. 752 (E.D.N.Y. 1973).
3. *Dixon v. Weinberger*, 405 F. Supp. 974 (D.D.C. 1975).
4. 446 F. Supp. 1295 (E.D. Pa. 1977); affirmed in part and modified in part, Nos. 78-1490, 1564, and 1602 (3rd Cir., December 13, 1979). The court of appeals' decision, handed down after this article was written, affirmed the basic thrust of the district court decision on statutory grounds, while modifying that decision with respect to the complete closing of Pennhurst. While recognizing that the Developmentally Disabled Assistance and Bill of Rights Act provides mentally retarded persons with a right to the least restrictive environment and that deinstitutionalization is the "clear preference of the Act," the appeals court held that in specific situations, "institutionalization might be appropriate once adequate habilitation and living conditions are established."
5. Since this article was written, the court has denied defendants' motion for modification. *Wyatt v. Ireland*, No. 3195-N (M.D. Ala., October 24, 1979).

27

"WILL THE CIRCLE BE UNBROKEN?"

Sterilizing the Genetically Impaired

NANCY S. WEXLER
Neurological Disorders Program
National Institute of Neurological, Communicative Disorders and Stroke
National Institutes of Health
Bethesda, Maryland 20205

> *Will the circle be unbroken,*
> *by and by, Lord, by and by?*
> *There's a better home awaiting*
> *In the sky, Lord, in the sky.*
> —FOLK SONG

I would like to quote an article from the *Miami Herald,* March 18, 1979, entitled "British Couple Wants Their Kids Sterilized."

London (AP). A British couple wants to have its three children sterilized because the mother suffers from a hereditary disease that they fear will be passed on to future generations. The decision has touched off a medical and ethical controversy.

Pamela Gooch, 33, has Huntington's chorea, a chronic disorder of the nervous system that causes jerking muscle spasms, slurred speech and premature aging. "I know what a dramatic step this is, but because of their genes, my children have a 50–50 chance of contracting it," Gooch said at the family's home in the Staffordshire coal-mining town of Cannock.

"I saw my father die in pain, and my children are seeing me going toward my death in the same way. Only by ending our line can we prevent them from watching their own children suffer."

Gooch and her husband said they plan to have Diane, 14, Tina, 11, and Scott, 9, sterilized when they reach the age of 16.

The eldest daughter said: "I would like to have children. But seeing the way Mum suffers, I would hate to have kids and pass it on. I understand what Mum and Dad are doing and I'll go along with it."

The father, Anthony, 40, who works an overnight shift in a coal mine so he can care for his housebound wife and the children during the day said: "Our eldest daughter does seem to understand and Tina is beginning to grasp the problem. . . . I know people might criticize us and say we have no right to make such a decision, but I disagree. We brought our children into the world. We're responsible for bringing them up, and we have to make decisions about their future too."

However, Dr. Brian Stott, the family physician said he was not happy about the parents' intentions. He said he does not believe any surgeon would be happy about performing the sterilization operations unless it could be proved which of the children were going to develop the disease. . . .

A spokesman for the British Medical Association said: "We hope no doctor would be prepared to carry out the sterilization. The thing to do is for the healthy parent and the doctor to ensure that the children obtain proper counseling when they grow up."[1]

THE REPRODUCTION LOTTERY: WHO WINS?

Mrs. Gooch's proposed solution to her problem is unambiguous and, within her own family sphere, effective. Perhaps many would sympathize, overtly or covertly, with her approach. As one gentleman said who testified at a public hearing before the Commission for the Control of Huntington's Disease and Its Consequences, "I've bred cattle and I've bred hogs and grown corn and beans. Now, I don't like to liken people to animals and plants, but if I had a cow that was throwing six-legged calves, I'd eliminate that line."[2]

Some recoil in horror, in vicarious identification with the young children who will be altered irrevocably at an early age. Sterilization is so final, so despairing of scientific progress. But is our horror only sentimentality? What if there were no Mrs. Gooch, feeling her disease working within her, who can insist from her own experience, "This illness should stop here"? Perhaps we should all become Mrs. Gooches, weighing, evaluating, mini-Mengeles dispatching some to the sterilization tables and others into the world of procreation. When God gave the command to "be fruitful and multiply," was it really meant only for a chosen people? If so, how does one choose?

To entertain the notion of sterilizing all the genetically impaired is to contemplate the end of civilization. Eugenic intervention is limited to those in whom one of the 4 to 10 deleterious genes that we all carry has had the bad grace to become obvious.[3] But what do we really mean when we speak of the sterilization of the genetically impaired? Certainly we mean the actual process of sterilizing people with reproductive capabilities. But sterilization can also be considered as a metaphor for the eugenic ideal—the systematic elimination of hereditary defects. For some disorders that are detectable prenatally, sterilization is obviously unnecessary, unless a couple refuses to avail themselves of screening, amniocentesis, or abortion.

There are several categories of individuals who are most often the eugenicists' targets. Historically, the bulk of discussion concerning involuntary sterilization has centered on the mentally incompetent, criminals, and the poor—the disadvantaged of our society who may not comprehend what is asked of them or who may be coerced into consenting. At stake in the sterilization of these individuals is not only their capacity to give informed consent, nor the questionable likelihood of genetically transmitting their own propensities, but also their ability to be adequate parents.[4–10]

Another group for whom mandatory sterilization is sometimes proposed consists of those people who are intellectually capable of making an informed decision, who know that they are or could be at greater than average risk for producing a child with a genetic abnormality, but for whom no *in utero* diagnostic test is yet available. These may be people who have already produced a genetically damaged child, the population that composes the bulk of our genetic counseling clientele, or who know that an illness "runs in the family."[11]

Thus the eugenic dream is fed primarily by two wishes: One is to persuade, coerce, cajole, or forcibly prevent from reproducing those more likely than average to produce a genetically handicapped child. The other is to eliminate from the reproductive ranks those individuals who by virtue of mental retardation, criminality, or poverty are considered constitutionally unfit for parenthood.

EUGENICISM ON THE RISE?

Since the days of Charles Benedict Davenport (1886–1944) and the Social Darwinists, the notion that all those whose likelihood of producing a defective offspring should be prevented from contributing to the gene pool has dropped from popularity.[12] Greater sophistication in our understanding of genetic transmission and an eyewitness observation of Hitler's Hereditary Health Laws has tarnished the popularity of eugenics.[13]

There are a number of pressures at work today, however, that could bring eugenic considerations back into the limelight. The first (although not in any order of priority) is economic. Health care costs have been rising exponentially and the dollars required to treat and care for the genetically impaired are in the billions. When indirect costs are included, the figures soar astronomically, Society may well ask why it should continue to absorb these costs when individuals can prevent some of them through abstention from procreation. From the White House to the mayor's office, conservation of dwindling resources is the watchword. Eugenicists ask, "Shouldn't these precious resources be reserved for those most able to use them?"[14]

The second pressure that will bring the issue of reproductive choices into greater prominence is our increasing ability to screen for hereditary conditions. Each year, we can detect more genetic diseases, and our technology is providing simpler and more accessible means for doing so. Sickle cell disease can now be spotted from an amniotic tap. And the recent discoveries of fetal cells in maternal blood may revolutionize the field of genetics and make amniocentesis a thing of the past.[15] Imagine being able to walk into a doctor's office and be screened with a simple blood test for a whole host of blighting diseases that could cripple or kill your children. What enormous freedom and what dread responsibility! Screening could become a routine part of obstetrical and gynecological care and even general medicine. But the decisions required when genetic abnormalities are found are never routine. The simplicity of the test and its lack of risk to the fetus may encourage many more to take it, including some who have not made a choice prior to the test (as most do who have amniocentesis) to abort if the findings are positive.

A third pressure toward a revival of eugenicism is the increasing availability of alternatives to one's own biologic offspring. Although the number of children available for adoption is declining, artificial insemination has been gaining popularity and acceptance. Couples in which both are carriers of a recessive disease or in which the man carries a dominant gene disorder can take advantage of this option. The work of Steptoe and Edwards may soon make it possible for women who carry a deleterious gene to give birth to children developed from *in vitro* fertilization of their husband's sperm with a donor egg.

A fourth eugenic push may come from a much more personal source than banks or laboratory benches. It is a voice that I suspect will become more vociferous and insistent as time passes. It is a voice of the children born destined for multiple blood transfusions, constant mucous-draining exercises, wheelchairs, and early death. It will come from children's advocate groups who will champion recent court decisions that every child has the right to be born healthy in mind and body. [16] We are seeing its tentative introduction now into the courts with the recent concept of "wrongful life" cases.

"Wrongful" Rights: Is Health a Birthright?

At the crux of the issue, apart from social burdens or rising costs, is the very personal confrontation between a parent's so-called right to have children and a child's right to be born healthy. Does the parent have a right to knowingly jeopardize the health and possibly life of offspring who have no choice but to be born? Should society passively observe parents' choices, watching as bystanders while the ranks of children file by, children gnawing on their own limbs, children twisted and deformed, children atrophying and dying in institutional beds? When does the natural exercise of "parental rights" become child abuse? What is the penalty for "accomplice to negligence" or "intent to jeopardize"?

In all the reading I did in preparation for this chapter, in all the discussions of economic burden and the protection of children and adults, welfare recipients or the mentally retarded from involuntary sterilization, never once did I read an interview with a child with muscular dystrophy, cystic fibrosis, thalassemia, sickle cell disease, spina bifida, or Friedreich's ataxia exploring how it feels to live with these diseases. There is a fair amount of literature on the emotional adjustment problems of handicapped children. But how often does an interviewer ask, "Would you rather never have been born?" One young woman of 20, at risk for Huntington's disease, did not wait for me to put the question. Shaking her fist and grimacing, she said with intense bitterness, "My Mom and Dad knew about HD and they had us kids anyway. I think it was the most selfish thing they could have done." Those who are most maimed, who die young or spend their lives in institutions, are least able to represent themselves.

To my knowledge, courts have not yet been confronted with the question of whether or not a child, born with a hereditary disorder of which the parents had knowledge prior to conception, has a cause of action against the parents. Dramatic changes in the law of torts over the last decade, however, make this suit a real possibility in the future. Reversals of the law in two areas have paved the way: in all jurisdictions, the infant's right to recover for prenatal injuries is uncontested; and in many jurisdictions, the tradition is crumbling of intrafamilial tort immunity that barred suits between parents and their children.[17,18]

The first case involving a cause of action for prenatal injuries was *Dietrich v. Inhabitants of Northampton* (1884).[19] Dietrich set the precedent, which prevailed until *Bombrest v. Katz* (1945),[20] for denying recovery for prenatal damage. In a startling reversal of the law, *Bombrest* ruled that a child is permitted to recover damages if born alive and if the fetus is viable at the time of injury. A series of subsequent decisions by courts in different jurisdictions soon eliminated the two requirements of fetal viability when damage is sustained, and live birth of the infant.[21]

In *Howard v. Lecher* (1976), the New York State Appellate Division of the Supreme Court denied a cause of action for Ashkenazi Jewish parents who sued their obstetrician for not warning them of their increased risk for Tay–Sachs disease. The defendant agreed to pay out-of-pocket costs for care of the genetically damaged child subsequently born, but contested paying for the parents' emotional distress. In one of the first so-called "wrongful life" cases, the *Howard* court ruled that the parents had no cause of action because it was the child that primarily suffered, and that it was not the public policy of the state to weigh the value of life itself.[22]

Only two months after the *Howard* case, the same appellate division of the Supreme Court decided in *Park v. Chessin* that the state should recognize an action for "wrongful life." It sustained a cause of action against an obstetrician for allegedly failing to warn a couple with one child dead of polycystic kidney disease that any subsequent children would have a 25% risk of the same disorder. In this instance, the parents sued on behalf of the deceased child for pain and suffering, rather than only on their own behalf as in *Howard*.[23] The couple ultimately lost their case when the Queens Supreme Court ruled that the obstetricians had, in fact, provided fair warning.[24]

Renslow v. Mennonite Hospital (1977) firmly established that there is a legal duty to an infant *prior* to conception.[25] In *Renslow*, a mother sued Mennonite Hospital and its director on her own behalf and as the next friend of her minor daughter.[26] Eight years prior to the birth of her daughter, the hospital had given her blood transfusions with the wrong Rh factor, neglecting to tell her. The child was born with severe damage due to the mother's prior Rh sensitization.[27] Of special interest in the *Renslow* decision is that it based its finding not only on the arguments of foreseeability and causation, which were easily sustainable in this case, but also on the grounds of sound public policy.[28] *Renslow* held that "public and social interests supported recovery for an injury of this nature."[29]

A number of different jurisdictions have recognized a child's right to begin life with a sound mind and body.[30] This is not equivalent to allowing an action in tort for "wrongful life." At issue in the "wrongful life" cases is the plaintiff's contention that because of extreme physical or emotional anguish, the plaintiff should never have been born. In most tort cases, courts award damages based on the difference between the condition of the plaintiff currently and the presumed condition had the negligence not occurred. If the negligence had been prevented in the "wrongful life" cases, there would be no plaintiff;

courts have found it difficult to give a monetary value to nonlife and have refused to permit actions for "wrongful life." Speculation on the quality of value of life is not in their purview, courts claim.[31]

The *Zepeda* court in particular, in denying recovery for social stigma due to illegitimacy, expressed concern that allowing recovery might encourage others to sue for being born under adverse circumstances, specifically mentioning those born with hereditary illnesses.[32]

CHILD VERSUS PARENT: A CAUSE OF ACTION?

In all cases in which infant plaintiffs have been allowed to recover damages for negligence committed prior to conception, there has been an issue of medical malfeasance. If children born with genetic impairments were to sue their parents for "wrongful life," at issue would be the wisdom of the parents' decision.

The earliest courts hearing suits of children against their parents refused to allow children even to maintain an action, and added insult to injury by holding that a child must aid and comfort the parent as long as the parent is obligated to support the child. *Hewlet v. George* (1891) ruled that actions by children against parents would disrupt peace in the family and were against public policy.[33] Parent–child tort immunity had four main justifications: (1) to preserve family harmony, (2) to maintain family coffers and not deprive other family members, (3) to protect against fraud and collusion, and (4) to restrain the judiciary from interfering in family affairs.[34]

An increasing number of jurisdictions, however, now permit tort actions to be maintained by children against their parents. Two main arguments have swayed the courts: (1) The child should not be denied redress just because of a relationship with the wrongdoer, and (2) parent–child immunity does not exist in property and contract rights law and surely the child's personal rights should be at least as well protected as property rights. The courts also concluded that family harmony can hardly be legislated, and if parent and child square off in court, the damage most likely already has been committed.[35]

In order for children to successfully sue their parents, they must demonstrate that the parent has a specific duty toward the child and has been negligent in fulfilling that duty. Parents have a legal responsibility to provide their children with necessary food, clothing, shelter, and medical care,[36] and to discipline their children as a "reasonable parent would in like circumstance."[37] If parents abuse their children to the extent that they cause considerable mental or physical injury, they may be held criminally liable.[38]

Courts that would allow children to maintain an action for negligence against their parents for exposing them to genetic jeopardy must first establish a standard of reasonable precaution. Negligence sometimes has been defined as a "failure to do what a reasonable person would do in like circumstances."[39] Courts have further specified that the amount of caution required in a particular situation is determined by three factors: "(1) the likelihood that the conduct will result in injury, taken with (2) the gravity of harm if the injury occurs, and balanced by (3) the interest which must be sacrificed to avoid risk."[40]

According to these definitions, a case for parental negligence in knowingly having a genetically damaged child could be made.

NIGHTMARES AND HYGIENE

Some authors have suggested that society may someday choose to mandate by statute parental responsibility for assuming genetic precaution.[41] But would society be so reckless as to pass a law nearly impossible to enforce? If we require mandatory genetic screening when couples apply for marriage licenses, how do we control how couples will act on the information provided? Should we institute economic sanctions against offending parents? Federal and private insurance companies could be encouraged to pay for screening, amniocentesis, and therapeutic abortion, and enjoined from paying maternity costs or medical coverage for children born to genetically high-risk couples. Midwifery would blossom if such a policy were enacted, babies would be born anyway, and most likely only the children would suffer because of inadequate care. Any legislation that proposed that people could not make claims on public assistance if they knowingly brought defective children into the world would work substantially to the detriment of the child and create endless lawsuits about the state of the parents' knowledgeability. Even fining parents would only take money away from the care of the handicapped child.

If economic sanctions are not invoked, what, realistically, are our options? Should gangs of medical thugs rove maternity wards looking for erring parents to sterilize? Shall we create a special class of genetic outlaws who come slouching through our streets, branded by yellow armbands with broken double helices advertising their defective status, or perhaps sporting scarlet letters on their breasts—HD, CF, MD? The sheer cost of enforcement would be prohibitive. And how would the legislators determine what constitutes a sufficiently severe genetic impairment to be banned? Achondroplasia? Multiple polyposis? Schizophrenia? No balance of justice is so exquisitely sensitive as to calibrate life and death by statute.

If courts of the future do consider individual common law tort cases, however, and children are able successfully to sue their parents for knowingly inflicting pain and suffering due to genetic insults, this may provide quite a sobering influence on parents' appreciation of their obligations as parents to do no harm to their children. Parents may be made by the courts to guarantee lifetime support or all medical costs for their handicapped children. The impact of such court decisions may be very good for encouraging parents to avail themselves of genetic services. But the entry of such cases into the court system is saddening. The litiginous nature of our society will have infected even the parent–child relationship.

SHOULD ADVOCATES BE LAWYERS OR COUNSELORS?

Courts that hear "wrongful life" cases in which children sue their parents will be asked to decide issues that genetic counselors have purposefully chosen not to decide but that they must confront daily. In recent years, genetic counselors (physicians, genetic associates, psychologists—whoever is doing the counseling) increasingly have assumed the position that prospective parents must decide for themselves about procreation. Even if a counselee asks what the counselor would do in his or her situation, the counselor is loath to respond. There are many excellent reasons for this nondirective stance of the counselor. Directive counseling may well drive away a client, and the client, not the counselor, must live with the decision. And yet, perhaps genetic counselors could provide couples with additional help that might forestall a later confrontation between parent and child in a court of law. In their efforts to be neutral, perhaps counselors are leaving their clients with too little guidance.

In order to help their clients, genetic counselors must be as fully informed as possible about the burden of the hereditary disease involved. They must understand the psychodynamics of reproductive decisions, in theory and in their own clients, and must be aware of their own internal reactions to clients' decisions. Counselors may feel strongly that a particular couple should not reproduce, while they are so moved by the despair and anguish of another couple that they secretly hope they will have children. Certainly the work on nonverbal communication done by Rosenthal and others would strongly suggest that the attitudes of counselors are conveyed to their counselees. Clients may be responding to a *sub rosa* dialogue with the counselor; if it is out of conscious awareness of the counselor, it cannot be adequately dealt with overtly in the counseling situation, but it can still exert a powerful covert influence.

The counselor has the unique task of representing not only the best interests of the client but also, ideally, those of any future children the client might have. Genetic counselors are among the few that can represent the unborn child, and the counselor's responsibility to this possible person may conflict directly with the needs and wishes of the living clients. Counselors must have the courage to push prospective parents to consider the reality of traumatic possibilities, to help them imagine how they would feel if the baby were born with a hereditary defect. Prospective parents in the 20s and 30s are not so far from the years of adolescent convictions of invulnerability, and the drive to deny and undo through acting normally is so strong that counselors may have a difficult time enabling parents to comprehend the emotional reality of what they learn. And the counselor must carefully balance pushing against the defenses of denial, isolation, and intellectualization while also shoaling them up to help a person cope with current pressures. It is often up to the counselor to emphasize to the couple the grave import of their reproductive decisions. If counselors are internally confused and conflicted, they may shy away from helping a client explore all the possible ramifications of a decision.

REFINING THE COUNSELING PROCESS

The timing of genetic counseling must be carefully considered. Couples are often counseled just after they have produced a damaged child. Although they may well understand and retain the risk figures of recurrence, they are not psychologically in a frame of mind to discuss their feelings about future children. How many of these couples return to the counselor when they are considering reproducing, particularly if amniocentesis is of no help for their particular genetic aberration? Should counselors insist on follow-up visits or strongly urge clients to return if they are contemplating additional children? Should one-year follow-up visits be as routine a part of genetic services as checkups postoperatively or after medication is prescribed? Certainly genetic information can be as potent as any drug or as scarring as any surgery.

How should genetic counselors convey the information they give? This is particularly important in working with clients who have or carry diseases for which prenatal diagnostic tests do not exist or who refuse to use them on religious grounds or otherwise. In many instances the client may have already seen the disease at its worst. But others may contemplate having additional children without knowing the full course of the disease to which they may be exposing that child. Should the counselor insist on showing a couple what the disease looks like? Since the couple might not return before deciding on more children, should they see the disease in the initial counseling sessions when the condition is first diagnosed? If it is too cruel to push such information on parents in their shock and grief over the diagnosis of someone close, is it not too cruel to allow them unknowingly to inflict such suffering on a subsequent child?

When counselors are working with a dominant gene disorder, the counseling situation may be all the more sensitive. I was once horrified to learn that a genetic counselor regularly takes her clients to the local VA hospital to see Huntington's disease patients on the wards. At first it seemed barbaric to me to force people to see a decaying vision of what they may become. It is no *Christmas Carol* where one can wake, like Scrooge, to learn that it was only a dream of Christmas future. And yet if couples are considering bequeathing this legacy to their children, is it not the least they can do to look?

Many Questions, Few Answers

There are so many questions about the impact of counseling for which we have no answers. We have some statistics about people's reproductive activity after genetic counseling, but we understand little about the motivations that led to this behavior. In Carter's study of 421 couples 3 to 10 years after they had received genetic counseling, two-thirds of the high-risk group (1 chance out of 10 or greater) reported that they decided not to have additional children, while in the low-risk group, one-fourth of the couples reported not wanting further children.[42] In looking at actual behavior, however, of the high-risk group, 46% had at least one additional child and nearly 70% of the low-risk group had another child or more. Curiously, of those who reported being attitudinally deterred from reproducing, 24% of the high-risk group who claimed not to want additional children went on to produce more offspring, while only 15% of the low-risk group who claimed that genetic counseling had persuaded them not to have more children actually produced one or more additional children.[43]

One wonders—what made almost a quarter of those who stated that their attitudes had been influenced by genetic counseling change their minds? How many of these children were planned? Were couples in the high-risk group who had decided not to have more children more prone to birth control "accidents" than couples in a low-risk group? Sorenson's conclusion from the Carter data is that "counseling seems to affect attitudes more than actual reproductive behavior."[44] If this is true, as it most likely is, what pushes people to violate their attitudes? What cognitive dissonance does this create for the couple? Surely people are ambivalent about the news they get from genetic counselors: do they act on one stream of the ambivalence and answer follow-up questionnaires on the other?

We know very little about the transference reactions of clients to their counselors. Shakespeare had a superb sense of the treacherous nature of truth-telling. In *Antony and Cleopatra* a messenger arrives with tidings for the Egyptian queen. She bids him, "Come hither sir/Though it be honest, it is never good/To bring bad news; give to a gracious message/An host of tongues, but let ill tidings tell/Themselves when they be felt." The messenger, not heeding the warning, relates that Antony has wed Octavia. The queen rails against the man, threatening him with pestilence, blinding, scalping, and stewing in brine like a pickle. In his defense, he protests: "Gracious madam,/I that do bring the news made not the match."[45]

In psychotherapy, the therapist is often imbued with all the attributes the client wishes to attain, and is sometimes envied and resented because of this fantasied perfection. How often do counselees presume the counselor to be "normal," with husband or wife and two smiling children at home? Do clients secretly discount what the counselor is saying? "Easy for you to talk," says the client, "you don't know what it feels like." Counselors must be alert to these feelings and be able to work them through with the client.

We know so pitifully little about the myriad of complex motivations that propel people to propagate their kind, particularly now that birth control methods are giving people such freedom of choice. Counting facts retained and babies born is not enough. How do people feel about their reproductive options? What goes into that crucial decision to create someone who will most likely be there for the rest of one's life? We know that some people have babies immediately after being counseled because to act as if everything were normal strengthens the denial that anything is potentially wrong. How do we help these individuals cope with the terror of their situation, help them come to terms with the reality of their risk, before they blindly jeopardize the life of some new person? What are people's secret "selling-out" prices? One 30-year-old man at risk for Huntington's disease, whose fiancé walked out on him when she learned of his risk, told me that he was firmly committed to not having children—unless it meant that he would never marry. Relief from loneliness was his price.

For some, the loss of the physiological experience of having one's own children is traumatic. Particularly for women, the monthly rehearsal for procreation runs longer than any Broadway play. We may be the "temporary repository for germ plasm,"[46] serving no more than as a way station for genetic messages from antediluvian times, but to store— and sometimes be captive to—an elegant apparatus that is never used can be a profound and intolerable loss. The throngs of women from all over the world who are sitting vigil in Dr. Patrick Steptoe's fertility clinic in Britain, willing to withstand arduous and painful procedures with slim chance for success, attest to some compelling urgency to have the experience of one's own biological child. Mrs. Brown, little Louise's mother, is barely articulate about why she so craved a child of her own body, but the saga of her several-years-long experience with Steptoe and Edwards, the physical pain and hardship she endured, is harrowing, impressive, and instructive.

In my study of people at risk for Huntington's disease, I did not dwell on people's motivations for having or refraining from having children. Of those that knew of their risk at the time their baby was conceived, most wanted to live normal lives and give a normal family life to their spouse, even if it meant risking the health of their child.[47] Several women who had decided not to have children reported feeling barren and desiccated. They felt stigmatized not only by the disease but by the ostracism that has been visited on sterile women down through the ages and across the world. Although they had no idea if they were fertile or not, the choice of remaining childless made them feel sterile. One woman in particular said that as a young child she had practiced for pregnancy by stuffing pillows under her shirt. She had been envious of her sister's heavier menstruation because her sister seemed more feminine and the rich padding of her womb would provide a better home for the developing baby.

Tomorrow: Chaos, Courts, or Counseling?

The fecundity of our population challenges our intellectual fertility in coping with genetic abnormalities so that harm to all concerned is minimized. Current practices and new options invite exploration. Perhaps genetic counseling clinics should use peer influence and support to a greater extent through ongoing "procreation" groups specifically designed for people grappling with the dilemma of whether or not to have a child. Couples in similar situations may be far better able to say strong truths, for and against reproduction, than the genetic counselor. Clients could be urged to come to the clinic to attend these groups prior to making procreative decisions, regardless of how long after the initial counseling session this may be. The group's composition would change and evolve as couples or individuals reached some resolution of the question. The group could also provide continuing support for couples who make difficult decisions, either to have or to refrain from having children, and who need occasional shoaling up and an opportunity to talk in order to abide by their choices.

The group or counselor or clinic should help couples explore, through lectures or workshop, alternatives to biologic children and the possibilities of life without children. Women may want information about careers other than mothering. The amount of attention and interest the counselor and clinic give to the question of childbearing will communicate in itself a message to the client to take time and care with these decisions.

We have many more creative possibilities open to us than helplessness, genocide, mourning, or legal entanglements. In a free society we should strive to maximize our spheres of freedom. Toward this end we encourage clients to make free and "informed" choices about children. Our best alternative to life under rule and regulation is information and education so that children, parents, and society can rejoice in these meaningful decisions. Let us not look ruefully back years from now to say, in the words of Alan Paton, "Cry the beloved country for the unborn child who is the inheritor of our fears."[48]

ACKNOWLEDGMENT

I would like to thank Mrs. Shirley Vinitsky for her valuable assistance in the research for and preparation of this chapter, and in sharing ideas.

REFERENCES AND NOTES

1. British couple wants their kids sterilized, *Miami Herald*, March 18 (1979).
2. Commission for the Control of Huntington's Disease and Its Consequences, Congressional Mandate, P.L. 94-63, Section 605(b), Report, Public Testimony **IV**(2):2.
3. Lipkin, M., Jr., and Rowley, P. T. (eds.), *Genetic Responsibility*, p. 81, Plenum Press, New York (1974).
4. Aptheker, H., Sterilization, experimentation and imperialism, *Political Affairs* **53**(1):37 (1974).
5. Babcock, R., Jr., Sterilization: Coercing consent, *The Nation* **218**(2):51 (1974).
6. Bass, M., Surgical contraception: A key to normalization and prevention, *Ment. Retard.* **16**(6):399 (1978).
7. Bayles, M., The legal precedents, *Hastings Cent. Rep.* **8**(3):37 (1978).
8. Neville, R., The philosophical arguments, *Hastings Cent. Rep.* **8**(3):33 (1978).
9. Thompson, T., The behavioral perspective, *Hastings Cent. Rep.* **8**(3):29 (1978).
10. Vitello, J., Involuntary sterilization: Recent developments, *Ment. Retard.* **16**(6):405 (1978).
11. Sorenson, J., Genetic counseling: Some psychological considerations, in: *Genetic Responsibility* (M. Lipkin, Jr., and P. T. Rowley, eds.), p. 62, Plenum Press, New York (1974).
12. Pickens, D., *Eugenics and the Progressives*, p. 37, Vanderbilt Press, Nashville (1968).
13. Although statutes permitting involuntary sterilization remain on the books in 21 states and over 70,000 known sterilizations of the retarded have occurred since 1907 when Indiana passed the first law legalizing such sterilizations, the mentally retarded are now in an anomalous position. Largey, G., Reversible sterilization: Socio-ethical considerations, *Soc. Biol.* **25**(2):139 (1978). Legally, they remain in some jeopardy. The new Department of Health, Education and Welfare regulations are problematic in that some claim they are too stringent, others too lenient. Ironically, in many states it is now exceedingly difficult for the mentally retarded to even take advantage of one of the most popular forms of birth control in our country if they are not able to give totally informed consent. DHEW proposes 30-day waiting period for sterilizations; no funds for under 21's, contraceptive hysterectomies, *Fam. Plann. Perspect.* **10**(1):39 (1978); Resolutions and policy statement: Voluntary sterilization, *Am. J. Public Health* **67**(1):105 (1977); Vining, E., and Freeman, J. M., Sterilization and the retarded female: Is advocacy depriving individuals of their rights?, *Pediatrics* **62**(5):850 (1978).
14. The insurance companies can effectively implement a eugenics policy, regardless of whether such a policy is legislated. Insurance companies may require that a couple be screened for genetic disease prior to selling a policy or paying for maternity benefits. If a couple has already had a child with a hereditary illness, the insurance companies may refuse to pay maternity benefits for any subsequent children born or to provide coverage for any such children who are genetically impaired. If National Health Insurance is ever enacted, the government must also decide if it chooses to enact a *de facto* eugenics policy through insurance reimbursement dollars. The government's decision to limit federally funded abortions may well increase the number of genetically handicapped babies being born.
15. *New York Times*, Apr. 15 (1979).
16. 31 N.J. 353, 364, 157 A2d 497, 503 (1960).

17. Chernaik, B., Recovery for prenatal injuries: The right of a child against its mother, *Suffolk U. L. Rev.* 10(3):582 (1976).
18. Hartye, F., Tort recovery for the unborn child, *J. Fam. L.* 15(2):276 (1977).
19. Chernaik, op. cit., p. 585.
20. Ibid., p. 587.
21. Veazey, L., Torts—An action for wrongful life brought on behalf of the wrongfully conceived infant, *Wake Forest L. Rev.* 13(3):712 (1977).
22. Curran, W., Tay–Sachs disease, wrongful life, and preventative malpractice, *Am. J. Public Health* 67(6):568 (1977).
23. Veazey, op. cit., p. 713.
24. *Medical World News*, p. 71, May 14 (1979).
25. *Renslow v. Mennonite Hospital*, 67 Ill.2d 348, 367, N.E.2d, 1250 (1977).
26. Roos, J., Torts—Negligence—Infant may maintain a cause of action for prenatal injuries resulting from a negligent action (prior to infant's conception), *Texas Tech. L. Rev.* 9(3):715 (1978).
27. Ibid.
28. Ibid., p. 726.
29. Ibid., p. 727.
30. See *Smith v. Brennan*, 31 N.J. 353, 364, 157 A.2d, 497, 503 (1960); *Sylvia v. Gobeille*, 101 R.I. 76, 78, 220 A.2d, 222, 223 (1966).
31. Chernaik, op. cit., p. 590.
32. Ibid.
33. Ibid., p. 591.
34. Ibid., p. 592.
35. Ibid., p. 592.
36. Ibid., p. 596.
37. Ibid., p. 597.
38. Ibid., p. 599.
39. Ibid., p. 606.
40. Ibid.
41. Shaw, M., Genetically defective children: Emerging legal considerations, *Am. J. L. Med.* 3(3):340 (1977).
42. Carter, C., Evans, K., Fraser-Roberts, J., *et al.*, Genetic clinic: A follow-up, *Lancet* 1:281 (1971).
43. Sorenson, J. R., Genetic counseling: Some psychological considerations, in: *Genetic Responsibility*, op. cit., p. 65.
44. Ibid., p. 65.
45. Shakespeare, W., *Antony and Cleopatra*, Act 2, lines 48–107.
46. Hardin, G., Parenthood: Right or privilege, *Science* 16(3944):31 (1970).
47. Wexler, N., Genetic Russian roulette, in: *Genetic Counseling and Psychological Considerations* (S. Kessler, ed.), Academic Press, New York (1979).
48. Paton, A., *Cry, the Beloved Country*, p. 80, Scribner's, New York (1948).

DISCUSSION

MR. G. MICHAEL SMITH (Attorney, Atlanta, Georgia): I would like to make one comment which might possibly open up a Pandora's box. First of all, I don't really see the likelihood of the breakdown of the intrafamilial protection in the family—the tort immunity that exists. The contract theory and the theory of property that attaches between parents and their children has been upheld. But we are talking about a tort action.

This type of tort has been a passive one, since parents are not really knowledgeable in most cases of the fact that they were bringing "a wrongful life" into existence.

However, where we have counseling, the parent may be knowledgeable. If there is going to be a breakdown, that parent will be the person who will be attacked as guilty of an active tort as opposed to a passive tort.

The problem is then parents will say, "I might be sued by my child if I bring a wrongful life into being." On the other hand, if we have a breakdown of this immunity, will this cause people ultimately to shy away from seeking counseling? I see a danger in giving the information on the one hand and placing the parent in a position of being an active as opposed to a passive tortfeasor. I think it needs to be thought about in terms of whether these types of immunity exist.

Realistically, I do not believe we are going to see much litigation of child against parent unless it is a situation where the parent actually has something, such as wealth.

PROF. SANFORD KATZ: Either that, or where the parent has some insurance. Actually, I haven't seen any cases where, in your terms, passive torts have been successful in courts. That is to say, where a child is suing a parent for some neglectful conduct, neglectful rearing. There was a recent suit in Colorado in which a child sued a parent for not raising the child properly and another suit in which a child sued a parent for not educating him properly.

None of these have been successful, as far as I can tell, for a couple of reasons. One is that courts are reluctant to find for the plaintiff in these cases, unless there is some reasonable opportunity for recovery monetarily. And as yet, I don't know of any insurance companies who insure an adult against these risks, namely, child-rearing.

However, the tort recoveries in the obvious areas of automobile accidents, and so on, are really based upon the likelihood of an insurance company paying off for the tortfeasors. In addition, I think one has to be realistic about the likelihood of success in these tort cases. To what extent do we allow the child to go back into the home in which he was raised and live happily ever after? I think there is something to the wisdom of the common law that talked about the ideal of family harmony and that one way of reaching family harmony was not to allow such suits.

PROF. RICHARD BERQUIST (College of St. Thomas, St. Paul, Minnesota): My point is a philosophic one. I can understand fairly well the idea of a child who is born handicapped suing his parents by reason of some negligent action by the parent during the pregnancy. At least I can understand it in a theoretical way. Whether it would be practical or not I guess is another problem. But if the mother took drugs deliberately and injured the child, I can see that, because the child is in existence and there is an action by the mother which does actually inflict injury on the child.

326

It is a good deal more difficult where the only thing the parent has done is to bring a child into existence and the injury is the result of a genetic problem which is presumably outside the parents' voluntary control. So to say that the parents have actually inflicted genetic damage on the child seems rather odd.

PROF. KATZ: These wrongful birth cases haven't been successful. The famous case of wrongful birth involved not a genetic problem but one of illegitimacy. I think the first case was in Illinois, *Peter v. Peter*, where a child sued his putative father. The mother and the child sued the putative father for what the Illinois lawyer, using creative imagination, called wrongful birth: allowing a child to be brought into the world with the social stigma of illegitimacy.

The Illinois Supreme Court thought it was a novel idea but denied recovery. They passed the buck to the legislature, noting that this was just a phenomenally new idea and if the legislature wished to enact a cause of action, they should do so. The judicial reluctance to carve out a new cause of action recognised that it would open the floodgates to litigation, with people suing their parents.

What good is it to sue a parent for negligent birth or wrongful birth if there is no monetary recovery? Is it just the principle?

PROF. BERQUIST: I wouldn't approve of a child suing the parents for wrongful birth even in an illegitimacy case. But at least it would be arguable in that case that there was something voluntary on the part of the parent. But in the case of the child who is born with a genetic disease, it doesn't seem to me there is the slightest color or slightest rationality to the argument that the parents have inflicted an evil on the child. That is the point I wanted to make.

RABBI GERALD WOLPE (Medical College and University of Pennsylvania): In these past three days I have felt a sense of unease which was somewhat exacerbated by Dr. Wexler's excellent presentation. We have referred to this problem in an oblique way. As long as our society had an overall concept of values, even though there were tensions, it was relatively easy to apply the tensions to the total concept. Today, when there is a bifurcation of values, a fragmenting of values, we have a very serious problem. And I am somewhat taken by the overall title of this conference and the insertion of ethics in certain sessions, such as yesterday and this morning.

Dr. Wexler, in your list of concerns for the genetic counselor there are certain things I found missing. I, for one, would be very interested in knowing why one would want to be a counselor in this field. What would he or she derive from counseling someone in this very difficult field? To use the old Moss concept, what is the prize they are receiving from this encounter? I think you hinted at it. But with all of the practical problems the counselor would have to face, such as the showing of a defective child to the parents, where does he or she derive a value system? If there is going to be nonverbal communication, then that counselor is going to give his or her values to the parent and those values are going to be part of the choice. At no time, very clearly, have these values been pointed out in this conference; they have been referred to as problems and I have heard of all the legal and medical problems that we are going to have to face. It is my contention that there is even a greater problem: the values have never been attacked, they have never been applied, and they have never been articulated. I think that is the greatest problem of all.

DR. NANCY S. WEXLER: I would invite you to see the "out-takes" on this speech. I think your questions are very well put. There are many issues which need to be addressed but time did not allow.

The values of the genetic counselor, the values of the physician, influence the work they do. It goes back to where we started this morning with Dr. Child's speech—what is the nature of education and how does it shape values? Certainly education is much more than reading, writing, and arithmetic. Education has to encompass and be guided by some value system: how you live, how you interact with people who are all variety of sizes, shapes, colors, and deformities.

I don't know what possesses somebody to be a genetic counselor or to do almost anything else that people do in the world. These are all questions that we need to spend much more time exploring.

I believe that when these wrongful life cases hit the courts, it is a sign of failure. It means that we haven't been successful either in defining our values or in communicating them, or in trying to work things out in a meaningful way between people. The court system can become a way of making a statement, seeking retribution through publicity. Whether or not anybody can recover any money from a case may be irrelevant in certain instances. I mean, if my parents knowingly risk having a child with a genetic defect and I am born with a defect and take my parents to court, their names would be in the headlines, they would be in the editorials of *Science* magazine, they would be publically shamed and humiliated or at least the objects of controversy. And everyone would know I was angry with them.

PROF. KATZ: That may be a misuse of the judicial process, though. We have laws against that. The judicial process is for resolving conflict. It isn't for making political statements.

DR. WEXLER: But you can't deny they are sometimes used for making political and ethical statements. People know that Zapeda had a bastard, whether he likes it or not. There is a lot of publicity which accrues to a case; I don't know if that is why the case was brought but it certainly is a consequence. And I think that making a statement may be very important for some people. It is an expression of anger when one may have no other way to express it.

PROF. KATZ: Maybe we ought to have Hyde Park here. That's serious. Our courts are clogged enough without having litigious people working out their psychological problems in a courtroom, although some say better there than the streets.

DR. LAWRENCE ULRICH (University of Dayton, Ohio): My comments are also addressed to Dr. Wexler's excellent presentation that serendipitously followed the speaker who just asked the question. I think you pointed out very nicely, Dr. Wexler, that the law is very inadequate in being able to enforce any kind of regulations requiring people not to reproduce if they are at risk.

The problem that arises, though, is that there may be things that are morally required that cannot be legally enforced. When it comes to this matter, I wonder if there is not a moral preference here that may not be just a preference but rather a moral requirement. We are quite capable in our society of instructing people about their moral obligations as members of society. That means that individuals may not capriciously follow their own wills and their own dispositions. Some of those moral requirements we can translate into laws, such as the requirement not to kill, the requirement not to steal, and so forth.

I am just wondering if this is not a situation where we have a moral requirement that at this point would be extremely cumbersome in legal circles but nonetheless that does not diminish the moral requirement of people with regard to their obligations toward their unborn—not their unborn children, but to the rest of society, to the species at large.

Before you answer that, I would just like to comment on the appalling sense of values that I see from the chairman, who thinks that the only points of law are points of law that can be decided in a monetary way.

MS. KATHI HANNA (Genetic Counseling Student, Sarah Lawrence College): I would like to answer the rabbi that one of the reasons I have gone into genetic counseling is because I feel I can play a positive role in what could be a negative experience, and I would also like to ask him why he has become a rabbi?

RABBI WOLPE: Well, I became a rabbi because six million of my people were destroyed and I wasn't going to give Hitler a victory.

Whether my answer is adequate or not, I certainly was not attacking the motive of anyone who is a genetic counselor. I would just be very interested in knowing why they wanted to be a counselor; because in a sense it would help me and perhaps Dr. Wexler to articulate, help shape, and teach the values which are inherent in that kind of counseling.

MS. TABITHA POWLEDGE: I want to ask the rabbi's question in a slightly different way. Your lovely presentation made it very clear that law is not the way to go. On the other hand, there is also a very strong eugenic thread in your espousal of counseling to achieve what are essentially the same ends. I want to ask a question that has troubled me as long as I have been looking at this field, which is now seven or eight years.

What is genetic counseling for? Is its purpose eugenic? And if so, maybe that ought to be explicity stated, which in fact is not the case now. That may be a rhetorical question. I don't expect you to answer it. But in fact that is not a question that this profession has addressed seriously and systematically, and it is an important question. You ought to be aware of what it is you want to do.

DR. WEXLER: May I try to answer? My personal belief is that genetic counseling should not be eugenic. Genetic counseling is similar to psychotherapy—psychotherapists can't require a patient to be healthy. You can't say, "Stop being neurotic or psychotic." Certainly therapists have some internal notion, albeit murky at times, of what it is to be psychologically healthy. Genetic counselors also have convictions about the "best outcomes" for parents and future children. Counselors should work with their clients to help them fully understand the technical, emotional, and financial information, and serve as "sounding boards."

Not being a eugenicist does not mean being totally passive. Counselors should actively pursue every alternative, should be frank if something does not make sense, should challenge defenses of withdrawal or denial if the birth of a damaged child may be the consequence, and must still maintain rapport and a working alliance with the client. That's a tall order. And I question whether all those who are doing genetic counseling are equipped by their training to provide what's needed.

The eugenics problem is a difficult one. The trouble I have when we talk about eugenics and morality is deciding who has the moral "upper hand"—the parent or the child. Many children who are born with genetic damage suffer physically and emotionally. Practicing eugenics may involve infringing on my freedom, but genetic responsibility must take into account the possibility of my imposing a question on some person who has no choice in the matter, and who has to live with the limitations or defects, for the rest of his or her life.

I do not want to say to anybody, "Do this" or "Don't do that." It probably wouldn't matter if I said it to them anyway. But we should remember that it is we who are adults and may want to have children who are talking to each other. There is no "fetal representative" for those who may yet be born with genetic damage.

PROF. Y. EDWARD HSIA: Sheldon Reed has expressed regret that the term *genetic counseling*, which he coined, has become misunderstood and excessively bandied about. Both Dr. Wexler and Dr. Childs have spoken very articulately on this issue and I suggest there is a common theme that they have which helps to resolve the legal and the moral aspects of the genetic counselor's responsibility.

I am a genetic counselor and I am a teacher. My responsibility is to try to confer knowledge, to try to convey understanding, and to encourage the people to whom I am trying to give genetic counseling to have wisdom. My responsibility is to try to get them to know, to understand, and to be wise. I am not responsible for directing what they are trying to do; my job is that of an educator.

There is one additional interesting point. I am delighted to be a genetic counselor when I have knowledge. But when I face a situation where I have no knowledge, and face a distressing situation without a diagnosis, I feel I'm in both a legal and a moral dilemma. A moral dilemma because I am unable to teach because of my ignorance, and a legal dilemma because whatever I say may turn out to be wrong and might have unforeseeable and unfortunate consequences.

MR. IAN KENNEDY (Attorney, British Broadcasting Corporation): Although Dr. Wexler referred to a newspaper report of parents seeking to have their children sterilized, it must be common knowledge to all of us that this would be unlawful in England—that is, sterilization of a minor without the consent of the court. There is a case decided by our only female judge, in which she said that a child could not be sterilized because she had a fundamental right to engage in reproduction. So the court's permission would be needed and would not normally be granted.

On the matter of the education of normal people to integrating handicapped people into ordinary schools, you may know that two years ago the British government produced a report which recommended the gradual dismantling of separate educational facilities so as to integrate the handicapped into the normal schools so as to overcome this separation, which creates later difficulties.

28

FATHERS ANONYMOUS

Beyond the Best Interests of the Sperm Donor

GEORGE J. ANNAS
Boston University Schools of Medicine and Public Health
Boston, Massachusetts 02118

Alex Haley concludes his international best seller, *Roots*, with the burial of his father in Little Rock, Arkansas. Walking away from the graveside he ponders the past generations, observing, "I feel that they *do* watch and guide." The book inspired whole industries devoted to the development of family trees, and locating one's "roots" has become somewhat of an obsession with many. Because of the current secrecy surrounding the practice of artificial insemination by donor (AID), there are an estimated 250,000 children conceived by AID (at the rate of 6000–10,000 annually in the United States) who will never be able to find their biological roots. There are almost no data available on these children, their psychological development, or their family life. The entire procedure has been shrouded in secrecy that is primarily justified by fear of potential legal consequences should the fact of AID be discovered.

It is the thesis of this brief chapter that most of the informal "policies" concerning AID as it is currently practiced in the United States have come about because of an exaggeration of potential legal pitfalls and a failure to pay sufficient attention to the best interests of the AID child. Accordingly, it is at least premature either to legislate "standards" or use AID as a "model" for *in vitro* fertilization (IVF). Most commentary on AID has concentrated on theoretical legal problems without paying attention to real psychological problems. Indeed, most of the legal literature reads like an answer to the following final exam question: "Review all of the case law and statutes relating to AID and discuss all possible lawsuits that any participant or product of AID might have against anyone. If time permits, suggest a statutory scheme that might minimize these problems." Rather than add another answer to this interesting but tangential question, this chapter will review the rationale for AID, the manner in which donors are selected, and the way records are kept, with a view toward developing policies and practices that maximize the best interests of the child.[1]

WHY USE AID?

The question of indications is almost never addressed in the medical or legal literature beyond assuming that it is almost exclusively a "treatment for husband infertility." To find a model for AID, one must consult the social satirists of the 20th century and the writings of philosophers. In George Orwell's *1984* reproduction by artificial insemination (although not necessarily by donor) was mandatory as part of a program to remove all pleasure from sexual intercourse. Other measures enacted toward this end were party approval of all marriages (always refused if the couple was physically attracted toward each other) and promotion of the view that sexual intercourse should be seen "as a slightly disgusting minor operation, like having an enema." All children were raised in public institutions.

Artificial insemination, however, is only one possible consequence, not a cause, of a totalitarian state of the type envisaged by Orwell. In this regard, Joseph Fletcher is quite correct in observing that in such a society "the modes of reproduction would be of a relatively minor concern... compared to the many human values certain to be destroyed."[2]

AID is taken more seriously, and viewed with more hope than fear, by Aldous Huxley. In *Brave New World Revisited* he writes that every new advance in medicine will "tend to be offset by a corresponding advance in the survival rate of individuals cursed by some genetic insufficiency... and with the decline of average healthiness there may well go a decline in average intelligence." Huxley presents one solution to this problem in his view of the ideal society, *Island*. In that society, AID is not mandatory but is in fact used by almost everyone—at least for the third child, and by most couples who decide to have only two, for their second child. The rationale is the one previously expounded by Huxley: to increase the general IQ of the population instead of allowing it gradually to decrease. In the words of his character, Vijaya:

> In the early days there were a good many conscientious objectors. But now the advantages of AI have been so clearly demonstrated, most married couples feel that it's more moral to take a shot at having a child of superior quality than to run the risk of slavishly reproducing whatever quirks and defects may happen to run in the husband's family... we have a central bank of superior stocks. Superior stocks of every variety of physique and temperament.[3]

The problems of selecting such "superior stock" have been discussed but not resolved. H. J. Muller, for example, argued in 1935 that no intelligent and morally sensitive woman would refuse to bear a child of Lenin,—while in a later version Lenin is omitted and Einstein, Pasteur, Descartes, Leonardo, and Lincoln are nominated.[4] Theodosius Dobzhansky has noted that "Muller's implied assumption that there is, or can be, *the* ideal human genotype which it is desirable to bestow upon everybody is not only unappealing but almost certainly wrong—it is human diversity that acted as a leaven of creative effort in the past and will so act in the future."[5] This is, of course, simply an axiom of evolution and natural selection. The problem of making conscious choices is that we cannot accurately predict what traits future generations will require for survival.

Sociobiologists have recently identified another genetic "truth" in the animal kingdom that may have relevance to the AID situation in man, i.e., that animals will try to maximize the spread of their genes. In the words of Richard Dawkins, "Ideally what an individual would 'like' (I don't mean physically enjoy, although he might) would be to copulate with as many members of the opposite sex as possible, leaving the partner in each case to bring up the children."[6] In this way the genes of the father are distributed maximally. AID and sperm banking remove the previous physical limitations of such a stretegy from the human animal.

On the moral plane, AID has been condemned by the Catholic church (primarily because masturbation is viewed as an unnatural and evil act), and by such writers as Paul Ramsey (on the basis that it is "an exercise of illicit dominion over man").[7] On the other hand, Joseph Fletcher has vigorously defended the morality of AID. He has argued, first, that there is ample precedent for the practice in the Old Testament (e.g., Deuteronomy 25:5-6, and Genesis 30:1-13) and that it is a licit means toward a highly desirable end (parenthood for the otherwise sterile couple). His conclusions are based on his belief that fidelity in marriage is a personal bond between husband and wife (not primarily a legal contract), and that parenthood is a moral relationship with children, not a material or merely physical one.[8]

Until recently it was impossible even to speculate with any authority on the indications for AID in contemporary medical practice. In March of 1979, however, Curie-Cohen, Luttrell, and Shapiro of the University of Wisconsin published their questionnaire survey of AID practitioners. They located 379 practitioners of AID who accounted for approximately 3576 births in 1977 and responded to a series of questions about their practices. The results of this survey provide the only data in existence on the current practice of AID in this country, and the survey, which will be referred to as the "Curie-Cohen survey," will be cited extensively in this chapter.[9]

As to indications, their survey findings are instructive: 95% of the respondents reported that their primary reason for using AID was for husband infertility. However, at least 40% had used AID for other reasons: one-third had used it for fertile couples when the husband feared transmission of a genetic disease (similar to Huxley's *Island* rationale), and almost 10% had used it to fertilize single women (removing sex from reproduction altogether, and highlighting the fears of many moralists). Therefore, whatever one views as society's rationale for permitting AID to continue, it must be recognized that a large percentage of practitioners are using it for eugenic purposes. In addition, those that use it to fertilize single women or members of lesbian couples are engaged in a practice that most of society would probably condemn because of its implications for the child and the family as a basic unit of society.

The issue of indications needs to be faced directly and clearly by commentators and practitioners alike so that an informed consensus can be reached. It is worth observing, however, that current rationales for servicing the infertile couple, the lesbian couple, and the single woman all rest primarily on one's definition of the best interests of the couple or prospective parent, and not on the best interest of the child. While many physicians "screen" recipients to determine if their motives are proper and their marriage "stable," there is no evidence that they are competent to make these judgments. This is not necessarily to say that AID should not be available to couples in which the husband is sterile; it is only to highlight the fact that we have no data concerning how children born into this situation fare, and to suggest that it is irresponsible to continue the practice of AID for this indication without attempting to gather such data.

DONOR SELECTION

Donor selection may be the most difficult issue in AID, but legal considerations are not controlling. First, it should be noted that "donor" is a misnomer. Virtually all respondents in the Curie-Cohen study paid for ejaculates, 90% paying from $20 to $35 per ejaculate, with 7% paying more, up to $100. Thus a more accurate term would be *sperm vendors*. While this distinction may seem trivial, it has legal consequences. For example, it makes no sense to designate the form signed by the "vendor" as a "consent form" since he is not a patient and isn't really consenting to anything. It is a contract in which the sperm vendor agrees to deliver a product for pay. We can debate the elements of the agreement, but most would probably agree that it should spell out the vendor's obligations in terms of his own physical and genetic health, including an accurate family history, the quality of the specimens he is required to produce, the necessity for complete and permanent anonymity of the recipient, and a waiver of any rights in any child resulting from the insemination. In return, the buyer agrees to pay the vendor and protect his anonymity.

The issue of *who* selects the sperm vendor has been given far too little attention. The Curie-Cohen study found two things of interest in this regard. First, 92% of practitioners never permit the recipient to select the donor, although the remainder do on rare occasion. Also, 15% used frozen semen obtained from sperm banks, and others used sperm from those selected by urologists or other personal associates. The point is that at least in a small minority of cases, someone other than the physician selects the source of the sperm. More significant, however, is the fact that almost all physicians make their own selection, most using medical students. Sixty-two percent used medical students or hospital residents; 10% used other university or graduate students; 18% used both; and the remaining 10% used donors from military academies, husbands of obstetric patients, hospital personnel, and friends.

Physicians in all of these situations are making eugenic decisions—selecting what they consider "superior" genes for AID. In general, they have chosen to reproduce themselves (or those in their profession), and this is what sociobiologists like Dawkins would probably have predicted. While this should not surprise us, it should be a cause of concern, since what may be controlling is more than just convenience. Physicians may believe that society needs more individuals with the attributes of physicians, but it is unlikely that society as a whole does. Lawyers would be likely to select law students; geneticists, graduate students in genetics; military personnel, students at the military academies; etc. The point is not trivial. Courts have found in other contexts that physicians have neither the training nor the social warrant to make "quality of life" decisions. In the *Houle* case, for example, a physician's decision not to treat a defective newborn was overruled on the basis that "the doctor's qualitative evaluation of the value of the life to be preserved is not legally within the scope of his expertise." Selecting donors in this manner, rather than matching for characteristics of the husband, for example, seems to be primarily in the best interests of the physician rather than the child, and can probably not be justified. Nor can the argument that medical students know more about genetics than other graduate students stand analysis. They are probably also just as susceptible to monetary influence as are some of the blood sellers described in Richard Tittmuss's, classic study, *The Gift Relationship*. Perhaps national guidelines, developed by a committee made up of a random sample of the population, would be more appropriate.

The Curie-Cohen survey also revealed that even on the basis of simple genetics, physicians administering AID "were not trained for the task" and made many erroneous and inconsistent decisions. Specifically, 80–95% of all respondents said they would reject a donor if he had one of the following traits, and more than 50% of all respondents would reject the same donor if one of these traits appeared in his immediate family: Tay–Sachs, hemophilia, cystic fibrosis, mental retardation, Huntington's, translocation or trisomy, diabetes, sickle cell trait, and alkaptonuria. This list includes autosomal recessive diseases in which carriers can be identified, and those in which they cannot, dominant, X-linked, and multigenic diseases.

The troubling findings are that the severity and genetic risk of the condition was not reflected in rejection criteria, and that genetic knowledge appears deficient. For example, 71% would reject a donor who had hemophilia in his family, even though this X-linked gene could *not* be transmitted unless the donor himself was affected. Additionally, although 92% said they would reject a donor with a translocation or trisomy, only 12.5% actually examined the donor's karyotype. Similarly, while 95% would reject a carrier of Tay–Sachs, fewer than 1% actually tested donors for this carrier state. In fact, only 29% performed any biochemical tests on donors other than blood typing, and these tests were primarily for communicable diseases. The conclusion must be that while prevention of genetic disease is a goal, it cannot be accomplished by the means currently in use. The findings also raise serious questions about the ability of these physicians to act as genetic counselors, and suggest that other nonmedical professionals may be able to do a better job in delivering AID services in a manner best calculated to maximize the interests of the child.

Since there is almost uniform agreement that certain genetic conditions contraindicate use of a person's sperm for AID, it is likely that a court would find a physician negligent in using such sperm even though few physicians actually test to make sure the sperm vendor is not affected.[10] "There are precautions so imperative that even their universal disregard will not excuse their omission."[11] This is an area in which uniform standards need to be developed within the profession.

Two other related issues concerning the donor or sperm vendor merit mention because they have apparently been dealt with strictly on the basis of fear of legal liability rather than any social or medical rationale or concern for the best interests of the child: consent of the donor's wife and record keeping.

The Donor's Wife

The American Medical Association, the British Medical Association, and authorities in Australia all agree, as do almost all legal commentators, that the wife of the sperm donor must sign the "consent" form "because marital interests are involved." None of these sources or commentators, however, provide any further explanation. This type of advice can be viewed as a paradigm of legalism based on fear and ignorance.

I do not know what the original source of this recommendation is, but it may be Joseph Fletcher's comments in 1954 that "it is clearly a requirement of personal integrity, of love and loyalty, that the donor's wife should be consulted by him (the donor) and agree to the role he plays."[12] Perhaps. But however one comes out on this pronouncement, it is not a legal requirement and does not seem to serve any useful social purpose. In terms of liability on the part of the physician, the potential grounds appear to be two: (1) an action in contract to recover a portion of the money received by the husband for his sperm on the grounds that the wife had a property interest in her husband's sperm; and (2) an action for alienation of affections by the wife against the physician on the basis that her husband prefers masturbation for pay to intercourse with her, or some other fantasy he may have developed that interferes with the marriage. Both of these strike me as being too silly to worry about, and any woman who would bring either action is not likely to be discouraged by the fact that she has signed a "consent" (read "contract") form. In addition, such a requirement is at odds with more recent United States Supreme Court decisions that refuse to permit one spouse to have veto power over procreation decisions made by the other spouse. Specifically, a husband may not be required to consent to his wife's abortion by state law because her right to make this decision is constitutionally protected.[13]

Record Keeping

While the Curie-Cohen survey found that 93% of physicians kept permanent records on recipients, only 37% kept permanent records on children born after AID (fewer than the 50% who provided obstetric care for their inseminated patients), and only 30% kept any permanent records on donors. Moreover, 83% opposed any legislation that would mandate the keeping of records because it would make protection of anonymity of the donor more difficult. The fear of record keeping seems to be based primarily on the idea, common in the legal literature, that if identifiable, the donor might be sued for parental obligations (e.g., child support, inheritance) by one of his "biological children" sired by the AID process, and that this suit might be successful. The underlying rationale is that without anonymity assured, there would be no donors. There are a number of responses to this argument.

1. It is important to maintain careful records to see how the sperm "works" in terms of outcome of the pregnancy. If a donor is used more than once, a defective child should be grounds for immediately discontinuing the use of the sperm for the protection of potential future children. Since the survey disclosed that most physicians have no policy on how many times they use a donor, and 6% had used one for more than 15 (with one using a donor for 50 pregnancies), this issue is much more likely to affect the life of a real child than the highly speculative lawsuit is to affect a donor.

2. No meaningful study of the characteristics of donors can ever be made if there are no records kept concerning them.

3. In those cases where family history is important (and it is important enough to ask *every* donor about his) the AID child will *never* be able to respond accurately.

4. Finally, and most importantly, if no records are kept, the child will *never*, under any circumstances, be able to determine its genetic father. Since we do not know what the consequences of this will be, it cannot be said that destroying this information is in the best interests of the child. The most that can be said for such a policy is that it is in the best interests of the donor. But this is simply not good enough. The donor has a choice in the matter, the child has none. The donor and physician can take steps to guard their own best interests, the child cannot.

Given the recent history of adopted children, it is likely that if AID children learn they are the products of AID, they will want to be able to identify their genetic father. It is now relatively accepted practice to tell adopted children that they are adopted as soon as possible, and make sure they understand it. This is because it is thought they will inevitably find out some day, and the blow will be a severe one if they have been lied to. In AID, the consensus seems to be not to tell on the basis that no one is ever likely to find out the truth, since to all the world it appears that the pregnancy proceeded in the normal course.

Moralists would probably agree with Fletcher that the physician should not accept the suggestion that a husband's brother be used as a donor without the wife's knowledge (his intent is to keep the bloodline in his children) because this is a violation of "marital confidence." It seems to me a similar argument can be made for consistently lying to the child—i.e., that it is a violation of parent–child confidence. There is evidence that AID children do learn the truth, and the only thing all 15 states with legislation on AID agree to is that it should legitimize the child—an issue that will never arise unless the child's AID status is discovered. If AID is seen as a loving act for the child's benefit, there seems no reason to taint the procedure with a lie that could prove extremely destructive to the child.

A number of policies would have to be changed to permit open disclosure of genetic parenthood to children. The first is relatively easy: a statute could be enacted requiring the registration of all AID children in a court in a sealed record that would only be available to the child; the remainder of the statute would provide that the genetic father had no legal or financial rights or responsibilities to the child. A variation on this would be to keep the record sealed until the death of the donor, or until he waived his right to privacy in this matter. In the long term, a more practical solution may lie in only using the frozen sperm of deceased donors. In this case full disclosure could be made without any possibility of personal or financial demands on the genetic father by the child.[14]

Worry about donors, in any event, is probably out of proportion to reality. There have been *no* suits against any donor by any child even though almost one-third of physicians engaging in AID keep permanent records of the donors. No matter what steps are taken to protect them, it seems essential to me that, in the potential best interests of the child, such records be kept and that their contents be based on the development of professional standards for such records.

Not keeping records can also lead to other bizarre practices. For example, some physicians use multiple donors in a single cycle to obscure the identity of the genetic father. The Curie-Cohen survey found that 32% of all physicians utilize this technique, which could be to the physical detriment of the child (and potential future of a donor with defective sperm) and cannot be justified on any genetic grounds whatsoever.[15]

SUMMARY AND CONCLUSIONS

Current AID practices are based primarily on consideration of protecting the interests of practitioners and donors rather than recipients and children. The most likely reason for this is found in exaggerated fears of legal pitfalls. It is suggested that policy in this area should be dictated by maximizing the best interests of the resulting children. The evidence from the Curie-Cohen survey is that current practices are dangerous to children and must be modified. Specifically, consideration should be given to the following: (1) removing AID from the practice of medicine and placing it in the hands of genetic counselors or other nonmedical personnel (alternatively, a routine genetic consultation could be added for each couple who request AID); (2) development of uniform standards for donor selection, including national screening criteria; (3) a requirement that practitioners of AID keep permanent records on all donors that they can match with recipients; I would prefer this to become common practice in the profession, but legislation requiring filing with a governmental agency may be necessary; (4) as a corollary, mixing of sperm would be an unacceptable practice, and the number of pregnancies per donor would be limited; (5) establishement of national standards regarding AID by professional organizations with input from the public; (6) research on the psychological development of children who have been conceived by AID and their families.

Dr. S. J. Behrman concludes his editorial on the Curie-Cohen survey by questioning the "uneven and evasive" attitude of the law in regard to AID, and recommending immediate legislative action:

> The time has come—in fact, is long overdue—when legislatures must set standards for artificial insemination by donors, declare the legitimacy of the children, and protect the liability of all directly involved with this procedure. A better public policy on this question is clearly needed.[16]

I have suggested that agreement with the need for "a better public policy" is not synonomous with immediate legislation. The problem with AID is that there are many unresolved problems with AID, and few of them are legal. There is no social or professional agreement on indications, selection of donors, screening of donors, mixing of donor sperm, or keeping records on sperm donations. Where there is agreement, such as in requiring the signature of the donor's wife on a "consent" form, the reasons for such agreement are unclear.

It is time to stop thinking about uniform legislation and start thinking about the development of professional standards. Obsessive concern with self-protection must give way to concern for the child.[17]

REFERENCES AND NOTES

1. Currently, at least 15 states (Alaska, Arkansas, California, Florida, Georgia, Kansas, Louisiana, Maryland, New York, North Carolina, Oklahoma, Oregon, Texas, Virginia, and Washington) have statutes on the books that mention AID. All of these statutes specifically provide that the resulting child is the natural child of the recipient's husband, provided he has consented to the procedure. Five states require that the consent be filed with a state agency (Kansas, Oklahoma, Georgia, Washington, and Oregon) and 6 states, either directly or by implication, limit the practice of AID to physicians (California, Oklahoma, Virginia, Washington, Alaska, and Oregon). Only 2 states, Washington and Texas, specifically provide that the sperm donor is not the father of the child. Oregon's is the only criminal statute, and makes it a Class C misdemeanor, punishable by 30 days in jail, for anyone but a physician to select sperm donors and for a donor to provide semen if he "(1) has any disease or defect known to him to be transmissible by genes; or (2) knows or has reason to know he has a venereal disease." The only state supreme court to ever rule on AID held a consenting husband liable for child support [*People v. Sorensen*, 437 P.2d 495 (Cal. 1968)]. An excellent overview of the law, which will not be repeated in this chapter, appears in the proceedings from the first conference, Healey, J., Legal aspects of artificial insemination by donor and paternity testing, in: *Genetics and the Law* (A. Milunsky and G. J. Annas, eds.), p. 203, Plenum, New York (1976).
2. Fletcher, J., *Morals and Medicine*, p. 134, Beacon Press, Boston (1954).
3. Huxley, A., *Island*, p. 193, Perennial Classic, New York (1972).
4. Quoted by Ramsey, P., *Fabricated Man*, p. 49, Yale University Press, New Haven (1970).
5. Ibid. at 53.
6. Dawkins, R., *The Selfish Gene*, p. 151, Oxford University Press, New York (1976).
7. Ramsey, P., *Fabricated Man*, supra note 4 at 48.
8. Fletcher, J., *Morals and Medicine*, supra note 2 at 116.
9. Curie-Cohen, M., Luttrell, L., and Shapiro, S., Current practice of artificial insemination by donor in the United States, *N. Engl. J. Med.* 300:585 (1979).
10. While there is not specific legal standard for screening sperm donors, when done by a physician the general law of specialists is applicable: "One holding himself out as a specialist should be held to the standard of care and skill of the average member of the profession practicing in the specialty, taking into account the advances in the profession" [*Brune v. Belinkoff*, 235 N.E. 2d 793 (1968)]. A recent analogous case involved an individual who had received two cornea transplants. The transplanted corneas turned out to be infected, and caused total and permanent blindness in the recipient. He sued the hospital and the resident who had removed the donor's eyes. The jury found in favor of the resident but against the hospital. In affirming the jury's verdict against the hospital, the court noted that while the hospital had "no printed or published checklist which could be used as a guideline for determining the suitability of a prospective donor," there was testimony that published criteria did exist and was "fairly uniform throughout the nation." The court further concluded that had these criteria been applied to the donor in this case, the jury could have rightfully decided that he would have been rejected:

    The jury heard expert testimony to the effect that cadavers with a history of certain types of illnesses are not generally wise choices for cornea donation. It follows that whoever may have had the responsibility of determining the suitability of the cornea for transplant would have been *required*, in the exercise of due care, *to review carefully and exhaustively the medical history of the propsoed donor*, ... The jury could have determined that Detroit General was negligent in failing to set up a procedure which would assure that the party responsible for determining the suitability of the cornea for transplant would have *access to all the relevant medical records of the proposed donor*. [emphasis added] [*Ravenis v. Detroit Gen. Hospital*, 234 N.W. 2d 411 (Mich. 1975)]

Applied to AID donors this case indicates that hospitals and physicians are responsible for determining the suitability of donors and, if they don't have a reasonable policy of their own, will be held to whatever policy has been accepted by other professionals engaged in the same activity.

11. T. J. Hooper, 60 F.2d 737,740 (2d Cir. 1932); and see *Helling v. Carey*, 519 P.2d 981 (Wash. 1974).

12. Fletcher, J., *Morals and Medicine, supra* note 2, 129.

13. For a fuller discussion of this issue see Chapter 18 by Leonard Glantz, this volume.

14. Sperm banks may soon begin marketing sperm directly to consumers, bypassing physicians and adding to current confusion in practice. See *Advertising Age*, p. 30, May 14 (1979).

15. A more encouraging finding was that only two physicians in the entire sample mixed donor sperm with the husband's semen. This apparently once-common practice has died, probably because it is now know to be medically contraindicated. See Quinlivan, W. L. G., and Sullivan, H., Spermatozoal antibodies in human seminal plasma as a cause of failed artificial donor insemination, *Fertil. Steril.* 28:1082 (1977).

16. Behrman, S. J., Artificial insemination and public policy, *N. Engl. J. Med.* 300:619 (1979).

17. The argument is based not on any alleged action for "wrongful life" that the child may have (I do not believe this would ever be recognized by a court in the absence of legislation) but on the theory that we should do what we can to protect the interests of "innocent" third parties whenever their interests are in conflict with those who have the ability to affect them.

Discussion

PROF. SANFORD KATZ: I am delighted to see this issue raised. Until very recently, the cases the courts have been asked to deal with in artificial insemination (AID) have dealt with two major issues. One is support and one is adultery. For the most part, the AID cases that arose in the courts did so where there was a dysfunctioning family, usually divorce, and where a child had been born through AID; there had been a disruption of the family, and then a father who was not the donor refused to continue support of the child. Some courts held that he didn't have to and some courts held that he was the sociological father and had, by his actions, brought the child into his home and that the child ought to be the subject of his father's support.

The other issue raised was whether it was a cause of action for a divorce—namely, that any woman who underwent AID would be committing adultery. I think there are some English cases addressing the question: Who is the one committing the act? Is it the physician or the donor, if you could find him? Interestingly enough, these kinds of cases form the jurisprudence for AID. For the longest time there was absolutely no concern for the questions that Professor George Annas raised in his paper.

I think they are extremely important and I think you are quite correct in linking it up with adoption. The point is that before we start legislating, we ought to be thinking about the social, philosophical, and theological questions this whole area raises.

PROF. MELVIN L. TAYMOR: Although I think Professor Annas has raised some very important issues and some I agree with, particularly in improving standards of selection of donors, I still feel on balance I have to disagree with his major tenet about opening up the whole area so that the secrecy is lost.

To me, the very essence of the success of AID is the secrecy for the child. I resent somewhat the implication that the secrecy is for the donor. The secrecy, as I see it, is first for the child; secondly, for the father. Many men would not agree, in our chauvinistic society, with AID if they thought it was going to be open. They do it because they know that it will not be opened up.

I would like to see more evidence that there is harm from the secrecy, or the actual incidence of genetic problems that might come up as a result of this. I either haven't seen them or am not aware of them. In AID, the doctor who makes the selection delves a little deeper than the usual wife does when she selects a husband. I mean, maybe you want every wife to select her husband on a genetic basis as well.

These are issues that I don't think are clear-cut. I think there are many things against opening them up.

PROF. KATZ: Professor Annas, was it that you were suggesting that the identity of the donor be named or merely the genetic characteristics?

PROF. GEORGE J. ANNAS: You could do it either way you wanted. Again, my primary suggestion is that we don't know the answer to your question about how much harm it is causing. We are assuming now it is not causing any. It seems to me there is no basis for that.

PROF. TAYMOR: I didn't say that. I say on balance, opening it up would cause tremendous harm and destroy the very way it works.

PROF. ANNAS: You *feel* that way. Again, you don't *know* that.

341

PROF. KATZ: That is true. I think that Professor Annas's linking it to the opening up of adoption records is very important. For many years we have been working on the assumption that everyone wants the records closed. It would be horrendous to have someone at your doorstep. Yet, more and more adoptees are asking, "Who was my mother; who was my father?" And also, on the other side, women who gave up their children for adoption twenty years ago are saying, "Now I would like to see the child I gave up" or the fathers who had fathered a child saying, "I would like to see my child."

This is unsettling to people working in the adoption field, particularly the social workers who have always said confidentiality is critical and we ought to protect the confidentiality of the biological mother and father.

Now we are seeing that maybe they don't want the secrecy and confidentiality that the professional people have been saying they ought to have.

PROF. TAYMOR: It is also unsettling to some of the parents who gave up their children and unsettling to the adoptive parents too. In adoption, you have to do it because the children know that they have a biological parent elsewhere. There is no need to do that in AID, other than just for the purpose of knowledge itself. Once you open the door, then they will have the right to seek out who the donor was and this will open up all kinds of problems and will probably destroy the treatment.

This is a very important treatment in the overall management of infertility for a large number of couples who do very well. The psychological implications have been studied.

MS. VIKKI ZEGEL (Library of Congress, Washington): It seems to me that these comments are in direct contradiction to the context of this conference on genetic responsibility. If you don't know the genes that have been contributed to at least one-half of your makeup, how can you possibly act responsibly? The mathematical possibilities for incest and expression of recessive genetic characteristics are amplified by something like AID, especially where you mix up half a dozen people. Especially where you have one donor who makes thirty-five dollars five or six times and is responsible for the births of X number of children, and moreover, in the immediate vicinity. That really opens up possibilities for incest, and the laws against incest are based on genetics. So this is totally unreasonable in the context of genetic responsibility.

PROF. TAYMOR: Will someone please explain to me exactly what incest would happen between a half brother and a half sister on an occasion, what that would accomplish once?

MS. ZEGEL: In the F_1 generation, half siblings both must have the same genetic makeup as the father.

PROF. TAYMOR: How does that increase the chances of a genetic abnormality? How many, what numbers, what percentages, what are you talking about?

MS. ZEGEL: The answer depends on the genetic abnormality. If the father is heterozygous for a recessive trait, the child might be a carrier for that trait. Then two of his children marrying might produce an affected child. It greatly increases the possibility of genetic expression of deleterious genes.

DR. THEODORE F. THURMON: The actual figure is twelve and one-half percent of the offspring have a polygenic or a recessive condition if half siblings mate.

PROF. TAYMOR: These apply to standards. I think the standards should be improved. I don't think that my talking about the secrecy is necessarily against the purpose of this meeting.

MS. ZEGEL: But standards don't mean anything if it is a recessive trait that hasn't been expressed in the father. You see, the fact is you are just throwing the door open for all kinds of possibilities for siblings unwittingly marrying and then expressing deleterious genes that may not have been expressed for generations. It really increases the probabilities enormously.

PROF. TAYMOR: But there could be some standard that a donor could not give more than an X number of times. I mean, not just throw it out without some firm mathematical basis.

MS. ZEGEL: That is why laws against polygamy were instituted.

DR. ALLEN BURDICK (University of Missouri): I think there are two population genetic considerations that are important to us. One is just the mention of the short-term population genetic consideration, and the other is long-term. If a donor is allowed to donate more than a few times, it tends to constrict (depending upon the mating system of the population) the numbers of people in that population who are reproducers.

The risk in early generations is the increased chance of consanguineous marriages in the first, second, or third generation beyond the time in which this donor contributed twice, three times, four times more than his share of offspring to the next generation. The primary population genetic consideration is that one individual should leave one offspring in the next generation, two on the average, two; unless that happens, the population is increasing in size.

The long-term consideration of the repeated use of an AID donor—not donor, vendor—is somewhat similar to the short-term consideration in that it makes the population the result of a more nearly polygamous mate system which is entirely sorted. It decreases the variation of the population, increases inbreeding of the population, and hastens evolution.

Evolution is something that one can decide on hastening in humans. Given the spread of the population, geneticists have considered that this would be desirable: to hasten evolution in evolution. A prominent geneticist, H. C. Muller, proposed sperm banks for propagating genes. But it does have short-term risks, even though the long-term might be something that may be desirable.

Ms. GINA KOLATA (*Science* magazine): I wanted to bring up the issue of who should be allowed to have AID by analogy with adoption. Not everybody can adopt. You have social workers who do a case study of a family. With AID, as I understand it, the doctor decides. I spoke to a doctor last fall who told me that he will not give AID to a lesbian woman but he will do it for certain single women if he thinks that they would be good mothers.

I wonder whether anybody has ever addressed the issue of whether there should be any standards. First, are the adoption standards any good? Should we do something analogous for AID, since it seems a bit unfair that a doctor should decide whether a woman would be any good as a mother?

PROF. ANNAS: I agree with you. We have no standards, and that is a problem, and you are precisely right. Right now the AID physician decides whether you are going to be a good mother or not with absolutely, as far as I can see, no basis in reality or predictive factors or psychological criteria to decide who is going to be a good mother.

29

CURRENT STATUS OF *in Vitro* FERTILIZATION AND REIMPLANTATION

MELVIN L. TAYMOR
Harvard Medical School
Boston, Massachusetts 02115

Division of Infertility and Reproductive Endocrinology
Beth Israel Hospital
Boston, Massachusetts 02115

The birth of baby Louise Brown in July 1978 was truly a dramatic event. Although it was the result of many years of intensive research, the event itself was a milestone in man's ability to work with nature. To some, *in vitro* fertilization and reimplantation represents a Pandora's box of "brave new world" reproduction: of oocyte donors and rental wombs. To others, it means a future chance, and I emphasize future and chance, of parenthood for the thousands of women who are sterile because of irrevocably damaged Fallopian tubes.

PHYSIOLOGICAL BACKGROUND

Oocyte Maturation

Knowledge of the maturation of the oocyte is a fundamental prerequisite for understanding the process of *in vitro* fertilization. The mammalian oocytes go through the initial stage of meiosis in fetal life and shortly after birth they become arrested in the diplotene stage, also called the dictyate stage.[1] It is believed that the arrest of maturation is brought about by some factor present in the granulosa cells, since release from the granulosa cells results in resumption of meiosis. The oocytes remain in the resting stage until an oocyte or oocytes in a mature follicle receive the stimulation of the mid-cycle surge of luteinizing hormone (LH). Meiosis then proceeds to metaphase 2, at which time ovulation occurs and meiosis again is halted. The impetus for the final resumption of meiosis is penetration of the oocyte by the sperm.

The above describes *in vivo* maturation. In 1935 Pincus and his co-workers showed

345

that the oocyte removed from the follicles and placed in a defined medium would resume meiosis.[2] This was later confirmed in humans.[3]

Cleavage and Tubal Transport

Edwards has described the timing of human preimplantation development.[4] Recently other studies have detailed the time element in the passage of these various stages throughout the Fallopian tube to the uterus.[5] Timing was related to ovulation by adding 36 hours to the beginning of the LH surge. The two pronuclei were seen 25 hours after ovulation, the two-stage zygote at 32 hours, four cells at 53 hours, and 16 cells at 79 hours. No cleavage beyond 16 cells was found in the Fallopian tube. Further stages of change, morula and blastocysts, were found in uterine washings approximately 100 hours after ovulation.

Thus it can be concluded that the zygote passes into the uterus at the 8- or 16-cell stage, approximately three days after ovulation, and remains in the uterus another day or two until the blastocyst forms and implantation occurs.

HISTORICAL REVIEW OF *in Vitro* FERTILIZATION AND REIMPLANTATION

Shettles was the first to discuss the possibility of *in vitro* fertilization and reimplantation for women with destroyed or absent oviducts in 1955.[6] His observations were based upon oocytes removed from follicles and matured *in vitro*. In 1966 Edwards further reported upon the maturation of oocytes *in vitro* but noted the inability to fertilize these oocytes.[7] Jagiello then described the resumption of meiosis *in vivo* with the injection of human chorionic gonadotropin.[8] Edwards and Steptoe, still working with oocytes matured *in vitro*, found an increase in the incidence of fertilization when capacitation of the sperm was carried out in the presence of follicular fluid.[9] However, it was not until they utilized oocytes matured *in vivo* with human menopausal gonadotropin and human chorionic gonadotropin that fertilization with cleavage to the 8- or 16-cell stage was noted.[10]

A number of years passed, and in 1976 Steptoe and Edwards reported their first successful reimplantation, which unfortunately ended in a tubal pregnancy.[11]

About this time, because of the low yield of successful reimplantation, they gave up superovulation and began to rely upon recovery of the single oocyte brought to maturation by the natural mid-cycle LH surge. In 1978 they reported upon the first successful live birth of an individual that was the result of *in vitro* fertilization and reimplantation.[12]

HOW THEY DO IT NOW

The details of the present techniques of Steptoe and Edwards have not been fully published, but Steptoe has given a number of presentations at national and international meetings so the details are now well known.

During the past three years or so they have not utilized gonadotropin therapy to stimulate or time ovulation but have relied upon the measurement of LH to time the laparoscopy to retrieve the single mature oocyte of the normal menstrual cycle. A daily assay of LH is carried out to determine the approximate day of the LH surge for a prospective patient. During the treatment cycle LH is assayed every three hours, starting a day or two before the anticipated LH peak. A urinary system based upon hemagglutination inhibition is utilized. It is their belief that repeated venipuncture may stress the patient and effect the time of ovulation. It is believed that ovulation occurs approximately 30 hours after the beginning of the LH surge. Laparoscopy can be carried out anywhere from 16 to 28 hours after the first rise in LH. Once maturation has been initiated *in vivo*, it apparently will continue *in vitro*. Special features related to the laparoscopy are, first, the use of a gas medium of 5% CO_2, 5% O_2, 90% N, which is replaced with CO_2 after follicle puncture. It is also important to enter the follicle from the side and to aspirate gently directly into the culture medium.

The oocyte is incubated for 12 to 18 hours in Bavister's medium along with spermatazoa diluted to 8×10^5 to 2×10^6 sperm per ml.

The zygote is then transferred to Ham's F-10 with 20% calf serum for further cleavage.

Reimplantation is carried out at the 8- to 16-cell stage. If a pregnancy occurs, careful monitoring is carried out with ultrasound and amniocentesis.

Reimplantation of the zygote is one of the most critical areas and probably one where the most losses occur at the present time. The zygote is drawn up in its medium into a fine plastic cannula that is inserted through the cervix. The tip should lie approximately 1 cm from the fundus, with the zygote at the end of the cannula. It is important that only a small amount of fluid, approximately 0.07 ml, be injected. Otherwise, there is a risk of the zygote migrating out the cervix before implantation.

SPECIAL PROBLEMS

Preliminary laparoscopy and laparotomy are required to assess the accessibility of the ovary to future laparoscopy. If at laparoscopy the ovary is not accessible, laparotomy will have to be carried out to improve this factor. Steptoe and Edwards strongly recommend preliminary occlusion of the Fallopian tubes in order to prevent an ectopic pregnancy. If laparoscopy is all that has been carried out, this can be achieved by cornual diathermy and, if laparotomy has been necessary, by salpingectomy.

Freedom from stress is considered to be an important element in successful *in vitro* fertilization and reimplantation. Stress at the time of mid-cycle may alter the temporal relationship of the LH surge to a previous cycle and even inhibit or modify the quality of the LH surge. This has been well documented by other observers. Less documented but probably of more importance may be the effect of stress on the success of the reimplantation. For this reason Steptoe and Edwards recommend that the entire procedure be carried out in a unit isolated from the confusion and stress of a general hospital, and that a separate clinic complete with comfortably furnished hospital rooms with nearby operating room and laboratory would be ideal.

The four successful implantations that occurred were in the last 21 attempts. All these occurred when reimplantation occurred somewhere between 8:00 P.M. and 12:00 midnight. This suggests some type of diurnal dependence on endometrial development. For this reason, they will ideally implant an 8-cell zygote between 8:00 and 12:00 P.M. If by midnight the 8-cell stage has not occurred, they will wait until the next morning, at which time it might be an 8- or 16-cell zygote.

The semen should be carefully handled. The husband should be checked before for bacteria and viruses, which should be eliminated if possible. Collection of the semen should be under as aseptic procedures as possible. The semen is diluted to prevent polyploidy. The laboratory should be as close to the operating room as possible, with filtered airflow directing bacteria away from all culture media and incubates.

RESULTS

In January 1979[13] Steptoe and Edwards reported upon the results of 79 patients admitted to the study (Table I).

The most frequent area of loss was at the level of failed implantation. Four pregnancies occurred. Two have resulted in full-term normal deliveries and two in spontaneous abortions. It is their belief that one of these latter occurred as a result of abnormal physical exertion in the second trimester, and that the loss was unnecessary. As indicated in the previous section, it is their belief that if the implantation is carried out between 8:00 and 12:00 P.M., the incidence of success would increase. It is possible that with attention to this detail and improved techniques the success rate might rise to as high as 30 to 50%. On the other hand, there may be limiting factors to ultimate success. One of these is the normal high death rate of embryos known from experimental observations on embryonic death. It is calculated that no more than 30% of fertilized gametes can go on to ultimate full development.[14] It is also possible that *in vitro* development proceeds at a different rate from *in vivo* development, and this too may be a limiting factor. However, if a success rate of even 20 to 30% can be obtained per procedure, a woman could undergo two or three laparoscopies with an expected success rate equal to, if not greater than, that achieved by tubal surgery.

TABLE I.   Results Reported to American
College of Obstetrics and Gynecology, 1979

Total patients in study	79
LH surge detected	68
Oocytes recovered	44
Fertilization	34
Cleavage to 8–10 cells and reimplanted	32
Pregnancies	4

QUESTIONS

Three levels of questions remain. The first of these center around the medical practice, or safety. Risks to the woman are minimal. Laparoscopy is a routine procedure. Whatever dangers do occur are far less than those related to many other procedures performed for infertility.

The dangers to offspring center about the ultimate normality of the individual. This is difficult to assess, and it probably will be decades before firm statistical figures are available. We can only theorize from animal data that such dangers are not overwhelming. Over 1000 animal offspring have been achieved from *in vitro* fertilization and reimplantation without any increase in congenital anomalies.[15] Furthermore, one could theorize that if genetic damage occurred at the early stages of cleavage, the end result would be pregnancy loss, rather than full-term development with congenital anomalies. Nevertheless, because of these factors, careful consideration must be given to this procedure and, if carried out, informed consent should be obtained. Even then, there are those who raise the question of whether it is possible to obtain consent from the offspring, those who are primarily involved.

Others are concerned about the fate of "embryos" that are not reimplanted. This was more of a potential problem when gonadotropins were used to induce ovulation and multiple oocytes were harvested. When *in vitro* fertilization and reimplantation is now utilized for clinical purposes, only one oocyte is obtained and fertilized, and then this objection becomes moot. However, the question of producing embryos for research still remains. The Ethics Advisory Committee of the NIH has recommended that when embryos are to be utilized for research they be kept in culture no more than 10 days.

The other problem area concerns the use of surrogate mothers and oocyte donors. There are some who find these possibilities extremely objectionable, while to others there are no objections. The advisory committee to the NIH has, for the present, recommended that when reimplantation is performed the gametes should be obtained from a married couple. It makes sense, for the beginning at least, to proceed along these lines so that the many women who might benefit from the procedure will not be denied the opportunity because of theoretical objections that can and should be controlled by society. The future course of the procedure, in terms of success and complications, will ultimately have an important bearing on society's view of these very issues.

REFERENCES

1. Baker, T. G., Oogenesis and ovarian development, in: *Reproductive Biology* (H. Balin and S. R. Glasser, eds.), p. 378, Excerpta Medica, Amsterdam (1972).
2. Pincus, G., and Enzman, E. V., The comparative behavior of mammalian eggs in-vivo and in-vitro I. The activation of ovarian eggs, *J. Exp. Med.* 62:665 (1935).
3. Pincus, G., and Saunders, B., Unfertilized human tubal ova, *Anat. Rec.* 69:163 (1939).
4. Edwards, R. G., Studies in human conception, *Am. J. Obstet. Gynecol.* 117:587 (1973).
5. Ortiz, M. E., and Croxalto, H. B., Observations on transport, aging, and development of ova in the human

genital tract, in: *Recent Advances in Reproduction and Regulation of Fertility* (G. P. Talwar, ed.), p. 307, Elsevier, Amsterdam (1979).

6. Shettles, L. B., Further observations on living human oocytes and ova, *Am. J. Obstet. Gynecol.* **69**:365 (1955).

7. Edwards, R. G., Donahue, R. P., Baraink, T. A., *et al.*, Preliminary attempts to fertilize oocytes matured in-vitro, *Am. J. Obstet. Gynecol.* **76**:192 (1966).

8. Jagiello, G., Karnicki, J., and Ryan, R. J., Superovulation with pituitary gonadotropins, *Lancet* **1**:178 (1968).

9. Edwards, R. G., Bavister, B. D., and Steptoe, P. L., Early stages of fertilization in-vitro of human oocytes matured in-vivo, *Nature* **221**:632 (1969).

10. Edwards, R. G., Steptoe, P. L., and Purdy, J. M., Fertilization and cleavage in-vitro of preovulatory human oocytes, *Nature* **227**:1307 (1970).

11. Steptoe, P. L., and Edwards, R. G., Reimplantation of a human embryo with subsequent tubal pregnancy, *Lancet* **1**:880 (1976).

12. Steptoe, P. L., and Edwards, R. G., Birth after the reimplantation of a human embryo, *Lancet* **2**:36 (1978).

13. Steptoe, P. L., and Edwards, R. G., Report to the Royal College of Obstetricians and Gynecologists, January 26, 1979.

14. Leridan, H., Démographie des échers de la reproduction, in: *Les Accidents Chromosomiques de la Reproduction* (A. Borie and C. Thibeault, eds.), p. 13, Centre International de L'Enfance, Paris (1973).

15. Biggers, J. D., In-vitro fertilization, embryo culture, and embryo transfer in the human, Report to Ethics Advisory Committee, Department of Health, Education and Welfare, p. 41, September 15 (1978).

30

Legal Implications and Regulation of *in Vitro* Fertilization

Barbara F. Katz
Office of General Counsel
Massachusetts Department of Public Health
Boston, Massachusetts 02111

Introduction

Recent events in England, which witnessed the birth of the world's first "test tube" baby, or baby conceived by means of *in vitro* fertilization (IVF), have elevated the serious medical, ethical, and legal issues surrounding this biomedical advance into the public forum.[1] Questions arise such as whether IVF is a nonhuman form of reproduction and is therefore immoral as a dehumanizing process; whether IVF is unethical and illegal experimentation with human beings; whether the state of science involved in IVF has not reached the point to warrant the participation of human beings; whether the potential danger of IVF children being born with physical abnormalities can be resolved; whether the law will create obstacles to the development of the process and/or to the individuals involved in it.

Resolution of most of the above problems is beyond the scope of this chapter. However, an examination of the role of the legal system in the IVF controversy will be undertaken, although many more questions will be raised than solved.

Background: The IVF Process

In order to understand the legal implications of IVF, it is first necessary to briefly summarize what the process itself involves, and the potential practical applications of that process.

The initial step in undertaking an IVF procedure is for the woman to be treated with hormones to stimulate maturation of eggs in the ovary. To locate the ovary, a laparoscope is inserted through an incision in the abdominal wall. Under direct vision, a needle is then inserted into the ovary to draw out several eggs. The eggs are placed in a dish containing blood serum and nutrients, to which sperm is added for fertilization. Once an egg is fertilized by one of the spermatozoa, it is then transferred to another dish of blood serum and sustaining nutrients. For the next three to six days, the fertilized egg divides, creating a cluster of cells called a blastocyst, or "conceptus." After the woman receives further hormone treatment to prepare the uterine lining, the conceptus is placed in the uterus, where it attaches to the wall and, ideally, normal embryo development proceeds. [2]

There are still several outstanding medical problems facing the evolution of IVF into generally accepted ordinary medical practice. At the present time it is considered an experimental technique, without assurances that an IVF conceptus will survive to term or that IVF infants will not face a heightened risk of having birth defects.

Thus, for example, only a small fraction of the eggs removed from the woman are at present able to be fertilized and grow, due to the difficulty of finding a culture medium that imitates the environment of the mother's body. [3] In addition, implantation can only occur during a short portion of the menstrual cycle, so that the timing of the transplantation creates a significant problem. [4] Finally, it may be possible that the unavoidable manipulation of the eggs and the conceptus during the IVF procedure may lead to severe birth defects in IVF infants. [5]

Numerous scenarios for the practical application of the IVF process are feasible, each step becoming progressively more controversial. First is the use currently envisioned, with a woman who is infertile by virtue of blocked or missing Fallopian tubes achieving pregnancy by having one of her own eggs fertilized in the laboratory with her husband's sperm, and thereafter having the conceptus implanted in her uterus.

Second, a woman capable of carrying a fetus to term but unable to produce normal egg cells will have the fertilized egg of another woman implanted in her uterus. There are several options here. In one variation, the egg is fertilized *in vivo*, in the uterus of the donor, via artificial insemination with the sperm of the recipient's husband, surgically removed from the donor's Fallopian tube at the appropriate moment, and then immediately implanted into the recipient's uterus. Another variation involves the *in vitro* fertilization of the donor's egg with the sperm of the recipient's husband, with subsequent implantation into the recipient's uterus.

Third, a woman with healthy tubes and ovaries, but with a condition that might be dangerously or even fatally aggravated by a pregnancy, has her egg, whether fertilized *in vitro* or *in vivo*, implanted into a second woman, who carries the fetus to term as a personal favor or in return for monetary compensation, after which she gives the baby back to its genetic mother.

Fourth, a woman who wants children, but for personal reasons, such as a desire not to disrupt career advancement, does not want to go through a pregnancy, arranges for a "surrogate mother" or "hostess" as described above.

Fifth, and certainly not within the foreseeable future, but a potential eventual application of the techniques involved nevertheless, women in general are able to select prepackaged embryos with clearly specified characteristics for implantation. This is not considered by all to be as farfetched as it sounds. The late Dr. H. J. Muller, Nobel Prize winner in physiology and medicine, seriously advocated that prospective parents forgo egotistical desires to reproduce their own genetic characteristics and, instead, that they help improve the human race by constructing their children from the "best" available egg and sperm cells. Dr. E. S. E. Hafez, an experimental biologist who has done animal research on embryo freezing, believes that there will come a time when parents will be able to select from one-day-old frozen embryos, guaranteed free of all genetic defects, with sex, eye color, probable IQ, and other traits described in detail on the label.[6]

FAMILY LAW RAMIFICATIONS OF IVF

IVF raises few issues of family law that artificial insemination has not already raised.[7] Therefore, a review of the judicial and legislative treatment of family law issues arising from the use of artificial insemination will help illuminate the probable treatment by the law of similar issues in the IVF area.

There are two types of artificial insemination. Homologous insemination (AIH) involves the injection by instrument of the husband's sperm into the woman's reproductive tract to induce pregnancy. Alternatively, heterologous insemination (AID) involves semen from one or more donors.[8] In general, the legal response to issues raised by artificial insemination has been inconsistent, and even contradictory at times, thereby making prediction extremely difficult.

There are very few legal problems with AIH, since the resulting child is the biological offspring of both husband and wife. On the other hand, AID presents a variety of legal problems. Early cases in the United States, England, and Canada held that the practice of AID was equivalent to adultery.[9] Such a holding provides grounds for divorce as well as the basis for possible criminal prosecution.[10] This issue generally arises as a defense in a divorce or support action.[12]

A modern trend may be discerned in cases such as *MacLennan v. MacLennan*.[13] This case considered the adultery question at length and presented a well-reasoned decision more in line with modern thinking. In an action for divorce brought by the husband on grounds of adultery, the wife claimed that her child was born as the result of AID, to which her husband had never consented. The court defined "adultery" as follows:

1. For adultery to be committed there must be the two parties physically present and engaging in the sexual act at the same time.
2. To constitute the sexual act there must be an act of union involving some degree of penetration of the female organ by the male organ.
3. It is not a necessary concomitant of adultery that male seed should be deposited in the female's ovum.
4. The placing of the male seed in the female ovum need not necessarily result from the sexual act, and if it does not, but is placed there by some other means there is no sexual intercourse.[14]

Under this definition of adultery, it seems unlikely that a present-day court would find that AID or IVF constituted adultery.

A related crucial question is the status of the child conceived as the result of AID. In those cases in which AID was held an adulterous act, the resulting child was found to be illegitimate. The question of illegitimacy has also arisen apart from any discussion of adultery. Indeed, many courts have held that a child born by AID is illegitimate without giving any regard to whether or not the husband gave his consent. [15]

However, recent cases are increasingly finding that an AID child is legitimate offspring of the husband. [16] For example, in *People v. Sorenson*, [17] a criminal case, the defendant, by written agreement, had consented to the artificial insemination of his wife. After a subsequent divorce, the wife, due to illness, became unable to support the child, and the district attorney of her county demanded child support payments under Section 270 of the Penal Code from the defendant. In the ensuing criminal prosecution for nonsupport, the defendant pleaded that he was not the father of the child.

The California Supreme Court held that the defendant was the lawful father of the child born to his former wife, that the child was conceived by artificial insemination to which the defendant had consented, and that his conduct carried with it an obligation of support within the applicable statutory meaning. [18] The court stated that the term father must be broadly interpreted. It should not for these purposes be limited to the biological or natural father, as those terms are generally understood, but rather tied to an evaluation of whether the legal relationship of father and child exists. Paternity, then, is established beyond a reasonable doubt when it is shown that a husband, unable to accomplish his objective of begetting a child, purchases semen from a donor and proceeds to use it to inseminate his wife. [19] The court said that, although both legitimate and illegitimate minors have a statutory right to support from their parents, "no valid public purpose is served by stigmatizing an artificially conceived child as illegitimate." [20] The court therefore held the defendant to be the lawful father of the child, liable for his support. It is possible that a similar analysis will be applied to the situation of an IVF child.

However, the mere adoption by some modern courts of a liberal interpretation does not remove artificial insemination and related reproductive technology from the context of present adultery and legitimacy laws. Yet, since these reproductive techniques do not belong in the "immoral" category of actions that these concepts are meant to address, it seems inappropriate to consider them as such. While there are dangers associated with these reproductive methods, they are substantially different from those involved in adultery and illegitimacy. Nevertheless, at the present time, these broad questions must be answered by the legislature.

A number of states have passed legislation to deal with artificial insemination, although the legislative response has been minimal. [21] In 1964 Georgia became the first state to pass a statute legitimizing children conceived by artificial insemination if both husband and wife consent in writing. [22] At least seven other states have legislation that deals with at least some aspect of the issue. [23] However, it is unlikely that any of these statutes would be interpreted to specifically apply to IVF.

To the extent that IVF does not differ significantly from artificial insemination, it is likely that family law will treat the two similarly. Legal clarification of artificial insemination thus seems a necessary prerequisite for regulating IVF.

When the husband provides the sperm and the wife the egg, the legal situation is analogous to that of homologous artificial insemination since the offspring would be the biological child of both husband and wife. When a donor provides the sperm used in the IVF process, the situation is similar to heterologous artificial insemination. Accordingly, the same laws and legal principles would most likely apply and govern the rights of the parties.

The situation becomes more complex if the egg also comes from a donor, or if the egg comes from a donor and the sperm from the husband. This would be the first time that there was any uncertainty concerning the identity of the mother of a child when it is born, since birth itself has generally been considered conclusive proof of motherhood. When the egg comes from another woman, is fertilized and implanted into the wife's uterus, the issue is whether providing the egg or carrying and giving birth to the child establishes that a particular woman is the mother. In order to avoid the problems that have been raised with AID, it may be necessary to have a state statute that provides that the wife will be deemed to be the mother of the child for all legal purposes. In general, though, it may be said that the situation is so similar to AID that the same rules and principles would apply.

The most difficult situation is that in which the wife's egg is fertilized by the husband's sperm *in vitro* and the embryo is implanted into another woman, or "surrogate" who carries the child for the remainder of the pregnancy and then gives birth to it. The child would be the biological offspring of the woman who provided the egg but gestationally the offspring of the "surrogate" who bore and delivered it.

Various legal problems would be raised by such a procedure. For example, if the "surrogate" were being paid for her services, what would happen if payments were missed? Would the "surrogate" then be able to claim the child as her own? Could the "surrogate" decide to abort in that situation? May the couple place restrictions on the "surrogate" concerning such things as diet, drugs taken, activities engaged in, physician? Could the couple require the "surrogate" to undergo amniocentesis? If a defect becomes known, could the couple require her to have an abortion? Could the couple decline to accept the child at birth? Could the "surrogate" decide to abort for health or any other reasons? What would happen if the "surrogate" refused to release the child to the couple at birth? Under what circumstances might the "surrogate" be left with the child and be liable for its support? What are the rights of the child, if any, against the "surrogate," such as right to support or right to inherit?

These issues indicate the necessity of, at the very least, having a clearly drawn contract that anticipates some of these problems and specifies the rights and duties of each of the parties involved. Nevertheless, it is possible that a court would not specifically enforce certain provisions of such a document as being against public policy, such as a clause requiring the "surrogate" to have an abortion if a defect in the fetus becomes known, if she decides at the time that she does not wish to undergo the procedure. In order to duly protect all interests, it may be the wisest course to enact a state statute that would indicate the status and responsiblities of each involved party.

The variance from the common conception of parenthood in these processes are similar to those of adoption. The couple who raise the child, care for him, and are legally responsible for him are his legitimate parents. As in other similar situations, the stigma of illegitimacy should not attach to the child. It is important that court challenges similar to those in AID cases be avoided with IVF. Accordingly, state statutory action is critical to legitimize children born of IVF, thereby resolving the basic family law problems. This is expecially important since IVF will eventually allow a departure from the traditional family concept even greater than artificial insemination, and the law must be ready to respond.

POTENTIAL CIVIL AND CRIMINAL LIABILITY

Liability for Birth of Defective Child

There is an as yet unknown possibility that IVF infants will be born with defects that could be attributable to the use of the reproductive technique.[24] There are several theories of tort law that could be brought to bear on the resolution of the problem of compensation for these injuries.

PRENATAL OR PRECONCEPTION INJURY. Although it is contrary to early tort law,[25] every state currently allows recovery in tort for prenatal injuries.[26] A number of recent cases also permit recovery for preconception infliction of personal injuries.[27] An obvious prerequisite to this type of action is the live birth of the child.[28]

The earliest cases in this area established viability as the test for recovery.[29] This seemed to be based on the evidence problem of proving that the defendant's actions actually caused the complained-of injuries. However, the modern trend is to permit recovery without regard to whether the infant was viable at the time of injury.[30] Yet the plaintiff is still faced with the difficult chore of proving the applicable standard of care and proximate cause.[31]

WRONGFUL LIFE. A severely defective IVF child could bring a "wrongful life" suit against his parents and/or physician for giving him life. However, recovery upon such grounds is usually denied.[32] Although a few cases have espoused this theory, as yet it has rarely been recognized as a viable basis for recovery.[33] Indeed, the wrongful life concept has been judicially accepted in only one decision.[34] It seems that the basic reason for denying recovery in this type of case is the impossibility of measuring damages. Such determination would require the court to make a value judgment that no life at all is better than a life with handicaps, which is a calculation the courts are reluctant to undertake.

There is some indication that the "wrongful life" theory may be on the verge of legal acceptance. This possibility has important implications for birth defects following the use of IVF. However, the plaintiff would still be faced with the difficult task of proving that such defect was caused by the use of the reproductive technology.

WRONGFUL BIRTH. Another line of cases that has possible ramifications for children born with defects as a result of IVF is that in which the woman who gave birth to the child brings the legal action. Cases falling into this category include a suit by the parents of a deformed child against the mother's physician for his failure to diagnose rubella during the course of the pregnancy,[35] a malpractice claim against a doctor for not diagnosing a pregnancy in time for an abortion,[36] an action by a woman against a pharmacist for mistakenly filling a prescription for the contraceptive Norinyl with a different drug, Nardil,[37] and a case brought by a woman against a physician who orally agreed to sterilize her husband so as to prevent procreation, the failure of which led to the birth of a third retarded child.[38]

In general, usual standards of medical malpractice would be applicable in this area. However, as discussed in reference to the wrongful life suit, the problem of proving causation is extremely complicated. A certain number of conventional births result in defective infants, and it might be difficult to prove that the IVF process itself was responsible. If the parents have been fully informed in advance of the procedure and the risks involved and have consented in an informed manner to the procedure in the face of those risks, the parents' action would be even harder to prove.[39]

Liability for "Termination" of IVF Conceptus

It is certain that a significant number of conceptus "terminations" will be attributable to the use of IVF. What are the potential legal ramifications of this occurrence?

CRIMINAL LAW.   Under present technology IVF involves the deliberate fertilization of a human egg, and in the course of the process the necessary "termination" of fertilized eggs.[40] The termination may occur a few moments after conception or a number of days thereafter. It may occur intentionally, by mistake, or by negligence. However, if work in the area of IVF is to continue, a large number of conceptuses will be destroyed. What are the criminal law implications of this?[41]

Infanticide has been defined as the killing of an infant after its birth.[42] It is the felonious taking of the life of a newborn child, which constitutes the crime of murder.[43] The crucial element in a case of infanticide is the birth of the child.

Determining exactly what constitutes the "birth" of a child has, however, been something of a problem. Precisely defining infanticide and accurately identifying the elements of that crime have significance when contemplating the criminal nature, if any, of the termination of an IVF conceptus, for the IVF conceptus *begins* life apart from the body of the mother. Yet it cannot be literally said that the conceptus has been "born alive." Nevertheless, the conceptus is living apart from the body of the mother and in this sense is somewhat analogous to a viable child.

A viable child is in essence a fetus that is capable of an existence independent of the mother. In the case of an IVF conceptus, "viability" occurs at conception. It is not only capable of an independent existence but maintains an independent existence. Yet it is still a fetus because it has never been "born alive."

However, in the final analysis, it is unlikely that the termination of an IVF conceptus would fall within a homicide statute. Such a statute requires the killing of a *person*.[44] It would be a strained interpretation for any court to consider an IVF conceptus to be a person within the meaning of these statutory schemes.

The viability analogy is also of importance due to the United States Supreme Court decision in *Roe v. Wade*.[45] In that case, the Court held unconstitutional statutes proscribing abortion at any stage of gestation. Basing its decision on the mother's right to privacy, the Court explained that the statutes, in restricting a woman's right to terminate her pregnancy at certain stages, violated the due process clause of the Fourteenth Amendment. The Court acknowledged, however, that the mother's right to terminate her pregnancy was not an unqualified right.[46] The decision differentiated the extent of this right during the three gestational trimesters. The decision to terminate the fetus during the first trimester must be solely that of the mother and her physician.[47] During the second trimester, a state may regulate the abortion procedure only to ensure maternal health.[48] During the final trimester of gestation, a state may protect fetal life by prohibiting abortion, except where it is necessary to preserve the health of the mother.[49] Thus the Supreme Court held that the state's police power could include a compelling interest in the protection of fetal life at the stage of viability, such that the state could regulate and even pass a statute proscribing the destruction of a fetus at that stage.

Yet the "viability" of an IVF conceptus does not encompass the concept or interest that the Supreme Court sought to protect in *Roe*. Thus it does not seem likely that termination of an IVF conceptus would be criminally prohibited by an abortion statute.

CIVIL LAW. What are the implications for potential civil liability for the termination of an IVF conceptus?[50]

First is the situation in which the researcher destroys the conceptus with the consent of the parents. As long as the parents have received sufficient information to give their informed consent to the termination, the parents would not be able to successfully sue the researcher.[51] It would also seem that there should not be any liability by the researcher or the consenting parents to the conceptus's estate for his termination. Considering the situation as analogous to abortion, it may be argued that, since the voluntary destruction of an *implanted* conceptus is abortion, the termination of an *unimplanted* conceptus may also come within the same "right of privacy" of the mother, again assuming that the viability distinction in *Roe*, discussed earlier, is not applicable here. Similarly, an analogy may be made to the use by a woman of an intrauterine device (IUD) for birth control purposes. While one difference is that the situation of an IUD takes place within a woman's body, it is nevertheless similar in that it deals with a fertilized egg that is voluntarily terminated by the woman by not permitting implantation. Both abortion, within certain legal limits, and the use of an IUD are rights that a woman may exercise without liability to the "entity" terminated in this manner. Accordingly, by analogy, it seems a logical extension to maintain that the parents and researchers should not face civil liability for the voluntary termination of the IVF conceptus.

However, there are situations in which the termination of the IVF conceptus may lead to civil liability. Such liability may theoretically be imposed on the researcher for the negligent termination of the conceptus. However, in a practical sense, a suit of this type would be extremely difficult for the plaintiffs to prove. Because of the experimental nature of the IVF procedure at the present time, a major obstacle to overcome would be the establishment of a standard of care by which to judge the actions of the physician.[52] In addition, it would also be difficult to establish the necessary causal link between the researcher's actions and the termination of the conceptus.[53]

Finally, liability to the parents may be possible for the intentional and non-consented-to termination of the IVF conceptus by another party. While the legal status of the preimplantation conceptus is unclear, it may be appropriate to consider it the "property" of the couple. Thus damages could still be awarded for its destruction if it were found to have occurred as the result of an intentional tort. Such a damage award would indicate that, although the conceptus is not a legal "person," its potential for development into a human being gives it a value greater than other tissue.

Alternatively, it may be possible to bring an action for wrongful death. Every state permits recovery if the child, injured as a result of some individual's wrongdoing, does not survive to bring his own action.[54] Neither the difficulty in determining damages nor the problem of proving causation has been considered sufficient to bar this particular type of action.

However, there is still disagreement among jurisdictions concerning whether or not a live birth is required in order to maintain a wrongful death action. Several states require a live birth, asserting that there has been no harm to a legally recognized "person" until the fetus is born alive.[55] Nevertheless, the modern trend is to permit an action for the wrongful death of a viable fetus regardless of whether it is born alive.[56]

However, even with the modern trend toward permitting recovery for the wrongful death of a viable fetus, it is still unlikely that parents of an IVF conceptus could make successful use of this cause of action, since the probable termination point of the conceptus would be prior to implantation, which at the present time is certainly before the stage of viability, as that term is used in wrongful death actions.

Finally, the parents could bring a suit for the intentional infliction of emotional distress. However, this type of action is unlikely to be successful, because of the difficulty in proving the "outrageousness" of the other party's conduct,[57] and in demonstrating that the parents have suffered harm from the emotional distress caused by the termination of their conceptus.[58]

Yet there has already been a successful IVF lawsuit based somewhat on a cause of action of this type recently decided in New York. John and Doris Del Zio brought a $1.5 million damage suit against Manhattan's Columbia–Presbyterian Medical Center and its chief of obstetrics and gynecology, Dr. Raymond Vande Wiele.[59] Despite several operations, Mrs. Del Zio had apparently been unable to become pregnant because her tubes had been blocked and partially destroyed by disease. Eggs were obtained from Mrs. Del Zio and bathed in follicular fluid that contained bits of tubal mucosa. They were exposed to Mr. Del Zio's sperm in a culture medium by Dr. Landrum Shettles.[60] However, prior to determination of whether the eggs had been fertilized, Dr. Vande Wiele destroyed the culture, contending that the procedure was risky, that an IVF child might be born with severe defects, that Dr. Shettles lacked the skills to undertake it, and that it had not been approved by the hospital's committee on human experimentation.[61] Mrs. Del Zio claimed that terminating the procedure without her consent and that of her husband denied them their last opportunity to have a child, damaged her both physically and psychologically, upset her sex life, and jeopardized her marriage.[62] A trial court awarded Mrs. Del Zio $50,000 in damages for emotional distress and awarded her husband nominal damages.[63]

GOVERNMENT REGULATION OF IVF

IVF is at present an experimental technique. Accordingly, current government regulations that govern human experimentation would also be applicable to research in the IVF area. However, there is a further complicating factor with IVF research that would seem to call for additional regulations being developed to prevent the unnecessarily hazardous use of the procedure.

Since the intent of the IVF procedure is to produce a healthy child, it would seem that the state may properly develop guidelines to protect the preimplantation conceptus. The government, under the mantle of its *parens patriae* power, may regulate a technology for the benefit of those affected by it. The structure of such a regulatory mechanism would be mandated by the concern for maternal and conceptus health, safety, and welfare.[64]

Risk to the IVF conceptus is not easily quantified. However, it seems clear that a certain number will be destroyed and that some will develop into defective children. The risks in the procedure become critical in the government's ultimate review and protection of the best interests of the conceptus.[65] Presumably the state could act to protect these interests and regulate the use of IVF.

Current HEW Regulations

Proposed regulations for the protection of human subjects to be generally applicable to all HEW grant-supported activity were published in November 1973.[66] This draft document proposed a moratorium on IVF research "until the safety of the technique has been demonstrated as far as possible in sub-human primates, and the responsibilities of donor and recipient 'parents' and of research institutions and personnel have been established."[67] It also recommended that all IVF proposals be approved by one of the proposed Ethical Review Boards (the current Ethics Advisory Boards).[68] However, the draft did not provide specific standards by which the board's decision was to be guided.[69]

A revised draft was published in August 1974,[70] with final regulations being promulgated in 1975.[71] The rules again do not provide specific regulations governing research with unimplanted IVF conceptuses. The document also does not establish the originally proposed moratorium on IVF research. It does follow the earlier drafts in leaving the resolution of issues in the IVF area to the Ethics Advisory Board.[72] It was noted that experimentation on implanted IVF conceptuses would be governed by the fetal research regulations.[73]

However, these fetal research regulations do not *per se* apply to IVF research, since the rules define fetus as "the product of conception *from the time of implantation*."[74] Thus, since many of the problems that are unique to IVF result from events occurring prior to implantation, current HEW regulations are of limited usefulness.[75]

It should also be noted that research on children born by means of IVF would be subject to the current regulations on experimentation with children. Again, these are of limited usefulness in considering the problems unique to IVF research.

State Laws

The only statutes states have that, by analogy, are closest to the IVF area are those that deal with fetal research. At least 15 states have such legislation.[76]

However, it is unlikely that any of these statutes have any effect on the IVF area, other than to regulate research with the implanted IVF conceptus, as is the case with the federal fetal research regulations. None of the state laws were drawn with regulation of IVF research in mind, as the issue was not then before the state legislatures. The only way in which these laws could be applicable would be if, for this purpose, the term *fetus* or the related term used in a particular statute were interpreted to include an *unimplanted* IVF conceptus, and such an interpretation would be a highly unusual and strained construction of the statutory language.

Possible Restrictions on Government Regulation of IVF

The state or federal government, in the exercise of its police power, may regulate the conduct of research with humans, and limit an individual's ability to participate in such an undertaking. Beyond that, it seems likely that, under certain circumstances, this regulation could extend to the point of prohibiting certain hazardous experimentation outright.[77] However, the constitutionality of any attempted governmental regulation of IVF is complicated by the fact that it involves a reproductive technology.

Any statute that impinges upon a "fundamental right" will be subject to the "strict scrutiny" test.[78] In essence, the burden will be on the state to show that the statute is necessary to protect a compelling state interest. There are two parts to this test. The state must establish a compelling interest that is advanced by the statute. In addition, the state must show that the statute is necessary to advance its interest and that there are no less restrictive alternatives. Under existing norms and case law, this burden of proof would be extremely difficult to satisfy.[79]

However, if it is determined that no "fundamental interest" is involved, the statute will be upheld if it is rationally related to a legitimate state objective. The state has broad powers in this regard, termed "police powers," enabling it to enact legislation for the health and general welfare of its citizens. This test, called the "rational basis" test, is relatively easy for the state to meet.

For this reason it is important for the person challenging a governmental regulation to show that a "fundamental right" is involved. Exactly what constitutes a fundamental right has been a source of continuing controversy. Basically, it is a matter for judicial interpretation and is therefore subject to change.[80]

In dealing with reproductive matters, the issue concerns the right to marry,[81] the right of privacy,[82] and the right to procreate.[83] Recent cases recognize the principle that marital and procreation decisions fall within a constitutionally protected zone of privacy.[84] As early as 1941 the United States Supreme Court stated that a person possesses the basic right to have offspring.[85] The Court in *Griswold v. Connecticut*[86] declared that the right to marital privacy was fundamental. In *Eisenstadt v. Baird*,[87] the Court struck down a statute that prohibited the dispensation of contraceptives to single women as violating a fundamental right of reproductive privacy.

The more recent case of *Roe v. Wade*,[88] discussed earlier, also dealt with a fundamental right to reproductive privacy. Such a right could be argued to include the right to reproduce by means of IVF. It is not clear, however, that the Court would extend *Roe* that far. In that case, the Court balanced fetal versus maternal rights. It found that the interests of the mother outweighed the interests of the fetus during the early stages of pregnancy. Yet this should not be interpreted to mean that the fetus, or in the case of IVF, the conceptus, is totally without interests that may deserve some measure of protection.[89] If the state's interest in protecting the IVF conceptus outweighs the mother's privacy right in having an IVF child, governmental regulation of IVF would be constitutionally valid. There may be sufficient state interest in preventing injury to a conceptus or fetus to justify regulation of a practice that may result in the birth of a defective child. The genetic control aspects of the procedure, in contrast to that involved in most other medical undertakings, may warrant added societal concern. Thus the state would seem to have an interest sufficient to justify regulating IVF because of these dangers and ethical issues raised by its use. Beyond that, since IVF experimentation occurs outside the woman's body, her right of privacy is less than that involved in abortion and diminishes in comparison to the state's interest.[90] Thus *Roe* should not present an overwhelming obstacle to governmental regulation of IVF.

Additionally, there are other legal precedents that uphold government restrictions on reproductive rights, indicating that it is possible to find certain state interests considered weighty enough to form a valid basis for some regulation or limitations on reproduction.[91] For example, in *Buck v. Bell*,[92] the Supreme Court upheld a state statute authorizing sterilization of inmates committed to state institutions who were found to have a "hereditary" form of insanity or imbecility.[93] Many states still have some form of compulsory sterilization legislation,[94] and courts typically uphold such statutes.[95]

Although this case was decided prior to those cases discussed above that have increasingly recognized the right to reproduce as being fundamental in nature, its underlying reasoning nevertheless remains as precedent for the view that it may be possible for a state to demonstrate a compelling justification to warrant some form of mandatory restriction on reproduction. It is possible that such justification may be found in society's interest in safeguarding the health and welfare of its citizens.[96]

Thus, given the hazards posed by IVF research to the general welfare, it would seem that a compelling state interest could be established for control. It should be noted, however, that if this interest is found to be sufficient to sustain such regulation, this would involve government regulation of an area of individual decision making in which the trend has been to recognize the claim of personal privacy. Accordingly, it does not appear that there would be a sufficient state interest to uphold an outright ban of the procedure.

Conclusion

In line with the above, the Ethics Advisory Board, discussed earlier, has recently completed its report and recommendations concerning IVF and has presented it to the secretary of Health, Education and Welfare.[97] Briefly, the major conclusions of the board are as follows:

1. The human embryo is entitled to profound respect, but this respect does not necessarily encompass the full legal and moral rights attributed to persons.
2. A broad prohibition of research involving human *in vitro* fertilization is neither justified nor wise.
3. The conduct of research involving human *in vitro* fertilization designed to establish the safety and effectiveness of the procedure is ethically acceptable under the following conditions: (a) the research complies with all appropriate provisions of the regulations governing research with human subjects; (b) the research is designed primarily to obtain important scientific information not reasonably attainable by other means; (c) the donors have given specific informed consent to the research; (d) no embryos will be sustained *in vitro* longer than 14 days after fertilization; (e) all interested parties and the general public will be advised if evidence begins to show that the procedure entails risks of abnormal offspring higher than those associated with natural human reproduction; (f) if the research involves implantation following human *in vitro* fertilization, the donors must be a lawfully married couple, with the embryo thereafter being implanted in the wife.
4. If the above conditions are met, it is acceptable from an ethical standpoint for the Department of Health, Education and Welfare (DHEW) to support or conduct research involving human *in vitro* fertilization and embryo transfer. However, the board did not address the question of the level of funding, if any, such research should receive.
5. DHEW should consider support of more animal research in order to assess the risks to both mother and offspring associated with the procedure.
6. DHEW should take the initiative in collecting, analyzing, and disseminating data from both research and clinical practice involving *in vitro* fertilization throughout the world.
7. DHEW should encourage the development of a Uniform Model Law to clarify the rights and responsibilities of donors, recipients, offspring, and other participants in the process, and should consider funding this drafting project.

Thus the board gives what could be considered guarded approval. At the present time there is no human IVF research being conducted in the United States. The board had before it one research application from Vanderbilt University that proposed studying chromosomal abnormalities but would not involve embryo transfer. However, the board referred the application back to the National Institutes of Health for further review in light of the board's report.

The board's recommendations must go through the entire federal rule-making process so as to amend the current human experimentation regulations. It is unclear what the outcome will be or what the final regulations will look like. However, it seems likely that the ultimate determination will be that it is ethical to go forward with human *in vitro* fertilization research. I believe this is as it should be, since it reflects the views of the majority of Americans, who in recent Harris and Gallup polls indicated that they are in favor of human *in vitro* fertilization.

REFERENCES AND NOTES

1. See, e.g., All about that baby, *Newsweek*, p. 66, Aug. 7 (1978); The first test-tube baby, *Time*, p. 58, July 31 (1978). According to news reports, the baby girl, born by Caesarean section, is in normal health. See, generally, In vitro fertilization: Four commentaries, *Hastings Cent. Rep.* **8**:7 (1978).
2. See, generally, Edwards, R. G., Bavister, B. D., and Steptoe, P. L., Early stages of fertilization in vitro of human oocytes matured in vivo, *Nature* **221**:632 (1969); Shettles, L. B., Human blastocyst growth in vitro in ovulation cervical mucus, *Nature* **229**:343 (1971); Steptoe, P. L., and Edwards, R. G., Laparoscopic recovery of preovulatory human oocytes after priming the ovaries with gonadotrophins, *Lancet* **1**:683 (1970).
3. Kass, L., Making babies—The new biology and the "old" morality, *Public Int.* **26**:18, 23 (1972); Soupart, P. J., and Morgenstern, L., Human sperm capacitation and in vitro fertilization, *Fertil. Steril.* **24**:462 (1973).
4. See Edwards, R., and Sharpe, D., Social values and research in human embryology, *Nature* **231**:87 (1974); Reilly, P., In vitro fertilization—A legal perspective, in: *Genetics and the Law* (G. Annas and A. Milunsky, eds.), pp. 359, 364 (1975).
5. See, e.g., Kass, *supra* note 3, at 27; Kass, L., Babies by means of in vitro fertilization: Unethical experiments on the unborn?, *N. Engl. J. Med.* **285**:1174 (1971).
6. Rorvik, D., The embryo sweepstakes, *New York Times* Magazine, pp. 17, 50, Sept. 15 (1974).
7. See Grad, F., Legislative responses to the new biology: Limits and possibilities, *UCLA L. Rev.* **15**:480, 501 (1968); Hudock, G., Gene therapy and genetic engineering: Frankenstein is still a myth, but it should be reread periodically, *Indiana L. J.* **48**:533 (1973).
8. E.g., Guttmacher, A., Artifical insemination, *DePaul L. Rev.* **18**:566 (1969); McLaren, Biological aspects of AID, in: *CIBA Foundation, Symposium in Legal and Other Aspects of Artificial Insemination by Donor- (A.I.D.) and Embryo Transfer* (Vol. 3) (1972). See L.v.L., 1 All. E.R. 141 (1949) (English); Kinney, L., Legal issues of the new reproductive technologies, *Calif. St. B. J.*, Nov./Dec.:514 (1977). See also Thies, W., A look to the future: Property rights and the posthumously conceived child, *Trusts and Estates* **110**:922 (1971).
9. E.g., *Orford v. Orford*, 49 Ont. L.R. 15, 58 D.L.R. 251 (1921). See Oakley, M. A., Test tube babies: Proposals for legal regulation of new methods of human conception and prenatal development, *Fam. L. Q.* **8**:385, (1974); Smith, G., Through a test tube darkly: Artificial insemination and the law, *Mich. L. Rev.* **67**:127, 135 (1968). See also *Doornbos v. Doornbos*, 23 U.S.L.W. 2308 (Super. Ct. Cook County, Ill., Dec. 13, 1954), *appeal dismissed on procedural grounds*, 12 Ill. App. 2d 473, 139 N.E. 2d 844 (1956); *Hoch v. Hoch* No. 44-C-8307 (Cir. A. Cook County, Ill., 1945); *Time*, p. 58, Feb. 26 (1945).
10. Hager, J., Artificial insemination: Some practical considerations for effective counseling, *N.C. L. Rev.* **39**:217, 232 (1961).
11. Plosowe, M., The place of law in medico-moral problems: A legal view II, *N.Y.U. L. Rev.* **31**:1238, 1242 (1956).
12. Hager, *supra* note 10, at 233.
13. (1958) Sess. Cas. 105, (1958) Scots L.T.R. 12. See *People v. Sorenson*, 66 Cal. Rptr. 7, 437 P.2d 495 (1968).
14. (1958) Sess. Cas. at 113.
15. E.g., *Gursky v. Gursky*, 39 Misc.2d 1083, 242 N.Y.S.2d 406 (Sup. Ct. 1963).
16. *In re Adoption of Anonymous*, 74 Misc.2d 99, 345 N.Y.S.2d 430 (Surrogate Ct. 1973); *Strnad v. Strnad*, 190 Mis. 786, 78 N.Y.S.2d 390 (Sup. Ct. 1948). See *People ex rel. Abajian v. Dennett*, 15 Misc.2d 260, 134 N.Y.S.2d 178 (Sup. Ct. 1958). See also *Levy v. Louisiana*, 391 U.S. 68 (1968), *rehearing denied* 393

U.S. 898; *Glona v. American Guarantee and Liability Ins. Co.*, 391 U.S. 73 (1968); *Re Estate of Jensen*, 162 N.W. 2d 861 (N.D. 1968); *Stor v. None*, 57 Misc. 2d 342, 291 N.Y.S. 2d 515 (1968); *Green v. Woodard*, 40 Ohio App.2d 101, 69 Ohio Ops.2d 130., 318 N.E.2d 397; C. Boardman, New York Family Law s.116 (Biskind ed.).

17. 66 Cal. Rptr. 7, 437 P.2d 495 (1968).
18. Ibid. at 10, 437 P.2d at 498.
19. Ibid. at 10, 437 P.2d at 498.
20. Ibid. at 13, 437 P.2d at 501.
21. See, generally, Note, Artificial insemination: A legislative remedy, *West St. U.L. Rev.* 3:48 (1975).
22. Ga. Code Ann. ss 74-101.1 (1973).
23. Okla. Stat. Ann. tit. 10, ss 551-53 (Supp. 1974). Kan. Stat. Ann. ss 23-128 to - 130 (1974). Ark. Stat. Ann. ss 61-141(c) (Supp. 1971). Md. Ann. Code art. 43, s 556E (Supp. 1974). N.C. Gen. Stat. s.49A-1 (Supp. 1974). N.Y. Dom. Rel. Law s73 (McKinney Supp. 1975). Cal. Civ. Code ss 195, 216 (West Supp. 1975). Cal. Penal Code s270 (West Supp. 1975).
24. Kass, *supra* note 3, at 28.
25. See, e.g., *Dietrich v. Inhabitants of Northampton*, 138 Mass. 14 (1884).
26. Prosser, W., *Handbook of the Law of Torts* (4th ed.), p. 336 (1971). See *Bonbrest v. Katz*, 65 F. Supp. 138 (D.D.C. 1946). See, generally, Wilson, Fetal experimentation: Legal implications of an ethical conundrum, *Denver L. J.* 53:581 (1976).
27. E.g., *Jorgensen v. Meade Johnson Laboratories Inc.*, 483 F.2d 237 (10th Cir. 1973); *Renslow v. Mennonite Hospital*, 67 Ill.2d 348, 367 N.W.2d 1250 (1977); *Park v. Chessin*, 88 Misc.2d 222, 387 N.Y.S.2d 204 (Sup. Ct. 1976), *aff'd*, 60 App. Div.2d 80, 400 N.Y.S.2d 110 (1977).
28. *Leccese v. McDonough*, 361 Mass. 64, 68, 279 N.E.2d 339, 342 (1972); *Keyes v. Construction Serv., Inc.*, 340 Mass. 633, 165 N.E.2d 912 (1960).
29. E.g., *Bonbrest v. Kotz*, 65 F. Supp. 138 (D.D.C. 1946).
30. E.g., *Smith v. Brennan*, 31 N.J. 353, 157 A.2d 497 (1960); *Daley v. Meier*, 33 Ill. App.2d 218, 178 N.E.2d 691 (1961); *Torigian v. Watertown News Co.*, 352 Mass. 446, 225 N.E.2d 926 (1967); *Kelly v. Gregory*, 282 App. Div. 542, 125 N.Y.S.2d 696 (1953).
31. Note, The "brave new baby" and the law: Fashioning remedies for the victims of in vitro fertilization, *Am. J. L. Med.* 4:319 (1978).
32. E.g., *Williams v. State*, 18 N.Y.2d 481, 223 N.E.2d 343 (1966); *Zepeda v. Zepeda*, 41 Ill. App.2d 240, 190 N.E.2d 849 (1963), *cert. denied*, 379 U.S. 945 (1964); *Nellis v. Chicago Wesley Memorial Hospital*, No. 701-15177 (Cir. Ct. Cook County, Ill., June 18, 1974).
33. See *Custodio v. Bauer*, 251 Cal. App.2d 303 (1967); *Gleitman v. Cosgrove*, 49 N.J. 22, 227 A.2d 689 (1967). See also Brodie, D., The new biology and the prenatal child, *J. Fam. L.* 9:391, 396 (1970); Gordon, The unborn "wrongful life," *N.Y.U. L. Rev.* 38:1078 (1963); Tedeschi, L., On tort liability for "wrongful life," *Israel L. Rev.* 513 (1966); Note, Fetal research: A view from right to left to wrongful birth, *Chi-Kent L. Rev.* 52:133 (1975). For example, the Supreme Court of Alabama recently refused to find a cause of action for "wrongful life" in a suit brought by a deformed child born despite her father's vasectomy. *Elliott v. Brown*, Dkt. No. 77-114 (Ala., Aug. 18, 1978).
34. See *Park v. Chessin*, 60 App. Div.2d 80, 400 N.Y.S.2d 110 (1977).
35. *Jacobs v. Theimer*, 519 S.W.2d 846 (Tex. 1975).
36. *Ziemoa v. Sternberg*, 45 A.D.2d 230, 357 N.Y.S.2d 265 (1974). Contra *Rieck v. Medical Protective Co.*, 64 Wis.2d 514, 219 N.W.2d 242 (1974).
37. *Troppi v. Scarf*, 31 Mich. App. 240, 187 N.W.2d 511 (1971). The case was subsequently settled.
38. *Doerr v. Villate*, 74 Ill. App.2d 332, 220 N.E.2d 767 (1966). See Robertson, J., Civil liability arising from "wrongful birth" following an unsuccessful sterilization operation, *Am. J. L. Med.* 4: 130 (1978).
39. See Note, *Park v. Chessin*: The continuing development of the theory of wrongful life, *Am. J. L. Med.* 4:211 (1978).
40. Kass, *supra* note 3, at 32.
41. Abel, K., The legal implications of ectogenetic research, *Tulsa L. J.* 10:243, 248 (1974).
42. *Black's Law Dictionary* (rev. 4th ed.), p. 917 (1968).
43. Abel, *supra* note 41, at 252. See *Gilpin v. Gilpin*, 94 N.Y.S.2d 706 (Dom. Rel. Ct. 1950).
44. See Annas, G., Glantz, L., and Katz, B., *Informed Consent to Human Experimentation: The Subject's Dilemma*, p. 200, Ballinger, Cambridge (1977).
45. 410 U.S. at 113.

46. Ibid. at 154.
47. Ibid. at 164.
48. Ibid.
49. Ibid.
50. See, generally, Note, *supra* note 31.
51. See, generally, Annas, G., Glantz, L., and Katz, B., *supra* note 44.
52. See Wilson, *supra* note 26, at 637.
53. Ibid. at 638.
54. Prosser, W., *Handbook of the Law of Torts* (4th ed.), p. 237 (1971).
55. *Norman v. Murphy*, 124 Cal. App.2d 95, 268 P.2d 178 (1954); *Drabbels v. Skelly Oil Co.*, 155 Neb. 17, 50 N.W.2d 229 (1951); *Graf v. Taggert*, 43 N.J. 303, 204 A.2d 140 (1964); *Carroll v. Skloff*, 415 Pa. 47, 202 A.2d 9 (1964); *Durrett v. Owens*, 212 Tenn. 614, 371 S.W.2d 433 (1963).
56. E.g., *Simmons v. Howard University*, 323 F. Supp. 529 (D.D.C. 1971); *Chrisafogeorgis v. Brandenberg*, 55 Ill.2d 368, 304 N.W.2d 88 (1973); *Britt v. Sears*, 150 Ind. App. 487, 277 N.E.2d 20 (Ct. App. 1971); *O'Neill v. Morse*, 235 Mich. 130, 188 N.W.2d 785 (1971); *Libbee v. Permanente Clinic*, 1268 Ore. 258, 518 P.2d 636 (1974); *Baldwin v. Butcher*, 184 S.E.2d 428 (W. Va. 1971). See *Eich v. Town of Gulf Shores*, 293 Ala. 95, 300 So.2d 354 (1974); *Porter v. Lassiter*, 91 Ga. App. 712, 87 S.E.2d 100 (1955).
57. See Prosser, W., *Handbook of the Law of Torts* (4th ed.), s 12, p. 56 (1971).
58. Ibid. at 59.
59. The first test tube baby, *Time*, pp. 58, 61, July 31 (1978).
60. Rorvik, *supra* note 6, at 55.
61. The first test tube baby, *supra* note 59.
62. Ibid.
63. Woman awarded $50,000 in suit on test-tube baby, *Boston Globe*, p. 2, Aug. 19 (1978).
64. Reilly, *supra* note 4, at 364.
65. Ibid. at 368.
66. 38 Fed. Reg. 31, 748 (1973).
67. Ibid.
68. Ibid.
69. Ibid. at 31, 743.
70. 39 Fed. Reg. 30, 648 (1974). See Martin, M., Ethical standards for fetal experimentation, *Fordham L. Rev.* 43:547 (1975).
71. 40 Fed. Reg. 33, 526 (1975) (Codified in 45 C.F.R. s 46). See Markey, K., Federal regulation of fetal research: Toward a public policy founded on ethical reasoning, *U. Miami L. Rev.* 31:675, 685 (1977).
72. 39 Fed. Reg. 30, 650 (1974).
73. Ibid. See, generally, Capron, A., The law relating to experimentation with the fetus, in: *Research on the Fetus*, 13-1, The National Commission for the Protection of Human Subjects of Biomedical and Behavioral Research, Appendix, DHEW Pub. No. (05) 76–128 (1975).
74. 40 Fed. Reg. 33, 529 (1975).
75. See Note, *supra* note 31.
76. See Annas, G., Glantz, L., and Katz, B., *supra* note 44, at 206.
77. See Annas, G., Glantz, L., and Katz, B., *supra* note 44, at 50.
78. See e.g., *Dandridge v. Williams*, 397 U.S. 471 (1970); *Shapiro v. Thompson*, 394 U.S. 618 (1969).
79. Note, Governmental control of research in positive eugenics, *J. L. Ref.* 7:615, 620 (1974).
80. E.g., *Prince v. Massachusetts*, 321 U.S. 158 (1944); *Jacobson v. Massachusetts*, 197 U.S. 11 (1905).
81. *Loving v. Virginia*, 388 U.S. 1 (1967).
82. *Griswold v. Connecticut*, 381 U.S. 479 (1965).
83. *Skinner v. Oklahoma*, 316 U.S. 535 (1941).
84. See, e.g., *Eisenstadt v. Baird*, 405 U.S. 438, 452–55 (1972). See also Smith, G., Manipulating the genetic code: Jurisprudential conundrums, *Georgetown L. J.* 64:697, 750 (1976).
85. *Skinner v. Oklahoma*, 316 U.S. 535, 541 (1941).
86. 381 U.S. 479, 495 (1965).
87. 405 U.S. 438 (1972).
88. 410 U.S. 113 (1973).
89. See Note, *supra* note 31; Green, H., Genetic technology: Law and policy for the brave new world, *Indiana L.J.* 48:559, 565 (1973). See, generally, Wilson, J., Fetal experimentation: Rights of the father and questions of personhood, *Villanova L. Rev.* 22:403 (1976–77).

90. See Kass, *supra* note 3, at 32.
91. See Golding, M., and Golding, N., Ethical and value issues in population limitation and distribution in the United States, *Vand. L. Rev.* **24**:495, 512 (1971).
92. 274 U.S. 200 (1927).
93. Ibid. at 207.
94. See Pate, R., and Plant, P., Sterilization of mental defectives, *Cumberland-Sanford L. Rev.* **3**:458 (1972).
95. E.g., *In re Moore*, 221 S.E.2d 307 (N.C. 1976).
96. See Vukowich, W., The dawning of the brave new world—Legal, ethical and social issues of eugenics, *U. Ill. L. F.* **1971**:189, 208 (1971).
97. Ethics Advisory Board, Report and Conclusions: HEW Support of Research Involving Human *In Vitro* Fertilization and Embryo Transfer, Department of Health, Education, and Welfare (May 4, 1979).

Discussion

Dr. J. Lorrin Lau (Johns Hopkins School of Medicine): Briefly, I would like to provide a few statistics which may aid your consideration and discussion of the subject. These are provided by Doctors Howard and Georgeanna Jones, who have retired from the Johns Hopkins Hospital but actively engage in research in Norfolk, Virginia, to correct one of the remaining unsolved problems generated and experienced in their practice.

Why? Why even talk about this subject? There are two reasons. One is that *in vitro* fertilization (IVF) is being practiced in the United States today and elsewhere, for example, Melbourne, Australia, by several physicians besides Steptoe and Edwards.

This second and more important realm is that this provides the only treatment to bypass obstructed tubes. It is unfortunate that the terms *test tube babies* or *fabricated babies* have been used. Instead, the terms *treatment to bypass obstructed tubes* or *assisted fertilization* should be used.

What is the problem? The problem is that fifteen percent of married women are infertile. Of these, twenty percent have obstruction of tubes. In turn, only fifteen percent of these can be helped by surgery, leaving in the United States approximately 500,000 to 600,000 couples without help. The aim is to help these couples have children which are genetically and developmentally their own, so I am not talking about surrogates.

What are the four difficulties?

The first difficulty is obtaining the egg by proper timing, by laparoscopy, and by use of hormones. The chance of success about one in ten.

The second difficulty is accomplishing fertilization. The chance of success of this is also one in ten.

The third is nurturing and maturing the embryo until it is twenty-three days. The chance of success of this is one in ten.

The fourth difficulty is placing and implanting the embryo, which is usually not visible, and the chance of this is one in two. The total chance of success is therefore about one in four hundred.

A fifth difficulty exists, which is funding. In the experience of the Joneses, it is not cost-effective to set up a clinic until funding can be guaranteed for a minimum of three years or a minimum of at least $100,000, which is a very conservative figure.

What are some of the benefits? Benefits are (1) relief of infertility; (2) applications to animal husbandry, i.e., getting better livestock and more food, especially in developing countries; (3) the provision of new information on development and differentiation; (4) the application of knowledge generated to problem areas such as cancer, which in a way is a genetic mutation.

What are some of the objections? The primary objection is that we cannot preclude a damaged child. Even without IVF, three to four percent of children have gross malformations, mental retardation, or genetic disease, and a half percent of all live-borns have some chromosomal disorder. Of aborted material inspected in the first trimester about fifty percent have chromosomal disorders.

The second objection is informed consent from the fetus. But this problem is not unique to IVF, since the problem of informed consent of the fetus exists even in normally conceived children.

The third objection is the inalienable right of couples to have children.

The fourth objection is that these techniques might be used for unmarried women, which once elicited the response, "Well, if unmarried women can't think of ways of procreating more easily, then something must be wrong."

The fifth objection is destruction of unused embryos, a topic covered by Professor Taymor. In most state legislatures the definition of viability has a limit after twenty-six weeks, therefore this may not be a problem.

The last objection is that of interfering with the family structure. This procedure as proposed at Norfolk is conducted only within the framework of marriage and the family.

Ms. BARBARA KATZ: My comment has to do with funding by government of research on IVF. I think it is interesting to note that there was only one proposal before the Ethics Advisory Board requesting approval for Department of Health, Education and Welfare funding. That involved fertilizing the egg in the laboratory and taking a look at chromosomal abnormalities. It didn't involve implantation at all.

The interesting thing is, first, that there was only one proposal before them and, second, that the board didn't approve it. They questioned the scientific validity of it and are sending it back to the National Institutes of Health (NIH) for a review. So at the present time, even with the Board going forward, there is still not any federally funded IVF research going on in the United States at the present time.

MR. JOHN HOAGLUND (Newport News, Virginia): One important question I think should be raised here is whether the material to make babies should be a marketable commodity. I don't believe that has been raised directly. It is a current question, apparently, in the case of your AID material, if frozen semen will soon be available through the mail. I think a situation that sharpens up the issue is when you are dealing with a possibility of IVF and surrogate motherhood. Suppose, for instance, I were unmarried and wanted a child. I could procure an ovum and hire the services of a woman who would be willing to serve as a surrogate mother and have the child herself on this basis.

If the materials for babies are to be regarded purely as marketable commodities, I don't think that there is anything that would rule out this situation. On the other hand, the situation doesn't seem to me to be right. So I think that there are probably some moral issues there that need to be addressed.

PROF. SANFORD KATZ: Yes, I think it is quite so. Maybe we should end the morning by saying this is the International Year of the Child and all over the world people are celebrating the rights of children.

I would just like to leave you with this one comment. In the discussion we just had, I wonder whether the products to make children really tell us more about our conception of children, namely, that the old view of the common law that children were chattel may be what we are all talking about. That the products of the human body that can make children mean that children are indeed chattel.

And we talk about surrogate mothers. I wonder whether the law itself shouldn't be looked into for answers. As Attorney Katz was talking, I was thinking of the surrogate mother as a simple case of bailment, where the owner transfers the personal property of his own to another for a specific purpose. After the purpose is achieved, there is a return of the property to the owner. That is the definition of bailment.

Doesn't that fit the case of the surrogate mother? And what we are talking about is ownership of property, which eventually turns into a youngster.

DR. CHARLES U. LOWE (Department of Health, Education and Welfare, Washington, D.C.): I wonder if I might make a comment and raise a question on the presentation by Attorney Katz.

At the time the original experiments by Steptoe and Edwards were conducted, one of the issues that emerged in Great Britain was whether the Medical Research Council would support the work. To the best of my knowledge, a decision was reached that the procedures undertaken were to be considered part of the health care of the mother and that therefore funding for the work or provision for meeting the costs of the special procedure would be met by the National Health Service and not by any research organization.

Now that is a comfortable position in Great Britain. But when this country was faced with the problem, they decided to sidestep it temporarily. With the success of the procedure in Great Britain, it seems to me there is a stronger argument to be made that this is health care.

I raise this point because the authority of the secretary to control this procedure is limited to two special situations. One is his authority to authorize the support of research in this area. The other is to determine whether or not any of the federally supported health delivery services will in fact compensate a provider for undertaking this practice.

The practice of medicine in this country as I understand it is controlled by the states, not by the federal government. I wonder whether Attorney Katz would be willing to comment on this issue. If this is not research and it is health care, what is the role of the states and where does the authority of the secretary cease?

Ms. KATZ: I think it is premature at this point to start talking about it in terms of nonresearch or not an experimental procedure. The fact that a few infants have been born, and luckily without defects at this point, still leaves a large question for research on the impact of this procedure. Also, I am not sure about whether we could really consider it to be solely just for the health care of the mother or the couple, but rather there are other interests involved: the interests of the conceptus and the eventual fetus.

So I am not sure that totally looking at it as a part of health care solely of the mother (as in treating a disease) would be appropriate. I certainly think it is premature at this point, anyway, to consider that. Again, any regulation that the government could come out with, I would want to make clear, only goes to regulation of federally funded research projects. It doesn't cover any other types of procedures.

PROF. MELVIN L. TAYMOR: I am not sure I would agree that because a procedure only has a certain level of success it necessarily is not health care. We don't gauge our various therapeutic maneuvers as being health care just on the basis of how successful they are. What level would you want to say passes from research to health care?

MS. KATZ: I think that is the whole issue.

PROF. TAYMOR: I think what is being done now by Dr. Steptoe is a mechanical series of steps that to me represent a step in health care. It is not very successful yet but we hope it will be better. There is research that maybe can be done aside from that as well, and I think Dr. Lau has made a very important point as far as the NIH is concerned. In our country, the NIH is for research and shouldn't necessarily be asked at this stage to provide money for care.

DR. LOWE: I think Attorney Katz has said something which is in a sense incorrect. The control in the current regulations does not require that the research be funded by the federal government—if the organization sponsoring research receives any federal funds, then the current regulations governing research apply to all research in that organization. The danger to the organization is that they lose federal funds if they undertake any research which is incongruent with regulations.

So the issue is broader, I think. than you allowed in your comments.

MS. KATZ: Yes.

DR. NEIL A. HOLTZMAN (Johns Hopkins, Baltimore): Dr. Lowe in his first comments raised a question that has probably not received adequate attention at this symposium, and that is the translation of new technologies into standard clinical practice and where one in fact draws the line between research and health care. This is not the first situation in which a government agency has ruled, I think fairly arbitrarily, on the question of what might be research and what might be clinical.

In the situation of trying to determine the efficacy of maternal serum alpha-feto protein screening through pilot programs, an agency that was dispensing funds for clinical projects indicated that this was still research, whereas in general the NIH have indicated that this is not research. The questions of research that come up here and also come up in alpha-feto protein screening, assuming that many of the technical aspects have been solved, are behavioral questions. How women, spouses, and others in society might respond to these new technological innovations. Until the research agencies in particular are prepared to expand their scope to provide investigators who are interested in these very important questions, I don't think we are going to successfully overcome the hurdle of ethically and morally applying the new technologies that are in hand.

PROF. SANFORD KATZ: Dr. Lowe, would you call adoption of children health care?

DR. LOWE: My perception is that it sits at the boundary between the responsibility that the state has for the care of all of its citizens and the rather more well-defined issue of health care. I would judge that this is a socially determined question rather than a professionally determined question.

PROF. KATZ: Suppose what we are talking about here didn't take place in a hospital but took place in a scientific laboratory. Would that be health care, this *in vitro* fertilization?

DR. LOWE: The site at which the application takes place I don't think determines quality.

PROF. KATZ: I am concerned about why you labeled this health care in contrast to something else. Is it because M.D.'s do it? Is it because it is done in a hospital? Is it because there is medication? What makes it health to nonhealth? What makes this transplant different from adoption? Is it because medical people do it, not social workers, not educators?

DR. LOWE: If you asked me to define research, I think I can do that more easily.

PROF. KATZ: No. I am just curious why this *in vitro* fertilization is a health problem for the mother, for the fetus. Who is getting healthy? Or who is getting unhealthy?

DR. LOWE: I do have to retreat into the definition of research to say that what is not research is health care. No hypothesis is being tested. The success of the procedure, for better or for worse, has been demonstrated. There are many examples of procedures which move into the domain of health care with rather less background than this procedure. I suspect I can give you a list of them.

PROF. TAYMOR: I think you are making a rather narrow limitation on health as being ill or not ill.

PROF. KATZ: Or diseased.

PROF. TAYMOR: And we are talking about the whole area. Is it the whole of infertility? Then this is part of an infertility problem. That is why it is health care.

PROF. KATZ: My question is, who is getting healthy from this? Is it the surrogate mother?

PROF. TAYMOR: No, not the surrogate. The husband and the wife are getting healthier from this procedure. They are curing their infertility, curing the emotional tensions that are involved with their infertility. It is health care. And if this is the procedure that is prescribed, you do it.

PROF. KATZ: Is it health care by adopting a child?

PROF. TAYMOR: We don't get involved with that. The social agencies get involved in that.

PROF. KATZ: That's all I wanted to know.

PROF. TAYMOR: If it cures their emotional illness, I suppose it is.

DR. MARK DINE (Pediatrician, Cincinnati, Ohio): Dr. Lowe has injected a very exciting note into this conference. Health care is a very difficult problem in definition in the study of bioethics. The thrust of medical education is the concept of treatment of the whole person. Therefore, family life, being part of a family, is part of health care in terms of creating the healthy family. Therefore, the direct answer to your question "Is adoption part of health care?" is certainly yes, perceived from this viewpoint. But that is absolutely the wrong way to perceive it, of course. Just because physicians think they should do everything doesn't mean everything they do is health care.

PROF. KATZ: That is for sure.

PROF. GEORGE ANNAS: But you said that what is taught in medical school is health care.

DR. DINE: It is the training of physicians, and physicians are created by society to perform a function, and society together with physicians decides what is disease.

PROF. ANNAS: But society hasn't done that yet. Physicians are not created by other physicians. I don't think there is any social concensus on how you define a disease, or on what is and is not "health care." "X" is not health care just because physicians say it is.

DR. DINE: Physicians are created by other physicians in the sense that their basic study, their basic work is understanding anatomy, physiology, and its application to nonphysicians; and they use physicians as their models.

PROF. ANNAS: Physicians are created by physicians? You can't support that.

GENETICS, LAW, AND ENVIRONMENTAL MUTAGENS/TERATOGENS

MODERATOR: GEORGE J. ANNAS

31

Issues in Human Mutagenesis and Teratogenesis

Arthur D. Bloom
College of Physicians and Surgeons
Columbia University
New York, New York 10032
Columbia-Presbyterian Medical Center
New York, New York 10032

There is perhaps no issue more critical for the survival and well-being of our species than the protection of mankind against those physical and chemical agents that damage the genetic material of our somatic and germ cells. In terms of the somatic cell effects of such agents, our concern here is with (1) their teratogenicity, i.e., their effects on the developing embryo or fetus; and (2) their carcinogenicity, i.e., their induction of neoplastic disease. Fully 250,000 children with birth defects are born each year in this country alone, and it is estimated that 25% of our population will develop cancer.[1,2]. Approximately 80–90% of these cancers are thought to be induced by environmental agents, while 35% of birth defects are caused by known genetic or environmental factors.[3] In terms of the germ cell effects of physical and chemical agents, documentation or proof of these in humans has been much more elusive, even when we are studying the effects of a known, potent mutagen to which a sizable human population has been exposed, at high doses. Nonetheless, concern for our species in terms of induced, heritable, deleterious mutations is the reasonable concern of our government and citizenry, since negligence in this area will doubtless lead, in our species as it does in experimental animals, to clinically significant genetic disease in generations to come.

I begin with a summary statement on the interrelationships of carcinogens, teratogens, and mutagens. First, it is now well established that 90% of agents that cause cancer when tested in bacterial assays, such as the Ames test, are mutagenic.[4] Second, the corollary is also, generally speaking true, namely, that compounds known for their ability to cause mutation tend also to be carcinogenic in experimental animal tests. Finally, the data on teratogenicity do not show as high a correlation with mutagenicity and carcinogenicity as mutagenicity and carcinogenicity do with each other. Nonetheless, if an agent is proven mutagenic and/or carcinogenic, it must be viewed with suspicion for its possible teratogenic effects, since all three of these kinds of effects are often brought about by the same basic mechanisms of action. The most generally accepted concept here, as described in more detail in a recent paper by Ames in *Science*,[5] is that chemical agents, at least, tend to exert their effects by damage to DNA.

In a recent issue of the *New Yorker*,[6] there is an amusing but disturbing piece on the concept of "risk–benefit assessment." Under discussion is the problem of the pentachlorophenols, chemicals manufactured to preserve wood and leather against slime molds. It seems that many of the pentachlorophenols are contaminated with the highly toxic dioxins and dibernofurans, and the Science Advisory Board of the U.S. Environmental Protection Agency has been considering how best to eliminate these contaminants. The article includes a discussion of the fact that the toxic contaminants may technically be readily separated from pentachlorophenol (PCP) during the manufacturing process. In the Appendix to the report of the Board are comments from several chemical manufacturers, including one that agreed that the dioxins and dibernofurans could, in fact, be separated off; but then the question would be what to do with the concentrated residual waste. The manufacturer suggests that there is no logical disposal site, and incineration is not feasible. The *New Yorker* concludes, in its usual ironic style: "Reading the proposal, . . . we felt we had never seen the issue of disposing of toxic wastes of a manufactured product put so plainly. Why spend all the money on and assume all the risks of separating and disposing of toxic wastes extracted from a product when the manufacturer has an alternative method that is both simple and inexpensive: to bypass such techniques and spread out the toxic wastes *by selling them right in the product itself?* How is that for creative chemistry?"

While it is, of course, true that we ourselves are responsible for the exposure of the species to such undesirable physical agents as the gamma radiation from atomic bombs and damaging chemical carcinogens as PCPs and the flame-retardant Tris-BP [tris (2,3-dibromopropyl) phosphate] found until recently in children's pajamas, it is clearly true that "natural" carcinogens over which we have little control also exist. The skin carcinogenicity of ultraviolet light is well known, and aflatoxin is a natural carcinogen produced by molds commonly found in a variety of foods (corn, peanuts). The identification of these more or less ubiquitous mutagens and carcinogens is an important current and future item of business for those who would protect the public health. It is clear that we shall not eliminate completely exposures to natural or man-made mutagens/teratogens/carcinogens; it is also clear that we should attempt to minimize such exposure. The problem is how best to do it.

There is now general agreement that screening for mutagens requires a series of tests, using both *in vitro* and *in vivo* systems, across several species.[7] The *in vitro* tests range from the so-called *Salmonella* (his^- → his^+) bacterial test of Ames *et al.*, to the use of yeasts like *Neurospora crassa*, to fruit flies like *Drosophila melanogaster*, to murine and human cells in culture. These and other *in vitro* systems allow for the study of many genetic phenomena: forward and reverse mutation, genetically controlled ability to repair damaged DNA, genetic recombination, chromosomal breakage and rearrangement, etc. When similar *in vivo* tests are used, mainly in rodents, we obtain a profile on the mutagenicity of a given compound and are able to classify it according to whether it is or is not a mutagen, and if it is, how strong a mutagen. We inevitably get into complex questions of dose, of drug interactions, maybe even of enhancement of the mutagenicity or carcinogenicity of a compound in the presence of vitamins (C and E, for example), and, of course, often return to the risk–benefit problem.

Most of us can comfortably do without saccharin and so will avoid its potentially carcinogenic effects. And yet, for diabetics and some others, it is a valuable compound, perhaps worth the risks it may pose. The recent episode at Three Mile Island (TMI), however, raises, in my view, a far more serious set of problems about the relationship between a known mutagen/teratogen/carcinogen, i.e., ionizing radiation, and the benefits and risks to society of the continued proliferation of nuclear reactors as a source of energy in our nation. I shall focus on the problem of radiation exposure in the remainder of this chapter.

Radiation is perhaps the best studied of all mutagens in terms of its effects, particularly in humans; the mechanisms of its action; and its dose–response relationships. We can control its use to a considerable extent and yet we have chosen to do so only modestly. I shall here try to summarize, in particular, the problems of radiation-induced mutation and teratogenesis in man, since these two problems, in addition to the carcinogenic effects of radiation, are very much upon us, and the approach may serve as a model for the problems of regulation of our exposure to chemical mutagens. It is clear that there exists at TMI and its environs, and perhaps across the country, an atmosphere of anxiety and doubt about nuclear reactors and radiation exposure generally. Every time an infant is born with a congenital malformation around Harrisburg, parents will question the etiology. Every time someone in the vicinity develops a cancer, the question of it being radiation-induced will be raised.

In terms of the direct teratogenic effects of radiation, most of what we know comes from animal experiments.[8] If the irradiation occurs preimplantation, depending on the dose, we may get death or failure of implantation; if the conceptus survives, we are likely to get little damage. If the irradiation occurs postimplantation, during organogenesis, we will obtain at high doses (at or above the LD_{50} for the species) either death of the conceptus, or slow growth persisting into postnatal life, or specific abnormalities of the central nervous system, eye, skeleton, or head. If the exposure dose is low (down to 5 rad in the mouse, 5–10 in the rat), these same kinds of abnormalities are seen, though at lower frequencies.

The human experience is largely that obtained from the Hiroshima and Nagasaki studies.[9] Above 50 rads, of largely gamma (Nagasaki), or gamma and neutron (Hiroshima) radiation, some mental retardation and "microcephaly" were described. Persistent reduction in body size has also been seen in these *in vitro* exposed, when studied at 17 years of age.

Our concern about intrauterine radiation exposure is mainly confined to early-in-gestation "embryonic" exposure, since radiation sensitivity in terms of birth defects is considerably reduced after organogenesis and in particular after eight weeks of gestation.

Overall, the figure recommended by the United Nations Scientific Committee on the Effects of Atomic Radiation (UNSCEAR) in its 1977 Report[10] is that of one abnormal child per 1,000 births per rad of exposure. I cannot extrapolate readily to the TMI situation since I have not as yet seen the estimated exposure doses, but if we accept impressions conveyed in the press, most of the approximately 150 pregnant women within five miles of the reactor had *at most* 5–10 mrem, and probably less. We estimate, then, based on these assumptions, that there is only a remote possibility of any *in utero* exposed embryo or fetus being born with a radiation-induced malformation. Nonetheless, we are in an area of scientific uncertainty, since as the UNSCEAR report says, "No satisfactory data are yet available for deriving reliable quantitative estimates of the risk from prenatal human irradiation at comparable developmental stages, particularly at the low doses and dose rates. The Committee emphasizes therefore the importance of further studies. . . ."[10]

In terms of the genetic effects of radiation, i.e., the effects on future generations, we are concerned with the production of different types of single-gene mutations (recessives, dominants), of multiple-gene effects, and of gross chromosomal abnormalities.[11,12] To illustrate the difficulties here: Even in the large F_1 generation of the Hiroshima and Nagasaki survivors, with the survivors exposed to high doses of radiation, no increase in genetic disease has been shown. And this in 3 or 4 F_1 cohorts totaling 17,000 each. Thus, if one cannot document genetic effects in a human population exposed at high doses to a known mutagen, the problems of weak mutagens at lower doses can be better appreciated.

And yet we know, based on experimental evidence, that mutations are induced at essentially all levels of radiation exposure. The concern for future human generations is then realistic. With the so-called direct method in mice (based on induced mutations per unit dose), the rate of recessives is about 60 per 10^6 gametes per rad*; of dominants, 20 per 10^6 mice per rad.† Extrapolating with this method to man, we estimate that a 1-rad parental exposure will result in 22–30 affected offspring per million live-borns (about 20 dominant mutations, 2–10 chromosomal mutations). Using the more indirect doubling-dose method, which assumes that the radiation-induced mutation rate is proportional to the spontaneous rate, and again extrapolating from mice, the UNSCEAR group[10] has estimated that with a 1-rad parental exposure, we expect 20 dominant or X-linked mutations that would be detectable, 38 chromosomal, and 5 of complex genetic inheritance, for a total of 63 per million offspring per rad. Including those recessives that would appear in later generations, we get 185 genetic abnormalities per million offspring, per rad. In TMI, at fractions of a rem, by all reasonable methods of estimating mutations, we would expect in the F_1 generation well under 1–10 abnormalities per million births.

Finally, in terms of cancer induction by ionizing radiation, a combination of data from animal studies and from human studies suggests that we may expect approximately 1–2 malignancies per 10,000 exposed persons per rad.[10] Numerous human populations have been exposed to ionizing radiation and have been followed over time or examined

---

* From mouse spermatogonial irradiations.
† Using dominant skeletal deformities.

retrospectively for cancer incidence. These include not only the Hiroshima and Nagasaki populations but the Marshall Islanders, the therapeutically X-irradiated ankylosing spondylitics, uranium miners, and numerous others. Again, for TMI and the Harrisburg area, the estimate of expected cancers resulting from the radiation exposure will be dose-dependent, and we expect few radiogenic cancers if the doses have been as low as we have been told.

In conclusion, I should like to emphasize the biological reality that when we consider protection of our population against radiation or other mutagens/teratogens/carcinogens, we must take particular cognizance of the fact that not all segments of the population are equally susceptible. At special risk might be individuals who are heterozygous for the genetically determined chromosome breakage syndromes (Bloom's syndrome, Fanconi's anemia, the Louis-Barr syndrome); individuals who are unable to repair DNA damage as efficiently as normals (as xeroderma pigmentosum patients); persons who are immunologically incompetent, for one reason or another; or persons from highly cancer-prone or birth-defect-prone families. Special protection efforts must be targeted at these persons, as well as at the society as a whole. Also, in terms of radiation exposure, from my three years of work in Hiroshima I can tell you that a scientifically well-founded negative statement has enormous impact on the population, and even when the exposure, as at TMI, is largely low dose, the studies ought to be done.

REFERENCES

1. McIntosh, R., Merritt, K. K., Richards, M. R., *et al.*, The incidence of congenital malformations: A study of 5,964 pregnancies, *Pediatrics* 7:505 (1954).
2. Fraumeni, J. F., Jr., Environmental and genetic determinants of cancer, *J. Environ. Patho. Toxicol.* 1:19 (1977).
3. Bloom, A. D., and Polin, R., Genetic diseases, in: *Neonatal-Perinatal Medicine* (R. E. Behrman, ed.), p. 171, Mosby, St. Louis (1977).
4. McCann, J., and Ames, B. N., Detection of carcinogens as mutagens in the *Salmonella*/microsome test: Assay of 300 chemicals: Discussion, *Proc. Natl. Acad. Sci. USA* 73:950 (1976).
5. Ames, B. N., Identifying environmental chemicals causing mutations and cancer, *Science* 204:587 (1979).
6. The talk of the town, *The New Yorker*, p. 25, May 21 (1979).
7. De Serres, F. J., Mutagenicity of chemical carcinogens, *Mutat. Res.* 41:43 (1976).
8. Brent, R. L., Radiation and other physical agents, in: *Handbook of Teratology* (Vol. I) (J. G. Wilson and F. C. Fraser, eds.), p. 153, Plenum Press, New York (1977).
9. Miller, R. W., Effects of ionizing radiation from the atomic bomb on Japanese children, *Pediatrics* 41:257 (1968).
10. Report of the United Nations Scientific Committee on the Effects of Atomic Radiation, Suppl. No. 40 (A/32/40), United Nations, New York (1977).
11. Neel, J. V., and Bloom, A. D., Detection of environmental mutagens, *Med. Clin. North Am.* 53:1243 (1969).
12. Muller, H. J., Radiation damage to the genetic material, *Am. Sci.* 38:25, 126, 399 (1951).

DISCUSSION

A PARTICIPANT: I don't want to detract in any way from the very real importance of everything that Dr. Bloom said. However, I think that we have to stick to reality. In almost every instance he mentioned syndromes due to mutagens and radiation. These syndromes have similar genetic forms. In almost every instance this has not been investigated.

I have personally had the experience of investigating a couple of them, such as the Hiroshima group, and some forms of microcephaly have been known in Japan for centuries and are inherited in the total population. I have also been associated with work on some chemical spills and found that the syndromes said to be due to chemical mutagens were also known for generations in the area and were familial.

I think until we remove this genetic background from studies of mutagenesis, we don't really have much more than anecdotal studies.

PROF. ARTHUR D. BLOOM: I did not pretend that the kinds of disorders that I was describing were in any way unique manifestations of radiation or chemical mutagen exposure. These are disorders which occur all the time and are due to mutant genes whether spontaneously present in the population or induced. Perhaps the groups I mentioned may be at increased risk. That is the concern. Certainly there is microcephaly throughout the world, but there is a statistically significant increase in microcephaly among those exposed *in utero* in Hiroshima.

32

Controlling Environmental Mutagens
Through Market Incentives or State Action?

Catherine J. Damme
Department of Community Medicine
University of Texas Medical School
Houston, Texas 77025

The infiltration of our environment by substances that mischievously and capriciously act on the human body is a subject for continuing public concern. That concern, which is not only for ourselves but for our offspring as well, has spawned calls for the control or elimination of these agents from our environment. I will discuss the need for control, and two ways control might be achieved: through market incentives or through state action.

The Need for Control

Before we talk about *how* society controls environmental mutagens, we must look at the more basic question: *Should* environmental mutagens be controlled? The decision of whether to control turns largely on the focus. A *macro* viewpoint might lead to the conclusion that *nothing* should be done. A *micro* viewpoint might dictate control at all costs.

For example, the proponent of a *macro* outlook would have as his objective the overall health of the gene pool. Although population geneticists have hypothesized many genetically harmful events, such as radiation exposure of varying degrees, most conclude that the effects on the gene pool would be imperceptible. But as to chemical mutagens present in the environment, virtually no work has been done to project effects on the gene pool; accordingly, one geneticist has stated that the only responsible advice his discipline can offer is to say they simply do not know; they can only be prepared to wait, see, and analyze.[1] At this time, no undue genetic load has been perceived, although such perception would be most difficult. After all, like the marketplace in classic economic theory, the gene pool, the population geneticist might say, is best left unmanipulated; it will do very well on its own; everything will work itself out. Thus, if society's goal is to maintain the vigor of the gene pool, we might easily justify doing nothing at all to control environmental mutagens.

Then there is the *micro* view, with a focus on the individual. This perspective would require controls on environmental mutagens, whatever the cost, regardless of effects on the gene pool, simply because some persons would be grievously injured. The advocate of the *micro* viewpoint would urge controls because society should not countenance sufferings that could be avoided. To stretch our economic analogy, our *micro* spokesman would take a Keynesian view: intervene, manipulate the system to further a societal goal—protection of health.

It is quite likely that a policy of "no control" of environmental mutagens would be acceptable to our society were there not such a high correlation between mutagens and carcinogens, a term that evokes fear, anxiety, and, in response to these feelings, demands for control. I suspect that those demands for regulation are really grounded in fear for ourselves and our vulnerability to cancer, *not* fear for the well-being of our offspring.

Therefore, the *micro* position that control is imperative becomes forceful and understandable when those two terms—*carcinogens* and *mutagens*, plus a third, *teratogens*—are linked together. Carcinogens, mutagens, and teratogens have taken on a highly charged meaning of their own; they have come to represent almost one term in the public lexicon, like Groucho, Harpo, and Chico—Groucho, of course, representing carcinogens. But the fears they evoke are real when society faces high cancer rates in urban areas,[2] when patterns of sterility and miscarriages appear in isolated work forces,[3] when estimates of occupationally related cancer fluctuate between 3% and 20%,[4] and when scientists, supposedly able (in the public mind) to quantify, vehemently disagree on risk assessments.

For those who value the individual, who keenly empathize with individual sufferings, and who do not look at the hazards of environmental toxins as an insignificant ripple in the gene pool or as a minor contribution to cancer rates in certain areas, the only course for society is to reduce human exposure to environmental dangers.

THE MEANS OF CONTROL

Through Market Incentives

One way to exert control with a minimum of state intervention or manipulation is through market incentives, or what Professor Guido Calabresi would call "general deterrence."[5] By this approach, we seek to control the production of a toxin by requiring its manufacturer (or others in the chain of distribution) to pay for injuries it actually causes. The control devices most readily available are claims by workers against employers (either under workers' compensation or tort law) and tort claims by members of the public who have been injured by exposure to a carcinogen, a mutagen, or a teratogen. If there are a large number of judgments against a producer of a toxin, it becomes theoretically less desirable to produce and thus, in a sense, it becomes "regulated" through the market.

Unfortunately, as I shall show, neither workers' compensation nor tort claims can be relied upon to place sufficient cost pressure on manufacturers. To understand the possible impact of these control devices, we must separate our triumvirate of disease-causing agents—carcinogens, mutagens, and teratogens—for the type of disease may dictate what legal theory would be used, the level of damages that may be awarded, and even who the plaintiff and defendant may be.

CARCINOGENS.   In the case of a carcinogen, two types of claimants may present themselves: a member of the work force or a member of the public. The worker may pursue a claim against his employer under the workers' compensation law. Workers' compensation systems offer employees certain advantages over tort law. It is much easier, for instance, to establish causation; instead of the complicated tort law requirement of proving proximate cause (about which I shall say more later), for the most part, the claimant need only show that the injury arose "out of employment."[6]

But there are drawbacks to the plaintiff as well. Under workers' compensation, the injured worker is compensated according to a schedule established by state law, which almost invariably provides much less than he could receive in a successful tort action. This means that even successful workers' compensation claims do not exert great cost pressure on employers, although they may force increases in his insurance premiums. The role of workers' compensation claims is problematic for another more fundamental reason: it is questionable whether certain cancers are covered, partly because establishing causation, although easier here than in tort cases, often is still quite difficult.

Historically, the vast majority of claims under workers' compensation were for injuries, not diseases. For many years, in fact, occupational disease was excluded from coverage altogether. Now, however, all 50 states cover occupational diseases to some extent. The difficulty for the victim of occupationally induced cancer lies in the very definition of "occupational disease." Of course, definitions vary among the states. But generally, an occupational disease has been defined as one that is peculiar to the employment, that is not easily developed in everyday life, that can be shown to have a causal connection with work, or, in some cases, that is not an ordinary disease of mankind.[7]

Meeting this definition presents little difficulty to the vinyl polymerization worker who is striken with angiosarcoma, a rare liver cancer closely associated with human exposure to polyvinyl chloride (PVC). But what if the same worker develops another disease that could be caused by PVC, such as lung cancer?[8] Because there are many causative factors in lung cancer, the employee's claim under workers' compensation may fail. For example, in *Amoroso v. Tubular and Cast Products Manufacturing Co.*,[9] a worker who was exposed to nitric, sulfuric, and chromic acid fumes at work was stricken with lung cancer. The workers' compensation board of New York denied compensation because the evidence revealed that the employee smoked over 40 cigarettes a day; therefore, a court concluded, there was no causal connection between the disease and the job. There are instances where the workers' compensation aegis will include cancers that *may* have been occupationally induced. In *Bolgor v. Chris Anderson Roofing Co.*,[10] where an employee alleged that development of cancer was caused by his work, the compensation board of New Jersey, while considering evidence that the employee was a heavy smoker, still allowed compensation, finding the type of employment to be a "major contributor" to his condition.

One additional note on occupational disease definitions: Boards have considered the issue of individual predispositions to diseases; the majority of jurisdictions have ruled that they are immaterial to granting workers' compensation.[11] Taking into account that we are *all* probably genetically predisposed to something, this is surely the sensible course.

There are other statutory vagaries that might exclude occupational cancers from workers' compensation coverage. One of the most prominent is the time limit for filing. Usually the disability or death must take place within a specified time after the last exposure or employment, and cancer typically has a long latency period. After recognition of radiation-induced diseases, however, the trend has been to liberalize these time limits.[12]

But an injured worker excluded from coverage is not out in the cold; indeed, he has incentives to avoid the workers' compensation route. Awards under workers' compensation are exceedingly low. The National Commission of State Workmen's Compensation Laws, which was created by the Occupational Safety and Health (OSH) Act, recommended that award schedules be revised.[13] Many commentators have suggested that workers' compensation systems by expanded to include family members, or even those residing in industrial neighborhoods[14]; other commentators have suggested other administrative systems be established to certify persons and classes of persons as victims of an environmental pollutant.[15] In addition to the paltry amounts of recovery they permit workers, many compensation statutes provide that a worker whose occupational cancer is covered by the act is foreclosed from suing his employer in a civil action (which might yield a much greater recovery). There have, however, been numerous attempts to give persons with occupational cancer remedies in civil law as well as workers' compensation, typically if they can show culpability. In *Reba Rudkin v. Johns-Manville Products Corporation et al.*,[16] for instance, a California Superior Court stated that "an employer who has fraudulently induced [an employee] to work in employment which is certain to injure him should not be allowed to assert his liability for the employee's injuries is limited by the workers' compensation statutes. . . . "[17] Clearly, a move is on to enlarge the scope of workers' remedies.

Exclusion from workers' compensation systems, then, allows the worker to sue his employer in tort; his position is then comparable to a member of the public who sues a manufacturer for causing an environmental cancer. And even if the employee is prohibited by workers' compensation laws from suing his employer, a number of courts have allowed him to sue third parties, e.g., manufacturers of the toxin,[18] regardless of workers' compensation coverage.

A tort is a legal wrong done by one person to another that is redressable in the courts by an award of damages. The most common type of tort is negligence. Under that theory, if A is injured by B, to establish his case A must show that (1) B owed him a duty of due care, (2) B breached that duty, (3) A was injured, and (4) B's breach was the proximate cause of A's injury. Damages in negligence normally are limited to those injuries that are foreseeable, not to every injury flowing from the act, however bizarre. Another tort action is strict liability; under this theory the courts impose liability without fault regardless of how careful or skillful the defendant was. It is a somewhat primitive doctrine and was originally applied to inherently dangerous activities such as using dynamite. Foreseeability is not a factor under strict liability; thus the tortfeasor is responsible for all injury flowing from his act, however unforeseen. Causation must be proven under strict liability as in negligence, and under both theories it represents the principal obstacle to recovery. If a straight negligence theory is used, the plaintiff will not have a difficult time establishing a duty of care or an injury; however, to satisfy the causation requirement, he must show at least that *but for* the negligent emission of the toxin, he would not have contracted cancer. This is a difficult burden for the plaintiff to sustain. It can be done, but the successes are rare and occur normally where the causal relation is obvious. For example, in *Heck v. Beryllium,*[19] the plaintiff, who lived near the defendant's beryllium plant, developed berylliosis. The court found that the beryllium-laden air the plaintiff breathed in that neighborhood was the proximate cause of his illness.

Both workers and affected members of the public have had more success using a strict products liability theory. Under this approach a manufacturer of a product that is inherently and unavoidably hazardous must, if the utility of the product still warrants its marketing, warn the consumer of the risks and let the consumer decide whether to use it. If the manufacturer fails to warn, he may be held strictly liable for injuries that result from the danger. Although the choice to use a product is a free one in theory, when the "consumer" is a worker, his choice may in fact be to use the product or to quit his job.

In an important case, *Borel v. Fibreboard Paper Products Corp.,*[20] the Fifth Circuit Court of Appeals allowed the plaintiff, an asbestos insulation worker who contracted asbestosis and died of mesothelioma, to sue the manufacturer of the asbestos. The worker's wife (who was substituted as plaintiff after his death) recovered on a strict products liability theory, with the court ruling that asbestos was an "unavoidably unsafe product." However, the court modified slightly the potential liability of defendants by placing a duty on them to warn of "only those dangers that are reasonably foreseeable."[21] Foreseeability is, of course, a negligence standard.

The *Borel* case also demonstrated that mulitple manufacturers could be held liable when it was unclear *which* supplier's asbestos actually started the irreversible carcinogenesis. And another similar case, *Karjala v. Johns-Manville Products Corp.*[22] dealt head on with the problem of statutes of limitation and diseases with long latency periods in that it allowed the use of the "discovery rule" in assessing whether the statute of limitations had run. The court in *Kerjala* said:

> There is rarely a magic moment when one exposed to asbestos can be said to have contracted asbestosis, the exposure is more in the nature of a continuing tort. It is when the disease manifests itself in a way which supplies some evidence of causal relationship to the manufactured product that the public interest in limiting the time for asserting a claim attaches and the statute of limitations will begin to run.[23]

So while a few courts have allowed claims from either a worker or a member of the public under tort theories, the results are still disappointing when one considers the numbers of victims of environmental or occupational cancers. Demonstrating causation, whether in negligence or strict liability, is still a major problem, particularly when the cancer in question is one of the prevalent types—lung, colon, breast—which have multiple causes. Add to that the time and expense of waging protracted litigation and it becomes clear that tort actions as they stand today will not have a very far-reaching effect as an indirect control factor. Generally, tort actions do a much better job of compensating victims and punishing wrongdoers than changing behavior on a broad social scale. Here, they do not do either very well. Furthermore, the slow, victim-by-victim, substance-by-substance process is not equitable, and nurtures a class-society mentality.

MUTAGENS AND TERATOGENS.   Mutagens and teratogens present all the problems of carcinogens, and more. Mutagenic agents that affect the germ cells of either males or females may cause birth defects in offspring. Teratogenic agents act directly on the fetus. The effects of mutagens, taken alone, are extremely difficult to detect in human populations. Unless the alteration of the germ cells causes unusual defects in offspring, chances are the event will go unnoticed. Teratogens echo this problem. Thalidomide caused the unusual "flippers" in affected offspring; diethylstilbesterol caused rare adenocarcinoma of the vagina in female offspring. But what of the substances that cause the more common birth defects that will remain undetected? Here, of course, the injured is the newborn. This often presents another set of problems. First of all, if the exposure to the toxic agent occurs in the workplace, but the only injury is to the offspring, no workers' compensation would be available. Another problem is that, with the case of mutagens, the tort takes place prior to conception. Although preconception torts are new, they are gaining acceptance in some jurisdictions.[24] In one of the leading cases, *Park v. Chessin*,[25] a couple consulted the defendant physician about future pregnancies after the death of their child from a polycystic kidney condition, an autosomal recessive disorder. The physician told them there was virtually no chance that future children would be affected. The couple had another child with polycystic kidneys who died. The parents and the child sued the physician. The deceased child's suit was eventually dismissed, but the court allowed the parents' preconception tort action. Furthermore, the court emphasized foreseeability in allowing an action based on prospective conditional liability. Any similar actions against a company that exposed a worker or a member of the public to a mutagen would also likely require a high degree of foreseeability before liability would result.

Teratogens have presented even more complicated problems. Because teratogens act only on the fetus, women in the workplace are vulnerable. There have been numerous cases of fertile women being foreclosed as a class from certain jobs because of the potential hazard to the fetus.[26] Companies fear liability for any birth defects that might be traced to the female worker's exposure to a teratogen. Although Title VII of the 1964 Civil Rights Act protects women from discrimination in employment,[27] many companies allege that harm to the fetus would justify exclusion of women under the bona fide occupational qualification exception.[28] The U.S. Equal Employment Opportunity Commission has failed to give a definitive ruling on this issue, and numerous women have been transferred to less hazardous (and often lower-paying) jobs. The unhappy result is that some young women, rather than lose their jobs, have elected to be sterilized in order to retain their positions.[29]

Thus the interests in fetal health (or the interests in avoiding liability) may come into conflict with the desires of women to enter or remain in job categories traditionally held by men. Can the interests of a person "not yet in being" have precedence over those already born? It is unclear from the disparate policies in various industries and the lack of policy from the federal government what the final outcome will be, but, clearly, the "voluntary" sterilization of fertile women is not acceptable.

A cause of action that merits slightly more than a footnote and that might be applicable to this context is the intrafamilial suit. Courts have a long-standing policy that precludes such suits. The immunity doctrine in actions between child and parent has roots in an 1891 decision, *Hewlett v. George.*[30] It was founded on the belief that to allow such suits would destroy parental authority, family tranquillity, family finances, and societal values, and would invite fraudulent suits.[31] Although there were some exceptions, such as those permitting suits by an emancipated minor[32] or against a deceased parent,[33] it was not until 1963 that a state clearly rejected the immunity doctrine.[34] A number of jurisdictions now allow such suits.[35] However, these decisions were made in cases that involved postnatal harms, usually child abuse. No successful case has occurred that has involved prenatal or preconception injuries. The closest is *Zepeda v. Zepeda,*[36] wherein an illegitimate child sued his father for damages stemming from his illegitimate status. The case failed, and courts have consistently disallowed such claims for reasons of public policy. It is conceivable that offspring injured by the mother's willful and knowing exposure of her fetus to a teratogen would attempt to sue the mother, as well as the employer, for damages. However, the chances for success of such a suit would be most remote.

The most difficult barrier in any of the tort actions described above, whether it is for carcinogenic, mutagenic, or teratogenic damage, is that of showing causation. The problem of proving causation, as well as the expense, time, and inefficiency of tort actions brought by either offspring, workers, or members of the public, will work to discourage such suits. Thus tort actions will do little to change the social policy or to prevent future injuries resulting from environmental poisons.

Through State Action

CONTROLLING SUBSTANCES. Another way to control environmental toxins is through direct governmental action to prohibit or restrict the manufacture or use of the substances (specific deterrence). There is a range of statutes on the state and federal level that have as their purpose the protection of the public against assaults on both person and the environment, but I will deal here with only federal statutes and actions taken pursuant to them.

There is an array of federal statutes that either implicitly or explicitly regulate human exposure to carcinogens, mutagens, or teratogens. For example, the Environmental Protection Agency (EPA) can regulate airborne carcinogens under the hazardous substances provision of the Clean Air Act.[37] Under the "unreasonable adverse affects" section of the registration provisions of the Federal Insecticide, Fungicide, and Rodenticide Act (FIFRA), EPA can require testing of substances for oncogenicity and mutagenicity.[38]

And the Occupational Safety and Health Administration has already regulated workers' exposure to some 23 carcinogens under the standard-setting authority of the OSH Act.[39] That agency has also proposed a generic carcinogen regulation, which, if adopted, will avoid relitigation of identical issues surrounding the carcinogenicity of a substance by classifying substances on the basis of available test data.[40] Carcinogenicity, mutagenicity, and teratogenicity are not specifically included in these statutory mandates, but, clearly, these laws permit such actions.

A few statutes explicitly include these specific human health hazards[41]; the most prominent among them is the Toxic Substances Control Act[42] (TSCA), which names carcinogenic, mutagenic, and teratogenic effects as those for which testing might be required to demonstrate the safety of a substance.[43]

TSCA is two-pronged in its approach to control of toxic chemicals. First, it has a data-generating function. Many of the sections of the act require testing and premarket notification procedures that force manufacturers of either new or in-use chemicals to produce information on the substances' safety to health and environment.[44] These tests may include those for the conditions we are concerned with in this chapter. Although TSCA has not yet reached its regulatory zenith, EPA, the administrating authority, has issued test guidelines for carcinogenic, mutagenic, and teratogenic effects.[45] The information developed from the testing requirements will enable the administrator of EPA to decide whether the substance should be allowed to remain in or (in the case of a new chemical) to enter the stream of commerce. The second means of control under TSCA is that of restriction; the administrator of EPA may take actions as mild as labeling or as stringent as prohibition.[46] Throughout the decision-making process, the administrator must, by law, make cost–benefit judgments that include economic and social factors.[47]

Even without the statutorily required cost–benefit calculus imposed under TSCA, federal government officials must be guided by inflationary and economic impact statements, and decisions made pursuant to all environmental laws are scrutinized by the Office of Management and Budget.

The result has been that cost–benefit calculations reign supreme. Unfortunately, even the term *calculation* reveals a myth. Certain costs can be readily toted up: jobs lost, business costs of complying with the measure, costs passed on to the consumer, and the like. But the best estimates of lives lost, cancers caused, and birth defects produced are oftentimes little better than guesses and tend, when quantified, to look quite small, and, indeed, acceptable to those who are not vulnerable to the dangers. But there are costs that cannot be measured, such as human suffering, which should not be forgotten in the pursuit of the perfect cost–benefit equation.

The consequence of this enforced balancing of costs and benefits has been a curious blend of action and paralysis in the federal government. There is great activity when a problem is perceived; but when the costs are reviewed, there is delay, often administratively, often in the courts. It is clear from various statutory mandates and from some early court decisions that government decision-makers need not, in fact must not, wait until the scientific community reaches agreement on whether a substance is a carcinogen, a mutagen, or a teratogen. The government decision-maker, after reviewing the literature, may declare that, as a matter of *policy*, not as a matter of scientific certainty, a substance is dangerous and must be regulated.

Most courts have upheld administrative rulings based on a reasonable assessment of available, but imperfect and conflicting, scientific data. For example, in *Ethyl Corp. v. Environmental Protection Agency*,[48] the Circuit Court of Appeals for the District of Columbia, construing the Clean Air Act, stated:

> Where a statute is precautionary in nature, the evidence difficult to come by, uncertain, or conflicting because it is on the frontiers of scientific knowledge, the regulations designed to protect the public health, and the decision that of an expert administrator, we will not demand rigorous step-by-step proof of cause and effect. Such proof may be impossible to obtain if the precautionary purpose of the statute is to be served. Of course, we are not suggesting that the Administrator has the power to act on hunches or wild guesses.
> ... The Administrator may apply his expertise to draw conclusions from suspected, but not from completely substantiated, relationships between facts, from trends among facts, from theoretical projections, from imperfect data, from probative preliminary data not yet certifiable as "fact," and the like.[49]

Squarely on the "frontiers of scientific knowledge" are the regulatory decisions that must be made on carcinogens, mutagens, and teratogens. When the Occupational Safety and Health Administration issued standards for various carcinogens present in the workplace, it based them on somewhat speculative information. And courts, echoing *Ethyl Corp.*, upheld the standards.[50] It was as though courts implicitly understood the administrator's dilemma and placed their imprimatur on decisions that might err on the side of safety.

Recently, however, there have been calls from various quarters to tighten up the decision-making process throughout government to include stricter cost–benefit analyses.[51] In addition, with each proposed regulation, the regulated and the regulators issue conflicting cost estimates of compliance.[52] The costs of human suffering, of medical care, and especially of further societal stratification between workers and the rest of the population are seldom considered.

The trend toward closer cost–benefit analyses is now reflected in the Fifth Circuit Court of Appeals' decision on the OSHA benzene standard,[53] which called for a reduction from 10 ppm to 1 ppm (averaged over an eight-hour day). The American Petroleum Institute (API), which challenged the standard, alleged that OSHA failed to show that the lowered standard was "reasonably necessary" to provide a safe workplace as specified under Section 652(8) of the OSH Act.[54] OSHA lawyers countered that Section 655b of the act[55] was dispositive in that it required that standards be enacted that would assure "that no employee will suffer material impairment of health ... from regular exposure ... for the period of his working life."[56] Thus, government lawyers argued, cost–benefit considerations could not negate their statutory charge. The court, in a departure from decisions on OSHA carcinogen standards in other circuits, concluded that OSHA's failure to demonstrate evidence of health benefits expected to be derived from reduction of the standard made it impossible for the court to assess the "reasonable necessity" for it as required under Section 652(8) of the Act.[57] Therefore, the standard was set aside. The U.S. Supreme Court agreed to review the decision; however, the Fifth Circuit action may presage a shift toward cost–benefit analyses in an area where all the factors simply cannot be quantified.

During the benzene standard proceedings, cost estimates were flying back and forth, and undoubtedly influenced the court. The plaintiffs, the American Petroleum Institute, stated that compliance with the standard would cost the industry between $187 million and $204 million in the first year.[58] The Manufacturing Chemists Association (MCA) agreed with that cost estimate but said it would be passed on to the customer "as the price of unnecessary and invalid government regulation."[59] The American Iron and Steel Institute (AISI) estimated that costs of compliance of coke-oven operators would be $60 million.[60] OSHA estimated that it would cost AISI members $19 million to comply.[61] And the AFL-CIO Industrial Union Department put the "economic impact" of the benzene standard at $34 million a year.[62]

CONTROLLING PERSONS. The cost estimates recited above underscore the danger of the cost–benefit mind-set: the ease of estimating economic costs and benefits, the difficulty of assessing human ones. As more decisions are pushed into the cost–benefit mode, as the cost of testing for chronic toxicities increases for each substance under each regulatory testing schema, and as the cost of developing and implementing control apparatus increases exponentially, a more insidious problem looms: If the combined social and economic costs of substance control are adjudged unacceptable, decision-makers in government may look to restrict not the substance but the person exposed to the substance. Such a stance would raise some serious constitutional law issues.

The U.S. Constitution, particularly the Bill of Rights, protects individuals against actions of the state that threaten to encroach on a citizen's freedoms. When a controversy that involves governmental restriction of a constitutional freedom comes before a court, the court balances the right of the individual against the state's interest in limiting those rights. Usually, the court examines a statute with a strong presumption of its legitimacy; it looks at the law only to make sure it bears a rational relationship to a proper state interest and is not arbitrary or capricious. But if the law will impinge upon what is called a fundamental right, the court will give it "strict scrutiny" and require that the state demonstrate a compelling interest in achieving its statutory objective and that the law accomplishes this end through the least restrictive means. Constitutional rights are not unbridled, but on occasion they must be sacrificed for the good of the whole.

As the flow toward quantification of costs of environmental controls on carcinogens, mutagens, and teratogens becomes a tide, and as the quantification of costs and benefits—particularly with substances that have no known safe dosages—becomes more and more ephemeral, a point will soon be reached at which the costs of reducing exposure levels of a toxin are deemed to be economically and socially unfeasible. Government, pursuant to its duty to protect public health, may decide to go beyond the scope of present environmental laws that restrict the toxin and instead restrict vulnerable persons.

When would it be permissible to control the person? And which persons would be restricted? If we take the example of a carcinogen, we would first have to determine that it was too vital to be banned and that its control would result in severe economic and social dislocation. Then we must identify which persons would be at risk from exposure to this carcinogen. They could be workers, their families, or persons who live in the neighbor-hoods surrounding the industrial site.

Not every person who has been exposed to asbestos will get mesothelioma; not every vinyl polymerization worker will get angiosarcoma; not every benzene worker will get leukemia. Each of us has preprogrammed genetic responses to the environmental assaults our bodies receive daily.[63] Pharmacogenetics has demonstrated that drug responses vary among individuals and that these responses are often genetically controlled. Perhaps the best-known example of monogenic control of drug metabolism is that of inactivation of isoniazid, a drug used to treat tuberculosis. Persons who are homozygous for the recessive gene are slow inactivators of the drug, and blood levels remain high for a significant time, often triggering a toxic reaction. A dose–response curve shows heterzygotes to be inter-mediate inactivators and homozygotes for the dominant gene to be fast inactivators.

Another drug with a known genetic variability is primaquine, an antimalarial com-pound. In sensitive individuals, the drug appears to cause destruction of red blood cells, which initiates hemolytic episodes of varying severity. Sensitivity to primaquine has been linked to G6PD enzyme deficiency, an X-linked recessive disorder.

If it can be argued that responses to environmental toxins are also genetically con-trolled, then perhaps screening tests could be devised to identify those with genetic predispositions to a carcinogen. Development of such tests would raise a multitude of legal issues. I will mention only a few of them.

As in other genetic screening programs, if the state establishes a voluntary program and builds in appropriate safeguards for protection of confidential data, there will not be severe barriers to justifying the program. However, voluntariness dilutes the program as a measure of control. For the state to require such tests, though, could spawn constitutional concern. Assuming a government program that compels testing, what if a genetic predis-position to an environmental carcinogen is shown to be race-linked, sex-linked, or ethni-cally linked? We need only look to G6PD deficiency to see that this is a realistic possibil-ity. G6PD deficiency is an X-linked recessive disorder; therefore, it affects only males. It is present in about 10% of U.S. blacks, is more common among black Africans, and is very common and more severe among Mediterranean peoples. Thus, with G6PD deficiency, we have demonstrated sex-, race-, and ethnically linked patterns. Courts would have to look very closely at a law that forbade certain persons and certain groups of persons from working and living in areas that exposed them, because of their genetic makeup, to grievous injury. The state would have to demonstrate a true health emergency to sustain a compel-ling interest and would also have to show that less restrictive measures would fail.

Questions of state action in the control of person instead of substance become even more complex if the substance at issue is a mutagen or a teratogen. Then the state's interest, which again must be compelling, lies in protecting those persons not yet in being—the gene pool in the case of mutagens, the fetus in the case of teratogens—rather than in protecting public health against cancer. To restrict the ability of fertile persons to decide where they work and live in the name of protecting future generations would be a drastic leap in state power and could only be justified by data showing a genetic catas-trophe of major dimensions.

These examples presuppose that the state would compel people to be protected. But perhaps the government's interest in health could be satisfied by less restrictive means, such as testing populations and informing them of their peril. The individual would then decide whether or not to accept the risk. On a smaller scale, this sort of choice is currently being made by states on the subject of motorcycle helmet laws. Some states have undertaken the duty to protect motorcyclists; others have allowed the risk to be a matter of choice. And courts have ruled on these laws with differing results.[64] However, if the state abdicated its duty to the "free" choice of the worker armed with knowledge of his vulnerability, our society would run the risk of creating a class of sacrificial workers who would work alongside the genetically resistant superworkers in high-paying jobs that exposed them to hazardous poisons.

But the choice may not be the worker's alone. Market pressures (liability under tort or workers' compensation law) may impel employers, not the state, to test current and potential employees (voluntarily or mandatorily) for genetic predispostions to toxins. Many tests already have been devised, if not applied, and some have been introduced. The first faltering steps were attempted with the investigation of aryl hydrocarbon hydroxylase (AHH) activity and its relationship to benzo(a)pyrene.[65] One report from the recent OSHA carcinogen regulation hearings tells of a test system that will measure DNA repair processes in industrial workers.[66] And some companies and universities are instituting programs that will monitor the numbers of chromosome breaks and sister chromatid exchanges (SCE) in certain worker populations.[67] In some settings, the tests are conditions of employment, yet in others, particularly university settings, the tests are considered so equivocal that the participants are classified as voluntary subjects of biomedical research, and the program is scrutinized by the institutional review boards. The testing programs for breaks and SCEs reveal a serious danger: It is very difficult to predict health risks from the results of the tests. Company officials are sometimes less than candid, therefore, in explaining the results to workers. Counseling a potential victim requires great skill since it is unlikely that the test results quantify risk as well as do present-day genetic screening tests for diseases with recessive, dominant, or X-linked patterns of inheritance. If genetic vulnerabilities to a carcinogen are uncovered in an individual, a counselor may wish to do additional studies on siblings and other relatives, particularly if they live and work in nearby or similar industrial areas. The company's and the counselor's duty (and potential liability for failure) to warn the potential proband or relatives might also become an issue. And, it may be controversial whether the results of such tests should be made part of the employee's medical records. Of course, social and even economic stigmatization may result from inclusion of the results in medical records. Moreover, many company officials feel that very few physicians could interpret or appreciate such information. But to keep such information from the employee or his physician would be unconscionable. From a legal standpoint, however, a history of breaks and SCEs many years prior to the development of cancer might prove to be a vital link in forging the chain of causation back to an environmental origin. Another problem might arise: company testing might reveal that a predisposition to an environmental carcinogen was linked to a racial, sexual, or ethnic group. Then, attempts to restrict employment might be subject to attack as violative of the Equal Employment Opportunity Act unless exempted by the bona fide occupational qualification provision.[68]

Although these issues of substance control versus person control appear to be far off in the future, it would be well to remember that as administrators, policymakers, legislators, and courts, as well as private sector officials, rely more heavily on cost–benefit analyses that too often quantify only economic and not human costs and benefits, as the inflationary spiral continues, as estimates of health effects remain conflicting and imperfect, and as test systems to identify the genetically weak are perfected, we may find a new urgency in devising ways to equitably protect the public and future generations without resorting to what might be a socially devastating genetic classification system.

References and Notes

1. Conversation with James Crow, Ph.D., Professor of Genetics, University of Wisconsin (May 10, 1979).
2. Mason, T., and McKay, F. W., *Atlas of Cancer Mortality for U.S. Counties: 1950–1969*, DHEW Publication No. (NIH) 75–780 (1975).
3. E.g., 1,2-dibromo-3-chloropropane has been highly suspect in causing sterility and reduced fertility in male workers exposed to it.
4. Bridbord, K., Decoufle, P., Fraumeni, J. F., Jr., *et al.*, *Estimates of the Fraction of Cancer in the United States Related to Occupational Factors*, Prepared by NCI, NIEHS, NIOSH (Sept. 15, 1978).
5. Calabresi, G., *The Costs of Accidents*, p. 68, Yale University Press, New Haven (1970).
6. Larson, A., *The Law of Workmen's Compensation*, Mathew Bender, New York § 6.00 (1975).
7. Larson, A., Occupational disease under workman's compensation laws, *U. Rich. L. Rev.* 9:87, 90 (1974).
8. Bridbord, *supra* note 4, at 32.
9. 13 N.Y.2d 992, 194 N.E.2d 694, 244 N.Y.S.2d 787 (1963).
10. 112 N.J. Super. 383, 271 A.2d 451 (1970), *aff'd.*, 117 N.J. Super. 497, 285 A.2d 228 (1971).
11. Larson, *supra* note 7, at 103.
12. Ibid. at 113.
13. The Report of the National Commission of State Workmen's Compensation Laws, 118 (1972). See also, S420, a bill introduced in February, 1979 by Sen. Jacob Javits (R-N.Y.) and Sen. Harrison Williams (D-N.J.) that would establish minimum federal standards for state workmen's compensation systems.
14. See, e.g., Toxic Sustance Pollution Victim Compensation Act, H.R. 9616, introduced October 18, 1977 by Rep. William M. Brodhead (D-Mich.). According to Brodhead, a bill such as this would "encourage manufacturers to develop innovative safety techniques to protect human health and, at the same time, provide victims of toxic substances poisoning with adequate compensation, expenses which properly should be borne by those who create the risks rather than by victims or society in general." Reported in *Chem. Reg. Rep. (BNA)* 33:1199 (1977). See also Asbestos Health Hazards Compensation Act (H.R. 2740) introduced by Rep. Millicent Fenwick (R-N.J.) and Rep. Edward P. Beard (D-R.I.) in August 1977. This proposed bill was attacked as an aid bill to the asbestos industry because it relieved them of liability. Reported in *Occup. Safety Health Rep. (BNA)* 8:183 (1978); *Occup. Safety Health Rep. (BNA)* 8:1723 (1979). Hearings on the subject of compensation for workers who are victims of asbestos-related maladies were held by the Subcommittee on Compensation, Health, and Safety of the House Committee on Education and Labor in October 1978. Further hearings and a bill were promised from that subcommittee. See *Occup. Safety Health Rep. (BNA)* 8:831 (1978); *Occup. Safety Health Rep. (BNA)* 8:1738 (1979).
15. For one such proposal, see Soble, S.M., A proposal for the administrative compensation of victims of toxic substance pollution: A model act, *Harv. J. Leg.* 14:683 (1977).
16. No. 159524 (Sup. Ct., Contra Costa Cty., Ca., 1979), reported in *Occup. Safety Health Rep. (BNA)* 8:1738 (1979).
17. Ibid. at 1738.

18. Rheingold, P. D., and Jacobson, L., The toxic tort cause of action: Law and procedure, in: *Toxic Torts* (P. Rheingold, N. Laudau, and M. Canavan, eds.), Association of Trial Lawyers of America, Washington, D.C. (1977).
19. 226 F.2d 87 (3rd Cir. 1966).
20. 493 F.2d 1076 (5th Cir. 1973), *cert. denied*, 419 U.S. 869 (1973).
21. Ibid. at 1088.
22. 534 F.2d 155 (8th Cir. 1975).
23. Ibid. at 160.
24. E. g., *Renslow v. Mennonite Hospital*, 40 Ill. App.3d 234, 351 N.E.2d 870 (1976).
25. *Park v. Chessin*, 387 N.Y.S.2d 204 (Sup. 1976), *aff'd.*, 400 N.Y.S.2d 110 (App. Div., 1977), *modified*, 386 N.E.2d 807 (1978).
26. See *Proceedings: Conference on Women and the Workplace* (SOEH 1976).
27. Equal Employment Opportunity Act, 42 U.S.C. § 2000e (1964).
28. Ibid. at § 2000e-2(e).
29. 4 Women assert jobs were linked to sterilization, *New York Times*, p. A21, col. 1 (Jan. 5, 1979).
30. 68 Miss. 703, 9 So. 885 (1891).
31. Parental liability for prenatal injury, *Col. J. L. Soc. Prob.* 14:45, 61 (1978).
32. Ibid. at 65.
33. Ibid. at 65.
34. Ibid. at 67.
35. Ibid. at 68.
36. 190 N.E.2d 849 (Ill. 1963).
37. The Clean Air Act, 42 U.S.C. § 7412 (as amended 1977).
38. The Federal Insecticide, Fungicide, and Rodenticide Act 7 U.S.C. § 136a (as amended 1978). Regulations issued under FIFRA for enforcement include the following as constituting "unreasonable risk":

    Chronic Toxicity. (A) Induces oncogenic effects in experimental mammalian species or in man as a result of oral inhalation, or dermal exposure; or induces mutagenic effects, as determined by multitest evidence.

    EPA Regulations for Enforcement of the Federal Insecticide, Fungicide, and Rodenticide Act., 40 C.F.R. § 162.11(a)(3)(ii)(1978). On July 12, 1977, EPA issued a preliminary draft of the Mutagenicity Testing Requirements Section of the FIFRA, Registration Guidelines for Hazard Evaluation of Humans and Domestic Animals. This document was revised and reissued on August 22, 1978. See EPA proposed guidelines for pesticide registration; hazard evaluation: Humans and domestic animals, *Fed. Reg.* 43:37,336 (1978) (to be codified in 40 C.F.R. §163).
39. *Occupational Safety and Health Standards; Subpart Z—Toxic and Hazardous Substances*, 29 C.F.R. § 1910.100 (1979). The toxic substances regulated include asbestos, 4-nitrobiphenyl, alpha-naphthylamine, methyl chloromethyl ether, 3,3-dichlorobenzidine (and its salts), bis-chloromethyl ether, betanaphthylamine, benzidine, 4-aminodiphenyl, ethyleneimine, beta-propiolactone, 2-acetylaminofluorine, 4-dimethylaminoazobenzene, n-nitrosodimethylamine, vinyl chloride, inorganic arsenic, lead, benzene, coke-oven emissions, cotton dust, 1,2-dibromo-3-chloropropane, acrylonitrile.
40. See OSHA: Identification, Classification, and Regulation of Toxic Substances Posing a Potential Occupational Carcinogenic Risk, *Fed. Reg.* 42:54,148 (1977) (to be codified in 29 C.F.R. § 1990).
41. E.g., Food, Drug, and Cosmetic Act, 21 U.S.C. § 348 (c)(3)(A)(as amended 1962).

    No such regulation shall issue if a fair evaluation of the data before the Secretary (A) fails to establish that the proposed use of the food additive, under the conditions of use to be specified in the regulation, will be safe: *Provided*, That no additive shall be deemed to be safe if it is found to induce cancer when ingested by man or animal, or if it is found, after tests which are appropriate for the evaluation of the safety of food additives, to induce cancer in man or animal, except that this proviso shall not apply with respect to the use of a substance as an ingredient of feed for animals which are raised. . . .
42. Toxic Substances Control Act 15 U.S.C. §2601 (Supp. 1977).
43. Ibid. at §2603(b)(1).
44. Ibid. at §2603, §2604.

45. EPA preliminary draft guidance for premanufacture notification (1978), *Chem. Reg. Rep. (BNA)* 2:656 (1978); EPA draft oncogenicity standards (1978), *Chem Reg. Rep. (BNA)* 2:466 (1978); EPA draft chronic testing standards (1978), *Chem. Reg. Rep. (BNA)* 2:468 (1978).

46. TSCA, *supra* note 42, at § 2605.

47. Ibid. at § 2601(c).

48. *Ethyl Corporation v. Environmental Protection Agency*, 541 F.2d 1 (D.C.Cir.), *cert. denied*, 426 U.S. 941 (1976).

49. Ibid. at 28 (citations omitted).

50. See *Industrial Union Department, AFL-CIO v. Hodgson*, 499 F.2d 467 (D.C. Cir. 1974) (asbestos standard); *Society of Plastics Industry, Inc. v. OSHA*, 509 F.2d 1301 (2nd Cir. 1975) (vinyl chloride standard); *American Iron and Steel Institute et al. v. OSHA*, 577 F.2d 825 (3rd Cir. 1978) (coke-oven emission standard).

51. Moves toward cost–benefit analyses have come from both the legislative and executive branches of the federal government. For example, the Senate Governmental Affairs Committee issued the final volume of its Study of Federal Regulation, which included an analysis of OSHA's performance. One recommendation was that legislation should be passed that would require an economic impact analysis for all significant regulations. Noting that it was difficult to make such calculations in the health and safety area, the report nonetheless stated that it was "essential to have more information on costs—and indeed the benefits—of both federal economic regulation and of federal health, safety, and environmental activities." Senators Abraham Ribicoff (D-Ct.) and Charles Percy (R-Ill.) plan to introduce such a statute. *Occup. Safety Health Rep. (BNA)* 8:1337 (1979). In April 1979, Sen. Thomas Eagleton (D-Mo.), a member of the same committee, stated during the hearings that perhaps legislation was in order to give the President specific authority to overrule a health or safety regulation if the costs would outweigh the benefits. *Occup. Safety Health Rep. (BNA)* 8:1650 (1979).

In January 1979, the Department of Labor issued guidelines for evaluating the impact of proposed regulatory actions. DOL guidelines will require the regulatory analysis when the economic impact of the action on the economy would be greater than $100 million per year, or $50 million or more costs increment in a segment of the economy, a dislocation of 10,000 jobs or more, or a limitation of competition or concentration in a market of $100 million or more. *Occup. Safety Health Rep. (BNA)* 8:1386 (1979).

52. See notes 58–62, *infra*.

53. *American Petroleum Institute et al. v. OSHA*, 581 F.2d 493 (5th Cir. 1978), *cert. granted, Marshall v. American Petroleum Institute et al.*, No. 78-1036 (February 21, 1979).

54. Occupational Safety and Health Act, 29 U.S.C. § 652(8) (1970): "The term 'occupational safety and health standard' means a standard which requires conditions, or adoption, or use of one or more practices, means, methods, operations, processes, reasonably necessary or appropriate to provide safe or healthful employment and places of employment."

55. Ibid. at § 655(b)(5): "The Secretary, in promulgating standards dealing with toxic materials or harmful physical agents under this subsection, shall set the standard which most adequately assures, to the extent feasible, on the basis of the best available evidence, that no employee will suffer material impairment of health or functional capacity, even if such employee has regular exposure to the hazard dealt with by such standard for the period of his working life."

56. Ibid.

57. *American Petroleum Institute et al. v. OSIA*, *supra* note 53, 581 F.2d at 501.

58. Industrial users challenge cost, feasibility of standard, *Chem. Reg. Rep. (BNA)* 2:44 (1978).

59. Ibid.

60. Fifth Circuit Court of Appeals stays standard pending review, *Chem. Reg. Rep. (BNA)* 2:73 (1978).

61. Ibid.

62. Court hears benzene arguments, asks for briefs on amended standard, *Chem. Reg. Rep. (BNA)* 2:483 (1978).

63. See, e.g., Strong, L. C., Genetic and environmental interactions, *Cancer* 40:1861 (1977); Knudsen, A.G., Strong, L. C., and Anderson, D. E., Heredity and cancer in man, in: *Progress in Medical Genetics* (Steinberg, A. G., Bearn, A. G., Motulsky, A., *et al.*, eds.), Saunders, Philadelphia (1973).

64. The courts in *American Motorcycle Association v. Davids*, 158 N.W.2d 72 (Mich. App. 1968); *People v. Fries*, 42 Ill.2d 441 (1969) struck down as unconstitutional state laws requiring motorcyclists to wear helmets. Similar laws were upheld in *Bisenius v. Karns*, 42 Wisc.2d 42, 165 N.W.2d 377 (1969); *State v. Odegaard*, 165 N.W.2d 677 (N.D., 1969).

65. Busbee, D. L., Shaw, C. R., and Cantrell, E. T., Aryl hydrocarbon hydroxylase induction in human leukocytes, *Science* 178:315 (1972); Kellerman, G., Luyten-Kellerman, M., and Shaw, C. R., Genetic variation of aryl hydrocarbon hydroxylase in human lymphocytes, *Am. J. Hum. Genet.* 25:327 (1973); Kellerman, G., Shaw, C. R., and Luyten-Kellerman, M., Aryl hydrocarbon hydroxylase inducability and bronchogenic carcinoma, *N. Engl. J. Med.* 289:934 (1973).
66. Ronald Hart of Ohio State University, appearing on behalf of the American Industrial Health Council (AIHC), an industry group, before the OSHA hearings on the proposed generic carcinogen standard, stated that at some time workers may be given badges to wear that would register DNA damage and would thus identify susceptible persons. Reported in *Occup. Safety Health Rep. (BNA)* 8:104 (1978).
67. The University of Texas Health Science Center at Houston and The University of Texas System Cancer Center are currently considering establishing such testing programs for selected at-risk worker populations. These programs will probably be voluntary and will be undertaken as experimentation under scrutiny of institutional review boards. Many petrochemical industries have similar programs that are carried out not as experimentation but rather as a condition of employment. One such program in cytogenetics had been in place at Dow Chemical Company, Freeport, Texas, for some 15 years. Workers exposed to vinyl chloride, benzene, and epichlorhydrin have been tested. The program has now been discontinued.
68. Equal Employment Opportunity Act, *supra* note 28.

DISCUSSION

DR. NEIL A. HOLTZMAN (Baltimore): The last part of your talk raised the question of genetic predisposition. Certainly there is (and has already been demonstrated) genetic variation due to mutagens. The question is: How much of the disease that will result from exposure to such chemicals can be expected on the basis of genetic variation? I think this goes back to something that Dr. Bloom talked about and that is the linear dose relationship. While there may be some individuals who have greater susceptibility, if there is no threshold to many of these compounds, then it means that virtually everyone is at risk.

I find working on that premise more satisfying, particularly in this society. Because when one begins to seek out those few instances where there may be a genetic predispostion, one is placing the burden on the individual and is suggesting that he or she make a choice of not seeking employment in dangerous areas. If this society was prepared to provide an alternative for those individuals, one that would be compatible with their economic viability, that would be appropriate. But we are already confronted with a situation where workers must make very difficult choices of seeking employment in hazardous circumstances because it is the only opportunity open to them.

MS. CATHERINE J. DAMME: As to the imperfect tests on genetically predisposed people, I agree with you. I think they are imperfect, although I do know of some that are being used and are being acted on (e.g., demonstrating chromosome breaks or sister chromatid exchanges). It is extremely hard on the worker when that worker must make a choice on whether to keep his job in a usually high-paying area or to leave.

Also, where these tests are being done, in many cases the workers simply aren't told of the results, but they are transferred to other areas.

MR. RIHITO KIMURA (Lawyer, Tokyo, Japan): I have a question about the role of scientists. You mentioned Professor Bloom that you have had experience in Hiroshima and Nagasaki. Scientists were rather too late to oppose the use of the atomic bomb, radium, and plutonium in Hiroshima and Nagasaki. But what about the case of defoliant use against Vietnam, which is also causally linked to genetic defects? Scientists are in some sense expected to work internationally against the policy of the U.S. government for this genocide by bombing.

I want to ask a question of Ms. Damme and Professor Bloom about the kind of model that could be effective if used by scientists and lawyers to avoid a scientific tragedy. We have seen the symbolic phenomenon already at Three Mile Island, but of course before that we had Hiroshima and Nagasaki. Do you have any thought of creating an ideal model in this case?

My second personal question to Professor Bloom is: As a genetic scientist when you were in Hiroshima did you have any feeling about ethical issues? You were there as a scientist and you did your work perhaps as a government officer. But you may have had some sort of feeling being in Japan as a scientist.

PROF. ARTHUR D. BLOOM: Obviously, I had, as all of us who spent time in Japan had, many moral and ethical questions about (a) what was done there, (b) what we were doing there. I think that one of the lessons that emerged was for scientists to collaborate with appropriate government people to study the exposed population. Because even though the studies that I cited earlier about the genetic effects in the F_1 generation were negative, it seems to me that that is an extremely reassuring statement to be able to make. Not that we think there were no effects. But it simply means that the magnitude of the effects was lower than we, with reasonable techniques, can perceive.

With reference to your first question in terms of a model, I think the best is an international scientific group meeting to discuss radiation effects at all levels from dosimetry to gene level effects. There is a model there that I think can be used for international cooperation, which is very important. I think that is beginning to happen in terms of chemical mutagenesis. But I think we are in the very early stages and obviously that needs to be encouraged.

MR. SCOTT M. THACHER (Harvard University): My feeling about the advantages of genetic screening in the workplace is that they have not often been well demonstrated. One example is alpha-1 antitrypsin deficiency, in which the affected person is much more susceptible to developing emphysema. Yet this covers a very small fraction of occupation-caused emphysema. My own personal concern, which you alluded to, is the legal or administrative barriers to determining or obtaining information of human cancer, especially occupational cancer. I think it is much easier to measure the level of a blood protein or blood enzyme than it often is to obtain hard information on human exposure or even human disease.

I wonder if you are willing to elaborate on some of the impediments to getting good human epidemiological data on occupational-related disease.

MS. DAMME: I agree. I have difficulty with the screening tests that are now in place. What worries me is that there seems to be a great dearth of good epidemiological work being done in this area. I am not sure why that occurs. Maybe there is reluctance on the part of workers. There is suspicion on the part of unions and there is great reluctance on the part of management. There is suspicion, obviously, on all sides. Therefore often nothing gets done. For example, if a company wants to institute a program, often they encounter great resistance on the part of workers safeguarding their privacy. Or, if the union wants to institute a program, the reverse may apply.

What frightens me is that many companies are latching onto these tests as though they were life preservers, and I am not sure how good the tests are. Also, I am not sure what is being done with the information. On the one hand, I feel that the tests should continue to be developed because knowledge about our genes is valuable. But I am not sure if I want everyone else to know or indeed anyone else to know.

Another problem is in waging a tort action many years down the line. If a company has done tests on someone, however imperfect those tests are, knowledge of that test and results of that test might prove a small link in the chain of showing causation down the line if the person developed cancer. So in that way it could work to the worker's benefit to have that test done.

33

Legal Regulation of Environmental Risk

Richard B. Stewart
Harvard Law School
Cambridge, Massachusetts 02138

I wish to address a question that has divided the two cultures of law and science: What should be the role of courts, lawyers, and litigation in the administrative regulation of environmental hazards?

Administrative Regulation

Over the past decade, there has developed within the federal government an extensive battery of administrative regulation of hazardous chemicals and radiation. Regulatory authority has been divided among many agencies, including the Consumer Product Safety Commission, the Occupational Safety and Health Administration, the Food and Drug Administration, the Environmental Protection Agency, and the Nuclear Regulatory Commission. Judicial review and lawyers' procedures are an integral and important part of this regulatory apparatus. I should like to outline the general legal principles governing administrative regulation and to evaluate their implications for control of environmental hazards.

Some introductory history may prove useful. The role of the courts and lawyers in administrative regulation goes back to the first extensive system of economic regulation in this country: the regulation of the railroads in the late 19th century. Rather quickly the courts developed a system of administrative law to control economic regulation by government agencies of private conduct. They required that agencies use formal, trial-type hearing procedures. The adversary hearing process was designed to elicit and resolve disputed factual issues relating to the agency's authority to regulate. Following such hearings, regulated firms were entitled to judicial review of the agency's fact-findings and determinations of law in order to ensure that the regulators were acting within their statutory authority to control private conduct.[1]

In the review by courts of environmental regulation, the administrative law tradition of protecting private autonomy has often been reflected in rather stringent requirements by courts that specific, concrete harm be shown as a basis for government intervention. Perhaps the most notable recent example of such a requirement was the initial decision in 1974 of the court of appeals in the *Reserve Mining* case, which denied the Environmental Protection Agency an injunction against the operation of an iron ore pelletizing facility on Lake Superior that discharged asbestoslike fibers into the air and water.[2] The judges denied relief because there had been no showing that anybody had yet been killed or injured as a result of these emissions, although there was disputed medical and scientific testimony that the fibers were carcinogenic[3] This evidence, the court found, was too speculative to justify shutdown of an important industrial facility.

However, several important developments during the past decade have modified traditional doctrines of administrative law and judicial attitudes. Traditionally, only persons whose liberty and property were subject to coercive regulation by government officials were entitled to formal procedures and judicial review. Judicial protections were not available to the beneficiaries of regulation—such as consumers or persons exposed to environmental hazards—if they complained that the administrative agency had failed to protect these beneficiary interests. In recent years, courts have become ircreasingly disturbed at the apparent failure of administrators vigorously to protect these beneficiary interests, and have extended procedural rights and judicial review to representatives of such interests. Accordingly, today adversary legal processes have not only the defensive function of protecting the private liberty and property of regulated firms but also the positive role of ensuring that the regulatory agency is taking affirmative action to protect those whom it is supposed to protect.[4]

A second important development has been a rather rapid growth in awareness by judges of the nature of environmental hazards. They have learned that causation is uncertain, that latency can extend over a long period of time, and that it is prudent and appropriate to take prophylactic measures before hazards result in discernible harm. This awareness was reflected in a second decision in the *Reserve Mining* case by the full court of appeals, which repudiated the notion that serious illness or death had to be shown before controls on hazardous activities could be imposed. More recently, the District of Columbia Court of Appeals has in a number of decisions upheld a variety of environmental regulations on the basis of rather sketchy or imcomplete data suggesting hazards of unknown severity or extent. The most notable of these is the *Ethyl* decision, sustaining EPA's phase-out of lead additives in gasoline.[5]

REGULATORY DILEMMAS

These recent developments have created a number of antinomies or dilemmas in the legal regulation of environmental risk:

First, we have a political and economic tradition of private autonomy and private enterprise, under which the government must affirmatively show substantial justification for intervention. But in the case of environmental hazards, there is recognized need for prophylactic intervention on the basis of incomplete and uncertain data and inference. These two principles are at war.

Second, there is a serious conflict between the necessity for delegating decisions to administrators and our democratic traditions. Given the pervasive nature of environmental hazards and the technical complexities in assessing risk, Congress obviously cannot deal with the problem through specific statutes directed at individual hazards. Nor can the courts serve as the "frontline" institution for dealing with such hazards through private lawsuits; there are serious problems of causation and proof that private adversary litigation is ill equipped to handle.[6] Legislative delegation of rather broad regulatory discretion to administrative agencies is inevitable. However, as some of the other chapters in this volume have indicated, questions of regulation of environmental hazards present very basic questions of economic trade-offs and social and moral choice. Traditionally, those basic social value choices are ones we want made by the legislature rather than by nonelected bureaucrats. So there is an incompatibility between the need for technically informed, specialized administrative decision and the need for politically accountable choice by a democratic decision-making process.[7]

Third, there is conflict between the need for good science and the need for administrative regulation. One of the reasons for having administrative agencies rather than courts is their specialized experience and superior technical capability. Yet there is evidence that many regulatory agencies do science poorly, in part because of the difficulty of providing attractive professional incentives to attract first-rate personnel, and in part because scientists in regulatory agencies may be subject to political or adversary pressures to slant research or analysis to fit agency policies dictated by nonscience concerns.[8]

Fourth, still another set of conflicts has been generated by the expanding goals of judicial review. In the last 10 or 15 years, there has been a developing perception that administrative agencies are not performing well, that they are often sluggish or unduly deferential to the interests of the regulated. As previously indicated, the courts in response have attempted to prod agencies to take affirmative action to protect those exposed to environmental hazards. This forces the courts to assess the justifications for initiating or refraining from regulatory intervention. But as they have assumed this more expansive role, courts have become increasingly aware of their limited competencies in technical analysis and social choice. Judges are not trained to understand scientific or economic complexities.[9] Moreover, for the courts to decide what degree of risk should trigger costly regulatory intervention is inconsistent with our political traditions.[10]

Fifth, the conception of the administrative agency itself has become divided. On the one hand, there is the tradition, going back to the New Deal, that the agency is an autonomous expert that has the self-starting authority to identify problems and take corrective measures in accordance with the overall public interest. On the other hand, we have become skeptical of agencies' ability to define and realize the "public interest," and increasingly tend to view administrators as arbiters (often one-sided) among the conflicting private interests of industry and environmental advocates.[11]

The Litigation of Risk

These various conflicts are well illustrated in our experience with the operation of the legal regulatory system in regard to hazardous chemicals. Most of the litigation that I will mention relates to the regulation of chemical substances on the ground that they are carcinogens. Evidence as to mutagenicity, including Ames test results, has sometimes been used as part of the effort to establish carcinogenicity. The Environmental Defense Fund, for example, relied upon evidence of mutagenicity in litigation before the Consumer Products Safety Commission seeking regulatory controls on Tris, a flame retardant used in children's sleepwear, and on hair dyes. Both EDF and EPA have relied on evidence with respect to mutagenicity in the regulation of pesticides. In the future, for the reasons adduced in this volume, mutagens and teratogens will become subject to regulation as such, rather than simply as surrogates for carcinogenicity. They may well represent the next frontier of regulatory control of environmental hazards.

What has been the experience to date in the legal regulation of such risks? The courts have basically seen their role in reviewing administrative determinations to regulate or not to regulate as a two-step process. First, the court defines the outer limits, as fixed by statute, of the agency's responsibility and authority. Second, within the often broad area of discretion that the agency enjoys under a statute, the court attempts to review whether the agency exercised that discretion in an informed and careful fashion.[12]

The Scope of Agency Regulatory Authority

Determining the bounds of the agency's statutory authority is a difficult task because Congress has given quite a variety of rather vague directives to various agencies to regulate hazardous chemicals. Just as a quick sampler, the Clean Air Act provision dealing with toxic chemical emissions says that the EPA is to set an emission limitation that provides an "ample margin of safety to protect the public health from such hazardous pollutants."[13] On the other hand, OSHA is required by statute to adopt standards for hazardous chemicals that "most adequately assure....to the extent feasible, on the basis of the best available evidence, that no employee will suffer material impairment of health or functional capacity...."[14] By contrast, the Federal Insecticide, Fungicide and Rodenticide Act authorizes EPA prohibition or regulation of pesticides involving any "unreasonable risk to man or the environment, taking into account the economic, social and environmental costs and benefits of [their] use...."[15] Depending on its form of use and location, the same chemical substance may be subject to regulation under different standards by different agencies.

In every one of these regulatory statutes the legislative directive is different. As a distinguished federal judge, Henry Friendly, said in a related context, "one would almost think there had been a conscious effort never to use the same phraseology twice."[16] The statutes vary in the specification of the harm or risk that justifies regulation. They also vary a good deal in the extent to which the benefits of the continued use of a substance or the costs of controlling its discharge may be taken into account.

However, the courts' decisions about the extent of agency authority to regulate have varied less than an examination of the statutory provisions might suggest. As a very rough generalization, I would say that we have a relatively risk-averse judiciary that gives great weight to possible adverse effects in defining the authority and the responsibility of the agency. In suits brought by environmentalists—and most of the decisions here are in the pesticides area—the courts have held that when there is some animal test evidence of carcinogenicity, the agency has a heavy burden to show why it should not take affirmative action to regulate. Conceding the difficulties in extrapolation from animal data and the problem of thresholds and the like, the courts have said, in essence, that if an environmental group can come in and show carcinogenic risks, the regulatory agency must take action or have a quite powerful justification for not doing so.[17]

In industry suits challenging regulatory controls, on the other hand, the courts have generally been deferential to the agency's conclusion that there are risks justifying intervention. In OSHA cases involving vinyl chloride and coke-oven standards[18] and in recent EPA decisions involving pesticides and lead additives in gasoline[19] the courts have concluded that if there is some clinical or epidemiologic evidence of some risk of serious health harm, and if the agency is not going to put an entire industry out of business, the agency has a very large degree of latitude.

So we have an essentially risk-averse judiciary in defining the boundaries of agencies' statutory responsibility to protect and authority to control. Whether that is because the courts see themselves as protecting basic human values by, if necessary, prodding sluggish bureaucracies or whether it is because judges reflect the proenvironmental bias of most upper-middle-class, well-to-do people is a question I shall not try to resolve here.

The recent decision of the Fifth Circuit in *American Petroleum Institute v. OSHA*,[20] which set aside OSHA's occupational exposure standard for benzene, is an apparent exception to this generalization. The court held that OSHA had to specify the health benefits that would be obtained by its action in lowering the standard from 10 parts per million to 1 part per million, and to determine that the benefit bore a "reasonable relationship" to the costs of compliance.

The agency relied on epidemiologic evidence of carcinogenicity (based on exposures of 100–150 ppm), an assumption that there were no thresholds, and the fact that the change would not put the industry out of business (although OSHA studies indicated compliance costs approaching half a billion dollars). OSHA declined to estimate what the benefits from the changed standard might be in terms of reduced mortality or morbidity, or to compare the benefits to the costs of compliance.

The Supreme Court has granted review of the Fifth Circuit's decision. That decision is a departure from the generally risk-averse attitude of the judiciary in reviewing environmental regulation. Yet it may reflect a new trend. Much of the regulatory reform legislation recently proposed by the president and by Senate committees would impose a general requirement that all regulatory agencies specify the benefits and burdens of regulation. Thus the Fifth Circuit's decision may be the wave of the future. I would differ with Professor Damme (see Chapter 32) as to whether such a shift would raise constitutional problems. In upholding the constitutionality of the Price–Anderson Act, limiting the liability of electric utilities and their suppliers for private damages resulting from nuclear power plant accidents, the Supreme Court recently asserted that environmental hazards are a matter of economic regulation and not individual liberty.[21] Under this approach, Congress will enjoy broad discretion to adopt a less risk-averse, more compliance-cost-sensitive approach to regulating environmental hazards.

Agency Exercise of Regulatory Discretion

Even given the risk-averse nature of the judiciary in defining agencies' statutory authority and responsibility, most statutory delegations of regulatory authority to administrators are so broadly drawn that most agencies have considerable discretion in dealing with environmental hazards, particularly in defining the extent of controls.

The courts have been unwilling to accord agencies *carte blanche* in the exercise of such discretion, in part because the stakes involved—both economic and health—are so great, and in part because courts perceive that agencies are not subject to any very clear system of incentives to make decisions that are either politically accountable or economically disciplined. Courts have attempted to control agencies' exercise of discretion by techniques that may be broadly characterized as procedural: devices to ensure that the agency has given reasoned consideration to relevant data, analysis, and choice values. The courts have not insisted, generally, on trial-type formalities, unless the statute requires it, but they have said that the agency must disclose all of the data and analysis on which it is relying in deciding whether to regulate and how stringently to regulate. Also, the agency must respond to criticisms of its position and to data and analysis submitted by outside parties—industry, environmental groups, or other governmental agencies. The courts have also developed a requirement that the agency rather extensively explain its decision through a reasoned discussion of the pertinent data, analysis, and consideration of policy.[22]

As a result, agencies write long opinions on the basis of an elaborate set of documentary materials. A year or more later the reviewing court will decide. It will rarely hold that it was impermissible for the agency to have reached the result it did. But it may well find the agency's explanation or documentary record inadequate, and remand for further proceedings. Some commentators have concluded that this approach is largely valueless, and that it simply promotes delay and the production of legal boilerplate by agency opinion-writers.[23] I believe that these criticisms are overdrawn, and that the courts' efforts to control the discretion of the agencies through essentially procedural requirments has been beneficial. My research as a member of several National Academy of Sciences studies of EPA decision making indicated that requiring agencies to justify their decisions in court on the basis of a documentary record tested by the adversary process had a distinctly beneficial inpact on the care and quality (in terms of information and analysis) of agency decisions.[24] I should point out that, at present, only the courts are institutionally positioned to review the quality of agency decision making on a regular basis. Review by Congress, the White House, or outside institutions is selective and fitful.

However, I would be quick to admit that courts, in assuming this reviewing function, have taken on a job they can't do very well, particularly in cases involving uncertainty and technical complexity. We need to develop other institutions to take on much of this task. I would suggest the development of additional nonjudicial review mechanisms within the executive branch of the government.[25] Such review is performed with respect to fiscal budgetary by the Office of Management and Budget. Why can't we have similar review mechanisms for regulatory decisions that pose technical issues on the frontiers of science and have far-reaching economic and welfare consequences? We have become a great regulatory state while relying on judicial mechanisms of oversight and control that are in part anachronistic.

The final point that I would raise with respect to legal regulation of environmental risk is the imperative need to improve the science upon which regulatory decisions are based. I have indicated earlier that there are serious difficulties in relying on the regulatory agencies to do their own science in-house. But there are equally serious difficulties in relying on the regulated firms to do all of the necessary data gathering and analysis, as the Toxic Substance Control Act experience is demonstrating. There are still other difficulties in relying on some other agency of government—such as the national laboratories or the National Cancer Institute—to develop the science.

These two problems—developing alternatives to courts for review of environmental regulation, and developing better science for regulation—will require the creative cooperation of both lawyers and scientists if our society is to succeed in coping with environmental hazards.

REFERENCES AND NOTES

1. See Stewart, R., The reformation of american administrative law, *Harv. L. Rev.* **88**:1667 (1975).
2. *Reserve Mining Co. v. United States*, 498 F.2d 1073 (8th Cir. 1974).
3. For discussion of the proof of harm or risk necessary to sustain legal remedies for environmentally hazardous activities, see Gelpe, M., and Tarlock, D., The uses of scientific information in environmental decisionmaking, *So. Cal. L. Rev.* **48**:371 (1974); Green, H., The risk–benefit calculus in safety determinations, *Geo. Wash. L. Rev.* **43**:808 (1975); Krier, J., Environmental litigation and the burden of proof, in: *Law and the Environment* (M. Baldwin and J. Page, eds.), pp. 105, 107, Waker, New York, (1970).
4. See Stewart, op. cit., note 1 *supra*.
5. *Reserve Mining Co. v. United States*, 514 F.2d 492 (8th Cir. 1975); *Ethyl Corp. v. EPA*, 541 F.2d 1 (D.C. Cir. 1976), *cert. denied*, 426 U.S. 941 (1977). See also *Hercules, Inc. v. EPA*, 12 ERC 1376 (D.C. Cir. 1978) (toxaphene and endrin); *Environmental Defense Fund v. EPA*, 549 F.2d 998 (D.C. Cir. 1976) (Heptachlor/chlordane); *Environmental Defense Fund v. EPA*, 510 F.2d 1292 (D.C. Cir. 1975) (Aldrin/Dieldrin).
6. See Stewart, R., and Krier, J., *Environmental Law and Policy* (2nd ed.), Chap. 4, Bobbs-Merrill, Indianapolis (1978).
7. See Stewart, R., Paradoxes of liberty, integrity and fraternity: The collective nature of environmental quality and judicial review of administrative action, *Environ. L.* **7**:463 (1977).
8. See National Academy of Sciences, *Decision Making in the Environmental Protection Agency*, p. 45, National Academy of Sciences, Washington (1977).
9. For discussion of the problems and some suggested solutions, see Bazelon, D., Coping with technology through the legal process, *Cornell L. Rev.* **62**:817 (1977); Leventhal, H., Environmental decision-making and the role of the courts, *U. Pa. L. Rev.* **122**:509 (1974).
10. See Stewart, op. cit. note 7, *supra*.

11. See Landis, J., *The Administrative Process*, Yale University Press, New Haven (1938); Stewart op. cit. note 1, *supra*.

12. See Stewart, R., The development of administrative and quasi-constitutional law in judicial review of administrative decision-making: Lessons from the Clean Air Act, *Iowa L. Rev.* **52**:713 (1977).

13. Clean Air Act Sec. 112 (b)(1)(B), 42 U.S.C. § 741 2(b)(1)(B).

14. 29 U.S.C. § 655(b)(5).

15. 7 U.S.C. § 136a(c)(5)(A) – (D).

16. *Associated Indus., Inc. v. Dep't of Labor*, 487 F.2d 342, 345 n.2 (2d Cir. 1973).

17. See *EDF v. Ruckleshaus*, 439 F.2d 584 (D.C. Cir. 1971) (DDT); *EDF v. EPA*, 465 F.2d 528 (1972) (Aldrin/Dieldrin). See also *NRDC v. Train*, 545 F.2d 320 (2d Cir. 1976) (airborne lead).

18. *Society of Plastics Indus. v. OSHA*, 509 F.2d 1301 (2d Cir. 1975); *American Iron and Steel Institute v. OSHA*, 577 F.2d 825 (3d Cir. 1978).

19. See decisions cited note 5, *supra*.

20. *American Petroleum Institute v. OSHA*, 581 F.2d 493 (5th Cir. 1978), *cert. granted*, 99 S. Ct. 1212 (1979).

21. *Duke Power Co. v. Carolina Environmental Study Group*, 98 S. Ct. 2620 (1978).

22. See Stewart, op. cit. note 12, *supra*.

23. Sax, J., The (unhappy) truth about NEPA, *Oklahoma L. Rev.* **26**:239 (1973).

24. See Stewart, op. cit. note 12, *supra*.

25. See Ackerman, B., Rose-Ackerman, S., Sawyer, J., *et al.*, *The Uncertain Search for Environmental Quality*, p. 147, Free Press, New York (1974) [proposal for policy review board].

34

Contamination of Breast Milk
A Clinical and Ethical Dilemma

Charles U. Lowe
National Institutes of Health
Bethesda, Maryland 20205

I would like to discuss an issue that has assumed increasing importance for the developed countries and may in the future create concern in countries now moving from a pastoral or farming economy to one with an industrial base. The matter for examination is the contamination of human milk with organohalides, products of the chemical industry. In this country, federal and state officials have discussed the variables at such length as to give the impression that the problem is unique to the United States. The truth may be, however, that we are merely more aware of the issue than others. There is every reason to believe, however, based on studies of one organohalide, DDT, that the problem is a general one and that we may expect breast milk contamination in any environment where organohalides are in common use.

A posture of concern appears justified for several reasons: (1) Breast-feeding is increasing dramatically in this country. I have learned of one community in which 100% of postpartum women are committed to breast-feeding. Nichols has reported that in Houston, 90% of newborns are feeding at the breast and nationally the figure is approaching 50%. Women starting with nursing generally continue for about three months. (2) The breast milk of the majority of American women is probably contaminated with detectable amounts of one or more organohalides. (3) The chemicals of immediate concern are carcinogenic and/or toxic in aninal systems; in addition, they appear to be toxic to human adults and probably more so to the human young.

I cite the following observations as the basis for my belief of special sensitivity of the young, though acknowledging that the evidence is generally circumstantial rather than derived fron planned experinentation.

1. When a mixture of organohalides was vented by an industrial explosion in a chemical factory in Seveso, Italy, the exposed children, not the adults, showed chloracne. This lesion seems to be the earliest hallmark of organohalide intoxication.

2. A virtual epidemic of intoxication occurred in Yusho, Japan, when a mixture of polychlorinated biphenyls (PCBs) was inadvertently leaked into rice oil later consumed by humans. Levels of PCBs ingested were probably quite high. Adults had vague conplaints of malaise and some exhibited cholracne. Neonates born to exposed mothers exhibited a variety of abnormalities, including low birth weight, hyperpigmentation, jaundice, and conjunctivitis. Children fed at the breast of exposed mothers had higher levels of PCB in their serum than did those in a control group. Perhaps most important is the observation that a study nine years after the episode found that the exposed children showed neurological and developmental impairment.

3. In one instance a young girl played with sand contaminated by dioxin and developed fleeting chloracne and a hemorrhagic cystitis. No adults in the same environment were reported to be affected.

4. Studies of PCB-fed pregnant rhesus monkeys revealed that their infants were small-for-dates and when breast-fed developed a lesion similar to chloracne and are now reported to show some developmental abnormalities. I hasten to add that these studies must be viewed with circumspection and translated to human infants with some hesitancy because of the strikingly different amounts of depot fat in primates in comparison with that in humans. Nonetheless, the findings are disquieting since the levels of PCB in the primate milk were not very different from those found in the milk of some American women.

Organohalides are a group of compounds used widely by industry for many different purposes. They serve as insecticides, herbicides, wood preservatives, coating for containers, a low dielectric filler for high-voltage transformers, and a fire retardant. The most widely known examples are DDT; 2,4,5-T; dieldrin; lindane; PCBs; PBBs; and benzenehexachloride. Related toxic compounds appear as intermediates in manufacture, as by-products of synthesis or as toxic degradation products. As a class they are highly lipophilic and accordingly, when ingested by animals or man, are stored in depot fat or circulate in blood in association with the lipid fraction. They resist enzymatic destruction in animal systems and nonenzymatic degradation in the biosphere. In the light of these properties, it is not surprising that they persist in the environment and are found in the body fat of many animals and of a large majority of Americans. Although manufacture of some of these has been discontinued, the environmental burden is so great that exposure of man and animals will probably continue for a long time.

A logical consequence of the presence of these compounds in human fat is the observation that as depot fat is released for conversions to milk fat, the lactating woman mobilizes organohalides, which are subsequently excreted in breast milk associated with the fat moiety.

Contamination of breast milk is widespread in America, and only rarely related to industrial exposure of the mother. For example, in a study of 1500 lactating women, dieldrin was found in 80% of the milk samples, heptachlorepoxide in 63%, oxychlordane in 73%, and detectable quantities of PCBs in 90%. Thus we are not dealing with a geographically isolated matter or unique industrial exposure or manufacturing error (as in Seveso, Italy, or the case of livestock feed contamination with PBBs in Michigan, or the

Yusho episode in Japan). Rather, we are dealing with a national phenomenon. Why, one might ask, is there uncertainty about the extent of the hazard or even whether any hazard exists?

I believe there are two reasons. First, the true burden of organohalides received by the breast-fed infant is unknown. The amount reported in breast milk varies over three orders of magnitude. Since the compounds are highly fat-soluble, they follow the levels of fat in breast milk. But the fat levels in milk are not constant. They are on average low in colostrum, rise in transitional milk, and are highest in mature milk, remaining relatively stable after 30 days of lactation. But, to complicate the matter, fat content changes during a single feeding and may in first milk be as low as 2% and rise when the breast is almost dry to levels approaching 8%. And the levels vary during the day and between the two breasts. So there is a sampling problem. The volatile nature of milk fat levels confounds any statement on the amount of organohalide transmitted to the nursing infant, since any analytic number depends in large measure upon when in the nursing cycle and when in the feeding pattern the sample was obtained. The true value for contamination could thus vary at least fourfold.

With this kind of uncertainty, it is difficult to decide whether we have a real health problem or simply a technological success—the ability to measure these organohalides at levels in parts per billion.

The second reason for uncertainty lies in the ethical constraint preventing conduct of experiments in human infants. Thus we cannot determine the true hazard. We must derive our information from accidents, the occurrence of which we have every responsibility to prevent, or extrapolate from animal models. In addition, of course, we are uncertain as to the lesion, if any, for which we should search.

Nonetheless, the available information is sobering. The few human data at our disposal indicate that we are dealing with substances that can have adverse effects on infants. The animal experiments demonstrate toxicity or death and, for dieldrin, carcinogenesis. The human milk findings, whether or not quantatively correct, are certainly qualitatively impeccable. We wish to encourage breast-feeding as the ideal method for nourishing the infant, but concomitantly we have a growing concern that in so doing, we may be advocating a subtle intoxication, the full expression of which may be delayed for decades.

Table I places the issue in stark relief. In it, I indicate the presumed exposure to these compounds of a breast-fed infant and then consider what a regulatory agency might have to do when faced with similarly contaminated market milk. It would probably take the milk off the market.

I close with a plea. We need much more knowledge of these matters. Research is the only way to obtain facts and guidance. At the very least we must determine the precise level of exposure. Second, we need to ascertain what, if any, are the adverse health effects of this exposure. Studies now under way will give us answers to the first question. Only in the next decade would we know the answer to the second if we were to initiate a comprehensive research effort right now.

As we endorse breast-feeding, and I include myself among those who do, we are

TABLE I. Estimated Intake and Advisory Levels for Contaminant Chemicals[a]

Chemical	"Approximate average" ppm in human milk (fat adjusted)	"Action level": ppm for condemnation of market milk (fat basis)[b]	Daily intake in breast-fed infant using high value average range[c] (μg/kg)	WHO-allowable intake, (μg/kg per day)[d]	Intake by 5-kg infant (μg/day)
Dieldrin	0.04–.024	0.3	0.8	0.1	4.0
BHC (total)	0.04–1.08	0.3	3.6	—	14.4
Lindane	0.04–1.4	0.3	5.0	—	25
DDT (total)	2–8	0.05[e]	28.0	5	140
Heptachlorepoxide	0.32–1.2	0.3	4.0	0.5	20
PCBs	1.6–4	2.5[f]	14.0	1	56

[a] SOURCE: Data made available by Dr. Walter Rogan, NIEHS, NIH.
[b] From an informal "action-level document" of FDA, cited courtesy of Mr. Ellis Gunderson.
[c] Calculated on the basis of a 5-kg infant and an intake of 700 ml milk/day.
[d] Based on animal studies. Dieldrin is a carcinogen. Other agents are toxins.
[e] Tolerance level, whole milk basis.
[f] Temporary tolerance level.

inadvertently encouraging the consumption of chemicals, about whose possible long-term health effects we are ignorant. Although among the general population in this country we have never identified an infant with a demonstrable health effect resulting from the exposures described above, our level of certainty in this matter remains low. Animal studies confirm toxicity and even genic effects. Since we have an implicit responsibility to protect the infant, what should we do pending completion of our research agenda? I feel we have an ideal dilemma for discussion in a volume on genetics and the law.

DISCUSSION

PROF. ARTHUR D. BLOOM: Why is it that ninety percent of babies in Houston breast-feed? That is much higher than the national average.

DR. CHARLES U. LOWE: I have no answer except that there has been a concerted effort by nutritionists and pediatricians in that community to encourage breast-feeding.

DR. MARK DINE (Pediatrician): There certainly is an encouragement related to bonding between the mother and child and knowledge along those lines as well as the desire to do things *au naturel*. I would like to address the question of feeding practices after nursing, recognizing that these deposits of the haloids into the fat are cumulative, since the only way of screening the haloids is by nursing.

Since these effects are to be noted ten to forty years later, how soon after nursing has a child in America ingested an amount equivalent to that if they had not ever nursed? At what point could one say, even without nursing, that enough has accumulated to make the percentage that came from nursing not persuasive?

DR. LOWE: Fortunately, there is only one predictably contaminated food in our environment at the present time, and that is freshwater game fish. The life expectancy of most food animals is relatively short and since the environmental protection activities we now are engaged in generally protect these animals, it is unlikely that an infant or an adult is going to consume significantly contaminated food.

DR. DINE: Then over a period of time nursing mothers will not have the problem that the nursing mothers of today have?

DR. LOWE: My first answer would be, I hope so.

The one study with which I am familiar covered a five-year period and showed a slow increase in the amount of PCBs in breast milk of groups of sixty to one hundred mothers. PCBs are a rather special situation, since the vast amount of PCB is still located in closed containers (transformers, for instance). It is only ten to thirty years from now, when these transformers are disassembled, that the major amount of PCB manufactured in this country years ago will be released to the environment. Now these substances have been discontinued and are no longer made.

So what I would anticipate is a fall in PCB and then, ten to fifteen years from now, a rise.

MS. TABITHA POWLEDGE (Hastings Center): The inference that I gleaned from your presentation is that we might be wise to recommend bottle-feeding meanwhile. I wonder if one couldn't perhaps make an analogy here with the use of ultrasound in prenatal diagnosis. We know next to nothing about the long-term effects of ultrasound and yet it is now a standard recommendation. We really do believe it increases the safety of the procedure.

Given what we know, particularly what we recently know about the qualities of breast milk, including its immunological properties, it seems to me that one might legitimately argue for encouragement of breast-feeding until the data are in, rather than what I thought I heard you saying.

DR. LOWE: I do not want to leave with you the thought that I do not advocate breast-feeding. That is not the point of my presentation. My hope is that those who do advocate it in their practice understand what some of the problems may be, are alerted to them, look for them. In fact, we might have to develop a sort of fallback position in which we limit breast-feeding for a period of time pending analysis of milk, when we really are sure how to determine what the burden is.

35

REGULATORY POLICY FOR MUTAGENS AND TERATOGENS

NICHOLAS A. ASHFORD
Center for Policy Alternatives
Massachusetts Institute of Technology
Cambridge, Massachusetts 02139

INTRODUCTION

Many geneticists perceive mutation to be the most serious health risk associated with exposure to chemicals. In addition, we are becoming increasingly aware of possible causal relationships between chemical agents and birth defects. Whatever the magnitude of the risk of chemically induced damage to the reproductive system and to the fetus, we have not as yet established an adequate regulatory policy to deal with the problems.

If an effective policy for regulating mutagens and teratogens is to be established, that policy must be coordinated with our approach to controlling carcinogens. This is the case for two reasons: (1) The regulation of a substance only as a carcinogen or only as a mutagen or only as a teratogen is likely to have significant consequences for the types of risks for which the substance is not regulated. (2) The regulatory policy that emerges for carcinogens will have important consequences for mutagens and teratogens. Both the outcome of the challenge to the OSHA benzene standard in the U.S. Fifth Circuit Court of Appeals and the fate of the OSHA generic cancer standard will have serious implications for shaping future regulatory policy for mutagens and teratogens. This chapter is intended to (1) summarize the crucial facts that are relevant for the design of regulatory policy; (2) survey the legislation as well as the other legal tools pertaining to the control of carcinogens, mutagens, and teratogens; (3) discuss the difficult legal questions in need of resolution and their significance for regulatory policy; and (4) explain how the screening of workers and use of medical removal protection as regulatory tools can produce undesired results.

SUMMARY OF CRUCIAL FACTS

The scientific basis for the chemical origin of mutagenesis and teratogenesis has been firmly established in specific cases. The variety of documented episodes include these experiences:

- Nurses working in operating theaters suffer seven times the miscarriage rates of other nurses in the same hospital.
- Male anesthesiologists father an excess number of malformed children.
- Dibromochloropropane (DBCP) causes sterility in male workers and cancer in animals.
- Vinyl chloride causes both cancer and chromosomal abnormalities in workers.
- Lead may cause both serious birth defects and damage to the germ cells.
- Strong evidence exists linking toxic wastes at the Love Canal with excess birth defects there.

Since 1945, the petrochemical industry has been growing at a rate of 15% per year. This growth means that there has been a quadrupling in per capita consumption of petrochemicals every 10 years. Thus, in 1975, each one of us may have experienced an exposure to chemicals that was 64 times greater, on the average, than the 1945 exposure. Furthermore, the nature of chemical production has shifted more toward the kinds of substances, e.g., chlorinated hydrocarbons, foreign to our "evolutionary soup."

In light of these changes in the levels and nature of chemical exposures, those mutagenic problems attended by long latency periods may be expected to rise dramatically over the next few decades.

As in the case of cancer, detection of mutagenesis and teratogenesis is beset by difficulties associated with multiple causation. Epidemiological investigations do not reveal small increases in events. They reveal *excess* events. If there is a large number of small exposures to different agents that make up the background, causal relationships will never be epidemiologically detected for any of them. The background can be quite large, and a large number of small exposures has the potential to do us in. Thus there are severe constraints to the utility of human epidemiology in assessing the extent and cause of the harm in multiple exposure, real-world situations.

Biological monitoring, such as that for chromosomal breaking, may reveal some mutation; however, it certainly will not reveal all types of harmful mutations. Birth defects are hard to detect ahead of time. Modeling techniques for predicting the number of mutagenic or teratogenic events at low exposures are in their infant stage and suffer from many of the same problems as those facing cancer researchers, e.g., the problems of multiple causation, extrapolation to low doses, and the validity of interspecies extrapolation.

In contrast to carcinogenic risks, there is far more agreement about the validity of the extrapolation of mutagenic risks across species. The fundamental mechanism of mutation is known, while the fundamental sequence of events in carcinogenesis is still uncertain. Thus it should appear somewhat surprising that mutagens as a class are not regulated. The reason for this is that mutagenic damage is often more difficult to detect in humans and has not yet been documented in ways that can galvanize strong public responses.

If substances were regulated as carcinogens on the basis of short-term mutagenicity tests alone, many human mutagens would probably be regulated. If this were the case, the cancer regulatory policy would be providing protection from mutagens as well as from carcinogens. However, the OSHA generic cancer standard does not propose this. Instead, it proposes to use the Bruce Ames short-term test as only one indication of carcinogenicity, not as the sole indication. Thus, where the only evidence for carcinogenicity is a short-term test, the substances would not be regulated, and significant mutagenic, as well as possible carcinogenic, risk would result.

So far, only OSHA is using the generic approach. In the future, other agencies may also follow. In regulating hazardous wastes under the Resource Conservation and Recovery Act and in administering the Toxic Substances Control Act, EPA envisions promulgating generic standards.

Bruce Ames in his May 11, 1979, *Science* article[1] states, "A high percentage of carcinogens is also likely to be able to reach and mutate the germ cell." Unless the short-term tests that detect mutagenicity can be used as the sole criteria for controlling carcinogens, cancer policy will not furnish an adequate defense against mutagens. In addition, it must be recognized that insufficiently stringent regulation of carcinogens will not provide sufficient protection against mutagens.

Occupational exposure to lead demonstrates the need for coordinated policy efforts in regulating carcinogens, mutagens, and teratogens. Regulating lead only as a teratogen suggests a policy of removing women of child-bearing age from the workplace. That policy falls far short of the mark from a public health viewpoint: removal of female workers would not prevent the possible lead-induced mutation of sperm cells in male workers.

Even if carcinogens are stringently regulated, resulting in the *de facto* control of many mutagens (this would happen if the Ames test or one of the related tests were one criterion), teratogens would still remain a problem.

LEGISLATION AND OTHER LEGAL TOOLS FOR CONTROL

The issues discussed above form the background against which regulatory efforts at control must be designed. The four most important pieces of legislation that form the existing base for regulatory action are the Occupational Safety and Health Act (OSH Act), the Consumer Product Safety Act (CPSA), the Toxic Substances Control Act (TSCA), and the Resource Conservation and Recovery Act (RCRA).

The OSH Act provides that

Each employer—
(1) shall furnish to each of his employees employment and a place of employment which are free from recognized hazards that are causing or are likely to cause death or serious physical harm to his employees;

> (2) shall comply with the occupational safety and health standards promulgated under this Act. [Sec 5(a)]

and that

> The Secretary [of Labor], in promulgating standards dealing with toxic materials or harmful physical agents under this subsection, shall set the standard which most adequately assures, to the extent feasible, on the basis of the best available evidence, that *no employee* will suffer *material impairment* of health or functional capacity even if such employee has regular exposure to the hazard dealt with by such standard for the period of his working life. [emphasis added] [Sec 6(b)(5)]

Thus the OSH Act clearly includes some mutagenic hazards in the definition "material impairment of health or functional capacity" of workers. DBCP exposure in Texas was regulated on the basis of its being a sterilant. The extent to which somatic mutagens or teratogens could be regulated is not yet clear.

CPSA requires that the public be protected from "unreasonable risks." Under the CPSA, there has been very little activity except for a handful of carcinogens, including Tris and vinyl chloride. RCRA intends a generic classification option as a method for controlling carcinogens, mutagens, and teratogens. Although RCRA is three years old, EPA has not yet issued hazardous waste regulations that are required to define the characteristics of these toxic substances. According to EPA, the criteria for stating what constitutes a mutagen, a teratogen, or even a carcinogen have not yet been sufficiently developed. Once again, one must wonder why EPA does not consider the Bruce Ames and other short-term tests as good social policy indicators of mutagenic risk.

TSCA, protecting consumers, workers, and the general public, is just gearing up. The act obligates EPA to regulate "unreasonable risks." From TSCA's legislative history, it is clear that the determination of unreasonable risk is to be established by balancing the economic consequences of regulation and the utility of the chemical product to society (costs) against the preceived environmental health and safety consequences (benefits). Since unreasonable risk is not explicitly defined in TSCA, considerable discretion is left to the EPA administrator in deciding how the various factors in the balancing process should be weighed. Clearly, this latitude in the law is an open invitation to lobbyists on both sides of the environmental health issue.

TSCA contains a special provision for hazards that are carcinogenic, mutagenic, or teratogenic. It provides that when

> there may be a reasonable basis to conclude that a chemical substance or mixture presents or will present a *significant* risk of serious or widespread harm to human beings from cancer, gene mutations, or birth defects, the Administrator shall... initiate appropriate action... to prevent or reduce to a sufficient extent such risk or publish in the Federal Register a finding that such risk is not unreasonable [Sec 4(f)(2)] (emphasis added).

The opportunity for bringing political pressure to bear for "significant" risks is waiting to be seized. There is potential for very strong social pressure to be generated for carcinogens, mutagens, and teratogens—possibilities that do not exist for many ordinary classical toxins.

The scope of protection intended by these laws is different. CPSA provides no special protection for carcinogens, mutagens, and teratogens beyond regulating "unreasonable risks." TSCA appears to provide a possibility for considerably more protection, but a large amount of discretion is left to the EPA administrator. Regulations under RCRA have not been formulated and here, too, EPA retains some latitude in how protective it will be.

The workplace, where possibly the most extensive exposure occurs, provides limited protection from mutagens and perhaps none for teratogens. Can (will) OSHA protect the fetus? Does "material impairment" to a worker include physic damage to the worker due to parenting a child with birth defects? For any of the legislation to provide protection, easily adminstered teratogenic screening tests need to be devised. This may be a critical bottleneck.

Aside from standards set under the various pieces of legislation, other legal tools are available to encourage protective activity. Workers have the right to refuse hazardous work both under labor law and under the OSH Act. It is encouraging that in the *Whirlpool* decision the U.S. Supreme Court upheld OSHA's regulations protecting this right as valid within the context of the act's authority. The worker's right to refuse hazardous work may prove the most crucial self-help technique offered in the legal arena, whether or not standards are set.

Labor law and collective bargaining contracts can be used as private enforcement mechanisms, if they begin incorporating, for example, lists of suspect agents regardless of whether or not they are listed as candidates for standard promulgation. If the National Institute for Occupational Safety and Health or the National Institute for Environmental Health Sciences believes a substance to be a potential teratogen or mutagen, and merely makes an announcement to that effect, the process of incorporating environmental health provisions into collective bargaining contracts would be expedited. This strategy might be especially effective for the 25% of the work force that are unionized.

Through the tool of injunctive relief, the state courts may be petitioned to order an employer or a manufacturer to stop exposing people to a substance, in and apart from the fact that it may be a regulated substance. Under the provisions of both TSCA and CSPA, once administrative remedies have been exhausted, affected parties can bring suit in federal court to prohibit further exposures. Suits for damages can be brought under common law and are specifically provided for in CPSA, though not in the OSH Act or TSCA. While the likelihood of these suits becoming commonplace is slim, they are more and more frequently directed toward supervisory personnel in the chemical and materials industries. A high rate of success with these suits is not needed to change behavior. When a person in a responsible company position has a chance of losing his house and assets, he or she will begin to manage the industrial process a bit differently. The products liability area is the most promising area today for changing the way business is managed in this country. Personal liability suits may prove to be an additional avenue for encouraging protective action by supervisory personnel, especially those with technical knowledge.

Unresolved Legal Issues

Of critical importance is the fate of OSHA's generic carcinogen standard. Proposed two years ago, this comprehensive cancer control strategy represents both an attempt to meet the urgent need for regulation of carcinogens and an ambitious effort to streamline bureaucracy. The standard sets some general criteria as to what constitutes a carcinogen and relies on human epidemiology, animal tests, and short-term mutagenicity tests. The major legal problem here is what constitutes the appropriate legal standard of proof, i.e., establishing causality. If there is to be any inroad to controlling chemical carcinogens, mutagens, and teratogens, it is essential that this policy survive.

Another unresolved issue is the quantification-of-benefits requirement for carcinogenic exposures that the U.S. Fifth Circuit Court of Appeals seems to think is required in order for OSHA to change the maximum permitted benzene exposure from 10 ppm to 1 ppm. I am cautiously optimistic about this case's reversal by the Supreme Court, not on the grounds that there should not be a cost–benefit analysis, but on grounds that have to do with the legal authority that the Fifth Circuit incorrectly drew upon in interpreting OSHA's actions. Regardless of this case's final outcome, the economic rationalists who insist that there be a benefit stream to compare against the cost stream in making these decisions will continue to press. Dealing with this pressure will present a serious problem as long as there is no consensus on the number of bodies that fall from a 1-ppm exposure to benzene. Comparing the soft incidence-of-cancer numbers against the hard compliance costs is not a fair game. If producing the numbers for a carcinogen like benzene seems tough, wait until we try it for mutagens and teratogens. Will modeling techniques ever be sufficiently developed?

Awareness of the dangers of mutagens and teratogens has had an insidious side effect. It has unwittingly cleared the way for a new opportunity for discrimination against women. Title VII of the 1964 Civil Rights Act outlaws discrimination in employment on the basis of race, color, religion, nationality, or sex. The practice of women undergoing hysterectomies in order to qualify for certain industrial jobs is being reported. A major battery plant routinely rejects all female applicants capable of bearing children for all jobs with lead exposure. Whether there is a legitimate business necessity exception that would successfully counter a claim of sex discrimination in this practice remains to be tested. Furthermore, if mutagenic as well as teratogenic damage can occur, a protective strategy based only on removing women of child-bearing capacity leaves other women and men unprotected from possible risks.

Medical Removal Protection: A Double-Edged Sword

In the OSHA context, there is serious consideration being given to a general requirement for medical removal protection (MRP). The theory behind MRP is to monitor individuals for biological changes relevant to specific workplace exposures. If monitoring detects these changes, workers are removed from the source of exposure, i.e., their jobs.

Employers are to pay workers for the time during which workers are removed so as not to create a disincentive for workers to take medical exams or submit to biological monitoring tests.

Although biological monitoring for reproductive system changes can be done, such monitoring is uncertain because existing epidemiology is very poor. Thus identifying target populations to monitor will be difficult. A serious problem with applying MRP as a disease-control strategy is that the result of removing workers who have chromosomal breakage is the substitution of other, healthy workers in their place. Depending on the shape of the dose–response curve, rotating workers can actually *increase* the total amount of genetic damage. In addition, if younger workers are substituted for older workers the probability of perpetuating damage due to greater reproductive activity may be expected to increase greatly.

The reduction of exposure via generic standards must be the primary control mechanism. To focus on the damaged worker is to emphasize the wrong pathology. The most serious pathology is poor and misguided industrial practices for the control of hazards that implicitly regard damaged workers as replaceable defective factors of production that can be recycled like metal parts. Rotating workers is a dangerous solution. There is no substitute for reducing exposure to these substances.

The existing legislation can be used aggressively to provide a greater degree of protection to workers, consumers, and the general public than is currently provided. Pressure must be generated to regulate carcinogens, mutagens, and teratogens in a comprehensive and correlated manner that befits their related biological damage mechanisms.

REFERENCES

1. Ames, B. N., Identifying environmental chemicals causing mutations and cancer, *Science* **204**(4393):587 (1979). See also references cited therein, including Wyrobeck, A. J., and Bruce, W. R., Chemical induction of sperm abnormalities in mice, *Proc. Natl. Acad. Sci. USA* **72**(11):4425–5529 (1975); Wyrobeck, A. J., and Bruce, W. R., The induction of sperm-shape abnormalities in mice and humans, in: *Chemical Mutagens: Principles and Methods for Their Detection* (Vol. 5) (A. Hollaender and F. J. DeSerres, eds.), p. 257, Plenum Press, New York (1978); Wyrobeck, A. J., and Gledhill, B. L., Human semen assays for workplace monitoring, in: *Proceedings of a Workshop on Methodology for Assessing Reproductive Hazards in the Workplace*, National Institute for Occupational Safety and Health, Rockville, Maryland (1980).

36

DEVELOPING A GOVERNMENT POLICY FOR THE REGULATION OF ENVIRONMENTAL MUTAGENS AND TERATOGENS

Shopping at the Grand Bazaar

GREGORY T. HALBERT
Enforcement Division
United States Environmental Protection Agency
New York, New York 10007

> Think of Washington as a grand bazaar, a steamy marketplace filled with tents and tables, an elegant bazaar where the din of buying and selling drowns out the patriotic music and the cologne of expensive lawyers mingles with the sweaty aroma of hard bargaining. This is what Washington is—not a government, but a sprawling marketplace.[1]

There is the popular belief that when an environmental health hazard is recognized, a new federal law is all that is needed to correct it. When the law is passed, an agency located in the executive branch will administer it and soon the hazard will be abated. Take the Toxic Substances Control Act (TSCA),[2] which became law in 1976. In signing the law, President Gerald Ford said:

> This... legislation provides broad authority to regulate any of the tens of thousands of chemicals in commerce.... Through the testing and reporting requirements of the law, our understanding of these chemicals should be greatly enhanced. If a chemical is found to present a danger to health or the environment, appropriate regulatory action can be taken before it is too late to undo the damage.[3]

The job seems almost automatic. The Environmental Protection Agency (EPA)[4] will simply test chemicals, review reports submitted by chemical manufacturers, and determine which substances pose an unreasonable danger of cancer, or mutagenic or teratogenic damage. Regulation of these dangerous chemicals will then follow to prevent the damage. It sounds simple—too simple. It is. The law doesn't describe what tests are to be performed. The law doesn't prescribe what information will be included in the reports.

The law also doesn't define what combination of facts will equal "a danger to health or the environment" that President Ford spoke of.

Environmental protection laws are general statements of what Congress wants accomplished—in this discussion, prevention of mutagenic and teratogenic damage—coupled with broad guidelines of how the agency is to proceed. The rest is up to EPA and the forces of the Grand Bazaar.

The agencies and departments that constitute the executive branch of the government operate like an open-air bazaar. Each agency is a stall where it barters, within the limits of the laws it has to work with. The Civil Aeronautics Board brokers airline fares and referees intramural battles over commercial airline routes. "The Department of Transportation dickers with Detroit over how safe to make automobiles. The Environmental Protection Agency dickers on how clean to make them."[5]

Money from the United States Treasury is the most obvious commodity that is bought and sold at the bazaar. Government procurement and grant-in-aid programs are a multi-billion-dollar business at the stalls. However, the bazaar also sells intangible commodities, less obvious than the cash transactions, but every bit as important. This is the activity EPA is largely involved in. These intangible commodities are the agency regulations that implement laws. The buying and selling of regulations is known as rule making.

There is no right or wrong, no good or bad at the bazaar. Barter is the rule, influence and persuasion the currency. The shoppers are the regulated and other parties interested in the subject matter that is to be regulated. With statutes intended to control environmental mutagens and teratogens, the regulated are the industries that manufacture, import, use, and dispose of chemicals and other substances that have mutagenic or teratogenic properties. Some state and local governments are in this category as well, either as disposers of these substances or as friends of industry. Their concern is that too much regulation will force industries to close, thus depriving their citizens of jobs. The trade associations of the regulated and their friends also barter for regulations they can live with.

Other shoppers should be active at the EPA stall to barter for the regulation of mutagens and teratogens. The public, those who will come in contact with these substances, are welcome participants. People who will eat food containing pesticide residues, use products made from chemicals, drink water that contains pollutants, and live near dumps where hazardous waste is disposed of all have a vital, personal interest in the regulations that emerge from the rulemaking process.

Citizens have an important role to play in helping the government establish a regulatory policy for environmental mutagens and teratogens. The environmental laws require, as all laws must, that the government accumulate data to support its regulatory actions. The public interest shopper can bring his demands and needs to the Grand Bazaar by adding data and reasons for strong, effective environmental regulations. EPA and other public health agencies can't do the job by themselves. The nature of the regulatory process and the demand for hard data will not allow that.

The magnitude of the rule making conducted in the Grand Bazaar can be appreciated by the fact that the District of Columbia, with a population of 700,000, has more practicing attorneys—over 30,000—than do 49 states.[6] Only California has more. A substantial number of these attorneys work in administrative law, either representing industry clients before agencies such as EPA or working for the agencies themselves. The ones who represent private clients have one goal: to get the best deal possible for their clients, in drafting regulations and in litigating cases brought for violations of the regulations. The public interest deserves, and needs, the same vigorous representation.

The Carter administration and EPA both welcome public interest shoppers to the bazaar. As will be discussed later, the president has issued an executive order and EPA has issued regulations designed to open the administrative process to citizen participation. EPA has also provided grants to national and local citizen groups to help them develop the organization necessary to bring their perspective to the government.[7] What follows in this chapter is an explanation of the process by which these all-important administrative regulations are developed, and how citizens can participate in the process. It is an explanation of how this corner of the Grand Bazaar functions: how policy is determined, and how regulations are drafted.

There are three steps, or activities, that lead to the government regulation of environmental mutagens and teratogens. First, EPA must develop a regulatory policy for assessing the risk posed by these substances. Once this is accomplished, specific regulations as authorized by the relevant laws may be written. At any point, before or after these regulations are finalized, citizens may file lawsuits against EPA, or in some cases against a suspected violator of the laws or regulations. The purpose of a suit against EPA is to compel the agency to take an action required by law. A suit against a suspected polluter has as its goal the abatement or cessation of the polluting acts.

The regulation of environmental mutagens and teratogens begins with the laws enacted by Congress for this purpose. The Federal Insecticide, Fungicide and Rodenticide Act (FIFRA)[8] regulates pesticides and required that all pesticides sold in the United States be registered with EPA. If it is determined that the chemicals in a pesticide will present an unreasonable risk of injury to health or to the environment, the pesticide may be denied a registration. If this determination of risk is made after a product is registered, the registration may be canceled. Millions of dollars in potential sales of a pesticide are at stake in a cancellation proceeding. As a result, a long time—in one case almost four years—is required to cancel a registration and to take a dangerous product off the market. In the interim, the EPA administrator may suspend the registration pending the termination of the administrative hearings necessary to cancel the registration.

A recent example of a suspension involved the pesticides 2,4,5-T and silvex.[9] Both products are herbicides used to kill weeds in pastureland and in forests. A contaminant in the pesticides, TCDD,[10] is one of the most deadly chemicals known. Studies in Oregon have demonstrated a high incidence of miscarriages in communities located near fields sprayed with 2,4,5-T.[11] A pesticide registration can also be canceled when tests of ingredients of a pesticide show that it induces mutagenic effects.[12]

A person seeking to discharge pollutants, from a factory or a mine for example, into navigable streams or rivers must obtain a discharge permit from EPA under the Federal

Water Pollution Control Act (FWPCA).[13] The permit will name the pollutants and the quantities that may be discharged. This system recognizes that a complete and instantaneous prohibition of pollutant discharge is not possible. That would cause most industries to suspend operations and bring the country to a halt. However, by combining discharge pretreatment requirements phased in over a period of years with the permit requirement, the amount of pollutants that enter the waters of the United States is reduced and water quality is improved. A potential discharger is not always given a permit to dump what he wants to dump, nor the quantities that he desires. The discharge permit is issued only after an administrative process—which can include a public hearing[14]—that considers the potential harm from the discharge as well as alternative disposal methods.

Land disposal of hazardous wastes has resulted in environmental crises in some parts of the country. Toxic chemicals in 55-gallon drums buried in open fields have escaped as the containers rusted and corroded. The vapors have risen through the ground to come in contact with people living in homes built nearby and have been linked to birth defects in children born to mothers who lived in these areas during their pregnancies.[15] Under the Resource Conservation and Recovery Act (RCRA),[16] disposal of hazardous waste will be allowed in the future only at those sites that have a permit issued by EPA for that purpose. This will keep hazardous substances away from existing residential areas. Records will be maintained to show the location of former disposal sites so that these areas can be avoided in future construction.

Public supplies of drinking water can be protected under the Safe Drinking Water Act (SDWA).[17] This law authorizes the EPA administrator to determine dangerous contaminants and to list by regulation their identity and the maximum levels of each that will be allowed in drinking water.

Under the Toxic Substances Control Act, the EPA administrator may require manufacturers of potentially harmful chemicals to conduct tests on the chemicals. Testing may be directed to clarify the health and environmental effects, including carcinogenic, mutagenic, or teratogenic effects. Based on the results of this testing, the administrator may find that there is a reasonable basis to conclude that a chemical presents an unreasonable risk of injury to health or the environment. In that case, he may issue an administrative rule to limit or prohibit entirely the manufacture and use of the offending chemical.

This law requires that the use of polychlorinated biphenyls (PCBs) be phased out. PCBs are widely used to insulate electrical equipment. Data have accumulated to show that PCBs are also a potent carcinogen and mutagen.[18]

Regulation of environmental mutagens and teratogens presents a new challenge to EPA. Historically, the agency has administered its laws on a media basis. The FWPCA regulates water pollution. The Clean Air Act (CAA)[19] governs the emission of pollutants into the air. The FIFRA regulates pesticides. Each program has been carried out independently of the others. However, regulation of mutagenic and teratogenic substances under TSCA and RCRA will transcend the traditional media lines. Also, regulatory efforts under one law must be coordinated with similar efforts under other laws to provide greater protection from these serious health hazards.

Other federal government agencies administer laws that apply to mutagenic and teratogenic substances. Each law contains different authorities and addresses health problems from separate perspectives. These regulatory activities must be coordinated. Three of these agencies have joined with EPA to form the Interagency Regulatory Liaison Group (IRLG).[20] They are the Food and Drug Administration (FDA), Consumer Product Safety Commission (CPSC), and the Occupational Safety and Health Administration (OSHA). This coordination effort was begun in 1977 and has provided for cross-training of enforcement personnel of each agency in the laws that other IRLG members enforce. Joint laboratory test protocols will also be developed.[21]

The task of regulating environmental mutagens and teratogens is further complicated by the very nature of the substances. They must be identified. To do this, test methods must be developed. It isn't necessary to prove to a scientific certainty that a substance has mutagenic or teratogenic properties before regulatory action can be taken. However, some evidence must be developed to support a conclusion that a substance has the potential to produce these effects.

EPA is now determining what test protocols will be acceptable to provide these data. When that task is complete, the 44,000 chemicals currently in commerce will have to be tested.[22] Only a small part of this testing can be performed by EPA. The agency does not have the laboratory resources to do the job. Nor the personnel. Toxicologists are in short supply. The passage of TSCA and RCRA has created a strong demand for people with toxicology training. The industries that will be subject to regulation under these laws are competing with EPA for their services. Many companies are able to bid higher than EPA. This again emphasizes the need for public input of scientific data, both laboratory and real-world experience, to EPA.

REGULATORY POLICY

The current approach employs the risk–benefit analysis. EPA developed this approach, with regard to carcinogens, in 1976.[23] It will be discussed in that context. However, the same principles will apply to mutagens and teratogens.

Until the late 1950s there were no regulatory actions taken against chemical carcinogens. However, in the 1950s there were studies of the health effects of radioactive fallout from the testing of nuclear bombs. As a result of these studies, standards were developed setting permissible exposure levels to ionizing radiation. These standards were somewhat arbitrary. They were set below levels of exposure that were known to have produced damaging health effects. These permissible exposure levels were not considered safe but were the maximum limits of exposure that would be allowed. Actual exposure was to be kept to a minimum. But, it was soon realized that there was no such thing as a safe dose. Any exposure, however small, will confer some risk of cancer on the people exposed to radiation. This finding led to the development of the no-threshold concept.

Evidence has also accumulated to indicate that the no-threshold concept is applicable to chemical carcinogens. The first chemical carcinogen regulatory measure was the

Delaney Clause that was added to the Federal Food, Drug and Cosmetic Act.[24] The Delaney Clause imposed a complete ban on any food additive that showed evidence of tumorigenic effect on humans or animals. This requirement is an attempt to eliminate all risk of cancer. However, it has become increasingly clear that all risk cannot be eliminated without serious social and economic consequences that are not acceptable. A recent example of this is the controversy surrounding the attempt by the Food and Drug Administration acting under the Delaney Clause to ban the use of saccharin in carbonated beverages. Saccharin is the only nonsugar sweetener easily available for "diet drinks." In a country where people want to combine slimness of figure with soft drinks, a nonsugar sweetener is essential.

The approach now taken by EPA in regulating carcinogens is to balance the risk of the disease against the benefits from use of the substance that presents the risk of cancer. The risk–benefit analysis involves a two-step decision-making process. The first decision is whether or not a particular substance presents any risk of cancer. This decision is made without regard to degree of risk and without regard to amount of exposure to the substance, whether by touch or by ingestion. If there is a risk of cancer, the second decision is what regulatory action, if any, should be taken to eliminate or reduce that risk.

With respect to the first determination, in many cases data accumulated from experiments conducted on animals will have to provide the basis for the decision. Environmental exposure of humans to a substance frequently requires so much time to result in cancer that timely decisions cannot be made. Moral principles will not permit experiments to be conducted on humans. Therefore, it will not be possible in many cases to "prove" that a substance is in fact a human carcinogen. When a substance "causes a statistically significant" incidence of tumors in test animals or in humans, it will be considered a possible carcinogen. This does not mean that EPA will take regulatory action, however.

The finding that a substance is a potential carcinogen will lead to a detailed risk–benefit analysis of the substance. Once this analysis is complete, EPA will decide what regulatory action to take. This decision will be reached after a consideration of the extent of the risk, the benefits conferred by the substance, the availability of substitutes, and the costs to control the substance. If the benefits clearly outweigh the risks, no action will be taken. Conversely, if the risks exceed the benefits, some regulatory action will be taken. This may range from complete ban on the use of the substance to limitations on the use or concentration of the substance that may be used.[25]

This risk–benefit analysis policy was adopted by EPA even though it was recognized that cancer research of the type necessary for this decision-making process was not well developed. Given the catastrophic effects that undetected carcinogenic substances can produce, it is the only viable approach.

Decision making regarding carcinogenesis, mutagenesis, and teratogenesis hovers at the frontiers of scientific knowledge. A well-stated description of the difficulty facing EPA was provided by the opinion in the *Ethyl Corp. v. Environmental Protection Agency*,[26] a landmark decision in environmental law:

> Man's ability to alter his environment has developed far more rapidly than his ability to foresee with certainty the effects of his alterations. It is only recently that we have begun to appreciate the danger posed by unregulated modification of the world around

us, and have created watchdog agencies whose task it is to warn us, and protect us, when technological "advances" present dangers unappreciated—or unrevealed—by their supporters. Such agencies unequipped with crystal balls and unable to read the future, are nonetheless charged with evaluating the effects of unprecedented environmental modifications, often made on a massive scale. Necessarily, they must deal with predictions and uncertainty, with developing evidence, with conflicting evidence, and, sometimes, with little or no evidence at all.[27]

In *Ethyl*, EPA regulations that required reduction in the lead content of leaded gasoline were challenged by manufacturers of lead additives and gasoline refiners. The regulations were promulgated pursuant to Section 211(c)(1)(A) of the Clean Air Act,[28] which authorizes the EPA administrator to regulate gasoline additives whose emission products "will endanger the public health or welfare...." One of the bases for the challenge was that the administrator did not have sufficient scientific evidence to support his decision that gasoline lead additives met the statutory criterion, "will endanger the public health or welfare." The court found that the administrator did have sufficient evidence, and upheld the regulation.

One of the arguments advanced by the Ethyl Corporation was that the administrator must find that the danger to health posed by lead additives was "probable."[29] The court disagreed, saying, "Danger depends upon the relation between the risk and harm presented by each case...."[30] Where the harm is severe, less showing of probability is necessary to produce a danger to public health.

Ethyl Corporation next contended that the administrator could not make a legislative policy judgment about the relative risk to health posed by gasoline additives, but could only make "a factual determination concerning the relative risks of underprotection as compared to overprotection."[31] Again the court rejected Ethyl's argument. "[T]he power to assess risks, without relying solely on facts, flows inexorably from the nature of the 'will endanger' standard."[32]

The court emphasized that in making public health regulatory decisions EPA is not bound by the same rigid rules of evidence that apply to a party in a courtroom trying to prove a fact: "... determination of endangerment to public health is necessarily a question of policy that is to be based on an assessment of risks that should not be bound by either the procedural or the substantive rigor proper for questions of fact."[33]

Regulators such as EPA must be accorded flexibility in making their determinations of risk to health because: "'reasonable medical concern' and theory long precede certainty. Yet the statutes—and common sense—demand regulatory action to prevent harm, even if the regulator is less than certain that harm is otherwise inevitable."[34]

The *Ethyl* opinion is important for it indicates the latitude the courts will allow EPA in making the necessary findings that a given substance poses an unacceptable risk of injury to health, be it cancer or mutagenic or teratogenic damage. The certainty with which this finding of risk or injury must be made is considerably less than the scientist requires before stating a scientific fact. The evidence that a danger to public health is present must be greater than the counterevidence that there is no endangerment. How much greater depends on the seriousness of the health hazard. Is it chronic, irreversible, fatal, affecting a large segment of the population? These are some factors that would add to the seriousness of the danger. Where the seriousness increases, the burden of proof that must be sustained decreases. The continuing battle at the Grand Bazaar centers over the questions of how serious a health hazard is and how reliable the data are that EPA is relying on.

The risk–benefit analysis is utilized with FIFRA and TSCA. These laws apply to pesticides and other chemicals that have both beneficial and harmful properties. The other laws, FWPCA, RCRA, and SDWA, regulate substances that are either discarded or serve no beneficial purpose. They contain their own standards, which must be met for regulatory action to be taken.

The focus of EPA in administering the laws that regulate mutagenic and teratogenic substances is the effect these substances have on drinking water supplies. That does not mean that other health threats are ignored. However, dangerous substances in drinking water can create the largest population-at-risk.

Drafting Regulations

The procedure that agencies such EPA must follow in promulgating regulations is set out in the Administrative Procedure Act (APA).[35] These procedures are intended to give the public an opportunity to provide information to agencies such as EPA with the goal that regulations accurately carry out the intent Congress had in enacting the law.

A *rule* or *regulation* (the terms are interchangeable) is defined in the APA[36] as any agency statement of general or particular applicability and future effect that implements, interprets, or prescribes a law. Rule making is the agency process for formulating or amending a regulation. A person includes an individual, partnership, corporation, association, public, or private organization. A party is a person appearing before an agency seeking relief of some type. The relief may include the drafting of a regulation to cover a certain subject, or the substantive content of the regulation.

A four-step process is followed in the promulgation of regulations. The first step, the Advance Notice of Proposed Rulemaking (ANPR), was established by Executive Order 12044,[37] which will be discussed later. This notice is published in the *Federal Register*, the daily newspaper of the executive branch of the federal government. It is published by the General Services Administration (GSA) and includes material submitted for publication by all executive branch agencies. The ANPR will state the legal authority for the proposed rule making and will describe the subject matter of the rule making. The notice will also solicit public comment as to the form the regulations should take, the scope, and the content of the regulations.

An example of an ANPR appeared in the *Federal Register* of February 16, 1979.[38] It stated that EPA was considering expansion of the criteria for designating hazardous substances under the Federal Water Pollution Control Act. The intent of the new regulations, in part, "is to consider substances which are carcinogenic, mutagenic, teratogenic, radioactive and which bioaccumulate in animal tissue to cause chronic toxic effects."[39]

Responses and comments to 10 specific questions were requested, and included: "(1) What is the appropriate basis for designating a carcinogen, mutagen, or teratogen as a hazardous substance? (2) How should reportable quantities be defined for carcinogens, mutagens or teratogens?"[40]

The public was given three months, until May 17, 1979, to submit comments on these questions to EPA.

The goal of an ANPR is to provide EPA with broad scientific input bearing on the subject matter of the proposed regulations. It is the opportunity for the public at large to make available to EPA knowledge gained through research and other studies of mutagenic and teratogenic substances that may be contained in waste water effluent, the subject of these regulations.

Following the ANPR, the APA procedures apply. A Notice of Proposed Rulemaking will be published in the *Federal Register*. This notice will state the legal authority for the proposed rules, citing the law and the relevant section(s). There will also be an identification of the subjects, in greater detail than the ANPR, that will be covered by the proposed rules. The public will be allowed 60 days to submit comments and suggestions to the agency. The name and address of the project officer is also given.

An example of this type of notice appeared in the *Federal Register*, March 11, 1977.[41] It pertained to proposed rulemaking under sections 8(c) through section 8(e) of TSCA.[42] These regulations were to set out the requirements on manufacturers of chemical substances to keep records of significant adverse reactions to health or the environment ascribed to their chemicals and to submit health and safety studies to EPA.

After public comments are received and analyzed by EPA, the proposed regulations will be drafted and published in the *Federal Register*. The public is again afforded 60 days in which to submit written comments to EPA concerning the regulations. Where there is a request, and if EPA determines that it would be helpful, public hearings will be held to receive oral comments to the proposed regulations. When the period for public comment closes, the comments are considered by the agency.

The final regulations are again published in the *Federal Register*. A preamble to the regulations is included that explains the scope and coverage of the component parts of the regulations. A reply to the public comments is also included in this publication. In this reply, EPA will summarize the comments received on each section of the proposed regulations, and explain how they were incorporated in the final version, or explain why certain comments were not considered. This preamble provides the "promulgative history" for the regulations in the same manner that congressional committee reports provide a legislative history for laws.

Citizen participation in the regulation drafting process is encouraged by a presidential executive order and by EPA regulations. President Carter issued Executive Order 12044 on March 23, 1978. The policy stated by this order was that agency regulations were to be "as simple and clear as possible. They shall achieve legislative goals effectively and efficiently."[43] To achieve these objectives, regulations are to be developed through a process that provides an opportunity for early participation and comment by individual members of the public.

To facilitate increased public participation in the regulation drafting process, the executive order grafted two procedures onto the steps already set out by the Administrative Procedure Act. An agenda of significant regulations under development or review must be published at least semiannually. On the first Monday in October each agency must publish in the *Federal Register* a schedule showing the times during the coming fiscal year when the agency's semiannual agenda will be published. Supplements to this agenda may be published at other times during the year as may be necessary. EPA scheduled four such agenda publications in fiscal year 1979.

The second procedural change provided by the executive order is to include opportunities for "early and meaningful" public participation in the development of agency regulations. People who commented on the draft of this executive order repeatedly stressed the need for early public participation in the rule-making process. It was felt that once a regulation is proposed, in the form the final version will take, it is quite difficult to change the language.

The order directs agencies to consider specific actions to encourage early public participation, and mentions as examples: "(1) publishing an advance notice of proposed rulemaking; (2) holding open conferences or public hearings; (3) sending notices of proposed regulations to publications likely to be read by those affected; and (4) notifying interested parties directly."[44]

As stated earlier, EPA issues advance notices of proposed rule making. Public meetings are also held at various cities across the country to solicit ideas and comments from the public. An example of this is the series of two-day public meetings that were held in each of the 10 EPA regions to discuss possible strategies for implementing the Resource Conservation and Recovery Act. After these meetings, 80 smaller workshops were held to discuss hazardous waste management.

Section 3 of Executive Order 12044 requires agencies to prepare an analysis of those regulations that have major economic consequences for the general economy, for individual industries, or for geographic regions. A Regulatory Analysis Review Group (RARG) was established in the Executive Office of the President. Chaired by the Council of Economic Advisors (CEA), this group coordinates the economic analysis by agencies of their regulations. This analysis is conducted after the proposed regulations are published and before they are put in final form.

The RARG review of regulations is a critical step. The industry that will come under the subject regulations can be expected to challenge them on the grounds they will be unreasonably costly.[45] This can be a difficult argument for EPA to meet. The cost of complying with a regulation lends itself to quantification, while the value of the benefit realized from the regulation does not. How is the cost of medical services avoided through regulation of a mutagenic or teratogenic substance determined? How many people would have been afflicted if the substances were not regulated?

This RARG regulation review should be seen for what it is—another stall at the Grand Bazaar. Concerned shoppers for effective regulation must be alert to this, and be prepared to make their argument before this group in support of proposed regulations.

EPA issued regulations February 16, 1979, to provide a mechanism for public participation in programs under RCRA, the Safe Drinking Water Act, and the Clean Water Act (another name for the FWPCA).[46] These laws all provide for the issuance of permits for polluting activities: hazardous waste disposal sites under RCRA, underground injection under SDWA, and pollution discharge into navigable waters under the CWA.[47]

The public participation regulations apply to EPA and state rule making, and the issuance and modification of permits, activities that may be delegated to state governments. The regulations also cover the development by EPA of strategy and policy guidance memoranda under these laws. Strategy and policy are important because they will influence the subsequent development of regulations and issuance of permits. The objectives in carrying out activities covered by these regulations are to promote public understanding of government programs and proposed actions, to consult with interested and affected segments of the public, and to encourage public involvement in implementing environmental laws.

EPA and state agencies are directed to conduct a continuing program for public information and participation in policy and regulation development. As part of the information requirement, each agency will maintain a list of persons and organizations who have expressed an interest in, or may be affected by, actions taken pursuant to these three laws. An example would be the permitting of a site for hazardous waste disposal. The permitting of these disposal sites is an activity that EPA must turn over to a state upon its request and the EPA determination that the state has an adequate program for this purpose. The permitting activity, also known as siting, has developed into a major political issue. No one wants a Love Canal in his or her neighborhood. The Love Canal is a former toxic chemical disposal site in Niagara Falls, New York. Fifty-five-gallon drums containing all types of lethal chemicals buried long ago have corroded and the contents have risen to the surface in the backyards of people who built homes there, unaware of what was buried in the area. Children born to mothers who lived in this area suffered birth defects that are linked to the seepage of the chemicals buried in the canal. The all-important decision of where to locate these toxic chemical disposal sites in the future will be made by state and substate, or regional, agencies.

Where possible, public hearings or meetings will be held before regulations are drafted or permits issued under these laws. In addition, state and local agencies are required to establish citizen advisory groups to make recommendations to the agencies regarding decision making. Such advisory groups must be representative of the interests in the area and may include members of public interest groups. However, an advisory group may not include a person who is likely to receive a special benefit from any decision made by the agency that the group advises.

Executive Order 12044 and the EPA regulations for public participation in programs under RCRA, SDWA, CWA, and the Administrative Procedure Act provide an opportunity and a challenge to the public. Citizens can have a valuable input to the making of important decisions that will greatly affect their lives. This work may be tedious, detailed, and time-consuming. But this is how the government develops its regulation of environmental mutagens and teratogens. The persons who have a financial interest in these regulations, the industries whose activities cause these serious health threats, are already active in the regulation-drafting and permit-issuance processes.

Up to this point the discussion has focused on how citizens can participate in the regulatory drafting process by reacting to proposals developed by EPA. However, citizens can also initiate administrative action. Section 4(e) of the Administrative Procedure Act[48] provides: "Each agency shall give an interested person the right to petition for the issuance, amendment, or repeal of a rule."

This is not an action that an individual can easily take. However, individuals acting through environmental organizations such as the Sierra Club, the Environmental Defense Fund, or the Natural Resources Defense Council, or acting through their state governments can, and have, submitted petitions to EPA to initiate administrative actions.[49]

The final regulations promulgated by EPA are subject to review in federal court. This procedure is called judicial review and illustrates the principle of checks and balances involving the executive, legislative, and judicial branches of the government. It allows the third branch of the government to determine whether the executive branch followed the Administrative Procedure Act and the environmental protection law enacted by Congress in drafting regulations.

Any person who suffers a "legal wrong" or is "adversely affected or aggrieved"[50] by the promulgation of an EPA regulation may file suit to challenge it. Such suits are frequently filed by regulated industries, environmental organizations, and public interest groups.[51] Only the final version of the regulation is subject to judicial review. This fact places added importance on participation in the rulemaking process at the administrative level. A record is maintained by EPA of all comments and hearing testimony received while the regulations are being developed. The more data a party has submitted to EPA in this process, the stronger their judicial challenge of the regulation can be.

The scope of the review of EPA regulations that a court may conduct is set out in Section 10(e) of the APA.[52] A court may set aside a regulation if it finds that it is arbitrary and capricious, is in violation of the Constitution, is in excess of statutory authority, or was promulgated without observance of procedure required by law.

One way of showing that a regulation is arbitrary and capricious is to show that EPA did not consider significant evidence presented to it through public comments or testimony. It must be kept in mind that "the public" includes the subjects of the regulations as well as public interest spokespersons. Neither side has a monopoly on the use of the term. This underscores the need EPA has for data to support strong regulation of environmental mutagens and teratogens. If the persuasive weight of the evidence presented to EPA during rulemaking supports the position taken by industry, a regulation would be vulnerable to attack on judicial review if this evidence were summarily ignored by the agency. It cannot be ignored. It must be balanced and outweighed by evidence that supports strong protection of health and environment.

CITIZEN SUITS

Another tool for shaping the government regulation of environmental mutagens and teratogens is the citizen suit. A citizen, or citizen group, such as an environmental public interest group, can file suit in federal court to compel EPA to take nondiscretionary action that is needed but for some reason is not being taken. There are an unlimited number of environmental protection needs. However, EPA resources to respond to those needs are severely limited. A citizen suit that is successful results in the entry of a court order compelling certain action by EPA before a deadline. This has the effect of forcing EPA's attention to a specific problem, such as mutagens or teratogens.

Many of the environmental protection laws contain deadlines established by Congress for the development of regulations. If they are not met, suit can be brought. Under 28 United States Code 1331 a suit can be brought in federal court over a federal question, that is, a legal dispute that arises out of federal law. Here the dispute would be over compliance by EPA with the statutory deadline.

All of the environmental protection laws that have been discussed here, except FIFRA, contain provisions to authorize citizen suits.[53] Before a suit can be brought under these provisions, the person must give EPA a 60-day notice of intent to sue. This will allow EPA an opportunity to take the action that is demanded without resort to litigation. In addition, these provisions in the environmental protection laws authorize the award of costs of the litigation to the person suing EPA if the court determines the suit has merit. These costs can include attorney and expert witness fees.

Under the specific provisions of these four environmental protection laws, suit can also be filed against a person (including the United States and any federal agency) that is alleged to be in violation of the law, or any permit issued pursuant to it. In these suits, EPA has the right to intervene, if it is not already a party. This type of citizen suit would seek an injunction to stop the alleged illegal activity.

An example of the first type of citizen suit is the case of *Illinois v. Costle*.[54] The state of Illinois sued the EPA administrator over the failure of EPA to meet the RCRA requirement that regulations governing disposal of hazardous wastes be promulgated by EPA within 18 months of enactment of the law.[55] The court was sympathetic to the complaint of Illinois and its coplaintiffs, various environmental public interest groups. Yet it pointed out its obvious limitations. The court could not appropriate funds to EPA to speed up this work, nor could it let consulting contracts to develop the necessary technical information to base the regulations on. What the court could do is use its influence to assure that EPA observes the congressional intent that the regulations be developed promptly to abate this serious health hazard.

The issue before the court was whether EPA was proceeding in good faith, given the resources it had available to it. The court found that EPA was acting in good faith, and accepted a new set of deadlines that EPA and the plaintiffs had agreed on, and noted the basis for this recurring problem of missed statutory deadlines: "Well-meaning statutes are not self-implementing. We need a national will to protect the environment from the threatening health and pollution hazards which this Act addresses. There is need of a massive commitment of funds, talent and purpose to these objectives."[56]

EPA is working to develop this "national will" that the court spoke of as essential if the environmental statutes are to be enforced to their full extent. There is no alternative to public participation—as individuals and as members of organizations—in the development of the regulation of environmental mutagens and teratogens. This participation is needed by EPA and by the state and substate agencies that will implement the laws Congress has enacted to protect the public health from these serious health hazards.

REFERENCES AND NOTES

1. *Washington Post*, p. 6, col. 1, Jan. 20 (1977).
2. 15 United States Code 2601 *et seq.*
3. Oct. 11, 1976, as reported in *EPA Journal* **Nov.–Dec.**: 2 (1976).
4. The Environmental Protection Agency (EPA) is a regulatory agency established by Reorganization Plan Number 3 of 1970. It is charged with the administration and enforcement of environmental protection laws. EPA is headed by an administration who is appointed by the president with the advice and consent of the United States Senate.
5. *Washington Post, supra,* note 1.
6. *District Lawyer* **Dec.–Jan.**: 4 (1978).
7. EPA has distributed $106,000 among 25 public interest organizations in New York and New Jersey. The groups will identify citizen roles in toxic substances issues, suggest ways the public can be educated about toxic pollution, and develop local-level action programs for dealing with toxic chemical problems. *Environmental Reporter*, p. 1423, Dec. 8 (1978). EPA has also made grants to the Natural Resources Defense Council (NRDC), the National Wildlife Federation, and the Izaak Walton League.
8. 7 U.S.C. 136 *et seq.*
9. Feb. 28, 1979. Published in *Fed. Reg.* **44**:15874 (Mar. 15, 1979).
10. 2,3,7,8-tetrachlordibenzo-p-dioxin.
11. Eight women informed EPA that they lived within 12 miles of Alsea, Oregon, where 2,4,5-T and silvex are used in forest management. They experienced a total of 13 miscarriages between 1972 and 1977. Most of the miscarriages occurred 8 to 10 weeks after conception and 4 to 6 weeks after spring application of 2,4,5-T in nearby forest areas. This was a rate almost three times greater than for control groups that were studied. *Fed. Reg.* **44**:15880 (1979).
12. Criteria for determining unreasonable adverse effects are contained in 40 CFR 162.11. Mutagenicity standard is given at 40 CFR 162.11(a)(3)(ii)(A).
13. 33 U.S.C. 1251 *et seq.* The National Pollutant Discharge Elimination System (NPDES) permit program is established uner 33 U.S.C. 1342.
14. 33 U.S.C. 1342(a)(1) and implementing regulations, 40 Code of Federal Regulations Part 125.
15. Preliminary data show an increased rate of miscarriages, birth defects, and low birth weights in pregnant women who live in areas surrounding the Love Canal in Niagara Falls, New York. The canal was long used as a dumping ground for toxic chemicals. *Environmental Reporter*, p. 2003, Feb. 23 (1979).
16. 42 U.S.C. 6901 *et seq.* Hazardous waste, solid waste that causes an increase in serious irreversible illness, is dealt with in sections 6921–6931.
17. 42 U.S.C. 300f *et seq.*
18. 15 U.S.C. 2605(e). PCBs have been detected in human breast milk. An estimated 300 million pounds of PCBs have escaped into the environment. Particularly tainted waters include Chesapeake Bay, the Hudson River, San Francisco Bay, the Gulf of Mexico, waters in Florida, and Puget Sound. A story is told about early-day PCB salesmen. The substance was then believed to be inert. Salesmen would pour themselves a shot glass of PCBs and drink it in front of customers to convince them of this fact.
19. 42 U.S.C. 7401 *et seq.*

20. Notice of formation of IRLG was published in *Fed. Reg.* **42**:54856 (Oct. 11, 1977). The Food Safety Quality Service of the United States Department of Agriculture (USDA) joined the IRLG in 1979.

21. The IRLG has published a list of 24 chemical substances or hazards that will be the focus of coordinated efforts by the member agencies. *Environmental Reporter*, p. 1422, Dec. 8 (1978). Included in the list of 24 are acrylonitrile (teratogen and mutagen), dibromochloropropane (sterilant), and ethylene oxide (mutagen).

22. Little is known about the chronic effects of these chemicals. Testing has been confined largely to acute toxic effects, and knowledge of environmental transport and fate, biological pathways, and chronic long-term effects is scant. A shortage of laboratory facilities and professionals trained in toxicology, industrial hygiene, pathology, and other such disciplines is a bottleneck in the flow of needed information. The lack of a comprehensive system for storing and analyzing information makes it difficult to use the data that do exist. *Environmental Quality, The Ninth Annual Report of the Council on Environmental Quality*, p. 178 (Dec. 1978).

23. *Fed. Reg.* **41**:21402 (may 25, 1976).

24. 21 U.S.C. 301 *et seq.* The Delaney Clause was added to section 348(c)(3)(A).

25. The weighing of risk and benefit of a substance inescapably is a comparison of dissimilar qualities. In the decision whether to suspend certain uses of 2,4,5-T and silvex, the risks were apparent increase in human miscarriages and fetotoxic effects in test animals. The benefits were largely economic. It was concluded that alternatives to these pesticides were available. Cost of agriculture production would increase only lightly from suspension of these products. Rangeland use of 2,4,5-T and silvex was not suspended because exposure to pregnant women was concluded to be minimal from this form of application. If a viable alternative to 2,4,5-T and silvex for weed control was not available, the registrations might not have been suspended. The administrator could have ordered changes to use directions to limit the concentration of the pesticide that could be used or to require communitywide warnings be given before application.

26. 541 F.2d 1 (D.C. Cir. 1976).

27. 541 F.2d at 26.

28. 42 U.S.C. 7545(c)(1)(A).

29. 541 F.2d at 18.

30. 541 F.2d at 18.

31. 541 F.2d at 20.

32. 541 F.2d at 20.

33. 541 F.2d at 24.

34. 541 F.2d at 25.

35. 5 U.S.C. 551 *et seq.*

36. 5 U.S.C. 551.

37. *Fed. Reg.* **43**:12661 (Mar. 24, 1978).

38. *Fed. Reg.* **44**:10270 (Feb. 16, 1979).

39. Ibid.

40. Ibid.

41. *Fed. Reg.* **42**:13579 (Mar. 11, 1977).

42. 15 U.S.C. 2607(c) through (e).

43. *Fed. Reg.* **43**:12661, *supra*.

44. *Fed. Reg.* **43**:12661, *supra*.

45. Minnesota Mining and Manufacturing Company (3M) has made great advances in pollution control that demonstrate that the real culprits are tradition-bound, weak corporate managers. Industrial pollution is often caused by inefficient processes. The 3M Pollution Prevention Pays (3P) program was started in response to the economic slowdown of 1974. In order to reduce operating costs while remaining in compliance with federal and state environmental protection laws, 3M shifted its focus from pollution control to resource conservation. The results have been dramatic: a cleaner environment plus an annual dollar saving of $10 million.

46. *Fed. Reg.* **44**:10286 (Feb. 16, 1979). These regulations will appear in 40 CFR Part 25.

47. EPA has also promulgated regulations to establish procedures for rule making under section 6 of TSCA, 40 CFR Part 750.

48. 5 U.S.C. 553(e).

49. The State of North Carolina petitioned EPA to amend the PCB marking and disposal regulations (40 CFR 761) issued pursuant to section 6(e) of TSCA. This petition was submitted pursuant to section 21 of TSCA, 15 U.S.C. 2620, which also provides authority for a citizen petition. *Fed. Reg.* **44**:13575 (Mar. 12, 1979). The City of East Orange, New Jersey submitted a petition under section 1424(e) of the Safe Drinking Water Act (SDWA) for a determination that a local aquifer is a sole or principal drinking water source for that area, and as such entitled to protection from underground injection of waste chemicals. *Fed. Reg.* **44**:13732 (Mar. 29, 1979).
50. 5 U.S.C. 702.
51. *Polaroid Corp. v. Costle*, 11 ERC 2134 (D. Mass. 1978) and *Manufacturing Chemists Assn. v. Costle*, 11 ERC 2014 (W.D. La. 1978). In *Polaroid* the court enjoined EPA from releasing information on chemicals that Polaroid claimed were trade secret and that it had submitted to EPA as required by the Toxic Substances Control Act (TSCA). EPA had planned to release the information to contractors working for the agency. The injunction was later lifted after EPA revised its regulations. In *MCA* the court struck down hazardous substance discharge regulations promulgated under section 311 of the FWPCA. The court found that the regulations were arbitrary and capricious. Section 311 was later amended by Congress.
52. 5 U.S.C. 706.
53. Section 30 of TSCA, 15 U.S.C. 2619; section 1449 of the SDWA, 42 U.S.C. 300j-8; section 7002 of the RCRA, 42 U.S.C. 6972; and section 505 of the FWPCA, 33 U.S.C. 1365.
54. 12 ERC 1597 (D.D.C. 1979).
55. Sections 3001 through 3006 of RCRA, 42 U.S.C. 6921 through 6926.
56. 12 ERC at 1599.

DISCUSSION

DR. THEODORE F. THURMON (New Orleans): Mr. Halbert, I think you lose credibility with the clinical geneticist when you report defects to be due to a teratogen when it turns out to be a known hereditary disorder. My advice would be to get clinical geneticists on your staff. You need people who can differentiate between inherited and environmentally caused defects.

I would invite you to study pedigrees before you decide that the defect is not hereditary. If you have found, as I have, that a disease has been occurring regularly, it is hard to attribute it to a genetic factor that just came on the scene. I think you will find on examination of many of the examples you used that just that will be found.

PROF. GEORGE J. ANNAS: Do you have any particular illustrative examples in mind?

DR. THURMON: Dioxin defects. Every one that I have seen reported may be an inherited defect. Now, some of them potentially could be mutations. But if you study the family and find that it has been occurring in the family from time to time, it is kind of hard to say it is a mutation.

MR. GREGORY T. HALBERT: You may well be proven correct. As I say, the matter is in litigation within the agency and probably will be appealed to the courts. There also are test data on laboratory animals showing damage to the skeletons of various test animals. I appreciate your comments and would appreciate if you and others here would direct the comments to the agency formally, either to our regional office or to the Office of Toxic Substances in Washington. Again, we need your input.

MR. LEONARD L. RISKIN (Attorney, University of Houston Law School): Professor Ashford, you mentioned that you thought suits against supervisors might be important in changing behavior. I am curious about the circumstances under which an employee might choose to sue a supervisor and not also the employer. If there are such circumstances, would not the employer be inclined somehow to cover the supervisor if there is a judgment against the supervisor? I guess I wasn't sure if you had suggested that there had been such cases or you thought they were coming.

MR. NICHOLAS J. ASHFORD: As you know, this is an area of state law, and the fifty states differ. It is a complex area.

First of all, let's take the issue of company doctors. They can be sued individually, even if they are working for the company, if they take upon themselves a duty to deliver medical care and they do that negligently. There is a movement in some states which allows direct suits against supervisory personnel such as production foremen, supervisory chemical engineers, or industrial hygienists, especially where they have superior knowledge and the ability to control toxic substance exposures.

In some cases the employer may decide they want to let the supervisory employee or safety specialist hang himself because he didn't pay attention to direction. In other cases the employer will provide assistance to his employee who is being sued. The employee can never be sure of his company's posture ahead of time.

I have observed one thing. Chemists in laboratories who manage laboratories and are research project leaders are very concerned about that small probability that they are going to get a lawsuit filed against them. Ultimately, it probably won't be successful. But they are going to have to get an attorney, and there is a chance that they will lose their house. It is not the court award alone that a successful plaintiff may collect that is the deterrent; it is the hassle factor that is going to cause these employees to say, "The hell with it, I don't want to use benzene as a solvent. I will use a safer substitute."

DR. DANIEL STOWENS: The efforts of EPA and the other agencies to protect us are laudable when you are talking of complex chemicals such as PCBs and all the others. But there has been work in teratogenesis for years with the most simple chemicals. One of the earliest experiments was inducing malformations in a little fish, the *Fundulus heteroclitus*, by changing the concentration of potassium. Now, this makes potassium a teratogen. And where do we stop in this business of regulation: how far does it go?

It is also almost axiomatic that anything that will kill a cell, if given in the proper dosage, will produce a malformation. And that covers just about everything. You can do it with rhubarb, you can do it with an extract of maple leaves. It goes on and on. So how far does the agency go and how far does the protection go: where does it all end?

MR. ASHFORD: First of all, it is not true that most substances given in enough concentration produce cancer or birth defects. We know that is not true. Take the organophosphate pesticides; less than ten percent of them turn out to be carcinogenic. So it is not true that if you ingest a substance in a high enough quantity it will cause cancer. That might be true for carcinogens placed on or under the skin, but not on ingestion or inhalation.

Secondly, the name of this game is to redirect the nature of industrial chemical production. It is not to trace the suspect chemicals one at a time until you find out which ones are bad. What this problem is about is the Halloween apple problem. Your child comes home on Halloween and brings back a bagful of apples. One in ten thousand contains a razor blade. Do you sit down and decide how much it costs to do X-ray analysis to determine which one has the razor blade or do you consider what the cost of corrective surgery is if he bites down on it and the probability of that occurring and do a cost–benefit analysis? No. You throw the bag out.

The point is, we don't want apples with chemical razor blades in them. Even if only five percent—and I am being deliberately conservative—five percent of the substances produced in the petrochemical industry are carcinogenic, mutagenic, or teratogenic, we have got to treat the entire set as if a significant risk exists and over the next decade or two change the nature of worker/consumer/general public exposure to these substances. We are not tracking down benzene today and toluene tomorrow. That is not what we are doing. We are trying to change the nature of industrial production.

MR. HALBERT: I agree personally with Attorney Ashford's comments. I think we have to keep in mind the realities, the real world EPA functions in. And it is this. In the District of Columbia, there are more than thirty thousand practicing attorneys and that is more attorneys than in forty-nine other states in this country. Only California, the most populous state, has more attorneys than D.C., a city of less than seven hundred thousand people. And what do you think most of these attorneys in private practice are doing? They are suing EPA and the other agencies, claiming that we are overstepping our scientific evidence in promulgating regulations.

Believe me, we would like to take stronger action in promulgating regulations and banning chemicals; but if we don't have the data to counterbalance the arguments of industry, we lose. And, as I say, we are working hard at it. We have more than enough chemical hazards and health hazards to keep us busy into the foreseeable future. And in response to the gentleman's comments from the floor, I don't think we are going to see the day when we are going to be overregulating substances that do not present a health hazard.

DR. NANCY S. WEXLER: It is a question both about the petrochemical industry and how to change medical practice. There is a great deal of concern now about unusual "slow" viruses such as those that cause Creutzfeld–Jacob disease, Kuru, scrapie in sheep, and others. There is very good evidence that these viral disorders can be transmitted through corneal transplants, etcetera.

My impression is that there has not been a great deal of action taken to protect people, and there are many areas in medical practice where people haven't even begun to think about changing ways. There is little discussion of precautions in dentistry during which viruses could be transmitted. We do not really know about the infectivity of Alzheimer's disease or other senile disease, or possible genetic susceptibilities.

The question I have is about protection. How can federal agencies protect the community against environmental threats of all types—petrochemical toxins, viruses, and so forth?

MR. HALBERT: EPA is the primary agency in controlling or regulating usuage of environmental mutagens and teratogens. Other agencies, though, have statutes which set up a subject area jurisdiction. What is sounds like here, the Food and Drug Administration and also the Occupational Safety and Health Administration would have jurisdiction. It is difficult sometimes to tell who has jurisdiction in some of these areas.

DR. AUBREY MILUNSKY: I speak as one member of the endangered species. We now know that there are some fifty-one thousand dumps of hazardous materials in this country. That figure comes from the EPA. The game maybe lost for this generation. We have all been exposed. We are all continuing to be exposed to the petrochemical and nuclear energy industry. I don't believe my colleague Mr. Ashford that the effective name of the game is to stop, to alter, or to change the direction of the chemical industry. I hope indeed he wins ultimately or that we all win. But what are we to do in the immediate future?

PROF. RICHARD B. STEWART: I think your comment on Mr. Ashford's approach raises a problem. As a sensible society, we are trying to shift the way we are doing business with a view to the hazards we are just now coming to realize. Much of the difficulty is in catching up. What do we do with established industries? What do we do with dumps? What do we do with the mistakes of the past while we are trying to correct them by decisions for the future? I share your skepticism about litigation in this regard.

I think if the public concern is generated enough, we may have a solution similar to one developing for nuclear waste. The federal government has accepted the responsibility or it is going to accept the responsibility very soon for complete control over nuclear waste, although one could argue that that should be borne by the industry as a cost that is foreseeable and which they should bear.

I think we may come to a similar conclusion with these toxic dumps because I don't think any amount of litigation is going to succeed in throwing the costs of dealing with them totally onto private industry. I think it is really going to be a decision at the political level on the part of the Congress and the executive branch whether similar commitments to mistakes of the past are going to be made in this regard as well.

MR. HALBERT: I don't think there is going to be one solution. It is going to be a combination of solutions, and some of the promising solutions that I am aware of now include a new form of fungicide which has been developed to consume a pesticide product.

MR. ASHFORD: It eats oil as well.

MR. HALBERT: Another possibility is incineration. Several chemical companies are developing incinerators that can maintain a temperature of about two thousand degrees Fahrenheit. Some chemicals can be destroyed in that way. Other chemicals can be reused through more efficient processing.

The Minnesota Mining and Manufacturing Company has pioneered a program called "Pollution Prevention Pays" and they have greatly reduced their costs of operation while remaining in compliance with environmental protection laws by coming up with new ways to use chemicals and also to improve the efficiency of their processing operations.

Ultimately, some chemicals will probably have to be stored if they are not disposed of in landfills. I don't see them ever leaving the plant or being injected into the ground. There are terrible problems with that. But it is still a combination of these. I am confident alternative waste disposal methods will be developed in the future.

MR. ASHFORD: I think both discussions are relevant. The adjustment problems aren't going to be solved by lawsuits. Future behavioral changes may very well be settled by lawsuits. What behavior you change in the management of the safe production of chemicals for workers and the production of safe materials for consumers is different from what you have to do to get the toxics problem solved, the problem which has been buried, pun intended, for years.

But for behavioral change, I have a lot more faith in the imposition of criminal penalties for midnight dumpers and for some good hard enforcement procedures as deterrents than the collection of fines would imply.

I find it offensive that the public should bear the cost of irresponsible past industrial practice.

BIBLIOGRAPHY OF
SELECTED RECENT BOOKS IN BIOETHICS

1. Reich, W. T. (ed.), *Encyclopedia of Bioethics*, Free Press, New York (1978).
2. Pastrana, G., *Medical Ethics: Ethical Reasoning in Medical Practice*, University of Santo Tomas, Faculty of Medicine and Surgery, Manila, Philippines (1979).
3. Munson, R., *Intervention and Reflection: Basic Issues in Medical Ethics*, Wadsworth, Belmont, Calif. (1979).
4. Fletcher, J., *Humanhood: Essays in Biomedical Ethics*, Prometheus, New York (1979).
5. Beauchamp, T. L., and Childress, J. F., *Principles of Biomedical Ethics*, Oxford University Press, New York (1979).
6. Curran, C. E., *Issues in Sexual and Medical Ethics*, University of Notre Dame Press, Notre Dame, Ind. (1978).
7. Belgum, D., *When It's Your Turn to Decide: A Hospital Chaplain Helps You Make Medical Decisions*, Augsburg, Minneapolis (1978).
8. Kieffer, G. H., *Bioethics: A Textbook of Issues*, Addison-Wesley, Reading, Mass. (1979).
9. Sollitto, S., and Veatch, R. M., *Bibliography of Society, Ethics and the Life Sciences: 1979–80*, Institute of Society, Ethics and the Life Sciences, Hastings-on-Hudson, N.Y. (1978).
10. Nevins, M. M. (ed.), *Annotated Bibliography of Bioethics: Selected 1976 Titles*, Information Planning Associates, Rockville, Md. (1977).
11. Sagen, H. B., *et al.*, *Teaching Biomedical and Health Care Ethics to Liberal Arts Undergraduates*, Associated Colleges of the Midwest, Chicago (1977).
12. May, W. E., *Human Existence: Medicine and Ethics: Reflections on Human Life*, Franciscan Herald Press, Chicago (1977).
13. Diamond, E.F., *This Curette for Hire*, ACTA Foundation, Chicago (1977).
14. Barber, B. (ed.), *Medical Ethics and Social Change*, American Academy of Political and Social Science, Philadelphia (1978).
15. Beauchamp, T. L., and Walters, L. (eds.), *Contemporary Issues in Bioethics*, Dickenson, Encino, Calif. (1978).
16. Taylor, N. K., *Bibliography of Society, Ethics and the Life Sciences: Supplement for 1977–78*, Institute of Society, Ethics and the Life Sciences, Hastings-on-Hudson, N. Y. (1977).
17. Duncan, A. S., Dunstan, G. R., and Welbourn, R. B., *Dictionary of Medical Ethics*, Darton, Longman and Todd, London (1977).
18. Marcolongo, F. J., *Moral Choices in Contemporary Society: A Study Guide for Courses by Newspaper*, Publishers Inc., Del Mar, Calif. (1977).
19. Hollis, H. N., *A Matter of Life and Death: Christian Perspectives*, Broadman Press, Nashville (1977).
20. Heyer, R. (ed.), *Medical/Moral Problems*, Paulist Press, New York (1976).
21. Zegel, V. A., and Parratt, D. (eds.), *CRS Bioethics Workshop for Congress, June, 1977*, Washington, D.C. (1977).

22. Sobel, L. A. (ed.), *Medical Science and the Law: The Life and Death Controversy*, Facts on File, New York (1977).

23. *A Comprehensive Study of the Ethical, Legal, and Social Implications of Advances in Biomedical and Behavioral Research and Technology*, National Technical Information Service, Springfield, Virginia (1977).

24. *Opinions and Reports of the Judicial Council, Including the Principles of Medical Ethics and Rules of the Judicial Council*, American Medical Association, Chicago (1977).

25. Spicker, S. F., and Engelhardt, H. T. (eds.), *Philosophical Medical Ethics: Its Nature and Significance. Proceedings*, D. Reidel, Boston (1977).

26. Wecht, C. H., *Medical Ethics and Legal Liability*, Practising Law Institute, New York (1976).

27. Rosen, J., *Medicine and Ethics: A Guide to the Issues and the Literature*, Michigan State University, Department of Human Development, East Lansing (1975).

28. Thomson, W. A., *A Dictionary of Medical Ethics and Practice*, John Wright, Bristol, England (1977).

29. Reiser, S. J., Dyck, A. J., and Curran, W. J. (eds.), *Ethics in Medicine: Historical Perspectives and Contemporary Concerns*, M.I.T. Press, Cambridge, Massachusetts (1977).

30. Veatch, R. M., *Case Studies in Medical Ethics*, Harvard University Press, Cambridge, Mass. (1977).

31. McElhinney, T. K. (ed.), *Human Values Teaching Programs for Health Professionals: Self-Descriptive Reports from Twenty-nine Schools* (3rd ed., October 1976), Society for Health and Human Values, Philadelphia (1976).

32. Hunt, R., and Appas, J. (eds.), *Ethical Issues in Modern Medicine*, Mayfield, Palo Alto (1977).

33. *Health Aspects of Human Rights, with Special Reference to Developments in Biology and Medicine*, World Health Organization, Geneva (1976).

34. *Development of Medical Technology: Opportunities for Assessment*, U.S. Government Printing Office, Washington, D.C. (1976).

35. *The Teaching of Bioethics: Report of the Commission on the Teaching of Bioethics*, Institute of Society, Ethics and the Life Sciences, Hastings-on-Hudson, N. Y. (1976).

36. Sollitto, S., and Veatch, R. M., *Bibliography of Society, Ethics and the Life Sciences. 1976–1977*, Institute of Society, Ethics and the Life Sciences, Hastings-on-Hudson, N.Y. (1976).

37. Shannon, T. E. (ed.), *Bioethics: Basic Writings on the Key Ethical Questions That Surround the Major, Modern Biological Possibilities and Problems*, Paulist Press, New York (1976).

38. Levinstein, J. L., *et al.*, *Biomedical Ethics and Jewish Law: A Symposium Sponsored by the Mount Sinai Hospital Medical Center Board of Trustees*, Mt. Sinai Hospital Medical Center, Chicago (1976).

39. Ostheimer, N. C., and Ostheimer, J. M. (eds.), *Life or Death—Who Controls?* Springer, New York (1976).

40. O'Donnell, T. J., *Medicine and Christian Morality*, Alea House, New York (1976).

41. McFadden, C. J., *The Dignity of Life: Moral Values in a Changing Society*, Our Sunday Visitor, Huntington, Indiana (1976).

42. Humber, J. M., and Almeder, R. F. (eds.), *Biomedical Ethics and the Law*, Plenum Press, New York (1976).

43. Gorovitz, S., *et al.* (eds.), *Moral Problems in Medicine*, Prentice-Hall, Englewood Cliffs, N. J. (1976).

44. Brody, H., *Ethical Decisions in Medicine*, Little, Brown, Boston (1976).

45. Milunsky, A., and Annas, G. J. (eds.), *Genetics and the Law*, Plenum Press, New York (1976).

46. *National Commission for the Protection of Human Subjects of Biomedical and Behavioral Research. Transcript of the Meeting Proceedings*, National Technical Information Service, Springfield, Virginia (1975).

47. *National Commission for the Protection of Human Subjects of Biomedical and Behavioral Research. Transcript of the Meeting Proceedings (8th)*, National Technical Information Service, Springfield, Virginia (1975).

48. Sollitto, S., and Veatch, R. M., *Bibliography of Society, Ethics and the Life Sciences*, Institute of Society, Ethics and the Life Sciences, Hastings-on-Hudson, N. Y. (1974).

49. Sollitto, S., and Veatch, R. M., *Bibliography of Society, Ethics and the Life Sciences*, Institute of Society, Ethics and the Life Sciences, Hastings-on-Hudson, N. Y. (1975).

50. Hiaring, B., *Ethics of Manipulation: Issues in Medicine, Behavior Control and Genetics*, Seabury Press, New York (1975).

51. Campbell, A. V., *Moral Dilemmas in Medicine: A Coursebook in Ethics for Doctors and Nurses* (2nd ed.) Churchill Livingstone, Edinburgh (1975).

52. Restak, R. M., *Premeditated Man: Bioethics and the Control of Future Human Life*, Viking Press, New York (1975).

53. Dedek, J. F., *Contemporary Medical Ethics*, Sheed and Ward, New York (1975).
54. *Religious Affects of Medical Care: A Handbook of Religious Practices of All Faiths*, Catholic Hospital Association, St. Louis (1975).
55. Bliss, B. P., and Johnson, A. G., *Aims and Motives in Clinical Medicine: A Practical Approach to Medical Ethics*, Pitman Medical, London (1975).
56. Gorovitz, S., *et al.*, *Medical Ethics Film Review Project*, University of Maryland, Council for Philosophical Studies, College Park, Maryland (1975).
57. Lobo, G. V., *Current Problems in Medical Ethics*, Allahabad Saint Paul Society, Allahabad, India (1974).
58. Jacobs, W., *The Pastor and the Patient. An Informal Guide to New Directions in Medical Ethics*, Paulist Press, New York (1973).
59. Jones, A., and Bodmer, W. F., *Our Future Inheritance: Choice or Chance? A Study by a British Association Working Part*, Oxford University Press, London (1974).
60. *Medicine and Morality: A Selected Bibliography*, Florida Technological University Library, Orlando, Florida (1974).
61. Carmody, J., *Ethical Issues in Health Services: A Report and Annotated Bibliography, Supplement 1, 1970-1973*, Department of Health, Education and Welfare, Washington, D.C. (1974).
62. Brown, B. S., *New Dimensions in Mental Health: Ethics and Values in Scientific Decisionmaking*, U.S. Government Printing Office, Washington, D.C. (1974).
63. Vaux, K., *Biomedical Ethics: Morality for the New Medicine*, Harper & Row, New York (1974).
64. Hunt, L. L. (ed.), *Institute on Human Values in Medicine: Human Values Teaching Programs for Health Professionals*, Society for Health and Human Values, Philadelphia (1974).
65. Nelson, J. B., *Human Medicine: Ethical Perspectives on New Medical Issues*, Augsburg, Minneapolis (1973).
66. Wogaman, J. P. (ed.), *The Population Crisis and Moral Responsibility*, Public Affairs Press, Washington, D.C. (1973).
67. Hiaring, B., *Medical Ethics*, Fides Publishers, Notre Dame, Indiana (1973).
68. Mertens, T. R., and Robinson, S. (eds.), *Human Genetics and Social Problems: A Book of Readings*, MSS Information Corporation, New York (1973).
69. Etzioni, A., *Genetic Fix*, Macmillan, New York (1973).
70. Veatch, R. M., Gaylin, W., and Morgan, C. (eds.), *The Teaching of Medical Ethics: Proceedings of a Conference Sponsored by the Institute of Society, Ethics and the Life Sciences and Columbia University, June 1-3, 1972*, Institute of Society, Ethics and the Life Sciences, Hastings-on-Hudson, N. Y. (1973).
71. Gorovitz, S., and Macklin, R., *Teaching Medical Ethics: A Report on One Approach*, Case Western Reserve University, Department of Philosophy, Cleveland (1973).
72. Sollitto, S., and Veatch, R. M., *Bibliography of Society, Ethics and the Life Sciences*, Institute of Society, Ethics and the Life Sciences, Hastings-on-Hudson, N. Y. (1973).
73. *Law and Ethics of A.I.D. and Embryo Transfer. CIBA Foundation Symposium 17*, Elsevier, New York (1973).
74. Marty, M. E., and Peerman, D. G. (eds.), *New Theology No. 10*, Macmillan, New York (1973).
75. Wertz, R. W. (ed.), *Readings on Ethical and Social Issues in Biomedicine*, Prentice-Hall, Englewood Cliffs, N. J. (1973).

Selected Recent Bibliography

Government Control of Science and Human Experimentation

1. Collins, F. L., Jr., Kuhn, I. F., Jr., and King, G. D., Variables affecting subjects' ethical ratings of proposed experiments, *Psychol. Rep.* **44**:155 (1979).
2. Neville, R., On the national commission: A Puritan critique of consensus ethics, *Hastings Cent. Rep.* **9**:22 (1979).
3. Levine, R. J., Regulation for the use of human tissues and body fluids as research materials—Current modifications, *Biochem. Pharmacol.* **28**:1893 (1979).
4. Broad, W. J., Proposals for ethics boards stir debate, *Science* **205**:285 (1979).
5. Recombinant DNA: Now is the time for Congress to act, *Nature* **279**:461 (1979).
6. Dickson, D., US drug companies push for changes in recombinant DNA guidelines, *Nature* **278**:385 (1979).
7. Recombinant DNA—How public, *Nature* **278**:383 (1979).
8. Guze, S. B., Federal regulation of psychiatric research, *J. Nerv. Ment. Dis.* **167**:265 (1979).
9. Ibrahim, M. A., and Spitzer, W. D., The case control study: The problem and the prospect, *J. Chronic Dis.* **32**:139 (1979).
10. Diamond, E. F., In vitro fertilization: A moratorium is in order, *Hosp. Prog.* **60**:66 (1979).
11. Rozovsky, L. E., A guide to influencing government policy, *Dimens. Health Serv.* **96**:26 (1979).
12. Rall, D. P., and Schambra, P. E., Environmental health research and regulation, *Environ. Health Perspect.* **30**:9 (1979).
13. Legislation and animal research: Is reform due? (Editorial), *Br. Med. J.* **1**:1035 (1979).
14. Downs, F. S., Whose responsibility? Whose rights? *Nurs. Res.* **28**:131 (1979).
15. Fuchs, V. R., Economics, health, and post-industrial society, *Milbank Mem. Fund Q.* **57**:153 (1979).
16. Dickson, D., Recombinant DNA research: Private actions raise public eyebrows (News), *Nature* **278**:494 (1979).
17. Curran, W. J., Sounding board. Reasonableness and randomization in clinical trial: Fundamental law and government regulation, *N. Engl. J. Med.* **300**:1273 (1979).
18. Copper, T., Shattuck lecture—In the public interest, *N. Engl. J. Med.* **300**:1185 (1979).
19. Brown, J. H., Schoenfeld, L. S., and Allan, P. W., The costs of an institutional review board, *J. Med. Educ.* **54**:294 (1979).
20. Barkes, P., Bioethics and informed consent in American health care delivery, *J. Adv. Nurs.* **4**:23 (1979).
21. Robertson, J. A., Ten ways to improve IRBs, *Inquiry* **9**:22 (1979).
22. Veatch, R. M., The national commission on IRBs: An evolutionary approach, *Inquiry* **9**:22 (1979).
23. Lipsett, M. B., Fletcher, J. C., and Secundy, M., Research review at NIH, *Inquiry* **9**:18 (1979).

*For a bibliography of earlier titles, see the first *Genetics and the Law* volume.

24. Dickson, D., MPs accuse health officials of unconcern over risks and benefits of DNA research (News), *Nature* 277:590 (1979).

25. Walgate, R., Genetic manipulation: Britain may exempt "self-cloning" (News), *Nature* 277:589 (1979).

26. A policy for biomedical and behavioral research, The Association of American Medical Colleges Executive Council, *J. Med. Educ.* 54:257 (1979).

27. Proposed rules for research on the institutionalized called restrictive by APA, *Hosp. Community Psychiatry* 30:285 (1979).

28. Paul, J., Gene cloning in cell biology, *Cell Biol. Int. Rep.* 2:311 (1978).

29. Watson, J. D., Let us stop regulating DNA Research (Editorial), *Nature* 278:113 (1979).

30. Dickson, D., DNA risks: Fears calmed, but doubts remain (News), *Nature* 277:505 (1979).

31. Genetic manipulation: New guidelines for UK, *Nature* 276:104 (1978).

32. King, J., New diseases in new niches, *Nature* 276:4 (1978).

33. Brenner, S., Six months in category four, *Nature* 276:2 (1978).

34. Lee, S. S., Regulation and academic medicine: The Quebec experience, *J. Med. Educ.* 54:12 (1979).

35. Belsey, A., Patients, doctors and experimentation: Doubts about the declaration of Helsinki, *J. Med. Ethics* 4:182 (1978).

36. Barclay, R. W., Legislative highlights, *J. Am. Diet. Assoc.* 74:57 (1979).

37. Cohen, M. E., The "brave new baby" and the law: Fashioning remedies for the victims of in vitro fertilization, *Am. J. Law Med.* 4:319 (1978).

38. Greenberg, D. S., Washington report. The frustrated reformers, *N. Engl. J. Med.* 300:211 (1979).

39. Galblum, T. W., Health care cost containment experiments: Policy, individual rights, and the law. *J. Health Polit. Policy Law* 3:375 (1978).

40. Guidelines for the use of human subjects in dental research, Council on Dental Research, *J. Am. Dent. Assoc.* 98:86 (1979).

41. McCartney, J. J., Research on children: National commission says "Yes, If . . . ," *Hastings Cent. Rep.* 8:26 (1978).

42. Powledge, T. M., A report from the Del Zio trial, *Hastings Cent. Rep.* 8:15 (1978).

43. Stetten, D., Jr., Valedictory by the chairman of the NIH Recombinant DNA Molecule Program Advisory Committee (Editorial), *Gene* 3:265 (1978).

44. Principal features of omnibus Public Law 95–622, which extends, amends and establishes authorities for Biomedical Research, including NCI, NHLBI, NRSA, and Ethics Commission. *Fed. Proc.* 38:1 (1979).

45. Singer, M. F., Spectacular science and ponderous process (Editorial), *Science* 203:9 (1979).

46. Marshall, E., Environmental groups lose friends in effort to control DNA research (News), *Science* 202:1265 (1978).

47. Dickson, D., US to increase public participation in regulation of DNA research (News), *Nature* 276:430 (1978).

48. Burkhardt, R., and Kienle, G. Controlled clinical trials and medical ethics, *Lancet* 2:1356 (1978).

49. Human experimentation: Human rights (Editorial), *Lancet* 2:1352 (1978).

50. May, S. C., Jr., An introduction to private enterprise perspectives on medicine, *J. Med. Assoc. Ga.* 68:45 (1979).

51. Brandt, A. M., Racism and research: The case of the Tuskegee syphilis study, *Inquiry* 8:21 (1978).

52. Science news, science fiction, and enticing readers into DNA (News), *Inquiry* 8:2 (1978).

53. The policy of the European association for the study of diabetes on human investigation (Editorial), *Diabetologia* 15:431 (1978).

54. Arpaillange, P., and Gaston, J. L., The point of view of the jurist on comparative trials, *Biomedicine* 28(Spec. No. 57) (1978).

55. Gordon, R. S., Jr., Issues in human experimentation, *Ann. Intern. Med.* 89:846 (1978).

56. Cerami, A., Drug development for "orphan" diseases (Editorial), *Am. J. Med.* 65:407 (1978).

57. Alexander, L. A., and Jabine, T. B., Access to social security microdata files for research and statistical purposes, *Soc. Secur. Bull.* 41:3 (1978).

58. Winton, R., The significance of the declaration of Helsinki. An interpretative commentary, *Med. J. Aust.* 2:78 (1978).

59. Sperber, P., Criteria for well-controlled device and diagnostic product investigations, *Med. Res. Eng.* 12:4 (1977).

60. Research on children (Editorial), *Br. Med. J.* 2:1043 (1978).

61. Smith, E. D., and Wieczorek, R. R., How is the topic of protection of human subjects taught in undergraduate and graduate nursing research courses, *Nurs. Res.* 27:328 (1978).

62. Jonsen, A. R., Research involving children: Recommendations of the national commission for the protection of human subjects of biomedical and behavioral research, *Pediatrics* **62**:131 (1978).
63. Watson, J. D., The case for expanding research into DNA, *N.Z. Vet. J.* **26**:182 (1978).
64. Wright, S., DNA: Let the public choose, *Nature* **275**:468 (1978).
65. Dickson, D., NIH confirms violation of recombinant DNA research guidelines (News), *Nature* **273**:5 (1978).
66. Dickson, D., Recombinant DNA risk-assessment studies to begin at Fort Detrick (News), *Nature* **272**:488 (1978).
67. Clouds over paediatric research (Editorial), *Lancet* **2**:771 (1978).
68. Dworkin, G., Legality of consent to nontherapeutic medical research on infants and young children, *Arch. Dis. Child* **53**:443 (1978).
69. Research involving children—Ethics, the law, and the climate of opinion, *Arch. Dis. Child* **53**:441 (1978).
70. Gray, B. H., Cooke, R. A., and Tannenbaum, A. S., Research involving human subjects, *Science* **201**:1094 (1978).
71. Hodgman, E. C., Student research in service agencies, *Nurs. Outlook* **26**:558 (1978).
72. Canone, F., and Guyot, J. C., Health policy in France: A major issue in the 1978 legislative elections, *Int. J. Health Serv.* **8**:509 (1978).
73. Wade, N., New rulebook for gene splicers faces one more test, *Science* **201**:600 (1978).
74. Mechanic, D., Prospects and problems in health services research, *Milbank Mem. Fund Q.* **56**:127 (1978).
75. Dickson, D., US proposes exemptions from DNA guidelines, *Nature* **274**:411 (1978).
76. Dickson, D., NIH relaxes recombinant DNA guidelines, *Nature* **274**:303 (1978).
77. Magraw, R. M., Fox, D. M., and Weston, J. L., Health professions education and public policy: A research agenda, *J. Med. Educ.* **53**:539 (1978).
78. Buntz, C. G., Macaluso, T. F., and Azarow, J. A., Federal influence on state health policy, *J. Health Polit. Policy Law* **3**:71 (1978).
79. Hales, D. R., Does biomedical research need an artificial watchdog, *Hospitals* **52**:176 (1978).
80. Cohen, C., Medical experimentation on prisoners, *Perspect. Biol. Med.* **21**:357 (1978).
81. Dickson, D., Supreme Court nullifies patent ruling on living organisms, *Nature* **274**:2 (1978).
82. Pritchard, R. H., Recombinant DNA is safe (Editorial), *Nature* **273**:696 (1978).
83. Muggia, F. M., and Devita, V. T., Moral dilemmas in clinical cancer experimentation (Letter), *Med. Pediatr. Oncol.* **4**:181 (1978).
84. Read, R. C., Criticising those in government (Letter), *Lancet* **2**:165 (1978).
85. Risk assessment protocols for recombinant DNA experimentation, *J. Infect. Dis.* **137**:704 (1978).
86. Snoke, A. W., Let's challenge the government for a change, *Hosp. Prog.* **59**:21 (1978).
87. Wales, J. B., and Treybig, D. L., Recent legislative trends toward protection of human subjects: Implications for gerontologists, *Gerontologist* **18**:244 (1978).
88. Berkowitz, S., Informed consent, research, and the elderly, *Gerontologist* **18**:237 (1978).
89. A fair trial (Editorial), *Br. J. Hosp. Med.* **19**:301 (1978).
90. Mcalister, W. H., Human experimentation: Regulations and ethics (Editorial), *Am. J. Roentgenol.* **130**:1200 (1978).
91. Svensson, L. G., Medical research in South Africa, *S. Afr. Med. J.* **53**:195 (1978).
92. Patten, S. C., Deceiving subjects (Letter), *Hastings Cent. Rep.* **8**:39 (1978).
93. Geison, G. L., Pasteur's work on rabies: Reexamining the ethical issues, *Hastings Cent. Rep.* **8**:26 (1978).
94. Comiskey, R. J., The use of children for medical research: Opposite views examined, *Child Welfare* **57**:321 (1978).
95. Levine, R. J., Metaethics and the new biology (Letter), *Pharos.* **41**:36 (1978).
96. Dickson, D., Friends of DNA fight back, *Nature* **272**:664 (1978).
97. Benedek, T. G., The "Tuskagee study" of syphilis: Analysis of moral versus methodologic aspects, *J. Chronic Dis.* **31**:35 (1978).
98. Brandt, A. M., Polio, politics, publicity, and duplicity: Ethical aspects in the development of the Salk vaccine, *Int. J. Health Serv.* **8**:257 (1978).
99. Comments on the report of the president's biomedical research panel (Letter), *Fed. Proc.* **37**:1992 (1978).
100. McMahon, F. G., The effects of new federal regulations on clinical investigation (Editorial), *Clin. Pharmacol. Ther.* **23**:495 (1978).

101. Price, S. L., The little beasts, *Pharos* **41**:2 (1978).
102. Horowitz, H. S., Ethical considerations in human experimentation, *J. Dent. Res.* **56**:154 (1977).
103. Califano, J. A., Jr., HEW's Califano urges debate on moral issues, *Hosp. Prog.* **59**:71 (1978).
104. Curran, W. J., Influence of the courts of law on biologicals development, regulation, and use, *Bull. WHO* **55**:53 (1977).
105. Fletcher, J., Ethical considerations in biomedical research involving human beings, *Bull. WHO* **55**:101 (1977).
106. Horowitz, H. S., Overview of ethical issues in clinical studies, *J. Public Health Dent.* **38**:35 (1978).
107. Vosberg, H. P., Molecular cloning of DNA. An introduction into techniques and problems, *Hum. Genet.* **40**:1 (1977).
108. Silverstein, A. M., Congressional politics and biomedical science, *Fed. Proc.* **37**:105 (1978).
109. Wortis, J., Noted psychiatrist scores government misuse of psychiatry (Editorial), *Biol. Psychiatry* **13**:1 (1978).
110. Statement by the American Society for Medical Technology in consideration of the proposed recombinant DNA legislation: Submitted to the Senate Subcommittee on Health of the Human Resource Committee and to the House Subcommittee on Health of the Interstate Foreign Commerce Committee, September 7, 1977, *Am. J. Med. Technol.* **44**:66 (1978).
111. The social implications of research on genetic manipulation, in: *Genetic Manipulation as It Affects the Cancer Problem* (Vol. 14) (J. Schultz and Z. Brada, eds.), p. 217, Academic Press, New York (1977).
112. Grobstein, C., What is the appropriate role of sponsoring institutions and of federal, state, and local governments in relation to recombinant DNA research? in: *Research with Recombinant DNA*, p. 193, National Academy of Sciences, Washington, D.C. (1977).
113. Meselson, M. S., and McCullough, J. M., The use of recombinant DNA research in biological warfare is ruled out for nations by the Biological Weapons Convention of 1972 and the Geneva Protocol of 1925: Is it conceivable that terrorists of nations may try to use this technology to develop weapons? in: *Research with Recombinant DNA*, p. 195, National Academy of Sciences, Washington, D.C. (1977).
114. Hubbard, R., and Robb, J. W., Ethical and moral issues of the research, in: *Research with Recombinant DNA*, p. 209, National Academy of Sciences, Washington, D.C. (1977).
115. Davis, B. D., Epidemiological and evolutionary aspects of research on recombinant DNA, in: *Research with Recombinant DNA*, p. 124, National Academy of Sciences, Washington, D.C. (1977).
116. Curtiss, R. B. D., and Clark-Curtiss, J., Effectiveness of physical and biological containment, in: *Research with Recombinant DNA*, p. 198, National Academy of Sciences, Washington, D.C. (1977).
117. Chakrabarty, A. M., and Ancker-Johnson, B., Can the results of basic research with recombinant DNA be transferred to industrial applications? What are the processes and problems involved in the application for patents involving recombinant DNA research? in: *Research with Recombinant DNA*, p. 202, National Academy of Sciences, Washington, D.C. (1977).
118. Barth, D., An environmental overview of research with recombinant DNA, in: *Research with Recombinant DNA*, p. 145, National Academy of Sciences, Washington, D.C. (1977).
119. Szybalski, W., Safety in recombinant DNA research (Editorial), *Gene* **1**:181 (1977).
120. Singer, M. F., A summary of the National Institutes of Health (USA) guidelines for recombinant DNA Research, *Gene* **1**:123 (1977).
121. Cowan, D. H., Ethical considerations of metabolic research in alcoholic patients, *Alcoholism* **1**:193 (1977).
122. Powledge, T. M., Recombinant DNA: Backing off on legislation, *Hastings Cent. Rep.* **7**:8 (1977).
123. Rachels, J., An overview of the scientist's responsibilities: Comments by an ethicist, *In Vitro* **13**:728 (1977).
124. Winslade, W. J., An overview of the scientist's responsibilities: Comments by an attorney, *In Vitro* **13**:712 (1977).
125. Nardone, R. M., An overview of the scientist's responsibilities: Comments by a scientist, *In Vitro* **13**:696 (1977).
126. Falk, H. L., The Toxic Substances Control Act and in vitro toxicity testing, *In Vitro* **13**:676 (1977).
127. Adams, E. M., The ethical responsibilities at issue, *In Vitro* **13**:595 (1977).
128. Hess, E. L., Wheel spinning in Washington, or another fizzle (Editorial). *Fed. Proc.* **36**:2647 (1977).
129. Use of ionizing radiation and radionuclides on human beings for medical research, training, and nonmedical purposes, Report of a WHO Expert Committee, *WHO Tech. Rep. Ser.* **611**:1 (1977).
130. Greenberg, D. S., Medicine and public affairs. Lessons of the DNA controversy, *N. Engl. J. Med.* **297**:1187 (1977).

131. Kupfer, S., Experimentation and ethics, *Mt. Sinai J. Med. N.Y.* **44**:648 (1977).
132. Ryan, K. J., Psychosurgery: Clarifications from the National Commission (Letter), *Hastings Cent. Rep.* **7**:4 (1977).
133. Milgram, S., Subject reaction: The neglected factor in the ethics of experimentation, *Hastings Cent. Rep.* **7**:19 (1977).
134. Davis, B. D., The recombinant DNA scenarios: Andromeda Strain, Chimera, and Golem, *Am. Sci.* **65**:547 (1977).
135. Curran, W. J., The confidentiality of researcher's notes and interviews in public policy and environmental health studies, *Am. J. Public Health.* **67**:1103 (1977).
136. Smith, R. J., SUNY at Albany admits research violations, *Science* **198**:708 (1977).
137. Weinberg, P., Research, regulation, and the public interest (Letter), *Science* **198**:668 (1977).
138. Kearney, P. J., Ethics, cancer and children, *Med. Hypotheses* **3**:174 (1977).
139. Skegg, P. D., English law relating to experimentation on children, *Lancet* **2**:754 (1977).
140. Ryan, K. G., Ethical issues in scientific research: Two views on recombinant DNA research, *Fed. Proc.* **36**:111 (1977).
141. Thomas, L., Ethical issues in scientific research: Two views on recombinant DNA research, *Fed. Proc.* **36**:1 (1977).
142. Duval, M. K., The provider, the government, and the consumer, *Daedalus* **185** (1977).
143. Federal health activities and current legislative priorities for the AFCR: Report of the Public Policy Committee, *Clin. Res.* **25**:213 (1977).
144. Axelsen, D., and Wiggins, R. A., An application of moral guidelines in human clinical trials to a study of a benzodiazepine compound as a hypnotic agent among the elderly, *Clin. Res.* **25**:1 (1977).
145. Berger, P. M., and Stallones, R. A., Legal liability and epidemiologic research, *Am. J. Epidemiol.* **106**:177 (1977).
146. Grobstein, C., The recombinant-DNA debate, *Sci. Am.* **237**:22 (1977).
147. Creighton, H., Legal concerns of nursing research, *Nurs. Res.* **26**:337 (1977).
148. Levine, R. J., Nondevelopmental research on human subjects: The impact of the recommendations of the national commission, *Fed. Proc.* **36**:2356 (1977).
149. Cohen, S. N., Development of drug therapy for children, *Fed. Proc.* **36**:2356 (1977).
150. Klerman, G. L., Development of drug therapy for the mentally ill, *Fed. Proc.* **36**:2352 (1977).
151. Lasagna, L., Prisoner subjects and drug testing, *Fed. Proc.* **36**:2349 (1977).
152. Lebacqz, K., The National Commission and research in pharmacology: An overview, *Fed. Proc.* **36**:2344 (1977).
153. Levine, R. J., Recommendations of the National Commission for the Protection of Human Subjects of Biomedical and Behavioral Research: Impact on research in pharmacology. Introduction, *Fed. Proc.* **36**:2341 (1977).
154. Dunsker, S. B., Health care delivery: The expensive role of government, *Ohio State Med. J.* **73**:501 (1977).
155. The National Cancer Act of 1971 with changes made by the National Cancer Act amendments of 1974, *J. Natl. Cancer Inst.* **59**:701 (1977).
156. Novak, E., Seckman, C. E., and Stewart, R. D., Motivations for volunteering as research subjects, *J. Clin. Pharmacol.* **17**:365 (1977).
157. Hershey, N., Research with human subjects: The medical staff's responsibility, *Hosp. Med. Staff* **5**:8 (1976).
158. Jonsen, A. R., Parker, M. L., Carlson, R. J., *et al.*, Biomedical experimentation on prisoners. Review of practices and problems and proposal of a new regulatory approach, *Ethics Sci. Med.* **4**:1 (1977).
159. Foley, T. S., A legislator's view on animal legislation affecting biomedical research, in: *The Future of Animals, Cells, Models, and Systems in Research, Development, Education, and Testing*, p. 171, National Academy Of Sciences, Washington, D.C. (1975).
160. Ladimer, I., Root and branch: Legal aspects of biomedical studies in man and other animals, in: *The Future of Animals, Cells, Models, and Systems in Research, Development, Education, and Testing*, p. 296, National Academy Of Sciences, Washington, D.C. (1975).
161. Schneider, H. A., in: *The Future of Animals, Cells, Models, and Systems in Research, Development, Education, and Testing*, p. 318, National Academy of Sciences, Washington, D.C. (1975).
162. Hendrix, T. R., Local institutional review boards (Editorial), *J. Med. Educ.* **52**:604 (1977).
163. Valid parental consent (Editorial), *Lancet* **1**:1346 (1977).

164. Krant, M. J., Cohen, J. L., and Rosenbaum, C., Moral dilemmas in clinical cancer experimentation, *Med. Pediatr. Oncol.* 3:141 (1977).

165. New York hospitals get new guidelines for patient research, *Urban Health* 5:9 (1976).

166. Reiss, D., Freedom of inquiry and subjects' rights: An introduction, *Am. J. Psychiatry* 134:891 (1977).

167. Musto, D. F., Freedom of inquiry and subjects' rights: Historical perspective, *Am. J. Psychiatry* 134:893 (1977).

168. Cole, J. O., Research barriers in psychopharmacology, *Am. J. Psychiatry* 134:896 (1977).

169. Meyer, R. E., Subjects' rights, freedom of inquiry, and the future of research in the addictions, *Am. J. Psychiatry* 134:899 (1977).

170. Gray, B. H., The functions of human subjects review committees, *Am. J. Psychiatry* 134:907 (1977).

171. Chalkley, D. T., Federal constraints: Earned or unearned, *Am. J. Psychiatry* 134:911 (1977).

172. Bean, W. B., The Fielding H. Garrison Lecture: Walter Reed and the ordeal of human experiments, *Bull. Hist. Med.* 51:75 (1977).

173. Annas, G. J., Psychosurgery: Procedural safeguards, *Hastings Cent. Rep.* 7:11 (1977).

174. Powledge, T. M., Recombinant DNA: The argument shifts, *Hastings Cent. Rep.* 7:18 (1977).

175. Callahan, D., Recombinant DNA: Science and the public, *Hastings Cent. Rep.* 7:20 (1977).

176. Dismukes, K., Recombinant DNA: A proposal for regulation, *Hastings Cent. Rep.* 7:25 (1977).

177. Wecht, C. H., Human experimentation and clinical investigation—Legal and ethical considerations, *Leg. Med. Annu.* 1976:299 (1977).

178. Horty, J. F., Patient's willingness cannot be presumed, *Mod. Health Care* 7:74 (1977).

179. Hershey, N., Research with human subjects: Trustees should ask questions, *Trustee* 29:7 (1976).

180. Anderson, F. O., and Abuzzahab, F. S., Demographic and psychometric features of anxious symptomatic volunteers (Proceedings), *Psychopharmacol. Bull.* 13:14 (1977).

181. Heller, P., Informed consent and the old-fashioned conscience of the physician-Investigator, *Perspect. Biol. Med.* 20:434 (1977).

182. Curran, W. J., Confidentiality of NIH research-project records, *N. Engl. J. Med.* 296:1273 (1977).

183. Toulmin, S., Ethical safeguards in research, *Conn. Med.* 42:235 (1977).

184. Hassar, M. Pocelinko, R. Weintraub, M., *et al.*, Free-living volunteer's motivations and attitudes toward pharmacologic studies in man, *Clin. Pharmacol. Ther.* 21:515 (1977).

185. Genest, J., Recombinant DNA experiments and universal guidelines (Letter), *Can. Med. Assoc. J.* 116:981 (1977).

186. Regulating research (Editorial), *Br. J. Hosp. Med.* 17:101 (1977).

187. Gordis, L., Gold, E., and Seltser, R., Privacy protection in epidemiologic and medical research: A challenge and a responsibility, *Am. J. Epidemiol.* 105:163 (1977).

188. Walters, L. R., Some ethical issues in research involving human subjects, *Perspect. Biol. Med.* 20:193 (1977).

189. Clothier, C. M., Consent to medical experiment, *Lancet* 1:642 (1977).

190. Laitinen, L. V., Ethical aspects of psychiatric surgery, *Neurosurgical Treatment In Psychiatry, Pain, And Epilepsy* (W. H. Sweet *et al.*, eds.), p. 483, University Park Press, Baltimore (1975).

191. Beyer, H. A., Changes in the parent–child legal relationship—What they mean to the clinician and researcher, *J. Autism Child. Schizo.* 7:84 (1977).

192. Simpson, G. H., and Coleman, M., Hopes and rights in conflict (Letter), *J. Autism Child. Schizo.* 7:103 (1977).

193. Branson, R., Prison research: National Commission says "No, Unless," *Hastings Cent. Rep.* 7:14 (1977).

194. An ex-convict testifies, *Hastings Cent. Rep.* 7:14 (1977).

195. Travitzky, D. E., Volunteering at Vacaville, *Hastings Cent. Rep.* 7:13 (1977).

196. Hatfield, F., Prison research: The view from inside, *Hastings Cent. Rep.* 7:11 (1977).

197. Estes, C. L., Revenue sharing: Implications for policy and research in aging, *Gerontologist* 16:141 (1976).

198. Marini, J. L., Sheard, M. H., and Bridges, C. I., An evaluation of "informed consent" with volunteer prisoner subjects, *Yale J. Biol. Med.* 49:427 (1976).

199. Smit, P. C., Clinical trials, children and the law, *S. Afr. Med. J.* 51:155 (1977).

200. Hofmann, F. G., The growing federal presence in human investigative studies (Editorial), *Fed. Proc.* 36:133 (1977).

201. Burger, E. J., Jr., Unmatured science and government regulation, *J. Toxicol. Environ. Health* 2:389 (1976).

202. Schrogie, J. J., Hensley, M. J., Digiore, C., *et al.*, Evaluation of the prison inmate as a subject in drug assessment, *Clin. Pharmacol. Ther.* **21**:1 (1977).

203. Almy, T. P., Therapeutic trials, town and gown, and the public interest (Editorial), *N. Engl. J. Med.* **296**:279 (1977).

204. Addae, S. K., Some problems of research design in the human experimental subject, *Ghana Med. J.* **14**:229 (1975).

205. Ingelfinger, F. J., Ethics of human experimentation defined by a national commission (Editorial), *N. Engl. J. Med.* **296**:44 (1977).

206. Bartholome, W. G., Parents, children, and the moral benefits of research, *Hastings Cent. Rep.* **6**:44 (1976).

207. McCormick, R. A., Experimentation in children: Sharing in sociality, *Hastings Cent. Rep.* **6**:41 (1976).

208. Robertson, J. A., Compensating injured research subjects: II. The law, *Hastings Cent. Rep.* **6**:29 (1976).

209. Rist, M., and Mohan, W. J., Wanted: Professional research subjects; rewards commensurate with risks, *Hastings Cent. Rep.* **6**:28 (1976).

210. Childress, J. F., Compensating injured research subjects: I. The moral argument, *Hastings Cent. Rep.* **6**:21 (1976).

211. Marks, P. A., Federal policy in biomedical research: The report of the President's Biomedical Research Panel, *Fed. Proc.* **35**:2536 (1976).

212. Falcone, D., and Jaeger, B. J., The policy effectiveness of health services research: A reconsideration, *J. Community Health* **2**:36 (1976).

213. Johnson, D. R., Clinical research—Some legal problems of the pharmaceutical manufacturer, *J. Clin. Pharmacol.* **16**:600 (1976).

214. Laves, B. S., Legal aspects of experimentation with institutionalized mentally disabled subjects, *J. Clin. Pharmacol.* **16**:592 (1976).

215. Curran, W. J., The proper and improper concerns of medical law and ethics, *N. Engl. J. Med.* **295**:1057 (1976).

216. Black, D., Looking at research, *Lancet* **2**:780 (1976).

217. Hopkins, E. L, McLean, L., and Rose, G. J., Issues related to the protection of the human fetus in biomedical research, *J. Natl. Med. Assoc.* **68**:340 (1976).

218. Perez, L. M., The National Health Planning and Research Development Act Pl 93, *J. Fla. Med. Assoc.* **63**:429 (1976).

219. Dimonda, R., Medical device legislation: More responsibility for hospitals, *Hospitals* **50**:105 (1976).

220. Schmidt, R. M., and Curran, W. J. A national genetic-disease program: Some issues of implementation, *N. Engl. J. Med.* **295**:819 (1976).

221. Randomized trials in preventive medicine and health service research. Report on a study group, in: *Randomized Trials in Preventive Medicine and Health Service Research*, p. 11.1, World Health Organization, Copenhagen (1976).

222. Harker, L. A., The need for human experimentation, in: *Animal Models of Thrombosis and Hemorrhagic Diseases*, p. 173, NIH, Bethesda (1975).

223. Aledort, L. M., and Grupenhoff, J. T., Legislative considerations for studies with animal models, in: *Animal Models of Thrombosis and Hemorrhagic Diseases*, p. 159, NIH, Bethesda (1975).

224. MBD, drug research and the schools. A conference on medical responsibility and community control, 1976, *Hastings Cent. Rep.* **6**:1 (1976).

225. Ramsey, P., The enforcement of morals: Nontherapeutic research on children, *Inquiry* **6**:21 (1976).

226. Biomedical ethics and the shadow of Nazism. A conference on the proper use of the Nazi analogy in ethical debate, 1976, *Inquiry* **6**:1 (1976).

227. Capron, A. M., Reflections on issues poised by recombinant DNA molecule technology, III, *Ann. N.Y. Acad. Sci.* **265**:71 (1976).

228. Mathews-Roth, M. M., The reporting of research on human subjects (Editorial), *Photochem. Photobiol.* **23**:381 (1976).

229. Connell, P. H., Christie, D. J., and May, W. W., Volunteers at risk, *J. Med. Ethics.* **2**:87 (1976).

230. Joubert, P., and Pannall, P., The selection of healthy volunteers for clinical investigation: The case for volunteer pools, *Curr. Med. Res. Opin.* **4**:192 (1976).

231. Diamond, E. F., Redefining the issues in fetal experimentation, *J. Am. Med. Assoc.* **236**:281 (1976).

232. Gorovitz, S., Research on captive populations (Letter), *J. Pharm. Sci.* **65**:IV (1976).

233. Row, C. F., At the heart of the review (Letter), *Hastings Cent. Rep.* **6**:43 (1976).

234. Delahunt, W. D., Biomedical research: A view from the state legislature, *Hastings Cent. Rep.* **6**:25 (1976).

235. Zimmerly, J. G., Clinical investigation in children, *Conn. Med.* **40**:320 (1976).

236. Pearse, W. H., The ethics of research, *Nebr. Med. J.* **61**:185 (1976).

237. Declaration of Helsinki. Recommendations guiding medical doctors in biomedical research involving human subjects, *Med. J. Aust.* **1**:206 (1976).

238. Revision of the Declaration of Helsinki (Editorial), *Med. J. Aust.* **1**:179 (1976).

239. McCormick, R. A., Experimental subjects. Who should they be, *J. Am. Med. Assoc.* **235**:2197 (1976).

240. Helmchen, H., and Muller-Derlinghausen, B., The inherent paradox of clinical trials in psychiatry, *J. Med. Ethics* **1**:168 (1975).

241. Cohen, E. S., Legal research issues on aging, *Gerontologist* **14**:263 (1974).

242. Michels, E., Research and human rights. Part 2, *Phys. Ther.* **56**:546 (1976).

243. Greenberg, D. S., Medicine and public affairs. Report of the President's Biomedical Panel and the old days at the FDA, *N. Engl. J. Med.* **294**:1245 (1976).

244. Regier, M., and Kurtz, N. R., Policy lessons of the Uniform Act; A response to comments, *J. Stud. Alcohol* **37**:382 (1976).

245. Little, J. W., Jr., An overview of the legal aspects of human experimentation and research, *J. Forensic Sci.* **21**:427 (1976).

246. Halper, T., Ethics and medical experimentation: Some unconfronted problems, *Conn. Med.* **40**:267 (1976).

247. Lowe, C. U., On legislating fetal research, in: *Genetics and the Law* (A. Milunsky and G. J. Annas, eds.), p. 351, Plenum Press, New York (1975).

248. Reilly, P., In vitro fertilization—A legal perspective, in: *Genetics and the Law* (A. Milunsky and G. J. Annas, eds.), p. 359, Plenum Press, New York (1975).

249. Capron, A. M., Experimentation and human genetics: Problems of "consent," in: *Genetics and the Law* (A. Milunsky and G. J. Annas, eds.) p. 319, Plenum Press, New York (1975).

250. Radiation dangers and volunteers (Editorial), *Br. Med. J.* **1**:672 (1976).

251. Rubin, J. S., Breaking into the prison: Conducting a medical research project, *Am. J. Psychiatry* **133**:230 (1976).

252. Declaration of Helsinki. Recommendations guiding medical doctors in biomedical research involving human subjects, *Ugeskr. Laeger.* **138**:399 (1976).

253. Fetal research (Letter), *N. Engl. J. Med.* **204**:849 (1976).

254. Dickens, B. M., The use of children in medical experimentation, *Med. Leg. J.* **43**:166 (1975).

255. Garnham, J. C., Some observations on informed consent in non-therapeutic research, *J. Med. Ethics* **1**:138 (1975).

256. Wecht, C. H., Medical, legal, and moral considerations in human experiments involving minors and incompetent adults, *J. Leg. Med.* **4**:27 (1976).

257. Varley, A. B., Protection of human research subjects: Are ethics necessary, *J. Leg. Med.* **4**:23 (1976).

258. Dornette, W. H., Whither goeth research (Editorial), *J. Leg. Med.* **4**:2 (1976).

259. Frankel, M. S., The development of policy guidelines governing human experimentation in the United States: A case study of public policy-making for science and technology, *Ethics Sci. Med.* **2**:43 (1975).

260. Trout, M. E., Ethics of clinical research, *Conn. Med.* **40**:201 (1976).

261. Capron, A. M., Social experimentation and the law, in: *Ethical and Legal Issues of Social Experimentation* (A. M. Rivlin and P. M. Timpane, eds.), p. 127, Brookings Institution, Washington, D.C. (1973).

262. Veatch, R. M., Ethical principles in medical experimentation, in: *Ethical and Legal Issues of Social Experimentation* (A. M. Rivlin and P. M. Timpane, eds.), p. 21, Brookings Institution, Washington, D.C. (1973).

263. Rivlin, A. M., and Timpane, P. M., Ethical and legal issues of social experimentation (Introduction and Summary), in: *Ethical and Legal Issues of Social Experimentation* (A. M. Rivlin and P. M. Timpane, eds.), p. 1, Brookings Institution, Washington, D.C. (1973).

264. Blomquist, C., Ethical guidelines for biomedical research, *Ann. Clin. Res.* **7**:291 (1975).

265. Barber, B., The ethics of experimentation with human subjects, *Sci. Am.* **234**:25 (1976).

266. Medicine and science in sports: Policy statement regarding the use of human subjects and informed consent, *Med. Sci. Sports* **7**:XIII (1975).

267. Reed, J., Knowledge, power, man and justice: Ethical problems in biomedical research, *Can. J. Genet. Cytol.* **17**:297 (1975).

268. Watts, M. S. The first decade of federal medicine (Editorial), *West. J. Med.* **123**:307 (1975).
269. Tiefel, H. O., The cost of fetal research: Ethical considerations, *N. Engl. J. Med.* **294**:85 (1976).
270. Tidd, M. J., and Ramsay, L. E., Safeguards for healthy volunteers in drug studies (Letter), *Lancet* **2**:875 (1975).
271. Feldmann, E. G., National research policy (Editorial), *J. Pharm. Sci.* **64**:1 (1975).
272. Holt, P. R., Digestive disease research and human experimentation (Editorial), *Gastroenterology* **69**:1357 (1975).
273. Katz, B. F., Children, privacy, and nontherapeutic experimentation, *Am. J. Ortho-psychiatry* **45**:802 (1975).
274. De Wet, B. S., Medical ethics and clinical therapeutic trials, *S. Afr. Med. J.* **49**:1849 (1975).
275. Spiro, H. M., Visceral viewpoints. Constraint and consent—On being a patient and a subject, *N. Engl. J. Med.* **293**:1134 (1975).
276. Foster, H. W., Jr., Biomedical research subjects: Who should be exempt (Editorial), *J. Med. Educ.* **50**:1069 (1975).
277. Fetal research: Response to the recommendations, *Hastings Cent. Rep.* **5**:9 (1975).
278. Veatch, R. M., Lay participation in medical policy making. II. Human experimentation committees: Professional or representative, *Hastings Cent. Rep.* **5**:31 (1975).
279. McHugh, J. T., Pursuit of fetal research guidelines must continue, *Hosp. Prog.* **57**:6 (1975).
280. Human experimentation in digestive disease research, *Gastroenterology* **69**:1165 (1975).
281. Dickens, B. M., What is a medical experiment, *Can. Med. Assoc. J.* **113**:635 (1975).
282. Curran, W. J., Ethical and legal considerations in high risk studies of schizophrenia, *Schizophrenia Bull.* **10**:74 (1974).
283. May, W. W., The composition and function of ethical committees, *J. Med. Ethics* **1**:23 (1975).
284. MacKay, R. C., and Soule, J. A. Nurses as investigators: Some ethical and legal issues, *Can. Nurse* **71**:26 (1975).
285. Taylor, J. L., Ethical and legal aspects of non-therapeutic clinical investigations, *Med. Leg. J.* **43**:53 (1975).
286. Smith, R. N., Safeguards for healthy volunteers in drug studies, *Lancet* **2**:449 (1975).
287. O'Rourke, K. D., Fetal experimentation: An evaluation of the new federal norms, *Hosp. Prog.* **56**:60 (1975).
288. An ethical consideration of large-scale clinical trials in cardiovascular diseases. Report of the Committee on Ethics of the American Heart Association, *Circulation* **52**:5 (1975).
289. Newman, S. H., and Beck, M. B., Abortion, obtained and denied: Research approaches, *Stud. Fam. Plann.* **53**:1 (1970).
290. Mattis, P. A., Standaert, F. G., D'Aguanno, W., *et al.*, Evaluation of safety of chemical agents in human subjects (Editorial), *Toxicol. Appl. Pharmacol.* **32**:449 (1975).
291. Kefauver, D. F., Federal regulations affecting clinical research, *Psychopharmacol. Bull.* **11**:15 (1975).
292. Blumberg, I., The new regulations on human experimentation—Civil liberties, emotions, and rational perspectives, *Psychopharmacol. Bull.* **11**:14 (1975).
293. Beaven, D. W., Morals and ethics in medical research, *N.Z. Med. J.* **81**:519 (1975).
294. Bernstein, J. E., Ethical considerations in human experimentation, *J. Clin. Pharmacol.* **15**:579 (1975).
295. Gaylin, W., and Steinfels, P. A letter on the Bauman Amendment (Letter), *Hastings Cent. Rep.* **5**:6 (1975).
296. Gaylin, W., Scientific research and public regulation, *Hastings Cent. Rep.* **5**:5 (1975).
297. Report of the National Commission for the Protection of Human Subjects of Biomedical and Behavioral Research: Commission recommendations, *Hastings Cent. Rep.* **5**:45 (1975).
298. Report of the National Commission for the Protection of Human Subjects of Biomedical and Behavioral Research: Deliberations and conclusions, *Hastings Cent. Rep.* **5**:41 (1975).
299. Toulmin, S., Exploring the moderate consensus, *Hastings Cent. Rep.* **5**:31 (1975).
300. McCormick, R. A., Fetal research, morality, and public policy, *Hastings Cent. Rep.* **5**:26 (1975).
301. Bok, S., Research, casual or planned, *Hastings Cent. Rep.* **5**:25 (1975).
302. Siegel, S., A bias for life, *Hastings Cent. Rep.* **5**:23 (1975).
303. Wasserstrom, R., The status of the fetus, *Hastings Cent. Rep.* **5**:18 (1975).
304. Walters, L., Fetal research and the ethical issues, *Hastings Cent. Rep.* **5**:13 (1975).
305. Steinfels, P., The National Commission and fetal research (Editorial), *Hastings Cent. Rep.* **5**:11 (1975).
306. Rozovsky, L. E., Medical experimentation and the law, *Dimens. Health Serv.* **52**:8 (1975).

307. Stryker, J., Research moratorium (Letter), *J. Leg. Med.* **3**:9 (1975).
308. Nelson, F. K., and Bernstein, J. E., Consent in pediatric medical experimentation, *J. Leg. Med.* **3**:15 (1975).
309. Eckstein, J. W., Public opinion: Effects on medical education and research, *J. Lab. Clin. Med.* **85**:8 (1975).
310. Bernstein, J. E., and Nelson, F. K., Medical experimentation in the elderly, *J. Am. Geriatr. Soc.* **23**:327 (1975).
311. Deriemer, T. A., Human research in connecticut prisons, *Conn. Med.* **39**:387 (1975).
312. Katz, J., Experimentation with human beings—Values in conflict, *Conn. Med.* **39**:375 (1975).
313. Robinson, D. S., Evaluation of drugs used in pregnancy and pediatric age groups (Editorial), *Ann. Intern. Med.* **82**:841 (1975).
314. Whalan, D. J., The ethics and morality of clinical trials in man, *Med. J. Aust.* **1**:1369 (1975).
315. Research on children (Editorial), *Lancet* **1**:1369 (1975).
316. Brody, E. B., The freedom of medical information, *J. Am. Med. Assoc.* **233**:145 (1975).
317. Oliver, T. K., and Morgan, T. E., Fetal research (Editorial), *J. Med. Educ.* **50**:484 (1975).
318. Anlyan, W. G., Quality, quantity and distribution of medical education and care: Regulation by the private sector or mandate by government, *Ann. Surg.* **181**:760 (1975).
319. Rada, R. T., and Jones, V. P., Human experimentation, informed consent, and the community mental health center, *Hosp. Community Psychiatry* **26**:305 (1975).
320. Hart, D. S., Fetal research and antiabortion politics: Holding science hostage, *Fam. Plann. Perspect.* **7**:72 (1975).
321. Rosner, F., Modern medicine, religion, and law; Human experimentation, *N.Y. State J. Med.* **75**:758 (1975).
322. Gray, B. H., An assessment of institutional review committees in human experimentation, *Med. Care* **13**:318 (1975).
323. Levine, R. J., Experiments on people. In Comment, *J. Am. Med. Assoc.* **232**:259 (1975).
324. Page, I. H., Experiments on people, *J. Am. Med. Assoc.* **232**:257 (1975).
325. Culliton, B. J., National Research Act: Restores training, bans fetal research, *Hastings Cent. Rep.* **4**:12 (1974).
326. Holzer, J. F., An analysis of the Edelin case, *Hosp. Prog.* **56**:20 (1975).
327. Carmody, D. P., The changing federal role in health, No. 52, NLN Publ. 19 (1974).
328. Hasselmeyer, E. G., Sudden Infant Death Syndrome Act of 1974, *Obstet. Gynecol.* **45**:214 (1975).
329. Greenberg, D. S., Medicine and public affairs. Biomedical research policy—A new inquiry begins, *N. Engl. J. Med.* **292**:489 (1975).
330. Research institutes and targeted research programs (Editorial), *J. Med. Educ.* **50**:301 (1975).
331. Trout, M. E., Should research in prisons be barred (Editorial), *J. Leg. Med.* **2**:6 (1974).
332. Stone, R. S., The rights of human beings participating as subjects in biochemical research (Guest Editorial), *J. Lab. Clin. Med.* **85**:183 (1975).
333. Ethics of human experimentation (Letter), *N. Engl. J. Med.* **292**:320 (1975).
334. Curran, W. J., Law-Medicine Notes. Experimentation becomes a crime: Fetal research in Massachusetts, *N. Engl. J. Med.* **292**:300 (1975).
335. Edwards, C. C., The federal involvement in health. A personal view of current problems and future needs, *N. Engl. J. Med.* **292**:559 (1975).
336. Arnold, L. C., Human experimentation, *Med. Trial Tech. Q.* **21**:121 (1974).
337. Ethical guidelines for the conduct of clinical investigation: Ethical policy of the American Society of Dentistry for Children pertaining to the use of human subjects in clinical research, *J. Dent. Child.* **42**:74 (1975).
338. Chalkley, D. T., Considerations in the protection of humans as subjects of research, *Bull. Pan Am. Health Organ.* **8**:289 (1974).
339. McCormick, R. A., Proxy consent in the experimentation situation, *Perspect. Biol. Med.* **18**:2 (1974).
340. Page, I. H., Growth—Magnitude and consequences in medicine, *Perspect. Biol. Med.* **18**:123 (1974).
341. Coleman, L. S., Perspectives on the medical research of violence, *Am. J. Orthopsychiatry.* **44**:675 (1974).
342. Schneiderman, M. A., and Mantel, N. The Delaney Clause and a scheme for rewarding good experimentation (Editorial), *Prev. Med.* **2**:165 (1973).
343. Schmahl, D., The Delaney Amendment (Editorial), *Prev. Med.* **2**:163 (1973).

344. Crampton, R. F., The Delaney Amendment (Editorial), *Prev. Med.* **2**:161 (1973).
345. Gilbert, J. P., Randomization of human subjects (Editorial), *N. Engl. J. Med.* **291**:1305 (1974).
346. Weinstein, M. C., Allocation of subjects in medical experiments, *N. Engl. J. Med.* **291**:1278 (1974).
347. Defelice, S. L., An analysis of the relationship between human experimentation and drug discovery in the United States, *Drug Metab. Rev.* **3**:167 (1974).
348. Duffy, J. C., Research with children: The rights of children (Editorial), *Child Psychiatry Hum. Dev.* **4**:67 (1973).
349. Ethical implications of investigations in seriously and critically ill patients. Committee on Ethics of the American Heart Association, *Circulation* **50**:1063 (1974).
350. Fantel, A. G., and Shepard, T. H., Legislative threats to research on human congenital defects, *Conn. Med.* **38**:535 (1974).
351. Bodies offered for research (Editorial), *Lancet* **2**:816 (1974).
352. Use of fetuses and fetal material for research, *Conn. Med.* **38**:539 (1974).
353. Pellegrino, E. D., Humanism in human experimentation: Some notes of the investigator's fiduciary role, *Tex. Rep. Biol. Med.* **32**:311 (1974).
354. Research on the fetus (Letter), *J. Am. Med. Assoc.* **230**:1124 (1974).
355. Fost, N., Our curious attitude toward the fetus, *Hastings Cent. Rep.* **4**:4 (1974).
356. Trout, M. E., Marihuana today, *J. Leg. Med.* **1**:44 (1973).
357. Warren, S., Informed consent in clinical investigation and practice, *J. Leg. Med.* **1**:21 (1973).
358. Perlman, J. L., Human experimentation, *J. Leg. Med.* **2**:40 (1974).
359. Spriesterbach, D. C., Hoppin, M. E., and McCrone, J., University research and the new federalism, *Science* **186**:324 (1974).
360. Medicolegal aspects of drug testing and systems of institutional review, *Am. J. Cardiol.* **34**:482 (1974).
361. Capron, A. M., Medical research in prisons. Should a moratorium be called, *Hastings Cent. Rep.* **3**:4 (1973).
362. Ethics and editors (Letter), *Am. J. Dis. Child.* **128**:422 (1974).
363. Shepard, T. H., and Fantel, A. G., Legislative threat to fetal research, *Am. J. Dis. Child.* **128**:295 (1974).
364. Veatch, R. M., and Sollitto, S., Human experimentation—The ethical questions persist, *Hastings Cent. Rep.* **3**:1 (1973).
365. Veatch, R. M., Experimental pregnancy: The ethical complexities of experimentation with oral contraceptives, *Hastings Cent. Rep.* **1**:2 (1971).
366. Perlman, D. A layman's look at human experimentation (Editorial), *Am. Rev. Respir. Dis.* **110**:387 (1974).
367. Pekar, P. P., Jr., Cancer—Attack or retreat, *J. Am. Med. Assoc.* **230**:555 (1974).

Genetic Counseling and Screening

368. Speaker, F., Board of trustees sets policy. Patient consent necessary for release of medical data, *Pa. Med.* **82**:40 (1979).
369. Bernstein, A. H., Incompetent's right to die: Who decides, *Hospitals* **53**:39 (1979).
370. Gray, C., and Garner, J., Health and the Law conference in Ottawa exposes pitfalls but offers few remedies, *Can. Med. Assoc. J.* **120**:1407 (1979).
371. Geekie, D. A., PEI doctors oppose government order breaching medical record confidentiality (News), *Can. Med. Assoc. J.* **120**:1404 (1979).
372. Foster, H. H., The devil's advocate, *Bull. Am. Acad. Psychiatry L.* **6**:475 (1978).
373. Jacknow, D., Access to medical records (Letter), *J. Am. Med. Assoc.* **2412701** (1979).
374. Goebert, II. W., Jr., Rx for malpractice—Informed consent, *Hawaii Med. J.* **38**:102 (1979).
375. McCormick, R. A., Bioethical issues and the moral matrix of U.S. health care, *Hosp. Prog.* **60**:42 (1979)
376. Medical records bills give government too much access, APA, other groups maintain (News), *Hosp. Community Psychiatry* **30**:497 (1979).
377. Stone, A. A., Informed consent: Special problems for psychiatry, *Hosp. Community Psychiatry* **30**:321 (1979).
378. Health and the law, *Dimens. Health Serv.* **56**:31 (1979).

379. Picard, E., A child's right to health care, *Dimens. Health Serv.* **56**:60 (1979).

380. Consent to medical treatment, *Br. Med. J.* **1**:1091 (1979).

381. Turner, J. H., Hayashi, T. T., and Pogoloff, D. D. Legal and social issues in medical genetics, *Am. J. Obstet. Gynecol.* **134**:83 (1979).

382. Pickens, A., Texas medical disclosure panel seeks comment on informed consent proposals, *Tx. Med.* **75**:38 (1979).

383. Berry, L. M., Confidentiality and ACC (Letter), *N.Z. Med. J.* **89**:97 (1979).

384. Trout, M. E., When can a child consent to treatment, *Leg. Aspects Med. Pract.* **6**:25 (1978).

385. Virginia court ruling stirs concern about confidentiality protections in group therapy (News), *Hosp. Community Psychiatry* **30**:428 (1979).

386. Pembrey, M. E., Genetic registers, *Arch. Dis. Child* **54**:169 (1979).

387. Horty, J. F., Health units sometimes can reveal private data, *Mod. Health Care* **9**:75 (1979).

388. Sadoff, R. L., Pennsylvania psychiatrist vindicated in refusing judge's request for records, *Leg. Aspects Med. Pract.* **7**:38 (1979).

389. Dickens, B. M., Legal protection of psychiatric confidentiality, *Int. J. L. Psychiatry* **1**:255 (1978).

390. Edelman, S., Confidential records are no secret (Letter), *Hospitals* **53**:50 (1979).

391. Supreme Court ruling to affect some hospitals (News), *Hospitals* **53**:24 (1979).

392. Wilson, R. G., Association keeps a careful watch on possible problems inherent in the Canadian Human Rights Act, *Can. Med. Assoc. J.* **120**:836 (1979).

393. Bunch, P. L., and Dowben, C., Status of the physician–patient privilege, *South. Med. J.* **72**:339 (1979).

394. Signed consent: Is it really the nurse's responsibility? (Letter), *Nurs. Outlook* **27**:154 (1979).

395. Connors, P. J., Malpractice and informed consent (Letter), *N. Engl. J. Med.* **300**:928 (1979).

396. Feinman, M. A., Getting along with the genetic genie, *Leg. Aspects Med. Pract.* **7**:37 (1979).

397. Chapman, S., What are your odds in the prenatal gamble, *Leg. Aspects Med. Pract.* **7**:30 (1979).

398. Genetic registers (Editorial), *Lancet* **1**:253 (1979).

399. Emery, A. E., Brough, C. Crawford, M., *et al.*, A report on genetic registers. Based on the report of the Clinical Genetics Society Working Party, *J. Med. Genet.* **15**:435 (1978).

400. Katz, E., Ontario Hospital Association Convention: Encouraging patients to donate organs, *Dimens. Health Serv.* **55**:33 (1978).

401. Andrews, R. G., Adoption: Legal resolution or legal fraud, *Fam. Process* **17**:313 (1978).

402. New regulations governing DHEW sterilization funding now in effect; Stress informed consent (News), *Fam. Plann. Perspect.* **11**:46 (1979).

403. Gilmore, A., Health records confidentiality: the rights of the individual and society, *Can. Med. Assoc. J.* **119**:1096 (1978).

404. Florey, C. Computers and confidentiality (Letter), *Br. Med. J.* **1**:346 (1979).

405. Curie-Cohen, M., Luttrell, and Shapiro, S., Current practice of artificial insemination by donor in the United States, *N. Engl. J. Med.* **300**:585 (1979).

406. Elliott, J., Controversy in medicine: Access to employee health records (News), *J. Am. Med. Assoc.* **241**:777 (1979).

407. Bernstein, A. H., Divulging hospital records: Need for access vs. confidentiality, *Hospitals* **53**:51 (1979).

408. Congress to examine confidentiality of medical records, *Hosp. Community Psychiatry* **30**:207 (1979).

409. Gargaro, W. J., Jr., Cancer nursing and the law. Informed consent: Update on the Tuma case, *Cancer Nurs.* **1**:467 (1978).

410. Kopelman, L., Genetic screening in newborns: Voluntary or compulsory, *Perspect. Biol. Med.* **22**:83 (1978).

411. Curran, W. J., Informed consent, Texas style: Disclosure and nondisclosure by regulation, *N. Engl. J. Med.* **300**:482 (1979).

412. Guidelines for the ethical, social and legal issues in prenatal diagnosis. A report from the genetics research group of the Hastings Center, Institute of Society, Ethics and the Life Sciences, *N. Engl. J. Med.* **300**:168 (1979).

413. Chiu, E., and Teltscher, B. Huntington's disease: The establishment of a national register, *Med. J. Aust.* **2**:394 (1978).

414. Brazier, M., Informed consent to surgery, *Med. Sci. Law* **19**:49 (1979).

415. Bernstein, A. H., Legal rights of mental patients, *Hospitals* **53**:49 (1979).

416. Bernstein, A. H., Current status of the law of consent to treatment, *Hospitals* **53**:83 (1979).

417. Wertheimer, E., Informed consent, *Conn. Med.* **43**:31 (1979).

418. Beigler, J. S., The APA model law on confidentiality, *Am. J. Psychiatry* **136**:71 (1979).
419. Model law on confidentiality of health and social service records, *Am. J. Psychiatry* **136**:137 (1979).
420. Regan, W. A., Legally speaking: When it comes to consent, empty gestures won't do, *R.N.* **42**:25 (1979).
421. Gilmore, M., and Shear, K., Case conferences from Cornell. Case 1: Ethical and legal considerations of confidentiality in the treatment of hospitalized health professionals, *Psychiatr. Q.* **50**:237 (1978).
422. Grant, M. A., Genetic control and the law, *Med. Trial Tech. Q.* **24**:306 (1978).
423. Ramsey, P., The Saikewicz precedent: What's good for an incompetent patient, *Inquiry* **8**:36 (1978).
424. Murphy, S. S., Privacy Commission report contains serious flaws, *Hospitals* **53**:65 (1979).
425. Bader, M., On "genetic counseling and wrongful life" (Letter), *Am. J. Public Health* **68**:1026 (1978).
426. Eaton, A. P., Ackerman, J. H., Linyear, A. S., *et al.*, Genetic services in Ohio. A regional approach, *Ohio State Med. J.* **74**:623 (1978).
427. Curran, W. J., Confidentiality of records in PSRO's, *N. Engl. J. Med.* **299**:1348 (1978).
428. Medical groups protest OSHA proposal regarding access to medical records. Statements of AOMA, AAOM and AMA, *J. Med.* **20**:769 (1978).
429. Foster, H. H., Confidentiality, *Bull. Am. Acad. Psychiatry L.* **6**:256 (1978).
430. Berger, C. J., "Minors may give own consent; no physicians liable for their care" (Letter), *Mich. Med.* **77**:488 (1978).
431. Mason, J. C., Backus, F. I., and Volberding, N. K., Obtaining informed consent for mental health treatment, *Hosp. Community Psychiatry* **29**:742 (1978).
432. Hansen, H. Brief Reports. Decline Of Down's syndrome after abortion reform in New York State, *Am. J. Ment. Defic.* **83**:185 (1978).
433. Schram, R. B., Kane, J. C., Jr., and Roble, D. T., Law-Medicine Notes. "No code" orders: Clarification in the aftermath of Saikewicz, *N. Engl. J. Med.* **299**:875 (1978).
434. Regan, W. A., Wrongful death alleged when psychiatric patient died from burns, *Hosp. Prog.* **59**:32 (1978).
435. Bernstein, A. H., Malpractice arbitration: Is it really binding, *Hospitals* **52**:46 (1978).
436. Massachusetts rule lets patients see medical files (News), *Hospitals* **52**:19 (1978).
437. Regan, W. A., OR nurses' problems and responsibilities in consent forms, *AORN J.* **28**:83 (1978).
438. Horty, J. F., Will HSA members disclose finances, *Mod. Health Care* **8**:45 (1978).
439. Gargard, W. J. Cancer nursing and the law. Informed consent. A specific case, *Cancer Nurs.* **1**:329 (1978).
440. Sklar, C. You and the law: Minors in the health care system, *Can. Nurse* **74**:18 (1978).
441. Disclosure of medical records, *Br. Med. J.* **2**:135 (1978).
442. Regan, W. A., Surgeon failed to disclose alternatives to sterilization, *Hosp. Prog.* **59**:98 (1978).
443. Kucera, W. R., Legal Briefs. Just an innocent slip of the tongue, *AANA J.* **46**:159 (1978).
444. Reilly, P., Government support of genetic services, *Soc. Biol.* **25**:23 (1978).
445. Horsley, J. E., How to keep consent forms from becoming hot potatoes, *R. N.* **41**:33 (1978).
446. Berger, L. R., Sounding Board. Abortions in America: The effects of restrictive funding, *N. Engl. J. Med.* **298**:1474 (1978).
447. Kidd, R., History of H.B. 1246—Confidential communications of medical records, *J. Med. Assoc. Ga.* **67**:275 (1978).
448. Regan, W. A., Chemotherapy withheld from retarded patient, *Hosp. Prog.* **59**:94 (1978).
449. Welch, L., Court determines that PSRO data must be given to consumer organization, *Hospitals* **52**:16 (1978).
450. Brown, J. H., Recent Canadian court decisions on consent letter, *Can. Med. Assoc. J.* **118**:896 (1978).
451. Curran, W. J., Genetic counseling and wrongful life, *Am. J. Public Health* **68**:501 (1978).
452. Drake, E. A., and Bardon, J. I., Confidentiality and interagency communication: Effect of the Buckley Amendment, *Hosp. Community Psychiatry* **29**:312 (1978).
453. Sklar,, C. L., Legal consent and the nurse, *Can. Nurse* **74**:34 (1978).
454. Lynch, P. M., and Lynch, H. T., Genetic counseling of high-cancer-risk patients: Jurisprudential considerations, *Semin. Oncol.* **5**:107 (1978).
455. Brearley, P., Confidentiality—The need for a code of practice, *Nurs. Times* **74**:505 (1978).
456. Martin, A. J. Confidentiality—Its nature in law, *Nurs. Times* **74**:503 (1978).
457. Shaw, M. W., The legal aspects of genetic screening and counseling, in: Armendares S., Lisker, R. *Human Genetics* (S. Armendares and R. Lisker, eds.), p. 3, Excerpta Medica, Amsterdam (1976).

458. Welsh, L., Concern expressed over data between PSRO's and HSA's, *Hospitals* **52**:48 (1978).

459. McKerrow, L. W., Policies for releasing psychiatric records, *Dimens. Health Serv.* **55**:38 (1978).

460. Sharpe, G., Consent and sterilization, *Can. Med. Assoc. J.* **118**:591 (1978).

461. Shaw, M. W., Genetically defective children: Emerging legal considerations, *Am. J. L. Med.* **3**:333 (1977).

462. Creighton, H., Law for the nurse supervisor. More about informed consent, *Superv. Nurse* **9**:84 (1978).

463. Horty, J. F., Oral okay upheld but written preferred, *Mod. Health Care* **8**:64 (1978).

464. Rosenstock, I. M., Public health aspects of Tay–Sachs screening, *Prog. Clin. Biol. Res.* **18**:329 (1977).

465. Shaw, M. W., Confidentiality and privacy: Implications for genetic screening, *Prog. Clin. Biol. Res.* **18**:305 (1977).

466. Horty, J. F., Court rules M.D.–parent talks are privileged, *Mod. Health Care* **8**:56 (1978).

467. Shaw, M. W., Perspectives on today's genetics and tomorrow's progeny, *J. Hered.* **68**:274 (1977).

468. Eberhard, L. S., Need for an interim regulation. Development of federal guidelines on disclosure of PSRO data, *QRB.* **3**:11 (1977).

469. Annas, G. J., Avoiding malpractice suits through the use of informed consent, *Leg. Med. Annu.* 217 (1977).

470. Sharpe, G., Recent Canadian court decisions on consent, *Can. Med. Assoc. J.* **117**:1421 (1977).

471. Gallop, M. R., Confidentiality of medical records, *Occup. Health Nurs.* **25**:13 (1977).

472. Westin, A. F., Medical records: Should patients have access, *Hastings Cent. Rep.* **7**:23 (1977).

473. Rozovsky, L. E., The patients' right to see the record, *Dimens. Health Serv.* **54**:8 (1977).

474. Terr, L. C., The Hearst trial and the confidentiality of residency records, *Am. J. Psychiatry* **134**:1283 (1977).

475. Bloom, A., The right of public access to information submitted under the requirements of the Health Planning Act, *Public Health Rep.* **92**:411 (1977).

476. Sheppard, G. A., Informed consent—Records, *Orange Cty. Dent. Soc. Bull.* **5**:5 (1977).

477. Thompson, C. E., Legal aspects of genetic screening, *JOGN Nurs.* **6**:34 (1977).

478. Creighton, H., Prediction of dangerousness in psychiatry. Law for the nurse supervisor, *Superv. Nurse* **8**:80 (1977).

479. Maroney, J. D., and Barber, W. G., Texas Supreme Court protects confidentiality of minutes of hospital meetings, *Tex. Med.* **73**:91 (1977).

480. Blume, S. B., Confidentiality of medical records: Legislation at last, *N.Y. State J. Med.* **77**:1328 (1977).

481. Stone, J. W., Advice about consent (Editorial), *J. Kans. State Dent. Assoc.* **60**:8 (1976).

482. Stone, A. A., The Tarasoff decisions: Suing psychotherapists to safeguard society, *Harv. L. Rev.* **90**:358 (1976).

483. Horty, J. F., Hospitals seen liable in consent case ruling, *Mod. Health Care* **7**:74 (1977).

484. Confidentiality and the three wise men, *Br. Med. J.* **1**:1291 (1977).

485. Harris, N. H., Confidentiality and the three wise men (Letter), *Br. Med. J.* **1**:1415 (1977).

486. Cooperman, E. M., Meningomyelocele: To treat or not to treat, *Can. Med. Assoc. J.* **116**:1339 (1977).

487. Chabon, R. S., The physician and parental consent, *J. Leg. Med.* **5**:33 (1977).

488. Annas, G. J., and Coyne, B. "Fitness" for birth and reproduction: Legal implications of genetic screening, *Leg. Med. Annu.* **1976**:273 (1977).

489. Kucera, W. R., What the nurse anesthetist should know about the law's view of informed consent, *AANA J.* **45**:309 (1977).

490. Yeaworth, R. C., Ethics: The agonizing decisions in mental retardation, *Am. J. Nurs.* **77**:864 (1977).

491. De Wet, B. S., Clinical trials, children and the law (Letter), *S. Afr. Med. J.* **51**:528 (1977).

492. Quimby, C. W., Jr., Informed consent: A dialogue, *Radiology* **123**:805 (1977).

493. Graham, J. B., Ethical and social issues posed by genetic studies of cardiovascular disease, *Perspect. Biol. Med.* **20**:260 (1977).

494. Protecting informants, *Br. Med. J.* **1**:521 (1977).

495. Horty, J. F., It isn't libel if there's privilege. No malice, *Mod. Health Care.* **7**:54 (1977).

496. Sandstrom, M. M., and Milunsky, A., Prenatal genetic diagnosis, *Am. Fam. Physician* **15**:121 (1977).

497. Jenkins, T., and Kromberg, J. G., Prenatal diagnosis and selective abortion, *S. Afr. med. J.* **50**:2091 (1976).

498. Informed consent suit against anesthesiologist, *Anesth. Analg.* **55**:906 (1976).

499. Milunsky, A., and Reilly, P., The "new" genetics: Emerging medicolegal issues in the prenatal diagnosis of hereditary disorders, *Am. J. L. Med.* **1**:71 (1975).

500. Fost, N., Ethical problems in pediatrics, *Curr. Probl. Pediatr.* **6**:1 (1976).

501. Hifman, L. F., III, Legal considerations and prenatal genetic diagnosis, *Clin. Obstet. Gynecol.* **19**:965 (1976).
502. Gardner, J. C., The Privacy Act and Social Security disability, *J. Tenn. Med. Assoc.* **69**:780 (1976).
503. Wiley, J., The nurse's legal responsibility in obstetric monitoring, *JOGN Nurs.* **5**:77S (1976).
504. Hansen, K. D., and Omenn, G. S., Genetic counseling: The search for the adopted child, *J. Leg. Med.* **4**:8AA (1976).
505. Boyarsky, S., Informed consent: Recent changes in the law, *Trans. Am. Assoc. Genitourin. Surg.* **67**:29 (1975).
506. Arnold, A., and Moseley, R., Ethical issues arising from human genetics, *J. Med. Ethics* **2**:12 (1976).
507. Annas, G. J., Legal aspects of medical confidentiality in the occupational setting, *J. Occup. Med.* **18**:537 (1976).
508. Morganstein, W. M., Informed consent—The doctrine evolves, *J. Am. Dent. Assoc.* **93**:637 (1976).
509. Knight, B., Forensic problems in practice. V. Medical Confidentiality, *Practitioner* **216**:588 (1976).
510. Rice, G. P., Jr., Freedom to access to information and the right of privacy, *Image* **8**:36 (1976).
511. Veatch, R. M., Ethical issues in genetics, *Prog. Med. Genet.* **10**:223 (1974).
512. Consent, *Pediatrics* **57**:414 (1976).
513. Annas, G. J., Problems of informed consent and confidentiality in genetic counseling, in: *Genetics and the Law* (A. Milunsky and G. J. Annas, eds.), p. 111, Plenum Press, New York (1975).
514. Callahan, D., On the necessity of legislating morality for genetic decisions, in: *Genetics and the Law* (A. Milunsky and G. J. Annas, eds.), p. 411, Plenum Press, New York (1975).
515. Hirschhorn, K., Medicolegal aspects of genetic counseling, in: *Genetics and the Law* (A. Milunsky and G. J. Annas, eds.), p. 105, Plenum Press, New York (1975).
516. Milunsky, A., Medico-legal issues in prenatal genetic diagnosis, in: *Genetics and the Law* (A. Milunsky and G. J. Annas, eds.), p. 53, Plenum Press, New York (1975).
517. Genetics and the Law. Selected recent bibliography, in: *Genetics and the Law* (A. Milunsky and G. J. Annas, eds.), p. 495, Plenum Press, New York (1975).
518. Genetic counseling—Mass population screening for homozygotes and heterozygotes. A discussion, in: *Genetics and the Law* (A. Milunsky and G. J. Annas, eds.), p. 123, Plenum Press, New York (1975).
519. Sorenson, J. R., From social movement to clinical medicine—The role of law and the medical profession in regulating applied human genetics, in: *Genetics and the Law* (A. Milunsky and G. J. Annas, eds.), p. 467, Plenum Press, New York (1975).
520. Paul, E. W., Pilpel, H. F., and Wechsler, N. F., Pregnancy, teenagers and the law, 1976, *Fam. Plann. Perspect.* **8**:16 (1976).
521. Castleman, N., A course on the social and ethical implications of new breakthroughs in genetics and medicine: Bioethics, *Ethics Sci. Med.* **2**:61 (1975).
522. Powledge, T. M., Genetic screening as a political and social development, *Birth Defects* **10**:25 (1974).
523. Fletcher, J., Roblin, R. O., and Powledge, T. M., Informed consent in genetic screening programs, *Birth Defects* **10**:137 (1974).
524. Lappe, M., Ethical, social and legal dimensions of screening for human genetic disease: Introduction, *Birth Defects* **10**:VII (1974).
525. Tietze, C., Human rights in relationship to induced abortion, *J. Sex Res.* **10**:89 (1974).
526. Bernstein, A. H., Contemporary consent cases, *Hospitals* **49**:101 (1975).
527. Urquhart-Hay, D., Ethics and vasectomy (Letter), *N.Z. Med. J.* **81**:568 (1975).
528. Skegg, P. D., "Informed consent" to medical procedures, *Med. Sci. L.* **15**:124 (1975).
529. Emery, A. E., Legal status of the unborn (Letter), *Lancet* **2**:715 (1974).
530. Goldsmith, L. S., The myth about informed consent, *J. Leg. Med.* **3**:17 (1975).
531. Lowry, R. B., Miller, J. R., Scott, A. E., *et al.*, The British Columbia Registry for Handicapped Children and Adults: Evolutionary changes over twenty years, *Can. J. Public Health*, **66**:322 (1975).
532. Reilly, P. Legislation on genetic screening (Letter), *Am. J. Hum. Genet.* **27**:120 (1975).
533. Basics of "informed consent" for Arizona physicians, *Ariz. Med.* **32**:426 (1975).
534. Nielsen, J., Chromosome examination of newborn children: Purpose and ethical apsects, *Humangenetik* **26**:215 (1975).
535. Reilly, P., Genetic screening legislation, *Adv. Hum. Genet.* **5**:319 (1975).
536. Ris, H. W., Further analysis of minor's consent bill (Letter), *Pediatrics* **54**:654 (1974).
537. Ashe, L., Doctors, defenses and the age of informed consent (second part), *Bol. Asoc. Med. Pr.* **65**:49 (1973).

538. Grunebaum, H., and Abernethy, V., Ethical issues in family planning for hospitalized psychiatric patients, *Am. J. Psychiatry* **132**:236 (1975).

539. Simonaitis, J. E., Faulty comparisons may constitute defective disclosure, *J. Am. Med. Assoc.* **231**:1086 (1975).

540. Shanklin, D. R., It is time to take a stand (Editorial), *J. Reprod. Med.* **14**:41 (1975).

541. Knapp, T. A., and Huff, R. L., Emerging trends in the physician's duty to disclose: An update of Canterbury v. Spence, *J. Leg. Med* **3**:41 (1975).

542. Age of Consent (Letter), *Can Med. Assoc. J.* **112**:420 (1975).

543. Newman, R. G., and Newman, T., Safeguarding confidentiality of methadone patient records, *Natl. Conf. Methadone Treat. Proc.* **2**:1431 (1973).

544. Bryant, T. E., Remarks on confidentiality, *Natl. Conf. Methadone Treat. Proc.* **2**:1427 (1973).

545. Feather, W., The physician and the law. Anesthesiologist fails to obtain informed consent, *Anesth. Analg. (Cleveland)* **53**:876 (1974).

546. Dornette, W. H., Informed consent and anesthesia, *Anesth. Analg. (Cleveland)* **53**:832 (1974).

547. Lieberman, E. J., Informed consent for parenthood, *Am. J. Psychoanal.* **34**:155 (1974).

548. Powledge, T. M., and Sollitto, S., Prenatal diagnosis—The past and future, *Hastings Cent. Rep.* **4**:11 (1974).

549. Pernoll, M. L., Prescott, G. H., and Hecht, F., Prenatal diagnosis: Practice, pitfalls, and progress, *Obstet. Gynecol.* **44**:773 (1974).

550. Reilly, P., Sickle cell anemia legislation, *J. Leg. Med.* **1**:36 (1973).

551. Reilly, P., Sickle cell anemia legislation, *J. Leg. Med.* **1**:39 (1973).

552. Salmon, E. L., Informed consent perplexity (Letter), *J. Leg. Med.* **1**:9 (1973).

553. Breckler, I. A., Price, E. M., and Shore, S., Informed consent: A new majority position, *J. Leg. Med.* **1**:37 (1973).

554. Rubsamen, D. S., Informed consent: New decisions create new impacts (Editorial), *J. Leg. Med.* **1**:6 (1973).

555. Consent held not informed, *J. Leg. Med.* **2**:54 (1974).

556. Autopsy consent at issue, *J. Leg. Med.* **1**:21 (1973).

557. Bernstein, A. H., Unauthorized disclosure of confidential information, *Hospitals* **48**:126 (1974).

558. Curran, W. J., The questionable virtues of genetic screening laws, *Am. J. Public Health* **64**:1003 (1974).

559. Orwoll, G., The doctor's duty of disclosure, *Anesth. Analg. (Cleveland)* **53**:759 (1974).

560. Powledge, T. M., New trends in genetic legislation, *Hastings Cent. Rep.* **3**:6 (1973).

561. Annas, G. J., Editorial correspondence: XYY and the law (Letter), *Hastings Cent. Rep.* **2**:14 (1974).

562. Powledge, T. M., Laws in question: Confusion over sickle cell testing, *Hastings Cent. Rep.* **3**:3 (1972).

563. Veatch, R. M., Brady, S. M., and Robitscher, J., Informed consent: When can it be withdrawn, *Hastings Cent. Rep.* **2**:10 (1972).

564. Committee on youth. The implications on minor's consent legislation for adolescent health care: A commentary, *Pediatrics* **54**:481 (1974).

565. Plaut, E. A., A perspective on confidentiality, *Am. J. Psychiatry* **131**:1021 (1974).

Genetics and Family Law

566. Hightower-Vandamm, M. D., Developmental Disabilities Act: An historic perspective, Part I, *AJOT* **33**:355 (1979).

567. Reilly, P., The law and artifical insemination (Letter), *N. Engl. J. Med.* **301**:588 (1979).

568. Smith-Hoffman, S. J., Abused children and developmentally disabled adults: The nurse's responsibility, *Wash. State J. Nurs.* **8**:4 (1978).

569. Curran, W. J., The Supreme Court and madness: A middle ground on proof of mental illness for commitment, *N. Engl. J. Med.* **301**:317 (1979).

570. Morris, G. H., Conservatorship for the "gravely disabled": California's nondeclaration of nonindependence, *Int. J. L. Psychiatry* **1**:395 (1978).

571. Diamond, E. F., In vitro fertilization: A moratorium is in order, *Hosp. Prog.* **60**:66 (1979).

572. McCormick, R. A., Bioethical issues and the moral matrix of U.S. health care, *Hosp. Prog.* **60**:42 (1979).

573. Copeland, L. S., Mandatory employment of the handicapped, *Soc. Secur. Bull.* **42**:23 (1979).

574. Shriner, T. L., Jr., Maternal versus fetal rights—A clinical dilemma, *Obstet. Gynecol.* **53**:518 (1979).

575. Flathouse, V. E., Multiply handicapped deaf children and Public Law 94–142, *Except. Child* **45**:560 (1979).

576. Dibden, W. A., Law as it relates to the mentally ill and the mentally handicapped, *Australas. Nurses J.* **8**:20 (1979).

577. Fitzgibbons, P. M., and Ferry, P. C., It's the law. Mandatory public education for handicapped children, *Am. J. Dis. Child.* **133**:476 (1979).

578. Jones, E. H., P. L. 94–142 and the role of school nurses in caring for handicapped children, *J. Sch. Health* **49**:147 (1979).

579. Kemp, J., Hospitals work to meet the letter and spirit of the law, *Hospitals* **53**:36 (1979).

580. Jonsen, A. R., Dying right in California—The Natural Death Act, *Clin. Toxicol.* **13**:513 (1978).

581. Wilson, R. G., Association keeps a careful watch on possible problems inherent in the Canadian Human Rights Act, *Can. Med. Assoc. J.* **120**:836 (1979).

582. Behrman, S. J., Artificial insemination and public policy (Editorial), *N. Engl. J. Med.* **300**:619 (1979).

583. Whelan, D., The law and artificial insemination with donor semen (AID), *Med. J. Aust.* **1**:56 (1978).

584. Cohen, S., Semmes, M., and Guralnick, M. J., Public Law 94–142 and the education of preschool handicapped children, *Except. Child* **45**:279 (1979).

585. Relman, A. S., The Saikewicz decision: A medical viewpoint, *Am. J. L. Med.* **4**:233 (1978).

586. Beck, W. W., Jr., Artificial insemination and preservation of semen, *Urol. Clin. North Am.* **5**:593 (1978).

587. Parker, E., and Tennent, G., The 1959 Mental Health Act and mentally abnormal offenders: A comparative study, *Med. Sci. L.* **19**:29 (1979).

588. Bernstein, A. H., Legal rights of mental patients, *Hospitals* **53**:49 (1979).

589. Soloff, P. H., Jewell, S., and Roth, L. H., Civil commitment and the rights of the unborn, *Am. J. Psychiatry* **136**:114 (1979).

590. Sachett, W. W., Jr., The right to die (Letter), *Leg. Aspects Med. Pract.* **5**:17 (1977).

591. Meisel, A., and Roth, L. H., Must a man be his cousin's keeper, *Hastings Cent. Rep.* **8**:5 (1978).

592. Rozovsky, L., Civil commitment of the mentally ill, *Dimens. Health Serv.* **55**:38 (1978).

593. Baron, C. H., Assuring "detached but passionate investigation and decision": The role of guardians ad litem in Saikewicz-type cases, *Am. J. L. Med.* **4**:111 (1978).

594. Artificial insemination by donor (Editorial), *S. Afr. Med. J.* **53**:1006 (1978).

595. Low, M. B., The Education For All Handicapped Children Act of 1975: A pediatrician's viewpoint, *Pediatrics* **62**:271 (1978).

596. Urbanus, P., Issues and opinions. II. Sterilization and the mentally retarded: HEW's new regulations, *J. Nurse Midwife* **23**:16 (1978).

597. Bradshaw, S., Disability without handicap: So you think you've got problems, *Nurs. Times* **74**:19 (1978).

598. Burton, C. V., The balance between regulation and freedom, *Neurosurgery* **1**:322 (1977).

599. Clark, J., Special educational needs of the young handicapped: After court, comes Warnock, *Nurs. Mirror* **147**:7 (1978).

600. Lesbian couples: Should help extend to aid? Clinical conference, *J. Med. Ethics* **4**:91 (1978).

601. Martin, R., The teacher's rights in PL 94–142, *J. Learn. Disabil.* **11**:331 (1978).

602. Michigan MH Department trying to eliminate abuse in mental retardation facilities, *Hosp. Community Psychiatry* **29**:621 (1978).

603. Huey, K., Conference report—Placing the mentally retarded: Where shall they live, *Hosp. Community Psychiatry* **29**:596 (1978).

604. Coye, J. L., and Clifford, D., A one-year report on rights violations under Michigan's new protection system, *Hosp. Community Psychiatry* **29**:528 (1978).

605. Kroeber, M. J., A fair deal for the handicapped, *Aust. Nurses J.* **7**:46 (1978).

606. Rothman, D. A., and Rothman, N. L., Has the Pennsylvania legislature over-reacted in setting standards for involuntary commitment of mentally ill persons, *Pa. Nurse* **33**:2 (1978).

607. Tiefel, H. O., The unborn. Human values and responsibilities, *J. Am. Med. Assoc.* **239**:2263 (1978).

608. Rothman, C., Clinical aspects of sperm bank, *J. Urol.* **119**:511 (1978).

609. Horan, D. J., Right-to-die laws: Creating, not clarifying, problems, *Hosp. Prog.* **59**:62 (1978).

610. Regan, W. A., Court blocked sterilization of a retarded minor, *Hosp. Prog.* **59**:96 (1978).

611. Wasow, M., For my beloved son David Jonathan: A professional plea, *Health Soc. Work* **3**:126 (1978).

612. Ramirez, B., and Smith, B. J., Federal mandates for the handicapped: Implications for American Indian children, *Except. Child* **44**:521 (1978).

613. Spencer, D. A., The revision of the Mental Health Act, 1959: Should the mentally handicapped be excluded, *R. Soc. Health J.* **98**:84 (1978).

614. Rees, F. W., Jr., and Gary, J. G., Compliance with 504 regulations need not require costly renovation, *Hospitals* **52**:78 (1978).

615. Rehabilitation Act: A matter of compliance (Editorial), *Hospitals* **52**:77 (1978).

616. Rosner, F. Sperm procurement and analysis in Jewish law, *Am. J. Obstet. Gynecol.* **130**:627 (1978).

617. Annas, G. J., The incompetent's right to die: The case of Joseph Saikewicz, *Hastings Cent. Rep.* **8**:21 (1978).

618. Ballard, J., and Zettel, J. J., Fiscal arrangements of Public Law 94–142, *Except. Child* **44**:333 (1978).

619. Harvey, J., Legislative intent and progress, *Except. Child* **44**:234 (1978).

620. Martin, A., A right to treatment, *Nurs. Times* **73**:1620 (1977).

621. Curran, W. J., Law-Medicine Notes. The freedom of medical practice, sterilization, and economic medical philosophy, *N. Engl. J. Med.* **298**:32 (1978).

622. Todd, M. C., The Jerome Cochran Lecture. Our federated individuality, *J. Med. Assoc. State Ala.* **44**:598 (1975).

623. Vaccaro, P. L., and Kaplan, R. S., Employing the handicapped: A look at the new regs, *J. Am. Health Care Assoc.* **3**:18 (1977).

624. Ballard, J., and Zettel, J. Public Law 94–142 and section 504: What they say about rights and protections, *Except. Child* **44**:177 (1977).

625. Abeson, A., and Zettel, J., The end of the quiet revolution: The Education for All Handicapped Children Act of 1975, *Except. Child* **44**:114 (1977).

626. Lewis, E. P., The right to inform (Editorial), *Nurs. Outlook* **25**:561 (1977).

627. Biklen, D., Myths, mistreatment, and pitfalls: Mental retardation and criminal justice, *Ment. Retard.* **15**:51 (1977).

628. Cull, J. G., and Levison, K. F., The rights of consumers of rehabilitation services, *J. Rehabil.* **43**:29 (1977).

629. Smith, G. P., 2d, The medicolegal challenge of preparing for a brave, yet somewhat frightening new world, *J. Leg. Med.* **5**:9 (1977).

630. Perlin, M. L., and Siggers, W. W., The role of the lawyer in mental health advocacy, *Bull. Am. Acad. Psychiatry L.* **4**:204 (1976).

631. Weitzel, W. D., Legislative liaison as critical intervention, *Bull. Am. Acad. Psychiatry L.* **4**:216 (1976).

632. Sullivan, J. L., Guardian's authority to involuntarily hospitalize the incompetent ward, *Bull. Am. Acad. Psychiatry L.* **4**:231 (1976).

633. Sandler, B., Donor insemination (Letter), *Br. Med. J.* **2**:188 (1977).

634. Langlois, L., The evolution of the legal status of mental patients in Quebec and Ontario—A comparison, *Can. Ment. Health* **24**:2 (1976).

635. Special legal implications of patient rights, *J. Am. Health Care Assoc.* **3**:19 (1977).

636. Fleming, B. A., and Hunt, J. F., Bartley v. Kremens: A step backward, *J. Natl. Assoc. Priv. Psychiatr. Hosp.* **8**:10 (1976).

637. Lax, B., and Foley, K. E., A lawsuit re: Legal rights The Retarded v. The People, *Ment. Retard.* **15**:3 (1977).

638. Templeton, A., AID (artificial insemination by donor)—What are the problems, *Midwife Health Visit. Community Nurse* **13**:208 (1977).

639. Macourt, D. C., and Jones, G. R., Artificial insemination with donor semen, *Med. J. Aust.* **1**:693 (1977).

640. Kraus, J., and Quinn, P. E., Human artificial insemination: Some social and legal issues, *Med. J. Aust.* **1**:710 (1977).

641. Chalkley, D. T., Federal constraints: Earned or unearned, *Am. J. Psychiatry* **134**:911 (1977).

642. Pennington, G. W., and Naik, S., Donor insemination: Report of a two-year study, *Br. Med. J.* **1**:1327 (1977).

643. Ficarra, B. J., History of legel medicine, *Leg. Med. Annu.* **1976**:3 (1977).

644. Gardner, D. C., Warren, S. A., and Gardner, P. L., Locus of control and law knowledge: A comparison of normal, retarded and learning diabled adolescents, *Adolescence* **12**:103 (1977).

645. Bazelon, D. L., Institutionalization, deinstitutionalization and the adversary process, *Columbia L. Rev.* **75**:897 (1975).

646. Stromberg, R. E., The right to die controversy: Physicians, not the courts, should resolve the issue, *Hosp. Med. Staff* 6:9 (1977).
647. Creighton, H., Law for the nurse supervisor. The right to refuse treatment, *Superv. Nurse* 8:13 (1977).
648. Scallett, L., What the Supreme Court said. And what it didn't, in: *Paper Victories and Hard Realities* (V. Bradley and G. Clarke, eds.), p. 8, Health Policy Center, Washington, D.C. (1975).
649. Stearns, J., Survey of state attorneys general, in: *Paper Victories and Hard Realities* (V. Bradley and G. Clarke, eds.), p. 118, Health Policy Center, Washington, D.C. (1975).
650. Lottman, M. S., Paper victories and hard realities, in: *Paper Victories and Hard Realities* (V. Bradley and G. Clarke, eds), p. 92, Health Policy Center, Washington, D.C. (1975).
651. Bradley, V., and Clarke, G., Introduction: Putting the decision in perspective, in: *Paper Victories and Hard Realities* (V. Bradley and G. Clarke, eds.), p. 1, Health Policy Center, Washington, D.C. (1975).
652. In the Supreme Court of the United States. No. 74–8. J. B. O'Connor, petitioner, v. Kenneth Donaldson, in: *Paper Victories and Hard Realities* (V. Bradley and G. Clarke, eds.), p. 106, Health Policy Center, Washington, D.C. (1975).
653. Creighton, H., Law for the nurse supervisor. Terminating life support, *Superv. Nurse* 8:66 (1977).
654. Ursin, K. G., Influence of litigation on the lives of developmentally disabled: A preliminary report, *Birth Defects* 12:137 (1976).
655. Wilson, G., Do patients have rights, too, *J. Long Term Care Adm.* 4:26 (1976).
656. Melcher, J. W., Law, litigation, and handicapped children, *Except. Child* 43:126 (1976).
657. Johan, M., Psychiatric society head critiques Mental Health Act, *Pa. Med.* 79:38 (1976).
658. Eaglstein, A. S., and Berman, Y., Legal enactment for the retarded in Israel, *Ment. Retard.* 14:34 (1976).
659. Garland, M., Politics, legislation, and natural death. The right to die in California, *Hastings Cent. Rep.* 6:5 (1976).
660. Collings, G. D., Barrier free access; Right or privilege, *Phys. Ther.* 56:1029 (1976).
661. Peckins, D. M., Artificial insemination and the law, *J. Leg Med.* 4:17 (1976).
662. Annas, G. J., Law and the life sciences. O'Connor v. Donaldson: Insanity inside out, *Inquiry* 6:11 (1976).
663. Wright, W. R., The Fair Labor Standards Act and wage payments as related to the handicapped in the Tennessee vocational training centers, in: *Rehabilitation of the Severely Disabled* (W. M. Jenkins *et al.*, eds.), p. 149, Kendall/Hunt, Dubuque, Iowa (1974).
664. Bush, E. S., Ramifications of the criminal justice system for the mentally retarded, in: *Rehabilitation and the Retarded Offender* (P. L. Browning, ed.), p. 88, Charles C Thomas, Springfield, Ill. (1976).
665. Morrow, C. C., An attorney's experiences in a legal center for retarded persons, in: *Rehabilitation and the Retarded Offender* (P. L. Browning, ed.), p. 81, Charles C Thomas, Springfield, Ill. (1976).
666. Ravenel, L. R., and Bush, E. S., A legal framework: An outsider's perspective, in: *Rehabilitation and the Retarded Offender* (P. L. Browning, ed.), p. 68, Charles C Thomas, Springfield, Ill. (1976).
667. Morrow, C. C., A legal framework: An insider's perspective, in: *Rehabilitation and the Retarded Offender* (P. L. Browning, ed.), p. 54, Charles C Thomas, Springfield, Ill. (1976).
668. Melia, R. P., Rights of handicapped Americans: A bicentennial review, *Am. Rehabil.* 1:2 (1976).
669. Harmon, J. C., Jr., Who speaks for the handicapped, *Am. Rehabil.* 1:19 (1976).
670. Laski, F. J., Legal advocacy, positive factor in rights for disabled people, *Am. Rehabil.* 1:12 (1976).
671. Dix, G. E., The Alabama "right to treatment" case: An opportunity not to be missed, *Community Ment. Health J.* 12:161 (1976).
672. Kucera, W. R., The right to privacy and the right to die, *AANA J.* 44:303 (1976).
673. Effland, R. W., Trusts and estate planning, in: *The Mentally Retarded Citizen and the Law* (M. Kindred *et al.*, eds.), p. 115, Free Press, New York (1973).
674. Johnson, R. H., and Wood, J. J., Sr., Judicial, legislative, and administrative competence in setting institutional standards, in: *The Mentally Retarded Citizen and the Law* (M. Kindred *et al.*, eds.), p. 528, Free Press, New York (1973).
675. Introduction, in: *The Mentally Retarded Citizen and the Law* (M. Kindred *et al.*, eds.), p. xxv, Free Press, New York (1973).
676. The fetus and the newborn. A discussion, in: *Genetics and the Law* (A. Milunsky and G. J. Annas, eds.), p. 45, Plenum Press, New York (1975).
677. Green, H. P., The fetus and the Law, in: *Genetics and the Law* (A. Milunsky and G. J. Annas, eds.), p. 19, Plenum Press, New York (1975).

678. Shaw, M. W., and Damme, C. Legal status of the fetus, in: *Genetics and the Law* (A. Milunsky and G. J. Annas, eds.), p. 3, Plenum press, New York (1975).

679. Healey, J. M., Jr., Legal aspects of artificial insemination by donor and paternity testing, in: *Genetics and the Law* (A. Milunsky and G. J. Annas, eds.), p. 203, Plenum Press, New York (1975).

680. Genetics and family law. A discussion, in: *Genetics and the Law* (A. Milunsky and G. J. Annas, eds.), p. 239, Plenum Press, New York (1975).

681. Disabled ratepayers, *Br. Med. J.* 1:406 (1976).

682. Gilbert, S., Artificial insemination, *Am. J. Nurs.* 76:259 (1976).

683. Beck, W. W., Jr., A criticial look at the legal, ethical and technical aspects of artificial insemination, *Fertil. Steril.* 27:1 (1976).

684. Dolan, W. V., Abortion vs. manslaughter (Letter), *Arch. Surg.* 111:93 (1976).

685. Schwitzgebel, R. K., A contractual model for the protection of the rights of institutionalized mental patients, *Am. Psychol.* 30:815 (1975).

686. Fitzgerald, J. L., Rights and birth by artificial insemination, *Med. Leg. Bull.* 24:1 (1975).

687. Stedman, D. J., State councils on developmental disabilities, *Except. Child* 42:186 (1976).

688. Peck, C. L., Current legislative issues concerning the right to refuse versus the right to choose hospitalization and treatment, *Psychiatry* 38:303 (1975).

689. Baucom, L. D., and Stedman, D. J., A survey of the membership of state and territorial developmental disabilities councils: Relevant characteristics and observations, 1974, *Ment. Retard.* 13:12 (1975).

690. Short, R. V., Focus: Current issues in medical ethics, *J. Med. Ethics* 1:56 (1975).

691. Cusine, D. J., Aid and the law, *J. Med. Ethics* 1:39 (1975).

692. Abeson, A., Bolick, N., and Hass, J., A primer on due process: Education decisions for handicapped children, *Except. Child* 42:68 (1975).

693. Artificial insemination (donor) (Editorial), *Br. Med. J.* 4:2 (1975).

694. Sterilization of minors, *Br. Med. J.* 3:775 (1975).

695. Aid and embryo transfer (Editorial), *Med. J. Aust.* 1:324 (1975).

696. Injury before birth (Editorial), *Lancet* 2:704 (1974).

697. Brenner, J., Jr., Commitment procedures: 1975 amendments, *J. Tenn. Med. Assoc.* 68:630 (1975).

698. Liss, R., and Frances, A. Court-mandated treatment: Dilemmas for hospital psychiatry, *Am. J. Psychiatry* 132:924 (1975).

699. Creighton, H. Moral and legal issues—Nursing judgments, No. 16–1551, NLN Publ. 16–1551:22 (1974).

700. Curran, W. J., Law-Medicine Notes. The right to psychiatric treatment: A "simple decision" in the Supreme Court, *N. Engl. J. Med.* 293:487 (1975).

701. Meisel, A., Rights of the mentally ill: The gulf between theory and reality, *Hosp. Community Psychiatry* 26:349 (1975).

702. Marsh, R. L., Friel, C. M., and Eissler, V., The adult MR in the criminal justice system, *Ment. Retard.* 13:21 (1975).

703. Beck, W. W., Jr., Artificial insemination and semen preservation, *Clin. Obstet. Gynecol.* 17:115 (1974).

704. Downey, G. W., "Crip lib": The disabled fight for their own cause, *Mod. Health Care* 3:21 (1975).

705. Whitten, E. B., The Rehabilitation Act of 1973 and the severely disabled (Editorial), *J. Rehabil.* 40:39 (1974).

706. Robitscher, J. B., The right to die, *Hastings Cent. Rep.* 2:11 (1972).

707. McNamara, V. P., Scoggin, W. A., and Raskin, F., Sterilization of the mentally retarded at the Medical College of Georgia, under Law No. 1288, *J. Med. Assoc. Ga.* 63:55 (1974).

708. Frankel, M. S., Artifical insemination: The medical profession and public policy, *Conn. Med.* 38:476 (1974).

BIOETHICS

709. Siegler, M., Clinical ethics and clinical medicine, *Arch. Intern. Med.* 139:914 (1979).

710. Dialysis for schizophrenia: Consent and costs, *Hastings Cent. Rep.* 9:10 (1979).

711. Stanley, A. T., Is it ethical to give hope to a dying person, *Nurs. Clin. North Am.* 14:69 (1979).

712. De Leon Siantz, M. L., Human values in determining the fate of persons with mental retardation, *Nurs. Clin. North Am.* **14**:57 (1979).
713. Davis, A. J., Ethics rounds with intensive care nurses, *Nurs. Clin. North Am.* **14**:45 (1979).
714. Aroskar, M. A., Ethical issues in community health nurses, *Nurs. Clin. North Am.* **14**:35 (1979).
715. Mahon, K. A., and Fowler, M. D., Moral development and clinical decision-making, *Nurs. Clin. North Am.* **14**:3 (1979).
716. Payton, R. J., Information control and autonomy: Does the nurse have a role, *Nurs. Clin. North Am.* **14**:23 (1979).
717. Lumpp, F., The role of the nurse in the bioethical decision-making process, *Nurs. Clin. North Am.* **14**:13 (1979).
718. Shriver, E. K., Symposium on bioethical issues in nursing, *Nurs. Clin. North Am.* **14**:1 (1979).
719. Horan, D. J., In vitro fertilization: Legal and ethical implications, *Hosp. Prog.* **60**:60 (1979).
720. Veatch, R. M., Professional medical ethics: The grounding of its principles, *J. Med. Philos.* **4**:1 (1979).
721. Brungs, R., Biomedical challenges: Should "human" be redefined, *Hosp. Prog.* **60**:55 (1979).
722. Butler, J. K., Is it ethical to conduct volunteer studies within the pharmaceutical industry, *Lancet* **1**:816 (1978).
723. Berman, L. B., and Bayley, C., Nephrology and bioethics, *J. Am. Med. Assoc.* **241**:1402 (1979).
724. Spitz, B., When a solution is not a solution: Medicaid and health maintenance organizations, *J. Health Polit. Policy L.* **3**:497 (1979).
725. Graber, G. C., and Marsh, F. H., Synthetic competence and consent. Ought a defendant be drugged to stand trial, *Inquiry* **9**:8 (1979).
726. Zeik, M., A condemned man's last wish: Organ donation and a "meaningful" death (Commentary), *Inquiry* **9**:17 (1979).
727. Bedau, H. A., A condemned man's last wish: Organ donation and a "meaningful" death (Commentary), *Inquiry* **9**:16 (1979).
728. Annas, G. J., All the president's bioethicists, *Inquiry* **9**:14 (1979).
729. Veith, F. J., Brain death and organ transplantation, *Ann. N.Y. Acad. Sci.* **315**:417 (1978).
730. Pollard, M. R., and Brennan, J. T., Jr., Disease prevention and health promotion initiatives: Some legal considerations, *Health Educ. Monogr.* **6**:211 (1978).
731. Toulmin, S., In vitro fertilization: Answering the ethical objections, *Hastings Cent. Rep.* **8**:9 (1978).
732. Ramsey, P., Manufacturing our offspring: Weighing the risks, *Hastings Cent. Rep.* **8**:7 (1978).
733. Soble, A., Deception in social science research: Is informed consent possible, *Hastings Cent. Rep.* **8**:40 (1978).
734. Levine, C., Can the fetus be an organ farm? (Commentary), *Hastings Cent. Rep.* **8**:24 (1978).
735. Maguire, D. C., Can the fetus be an organ farm? (Commentary), *Hastings Cent. Rep.* **8**:24 (1978).
736. Warren, M. A., Can the fetus be an organ farm? (Commentary), *Hastings Cent. Rep.* **8**:23 (1978).
737. Robertson, J. A., In vitro conception and harm to the unborn, *Hastings Cent. Rep.* **8**:13 (1978).
738. Lappe, M., Ethics at the center of life: Protecting vulnerable subjects, *Hastings Cent. Rep.* **8**:11 (1978).
739. Report of the Pope John XXIII Center Consultation and Board Meeting: Center's board identifies key issues, reorganizes, starts task forces, Hosp. Prog. **60**:28 (1979).
740. Fabro, F., Ethics of in-vitro human fertilization (Editorial), *Conn. Med.* **42**:808 (1978).
741. Tsuang, M. T., Genetic counseling for psychiatric patients and their families, *Am. J. Psychiatry* **135**:1465 (1978).
742. Murphy, E. A., Eugenics. An ethical analysis, *Mayo Clin. Proc.* **53**:655 (1978).
743. Culliton, B. J., and Waterfall, W. K., Flowering of American bioethics, *Br. Med. J.* **2**:1270 (1978).
744. Agassi, J., Liberal forensic medicine, *J. Med. Philos.* **3**:226 (1978).
745. Sanders, C. A., Medical technology: Who's to say when we've had enough, *Hospitals* **52**:66 (1978).
746. Jonsen, A. R., Books on bioethics—A commentary, *Pharos* **41**:39 (1978).
747. Newton, L., Is abortion a religious issue? 3. The irrelevance of religion in the abortion debate, *Hastings Cent. Rep.* **8**:16 (1978).
748. Jaffe, F. S., Is abortion a religious issue? 2. Enacting religious beliefs in a pluralistic society, *Hastings Cent. Rep.* **8**:14 (1978).
749. Siemsen, A. W., Beauchamp, T. L., and Robertson, J. A., Medical, moral and legal aspects of renal replacement therapy, *Proc. Clin. Dial. Transplant Forum* **7**:185 (1977).
750. McCarthy, D. G., Positive view of church's stand on artificial fertilization (Letter), *Hosp. Prog.* **59**:6 (1978).

751. Science outpacing ethical studies in genetics, *Can. Med. Assoc. J.* **119**:78 (1978).
752. National Society of Professional Engineers v. United States (April 25, 1978), *ASHA Rep.* **20**:542 (1978).
753. Mellinger, G., Balter, M., and Huffine, C. L. Progress report on a study of public judgments regarding ethical issues in biomedical reports proceedings, *Psychopharmacol. Bull.* **14**:29 (1978).
754. Tiefel, H. O., The unborn. Human values and responsibilities, *J. Am. Med. Assoc.* **239**:2263 (1978).
755. Altman, A., Abortion funding: Legal and moral questions (Letter), *Hastings Cent. Rep.* **8**:4 (1978).
756. Bishop, V. A., A nurse's view of ethical problems in intensive care and clinical research, *Br. J. Anaesth.* **50**:515 (1978).
757. Payne, J. P., Ethical problems in clinical research and intensive care (Editorial), *Br. J. Anaesth.* **50**:413 (1978).
758. Mnookin, R. H., Children's rights: Legal and ethical dilemmas, *Pharos* **41**:2 (1978).
759. Ravin, A. W., Genetics in America: A historical overview, *Perspect. Biol. Med.* **21**:214 (1978).
760. Smith, G. P., 2nd, Uncertainties on the spiral staircase: Metaethics and the new biology, *Pharos.* **41**:10 (1978).
761. Refshauge, W., The place for international standards in conducting research on humans, *Bull. WHO* **55**:133 (1977).
762. Ladimer, I., Biological research for the community: Legal and ethical perspectives, *Bull WHO* **55**:111 (1977).
763. Cook, J. W., Altman, K., and Haavik, S., Consent for aversive treatment: A model form, *Ment. Retard.* **16**:47 (1978).
764. Steinfels, M. O., New childbirth technology: A clash of values, *Hastings Cent. Rep.* **8**:9 (1978).
765. Pauli, R. M., and Cassell, E. J., Nuturing a defective newborn, *Hastings Cent. Rep.* **8**:13 (1978).
766. Creighton, H., The diminishing right of privacy: Computerized medical records. Law for the nurse supervisor, *Superv. Nurse* **9**:58 (1978).
767. Norton, M. L., Ethics in medicine and law: Standards and conflicts, *Leg. Med. Annu.* 201 (1977).
768. Rheingold, P. D., and Goldhirsch, L., Doctors' countersuits, *J. Leg. Med.* **5**:32 (1977).
769. Hartung, J., and Ellison, P., A eugenic effect of medical care, *Soc. Biol.* **24**:192 (1977).
770. Kaufman, A. L., Law and ethics, *Hastings Cent. Rep.* **7**:7 (1977).
771. Veatch, R. M., Medicine, biology, and ethics, *Hastings Cent. Rep.* **7**:2 (1977).
772. Blackstone, W. T., The environment and ethics, *Hastings Cent. Rep.* **7**:16 (1977).
773. Stycos, J. M., Desexing birth control, *Fam. Plann. Perspect.* **9**:286 (1977).
774. Brams, M., Transplantable human organs: Should their sale be authorized by state statutes, *Am. J. L. Med.* **3**:183 (1977).
775. Graham, L. R., Political ideology and genetic theory: Russia and Germany in the 1920's, *Hastings Cent. Rep.* **7**:30 (1977).
776. Buss, A. R., Galton and the birth of differential psychology and eugenics: Social, political, and economic forces, *J. Hist. Behav. Sci.* **12**:47 (1976).
777. Lesse, S., The law and our future society (Editorial), *Am. J. Psychother.* **31**:341 (1977).
778. Malter, S., Genetic counseling: A responsibility of health-care professionals, *Nurs. Forum* **16**:26 (1977).
779. Veith, F. J., Fein, J. M., Tendler, M. D., *et al.*, Brain death. 1. A status report of medical and ethical considerations, *J. Am. Med. Assoc.* **238**:1651 (1977).
780. Smith, G. P., 2nd, The medicolegal challenge of preparing for a brave, yet somewhat frightening new world, *J. Leg. Med.* **5**:9 (1977).
781. Riis, P., Letter from Denmark. Planning of scientific-ethical committees, *Br. Med. J.* **2**:173 (1977).
782. Hummer, R. L., The era of humane awareness, in: *The Future Of Animals, Cells, Models, and Systems In Research, Development, Education, and Testing*, p. 25, National Academy of Sciences, Washington, D.C. (1975).
783. Donachie, E., Dilemmas and pressures. Part 2. On patients and nurses—Abortion legislation, *World IR Nurs.* **6**:1 (1977).
784. Etzioni, A., Public policy issues raised by a medical breakthrough, *Policy Anal.* **1**:69 (1975).
785. Macklin, R., Consent, coercion, and conflicts of rights, *Perspect. Biol. Med.* **20**:360 (1977).
786. Minkler, M., "Thinking the unthinkable": The prospect of compulsory sterilization in India, *Int. J. Health Serv.* **7**:237 (1977).
787. Livingstone, F. B., A problem in determining fitness differences, *Ann. Hum. Genet.* **40**:367 (1977).
788. ABA relaxes ethics code on practice information, *J. Mercer Dent. Soc.* **30**:10 (1976).

789. Powell, C. H., Professionalism and a code of ethics—The extra mile (Editorial), *Am. J. Hyg.* **38**:4 (1977).
790. Kurlantzick, L., "Medical story": Some (mis?) conceptions about law, *Med. Sci. L.* **17**:71 (1977).
791. Lawler, J., The ethical dilemma of resuscitation, *Lamp* **33**:27 (1976).
792. Curran, W. J., Direct advertising to the public: Free speech, risks, and benefits, *Am. J. Public Health* **67**:56 (1977).
793. Rickham, P. P., The swing of the pendulum: The indications for operating on myelomeningoceles, *Med. J. Aust.* **2**:743 (1976).
794. Shephard, D. A., The 1975 Declaration of Helsinki and consent (Editorial), *Can Med. Assoc. J.* **115**:1191 (1976).
795. Sabin, N. M., President's message: FTC challenges codes of ethics, *J. Wis. Dent. Assoc.* **52**:379 (1976).
796. Schnaper, N., Schnaper, H. W., and Schnaper, L. A., Euthanasia: Is there an answer, *Md. State Med. J.* **25**:66 (1976).
797. Isely, D. C., The control and propagation of a genetic disease, *ISA Trans.* **15**:115 (1976).
798. Biomedical research, *WHO Chron.* **30**:360 (1976).
799. Tanay, E., Money and the expert witness: An ethical dilemma, *J. Forensic Sci.* **21**:769 (1976).
800. Gillette, R. D., Malpractice: Why physicians and lawyers differ, *J. Leg. Med.* **4**:9 (1976).
801. Macklin, R., Moral concerns and appeals to rights and duties. Grounding claims in a theory of justice, *Hastings Cent. Rep.* **6**:31 (1976).
802. Henry, M., Inevitable failure of a shortcut policy. Compulsory sterilization in India: Is coercion the only alternative to chaos, *Hastings Cent. Rep.* **6**:14 (1976).
803. Beckwith, J., Social and political uses of genetics in the United States: Past and present, *Ann. N.Y. Acad. Sci.* **265**:46 (1976).
804. Twiss, S. B., Jr., Ethical issues in priority-setting for the utilization of genetic technologies, *Ann. N.Y. Acad. Sci.* **265**:22 (1976).
805. Hummel, R. F., Death with dignity legislation: A foot in the door, *Hosp. Prog.* **57**:50 (1976).
806. Bagley, C., On the sociology and social ethics of abortion, *Ethics Sci. Med.* **3**:21 (1976).
807. Zucker, A., Research and other freedoms (Letter), *Genetics* **82**:559 (1976).
808. Neel, J. V., Human genetics, in: *Advances in American Medicine: Essays at the Bicentennial* (Vol. L) (J. Z. Bowers and E. F. Purcell, eds.), p. 39, Josiah Macy, Jr., Foundation, New York (1976).
809. Lappe, M., Why shouldn't we have a eugenic policy?, in: *Genetics and the Law* (A. Milunsky and G. J. Annas, eds.), p. 421, Plenum Press, New York (1975).
810. Veatch, R. M., Sterilization: Its socio-cultural and ethical determinants, in: *Advances in Voluntary Sterilization* (M. E. Schima *et al.*, eds.), p. 138, Excerpta Medica, Amsterdam (1973).
811. Lederberg, S., State channeling of gene flow by regulation of marriage and procreation, in: *Genetics and the Law* (A. Milunsky and G. J. Annas, eds.), p. 247, Plenum Press, New York (1975).
812. Lederberg, S., Law and cloning—The state as regulator of gene function, in: *Genetics and the Law* (A. Milunsky and G. J. Annas, eds.), p. 377, Plenum Press, New York (1975).
813. Curran, W. J., Public health and the law. The European scene, *Am. J. Public Health* **66**:178 (1976).
814. Redlich, F., and Mollica, R. F., Overview: Ethical issues in contemporary psychiatry, *Am. J. Psychiatry* **133**:125 (1976).
815. Nolan-Haley, J. M., Defective children, their parents, and the death decision, *J. Leg. Med.* **4**:9 (1976).
816. Etzioni, A., Amniocentesis: A case study in the management of "genetic engineering," *Ethics Sci. Med.* **2**:13 (1975).
817. Ethical, social and legal dimensions of screening for human genetic disease, *Birth Defects* **10**:1 (1974).
818. Muller, S. A., The treatment of genetic disorders, *Arch. Dermatol.* **3**:1620 (1975).
819. Tarjan, G., Sex: A tri-polar conflict in mental retardation, *Monogr. Am. Assoc. Ment. Defic.* **1**:175 (1973).
820. Goldberg, A. J., Ethics in the professions, *J. Am. Coll. Dent.* **42**:218 (1975).
821. Gordon, L., The politics of birth control, 1920–1940: The impact of professionals, *Int. J. Health Serv.* **5**:253 (1975).
822. Brill, H., Presidential address: Nature and nurture as political issues, *Proc. Am. Psychopathol. Assoc.* **63**:283 (1975).
823. Ethics and the professions, *J. Med. Ethics* **1**:2 (1975).
824. Sterilization of minors, *Br. Med. J.* **3**:775 (1975).

825. Bruce, E., and Vodicka, J. D., Medical discipline. Part II. Constitutional consideration—The police power, *J. Am. Med. Assoc.* **233**:1427 (1975).

826. Frohman, I. G., Lawyers' ethics (Letter), *J. Leg. Med.* **3**:8 (1975).

827. Allen, G. E., Genetics, eugenics and class struggle, *Genetics* **79**:29 (1975).

828. Beauchamp, J. M., Euthanasia and the nurse practitioner, *Nurs. Forum* **14**:56 (1975).

829. Mass, B., An historical sketch of the American population control movement, *Int. J. Health Serv.* **4**:651 (1974).

830. Holt, N. R., Science as social and political reform. Monists and Nazis: A question of scientific responsibility, *Hasting Cent. Rep.* **5**:37 (1975).

831. Hanid, T. K., Ethical dilemmas in obstetric and newborn care, *Midwife Health Visit.* **11**:9 (1975).

832. Murphy, E. A., The normal, eugenics and racial survival, *Johns Hopkins Med. J.* **136**:98 (1975).

833. Beck, W. W., Jr., Artificial insemination and semen preservation, *Clin. Obstet, Gynecol.* **17**:115 (1974).

834. Lipman, M., Defamation: A rash comment could get you sued, *R.N.* **38**:48 (1975).

835. Coffey, P. G., Therapeutic abortion (Letter), *Can. Med. Assoc. J.* **112**:283 (1975).

836. Ethics of selective abortion (Editorial), *Br. Med. J.* **4**:676 (1974).

837. Cattell, R. B., Differential fertility and normal selection for IQ: Some required conditions in their investigation, *Soc. Biol.* **21**:168 (1974).

838. Osborn, F., History of the American Eugenics Society, *Soc. Biol.* **21**:115 (1974).

839. Bernstein, A. H., Judging hospital trustees, *Hospitals* **49**:83 (1975).

840. Rozovsky, L. E., Conflict of interest on hospital boards, *Dimens. Health Serv.* **51**:8 (1974).

841. Hogshead, H. P., Responsibility: A modality for the next decade, *Phys. Ther.* **54**:588 (1974).

842. Stewart, D. B., Musings: Side glances at obstetrics. II, *Can. Med. Assoc. J.* **111**:918 (1974).

843. Twiss, S. B., Jr., Examining the pros and cons of parental responsibility for genetic health, *Hastings Cent. Rep.* **4**:9 (1974).

844. Hilton, B., Will the baby be normal? And what is the cost of knowing, *Hastings Cent. Rep.* **2**:8 (1972).

845. Lasch, C., Birth, death and technology: The limits of cultural laissez-faire, *Hastings Cent. Rep.* **2**:1 (1972).

846. Etzioni, A., and Castleman, N., Amniocentesis, a forerunner of the genetic fix, *Conn. Med.* **38**:487 (1974).

Genetics and Environmental Mutagens; Related Topics

847. Moon, B. F., Bureaucracy vs. the elders' counsel: The people's doubts about the regulatory process, *Tex. Rep. Biol. Med.* **37**:217 (1978).

848. Giblin, P. M., The advocacy process, *Tex. Rep. Biol. Med.* **37**:212 (1978).

849. Smith, C. W., Industrial views and concerns with regard to regulatory process, *Tex. Rep. Biol. Med.* **37**:204 (1978).

850. Douglas, D. B., Implications of health and safety legislation, *IARC Sci. Publ.* **74**:81 (1979).

851. Higginson, J., Carcinogenic risks. Strategies for intervention. Introduction, *IARC Sci. Publ.* **74**:3 (1979).

852. Carcinogenic risks. Strategies for intervention, *IARC Sci. Publ.* **74**:3 (1979).

853. Kiss, A. C., Legal forms of intervention against carcinogenic risks, *IARC Sci. Publ.* **74**:241 (1979).

854. Latarjet, R., Regulation of radiation pollution: Its possible usefulness in strategy for intervention against chemical mutagens, *IARC Sci. Publ.* **74**:207 (1979).

855. Pitts, J. N., Jr., Chemical and biological implications of organic compounds formed in simulated and real urban atmospheres, *Ecotoxicol. Environ. Safety* **2**:199 (1978).

856. Upholt, W. M., Philosophies of the Environmental Protection Agency in regulating human exposure to carcinogens, *Ann. N.Y. Acad. Sci.* **298**:583 (1978).

857. Lord, M. W., Aquatic pollutants and biologic effects with emphasis on neoplasia: Keynote address, *Ann. N.Y. Acad. Sci.* **298**:201 (1978).

858. Fairchild, E. J., and Ede, L., Occupational safety and health: United States of America, *Soz. Praeventivmed.* **23**:20 (1978).

859. Culliton, B. J., Toxic substances legislation: How well are laws being implemented? (News), *Science* **201**:1198 (1978).

860. Quarles, J. R., Existing legislation and government regulatory agencies, *Bull. N.Y. Acad. Med.* **54**:442 (1978).

861. Damme, C., Diagnosing occupational disease: A new standard of care, *J. Med.* **20**:251 (1978).

862. Falk, H. L., The Toxic Substances Control Act and in vitro toxicity testing, *In Vitro* **13**:676 (1977).

863. Schmidt, B. C., Five years into the national cancer program: Retrospective perspectives—The National Cancer Act of 1971, *J. Natl. Cancer Inst.* **59**:687 (1977).

864. Westerholm, P., Administrative aspects on regulation of carcinogenic hazards in occupational environment, *Bull. Inst. Mar. Trop. Med. Med. Acad. Gdansk* **28**:57 (1977).

865. Rose, V., Why the delay in cancer-protection standards, *Occup. Health Saf.* **46**:49 (1977).

866. Catton, J. A., Legislative framework of control of occupational carcinogens in the United Kingdom, *IARC Sci. Publ.* **13**:435 (1976).

867. Munn, A., Control of carcinogenic hazards in industry, *IARC Sci. Publ.* **13**:417 (1976).

868. Westerholm, P., Administrative aspects on regulation of carcinogenic hazards in occupational equipment, *IARC Sci. Publ.* **13**:403 (1976).

869. Epstein, S. S., Regulatory aspects of occupational carcinogens: Contrasts with environmental carcinogens, *IARC Sci. Publ.* **13**:389 (1976).

870. Montesano, R., and Tomatis, L. Legislation concerning chemical carcinogens in several industrialized countries, *Cancer Res.* **37**:310 (1977).

871. Gauvain, S., and Pittom, A., Governmental action in Britain, *Ann. N.Y. Acad. Sci.* **271**:220 (1976).

872. Fairchild, E. J., I. I. Guildelines for a NIOSH policy on occupational carcinogenesis, *Ann. N.Y. Acad. Sci.* **271**:200 (1976).

873. Weaver, N. K., Measures for the prevention of occupational cancer, *J. Occup. Med.* **18**:607 (1976).

874. Rauscher, F. J., Jr., The federal role in the national cancer effort. Address to the American Cancer Society National Conference on Gynecologic Cancer, *Cancer* **38**:541 (1976).

875. Shephard, D. A., Genetic hazards to man from environmental agents, *Can. Med. Assoc. J.* **112**:1460 (1975).

MISCELLANEOUS BIBLIOGRAPHY, INCLUDING ADOPTION, RIGHTS OF CHILDREN, ABORTION, THE FETUS, STERILIZATION, SEX SELECTION, GENETIC ENGINEERING, RIGHTS OF INSTITUTIONALIZED PERSONS

876. Callahan, D., The moral career of genetic engineering, *Hastings Cent. Rep.* **9**:9 (1979).

877. Horan, D. V., An analysis of Bellotti v. Baird, *Hosp. Prog.* **60**:18 (1979).

878. Derdeyn, A. P., Adoption and the owership of children, *Child Psychiatry Hum. Dev.* **9**:215 (1979).

879. Kucera, W. R., Feminist Women's Health Center v. Mohammad: A limitation on medical review committees, *AANA J.* **47**:330 (1979).

880. Amstrong, B., A question of abuse: Where staff and patient rights collide, *Hosp. Community Psychiatry* **30**:348 (1979).

881. Kucera, W. R., The dead shall not speak, *AANA J.* **47**:72 (1979).

882. Shriner, T. L., Jr., Maternal versus fetal rights—A clinical dilemma, *Obstet. Gynecol.* **53**:518 (1979).

883. Jowers, L. V., MD's prosecution for abortion blocked (News), *Leg. Aspects Med. Pract.* **6**:29 (1978).

884. The hazard that disappeared (Editorial), *Lancet* **2**:1123 (1979).

885. Gudeman, H. E., Nelson, M. I., Kux, L. J., et al., Chancing admission patterns at Hawaii State Hospital following the 1976 revision of the Hawaii Mental Health Statutes, *Hawaii Med. J.* **38**:65 (1979).

886. Swadron, B. B., and Himel, S. G., Legal opinion on position paper: Withholding treatment, *Can. J. Psychiatry* **24**:81 (1979).

887. Garner, J., Swadron discusses ethics of "Dickering with death," *Can. Med. Assoc. J.* **120**:718 (1979).

888. Relman, A. S., Michigan's sensible "living will" (Editorial), *N. Engl. J. Med.* **300**:1270 (1979).

889. Wald, P. M., and Friedman, P. R., The politics of mental health advocacy in the United States, *Int. J. L. Psychiatry* **1**:137 (1978).

890. Goldstein, J., Psychoanalysis and a jurisprudence of child placement—With special emphasis on the role of legal counsel for children, *Int. J. L. Psychiatry* **1**:109 (1978).

891. Twelftree, H., Abortion law and problems experienced by operating nursing staff, *Australas. Nurses J.* **8**:3 (1978).

892. Keene, B., The Natural Death Act: A well-baby check-up on its first birthday, *Ann. N.Y. Acad. Sci.* **315**:376 (1978).

893. Morris, P., The balance of injustice, *Nurs. Mirror* **148**:16 (1979).

894. Swan, M., and Dipert, D., A survey of the impact of a patients' rights law on state facilities in Indiana, *Hosp. Community Psychiatry* **30**:234 (1979).

895. Gunby, P., Sex selection before child's conception (News), *J. Am. Med. Assoc.* **241**:1220 (1979).

896. Rozovsky, L., The patient's right to die, *Dimens. Health Serv.* **56**:38 (1979).

897. Relman, A. S., The Saikewicz decision: A medical viewpoint, *Am. J. L. Med.* **4**:233 (1978).

898. Leder, P., Recombinant DNA technology: Prologue and promise, *Radiology* **130**:289 (1979).

899. Trebilcock, A. M., OSHA and equal employment opportunity laws for women, *Prev. Med.* **7**:372 (1978).

900. Laking, G. R., The ombudsman and medicine, *N.Z. Med. J.* **88**:410 (1978).

901. Horan, D. V., "Viability" revisits the court in Colautti v. Franklin, *Hosp. Prog.* **60**:18 (1979).

902. Coye, J. L., Children's rights in Michigan: New law and old paternalism, *Hosp. Community Psychiatry* **30**:132 (1979).

903. Soloff, P. H., Jewell, S., and Roth, L. H., Civil commitment and the rights of the unborn, *Am. J. Psychiatry* **136**:114 (1979).

904. Stevens, G. E., Medical photography, the right to privacy and privilege, *Med. Trial Tech. Q.* **24**:456 (1978).

905. Hayes, J., The patient's right of access to his hospital and medical records, *Med. Trial Tech. Q.* **24**:295 (1978).

906. Williamson, N. E., Lean, T. H., and Vengadasalam, D., Evaluation of an unsuccessful sex preselection clinic in Singapore, *J. Biosoc. Sci.* **10**:375 (1978).

907. Bhalla, C. K., Adoption: Medical and legal aspects, *Indian Pediatr.* **15**:523 (1978).

908. Lawrence, G. J., Jr., Another instance of harm by regulation of the practice of medicine (Editorial), *N.Y. State J. Med.* **78**:13 (1978).

909. Winton, R., Father has no say (Editorial), *Med. J. Aust.* **2**:58 (1978).

910. Rozovsky, L., Civil commitment of the mentally ill, *Dimens. Health Serv.* **55**:38 (1978).

911. Neri, G., and Serra, A. Italy's abortion law (Letter), *Lancet* **2**:895 (1978).

912. Hogan, N. S., Patient's rights: Voluntary or mandatory, *Hospitals* **52**:113 (1978).

913. Regan, W. A., Class action brought against hospital to allow abortions, *Hosp. Prog.* **59**:34 (1978).

914. Despite church opposition, Italy adopts new law providing free abortion for a variety of indications (News), *Fam. Plann Perspect.* **10**:241 (1978).

915. Dembitz, N., The family: Changing values. Family law in flux, *N.Y. State J. Med.* **78**:1410 (1978).

916. Smith, T., Abortion and pregnancy screening, *J. Med. Ethics* **4**:99 (1978).

917. Coye, J. L., and Clifford, D., A one-year report on rights violations under Michigan's new protection system, *Hosp. Community Psychiatry* **29**:528 (1978).

918. Shettles, L., Sex preselection (Letter), *Obstet. Gynecol.* **51**:513 (1978).

919. Krass, M. E., Operation of the abortion law (Letter), *Can. Med. Assoc. J.* **118**:1362 (1978).

920. Bury, J., Abortion law reform (Letter), *Br. Med. J.* **1**:1698 (1978).

921. Payson, H. E., The professional guardian: A new type of non-institutional administrator, *Bull. Am. Acad. Psychiatry L.* **5**:318 (1977).

922. White, N. S., ERA: The chance for equality, *Nurs. Adm. Q.* **2**:79 (1978).

923. Horan, D. J., Illinois court requires physician to try to save viable fetus (Letter), *Hosp. Prog.* **59**:6 (1978).

924. Curtiss, R., 3d, Clark, J. E., Goldschmidt, R. et al., Biohazard assessment of recombinant DNA molecule research, in: *Plasmids* (S. Mitsuhashi et al., eds.), Avicenum, Prague (1976).

925. A lawyer looks at medical ethics, *Med. Leg. J.* **45**:104 (1977).

926. Castelli, J., Supreme Court to review cases on fetal viability, *Hosp. Prog.* **59**:18 (1978).

927. Shires, D. B., and Duncan, G., Freedom of information legislation and medical records (Editorial), *Can. Med. Assoc. J.* **118**:343 (1978).

928. Creighton, H., Action for wrongful life. Law for the nurse supervisor, *Superv. Nurse* **8**:12 (1977).

929. Vear, C. S., Preselective sex determination, *Med. J. Aust.* **2**:700 (1977).

930. Contraception, sterilization and abortion in New Zealand, *N.Z. Med. J.* **85**:428 (1977).

931. Tucker, P., Adoption problems stimulate public interest, *Midwife Health Visit. Community Nurse* **14**:14 (1978).

932. Kornitzer, M., New adoption legislation and the health visitor, *Health Visit.* **51**:4 (1978).

933. Baltimore, D., Case analysis 5. Genetic engineering: The future. Potential uses, in: *Research with Recombinant DNA* p. 237, National Academic of Sciences, Washington, D.C. (1977).

934. Lee, L. T., and Paxman, J. M., Legal aspects of menstrual regulation, *Stud. Fam. Plann.* **8**:273 (1977).

935. Martin, A., A right to treatment, *Nurs. Times* **73**:1620 (1977).

936. Hegde, U. C., Shastry, P. R., and Rao, S. S., A simple and reproducible method for separating Y-bearing spermatozoa from human semen, *Indian J. Med. Res.* **65**:738 (1977).

937. Stroud, J., Adoption counselling and the new Children's Act, 1975, *Midwife Health Visit. Community Nurse* **13**:308 (1977).

938. Ballard, J., and Zettel, J., Public Law 94–142 and section 504; What they say about rights and protections, *Except. Child* **44**:177 (1977).

939. Affleck, G., and Thomas, A., The Edelin decision revisisted: A survey of the reactions of Connecticut's ob/gyns, *Con. Med.* **41**:637 (1977).

940. Hollowell, E. E., The right to die. How legislation is defining the right, *J. Pract. Nurs.* **27**:20 (1977).

941. Lincoln, R., Doring-Bradley, B., Lindheim, B. L., *et al.*, The court, the Congress and the president: Turning back the clock on the pregnant poor, *Fam. Plann. Perspect.* **9**:207 (1977).

942. Primrose, S. B., Genetic engineering, *Sci. Prog.* **64**:293 (1977).

943. Lewis, E. P., The right to inform (Editorial), *Nurs. Outlook* **25**:561 (1977).

944. Tuma, J. L., Professional misconduct? (Letter), *Nurs. Outlook* **25**:546 (1977).

945. Gaylin, W., Sounding Board. The frankenstein factor, *N. Engl. J. Med.* **297**:665 (1977).

946. Mason, A., and Bennett, N. G., Sex selection with biased technologies and its effect on the population sex ratio, *Demography* **14**:285 (1977).

947. Williamson, R., Genetic engineering, *Nurs. Mirror* **145**:23 (1977).

948. Seidel, H. M., Legal change for child health: Report on a conference (Editorial), *Pediatrics* **60**:251 (1977).

949. Abortion (Letter), *Lancet* **2**:247 (1977).

950. U.S. District Court orders payment of maternity benefits to unwed mothers under Title VII maternity issues, *Employ. Benefit Plan Rev.* **29**:24 (1975).

951. Page, S., Toward evaluating the meaninfulness of legal counsel for psychiatric patients, *Can. Ment. Health* **24**:6 (1976).

952. Cook, R. J., Abortion laws in Commonwealth countries, *IPPF Med. Bull.* **10**:1 (1976).

953. Robertson, J. A., After Edelin: Little guidance, *Hastings Cent. Rep.* **7**:15 (1977)

954. Special legal implications of patient rights, *J. Am. Health Care Assoc.* **3**:19 (1977).

955. Fleming, B. A., and Hunt, J. F., Bartley v. Kremens: A step backward, *J. Natl. Assoc. Priv. Psychiatr. Hosp.* **8**:10 (1976).

956. Smith, T., The parliamentary scene, *J. Med. Ethics* **3**:100 (1977).

957. Wee, Kim Seng K., Some laws relating to population growth: Laws directly affecting fertility, *Nurs. J. Singapore* **17**:20 (1977).

958. Yankauer, A., Abortions and public policy (Editorial), *Am. J. Public Health* **67**:604 (1977).

959. Chabon, R. S., The legal status of the unborn child, *J. Leg. Med.* **5**:22 (1977).

960. Mattingly, R., Adoption and the Children Act. Part I: Progress in implementing the act, and the right of adopted people to obtain their birth certificate, *Nurs. Mirror* **144**:63 (1977).

961. Lipton, G. I., Mother is a bad word: The rights of children in law in two custody cases, *Aust. N.Z. J. Psychiatry* **11**:19 (1977).

962. Thomas, W. D., The Badgley report on the abortion law (Editorial), *Can. Med. Assoc. J.* **116**:966 (1977).

963. Kemp, K. A., Carp, R. A., and Brady, D. W., Abortion and the law: The impact on hospital policy of the Roe and Doe decisions, *J. Health Polit. Policy L.* **1**:319 (1976).

964. Shephard, D. A., Genetic intervention: The modern chimera (Editorial), *Can. Med. Assoc. J.* **116**:705 (1977).

965. Goodman, J. D., On unsealing the records in adoption (Letter), *Am. J. Psychiatry* **134**:95 (1977).

966. Creighton, H., Law for the nurse supervisor. The right to refuse treatment, *Superv. Nurse* **8**:13 (1977).

967. Abortion (amendment) bill (Letter), *Lancet* **1**:606 (1977).

968. Weissmann, G., Experimental enzyme replacement in genetic and other disorders, *Hosp. Pract.* 11:49 (1976).

969. Pakalnis, L., and Makoroto, J., Reproduction and the test tube baby, *Can. Nurse* 73:34 (1977).

970. Blau, S., Not in the best interests of the child (Letter), *Am. J. Psychiatry* 134:210 (1977).

971. Glass, R. H., Sex preselection, *Obstet. Gynecol.* 49:122 (1977).

972. Leete, R., Some comments on the demographic and social effects of the 1967 Abortion Act, *J. Biosoc. Sci.* 8:229 (1976).

973. Derdeyn, A. P., Child custody contests in historical perspective, *Am. J. Psychiatry* 133:1369 (1976).

974. Grewall, S., Medical termination of pregnancy—Its status, achievements and lacunae, *J. Indian Med. Assoc.* 66:269 (1976).

975. Parents, children, and due process: The case of Kremens v. Bartley, *Hosp. Community Psychiatry* 27:705 (1976).

976. Bryant, M. D., Jr., State legislation on abortion after Roe v. Wade: Selected consitutional issues, *Am. J. L. Med.* 2:101 (1976).

977. Outman, J. B., Revision of Georgia's adoption laws, *J. Med. Assoc. Ga.* 65:333 (1976).

978. Miller, E. Psychotherapy of a child in a custody dispute, *J. Am. Acad. Child Psychiatry* 15:441 (1976).

979. Annas, G. J., Law and life sciences. Abortion and the Supreme Court: Round two, *Hastings Cent. Rep.* 6:15 (1976).

980. Somers, R. L., and Gammeltoft, M., The impact of liberalized abortion legislation on contraceptive practice in Denmark, *Stud. Fam. Plann.* 7:218 (1976).

981. Ellis, J. W., Law: Mental health rights of children, *Ment. Health* 60:20 (1976).

982. Draconian images and the new biology, in: *The Post-Physician Era* (J. S. Maxmen, ed.), p. 176, Wiley, New York (1976).

983. Rogers, S., Reflections on issues posed by recombinant DNA molecule technology. II, *Ann. N.Y. Acad. Sci.* 265:66 (1976).

984. Green, H. P., Law and genetic control: Public-policy questions, *Ann. N.Y. Acad. Sci.* 265:170 (1976).

985. Neville, R., Gene therapy and the ethics of genetic therapeutics, *Ann. N.Y. Acad. Sci.* 265:153 (1976).

986. Morrow, J. F., The prospects for gene therapy in humans, *Ann. N.Y. Acad. Sci.* 265:13 (1976).

987. Hull, R. T., Philosophical considerations in the growing potential for human genetic control, *Ann. N.Y. Acad. Sci.* 265:118 (1976).

988. Sorosky, A. D., Baran, A., and Pannor, R., The effects of the sealed record in adoption, *Am. J. Psychiatry* 133:900 (1976).

989. Select committee on abortion (Letter), *Lancet* 2:306 (1976).

990. Lappe, M., Realities of "Genetic engineering," *Med. Res. Eng.* 12:25 (1976).

991. Bryn, R. M., The new jurisprudence, *J. Am. Med. Assoc.* 236:359 (1976).

992. Bernstein, A. H., An abortion law update, *Hospitals* 50:90 (1976).

993. Lorrain, J., Pre-conceptional sex selection, *Int. J. Gynaecol. Obstet.* 13:127 (1975).

994. Dix, G. E., The Alabama "right to treatment" case: An opportunity not to be missed, *Community Ment. Health J.* 12:161 (1976).

995. Gene cloning: One milestone on a very long road (Editorial), *Lancet* 1:893 (1976).

996. Dunlop, J. M., Genetic engineering, *Nurs. Times* 72:169 (1976).

997. Cohen, M., Rapson, L., and Watters, W. W., The Canadian abortion law (Letter), *Can. Med. Assoc. J.* 114:593 (1976).

998. Glantz, L. H., The legal aspects of fetal viability, in: *Genetics and the Law* (A. Milunsky and G. J. Annas, eds.), p. 29, Plenum Press, New York (1975).

999. Green, H. P., The fetus and the law, in: *Genetics and the Law* (A. Milunsky and G. J. Annas, eds.), p. 19, Plenum Press, New York (1975).

1000. Shaw, M. W., and Damme, C., Legal status of the fetus, in: *Genetics and the Law* (A. Milunsky and G. J. Annas, eds.), p. 3, Plenum Press, New York (1975).

1001. Singer, K., Psychiatric aspects of abortion in Hong Kong, *Int. J. Soc. Psychiatry* 21:303 (1975).

1002. Shettles, L. B., Sex selection (Letter), *Am. J. Obstet. Gynecol* 124:441 (1976).

1003. Playing with genes (Editorial), *Br. Med. J.* 1:302 (1976).

1004. Wolfe, B. E., The Donaldson decision, *Schizophr. Bull.* 13:4 (1975).

1005. Dolan, W. V., Abortion vs. manslaughter (Letter), *Arch. Surg.* 111:93 (1976).

1006. Schwitzgebel, R. K., A contractual model for the protection of the rights of institutionalized mental patients, *Am. Psychol.* 30:815 (1975).

1007. Mattingly, R., Adoption and the Children Bill, *Nurs. Mirror* **141**:69 (1975).

1008. Restrictions in Missouri abortion statute upheld, *J. Miss. State Med. Assoc.* **16**:382 (1975).

1009. Evidence of the Royal College of General Practitioners to the Select Committee of Parliament on the Abortion (amendment) bill, *J. R. Coll. Gen. Pract.* **25**:774 (1975).

1010. Gallagher, U. M., and Katz, S. N., The model State Subsidized Adoption Act, *Child Today* **4**:8 (1975).

1011. Walter, S. D., The transitional effect on the sex ratio at birth of a sex predetermination program, *Soc. Biol.* **21**:340 (1974).

1012. Blackwood, R., What are the father's rights in abortion, *J. Leg. Med.* **3**:28 (1975).

1013. Mukherjee, J. B., The Medical Termination of Pregnancy Act, 1971 and the registered medical practitioners, *J. Indian Med. Assoc.* **65**:13 (1975).

1014. Ament, M., The right to be well born, *J. Leg Med.* **2**:24 (1974).

1015. Hirsh, H. L., Legal guidelines for the performance of abortions, *Am. J. Obstet. Gynecol.* **122**:679 (1975).

1016. Creighton, H., Moral and legal issues—Nursing judgments, SO NLN Publ. 16–1551, p. 22 (1974).

1017. Ryan, K. J., The legitimacy of a diverse society, *J. Am. Med. Assoc.* **233**:781 (1975).

1018. Abortion (amendment) bill (Letter), *Br. Med. J.* **2**:558 (1975).

1019. Berg, P., Baltimore, D., Brenner, S., *et al.*, Asilomar Conference on Recombinant DNA Molecules, *Science* **188**:991 (1975).

1020. Pilpel, H. F., Abortion: U.S.A. style, *J. Sex Res.* **11**:113 (1975).

1021. Gibbs, R. F., Commonwealth v. Edelin (Editorial), *J. Leg. Med.* **3**:6 (1975).

1022. Rinehart, W., Sex preselection—Not yet practical, *Popul. Rep.* **1**:121 (1975).

1023. Glowienka, E., A brighter side of the new genetics, *Bioscience* **25**:94 (1975).

1024. Ward, R., Paul, E. W., and Pilpel, H. F., Pregnancy, teenagers and the law (Letter), *Fam. Plann. Perspect.* **7**:97 (1975).

1025. Kramer, M. J., Legal abortion among New York City residents: An analysis according to socioeconomic and demographic characteristics, *Fam. Plann. Perspect.* **7**:128 (1975).

1026. Tietze, C., The effect of legalization of abortion on population growth and public health, *Fam. Plann. Perspect.* **7**:123 (1975).

1027. Paul, E. W., Pilpel, H. F., and Wechsler, N. F., Pregnancy, teenagers and the law, 1974, *Fam. Plann. Perspect.* **6**:142 (1974).

1028. Abortion (amendment) bill (Letter), *Br. Med. J.* **3**:99 (1975).

1029. Stubblefield, P. G., Abortion vs. manslaughter, *Arch. Surg.* **110**:790 (1975).

1030. Rozovsky, L. E., Morgentaler vs. the Queen, *Dimens. Health Serv.* **52**:8 (1975).

1031. Knowles, J. H., The health system and the Supreme Court decision; an affirmative response, *Fam. Plann. Perspect.* **5**:113 (1973).

1032. Tietze, C., Two years' experience with a liberal abortion law: Its impact on fertility trends in New York City, *Fam. Plann. Perspect.* **5**:36 (1973).

1033. Abortion: The high court has ruled, *Fam. Plann. Perspect.* **5**:1 (1973).

1034. McLaren, H. C., Abortion (amendment) bill (Letter), *Br. Med. J.* **2**:613 (1975).

1035. Walter, S. D., Sex predetermination and epidemiology, *Soc. Sci. Med.* **9**:105 (1975).

1036. Lappe, M., The human uses of molecular genetics, *Fed. Proc.* **34**:1425 (1975).

1037. Roblin, R., Ethical and social aspects of experimental gene manipulation, *Fed. Proc.* **34**:1421 (1975).

1038 Eisinger, J., The ethics of human gene manipulation. Introductory remarks, *Fed. Proc.* **34**:1418 (1975).

1039. Pilpel, H. F., The fetus as person: Possible legal consequences of the Hogan-Helms Amendment, *Fam. Plann. Perspect.* **6**:6 (1974).

1040. Potts, M., The world abortion picture—1975, *IPPF Med. Bull.* **9**:2 (1975).

1041. Creighton, H., Law for the nurse supervisor: More about abortion decisions, *Superv. Nurse* **6**:10 (1975).

1042. Introcaso, D. A., Abortion (Letter), *Obstet. Gynecol.* **45**:234 (1975).

1043. Roghmann, K. J., The impact of the New York State abortion law on black and white fertility in upstate New York, *Int. J. Epidemiol.* **4**:45 (1975).

1044. Dunlop, J. M., Genetic engineering—A waste of valuable resources, *Public Health* **89**:13 (1974).

1045. Edelin supported (Letter), *N. Engl. J. Med.* **292**:705 (1975).

1046. Ingelfinger, F. J., The Edelin trial fiasco (Editorial), *N. Engl. J. Med.* **292**:697 (1975).

1047. Weinstock, E., Tietze, C., Jaffe, F. S., *et al.*, Legal abortions in the United States since the 1973 Supreme Court decisions, *Fam. Plann. Perspect.* **7**:23 (1975).

1048. Lewis, D. O., Balla, D., Lewis, M., *et al.*, The treatment of adopted versus neglected delinquent children in the court: A problem of reciprocal attachment, *Am. J. Psychiatry* **132**:142 (1975).

1049. Iffy, L., Abortion laws in Hungary (Letter), *Obstet. Gynecol.* **45**:115 (1975).

1050. Gibbs, R. F., Therapeutic abortions and the minor, *J. Leg. Med.* **1**:36 (1973).

1051. Thomas, L., Notes of a biology-watcher. On cloning a human being, *N. Engl. J. Med.* **291**:1296 (1974).

1052. Reilly, P., Legal status of the unborn (Letter), *Lancet* **2**:1207 (1974).

1053. David, H. P., Abortion and family planning in the Soviet Union: Public policies and private behaviour, *J. Biosoc. Sci.* **6**:417 (1974).

1054. Vaux, K., Generating man, *Tex. Rep. Biol. Med.* **32**:351 (1974).

1055. Phillipson, D., Supreme Court considers Morgentaler abortion case, *Can. Med. Assoc. J.* **111**:872 (1974).

1056. Simms, M., How do we judge the Abortion Act/Reflections on the Lane Committee and the 1967 Abortion Act, *Public Health* **87**:155 (1973).

1057. Lappe, M., Genetic counseling and genetic engineering, *Hastings Cent. Rep.* **3**:13 (1971).

1058. Hazards of genetic experiments (Editorial), *Br. Med. J.* **3**:483 (1974).

1059. Davis, B. D., Genetic engineering: How great is the danger (Editorial), *Science* **186**:309 (1974).

1060. Rozovsky, L. E., Canadian law vs. Catholic hospitals, *Dimens. Health Serv.* **51**:12 (1974).

INDEX

Page numbers in *italic* type represent citations in Discussions.